Theoretical Approach to Mathematics

Volume I

Theoretical Approach to Mathematics
Volume I

Edited by **Matt Ferrier**

NY RESEARCH PRESS
New York

Published by NY Research Press,
23 West, 55th Street, Suite 816,
New York, NY 10019, USA
www.nyresearchpress.com

Theoretical Approach to Mathematics: Volume I
Edited by Matt Ferrier

International Standard Book Number: 978-1-63238-443-0 (Hardback)

Printed in China.

Contents

Preface

Mathematics is a branch which deals with topics related to quantity, space and their related changes. A system of numeration is a system which is used for expressing numbers in a simplified manner. A mathematical notation which uses digits or other symbols in a consistent manner to represent numerals of a given set constitutes the field of Mathematics.

A numeral system must essentially represent a useful set of numbers such as rational numbers, or integers and a unique representation which reflects the algebraic and arithmetic structure of numbers. These numbers set the base of the most important field of study in human life which is called mathematics.

Like all other human invention, mathematics also originated in the ancient times, due to human need. The basis of Mathematics was the need of society to calculate and keep track of things. So the more complex the society grew, the more complex the needs became. This need of ever growing social requirements of humans gave birth to most of the advances in the field of mathematics.

I would like to express my thanks to the authors who contributed to this book. Without their timely submissions, this publication wouldn't have been possible. I thank them for allocating much of their valuable time to this project. I would also like to thank the organizers and the staff of the publishing team for their constant support at every step. Last but not the least, I wish to thank my family and friends, who have supported me in my life through everything. I do believe that the contents in this book will be helpful to many researchers in this field around the world.

Editor

The Fractional S-Transform on Spaces of Type S

Sunil Kumar Singh

Department of Mathematics, Rajiv Gandhi University, Doimukh, Arunachal Pradesh 791112, India

Correspondence should be addressed to Sunil Kumar Singh; sks_math@yahoo.com

Academic Editor: Peng-Yeng Yin

Certain properties of the S-transform are given and composition of S-transform is defined. We study the fractional S-transform on spaces of type S. It is shown that the fractional S-transform is a continuous linear map of the space of type S defined on $\mathbb{R}$ into the space of type $\widetilde{S}$ defined on $\mathbb{R} \times (\mathbb{R} \setminus \{0\})$.

1. Introduction

In this section we give the definition and some basic properties of the S-transform. The S-transform was first introduced by Stockwell et al. [1] in 1996 as invertible time-frequency spectral localization technique. The S-transform is an extension to the ideas of continuous wavelet transform and is based on a moving and scalable localizing Gaussian window and has characteristics superior to both of the Fourier transform and the wavelet transform [2–5].

The one-dimensional continuous S-transform of $u(t)$ is defined as [4]

$$\begin{aligned}(Su)(\tau, f) &= S(u(t))(\tau, f) \\ &= \int_{\mathbb{R}} u(t)\, \omega(\tau - t, f)\, e^{-i2\pi f t} dt,\end{aligned} \tag{1}$$

where the window ω is assumed to satisfy the following:

$$\int_{\mathbb{R}} \omega(t, f)\, dt = 1 \quad \forall f \in \mathbb{R} \setminus \{0\}. \tag{2}$$

The most usual window ω is the Gaussian one

$$\omega(t, f) = \frac{|f|}{k\sqrt{2\pi}} e^{-(f^2 t^2/2k^2)}, \qquad k > 0, \tag{3}$$

where f is the frequency, t is the time variable, and k is a scaling factor that controls the number of oscillations in the window.

Then, (1) can be rewritten as a convolution

$$(Su)(\tau, f) = \left(u(\cdot)\, e^{-i2\pi f \cdot} * \omega(\cdot, f)\right)(\tau). \tag{4}$$

Applying the convolution property for the Fourier transform, we obtain

$$(Su)(\tau, f) = \mathscr{F}^{-1}\{\widehat{u}(\cdot + f)\, \widehat{\omega}(\cdot, f)\}(\tau), \tag{5}$$

where $\mathscr{F}^{-1}$ is the inverse Fourier transform. For the Gaussian window case (3),

$$\widehat{\omega}(\alpha, f) = e^{-2(\pi k \alpha/f)^2}. \tag{6}$$

Thus we can write the S-transform in the following form:

$$(Su)(\tau, f) = \int_{\mathbb{R}} \widehat{u}(\alpha + f)\, e^{-2(\pi k \alpha/f)^2} e^{i2\pi\alpha\tau} d\alpha. \tag{7}$$

Also, if $\widehat{u}(f)$ and $(Su)(\tau, f)$ are the Fourier transform and S-transform of u, respectively, then

$$\widehat{u}(f) = \int_{\mathbb{R}} (Su)(\tau, f)\, d\tau, \tag{8}$$

so that

$$u(t) = \mathscr{F}^{-1}\left(\int_{\mathbb{R}} (Su)(\tau, \cdot)\, d\tau\right)(t). \tag{9}$$

2. Some Important Properties of S-Transform

Some basic properties of S-transform can be found in [6, 7] and certain properties of S-transform are obtained in this section.

Theorem 1. *If* $u(t) \in L^1(\mathbb{R})$, *then* $\lim_{|\tau|\to\infty}|(Su)(\tau, f)| = 0$.

Proof. We have by definition

$$\begin{aligned}(Su)(\tau, f) &= \int_{\mathbb{R}} \widehat{u}(\alpha + f)\, e^{-2(\pi k\alpha/f)^2} e^{i2\pi\alpha\tau} d\alpha \\ &= -\int_{\mathbb{R}} \widehat{u}(\alpha + f)\, e^{-2(\pi k\alpha/f)^2} e^{i2\pi(\alpha+1/2\tau)\tau} d\alpha \\ &= -\int_{\mathbb{R}} \widehat{u}\left(x + f - \frac{1}{2\tau}\right) e^{-2[\pi k(x-1/2\tau)/f]^2} e^{i2\pi\tau x} dx.\end{aligned} \tag{10}$$

Thus

$$\begin{aligned}&(Su)(\tau, f) \\ &= \frac{1}{2}\left[\int_{\mathbb{R}} \widehat{u}(\alpha + f)\, e^{-2(\pi k\alpha/f)^2} e^{i2\pi\alpha\tau} d\alpha \right.\\ &\left. \quad - \int_{\mathbb{R}} \widehat{u}\left(\alpha + f - \frac{1}{2\tau}\right) e^{-2[\pi k(\alpha-1/2\tau)/f]^2} e^{i2\pi\tau\alpha} d\alpha\right].\end{aligned} \tag{11}$$

Clearly,

$$\begin{aligned}&\lim_{|\tau|\to\infty} |(Su)(\tau, f)| \\ &\le \frac{1}{2}\lim_{|\tau|\to\infty} \int_{\mathbb{R}} \left|\widehat{u}(\alpha + f) - \widehat{u}\left(\alpha + f - \frac{1}{2\tau}\right)\right| d\alpha = 0.\end{aligned} \tag{12}$$

□

Theorem 2. *If* $u(t) \in L^1(\mathbb{R})$, *then* $\lim_{|f|\to 0}|(Su)(\tau, f)| = 0$.

Theorem 3. *If* $e^{-2(\pi k\alpha/f)^2}\widehat{u}(\alpha + f) \in L^2(\mathbb{R})$, *then* $\|(Su)(\tau, f)\|_{L^2} \le \|u\|_{L^2}$.

Theorem 4. *If* $u(t) \in L^1(\mathbb{R})$, *then* $(Su)(\tau, f)$ *is continuous with respect to the variables* τ *and* f.

Proof. By the definition

$$\begin{aligned}&|(Su)(\tau + h, f) - (Su)(\tau, f)| \\ &= \left|\int_{\mathbb{R}} \widehat{u}(\alpha + f)\, \widehat{\omega}(\alpha, f)\left(e^{i2\pi\alpha h} - 1\right) e^{i2\pi\alpha\tau} d\alpha\right| \\ &\le \int_{\mathbb{R}} |\widehat{u}(\alpha + f)| \left|e^{i2\pi\alpha h} - 1\right| d\alpha, \\ &\longrightarrow 0 \quad \text{as } h \longrightarrow 0.\end{aligned} \tag{13}$$

Hence the S-transform is continuous with respect to the first variables τ. Similarly we can show that it is also continuous with respect to the second variables f. □

2.1. Composition of S-Transforms. Let S_1 and S_2 be two S-transforms defined as

$$\begin{aligned}(S_1u)(\tau_1, f_1) &= \int_{\mathbb{R}} \widehat{u}(\alpha + f_1)\, \widehat{\omega_1}(\alpha, f_1)\, e^{i2\pi\alpha\tau_1} d\alpha, \\ (S_2u)(\tau_2, f_2) &= \int_{\mathbb{R}} \widehat{u}(\alpha + f_2)\, \widehat{\omega_2}(\alpha, f_2)\, e^{i2\pi\alpha\tau_2} d\alpha,\end{aligned} \tag{14}$$

provided the integrals exist. Then their composition $S_1 \circ S_2$ is defined by

$$\begin{aligned}&\left(\overset{*}{S} u\right)(\tau_1, f_1, f_2) \\ &= ((S_1 \circ S_2)u)(\tau_1, f_1, f_2) \\ &= \int_{\mathbb{R}} \mathcal{F}\{(S_2u)(\tau_2, f_2)\}(\alpha + f_1)\, \widehat{\omega_1}(\alpha, f_1)\, e^{i2\pi\alpha\tau_1} d\alpha \\ &= \int_{\mathbb{R}} \widehat{u}(\alpha + f_1 + f_2)\, \widehat{\omega_2}(\alpha + f_1, f_2)\, \widehat{\omega_1}(\alpha, f_1)\, e^{i2\pi\alpha\tau_1} d\alpha \\ &= \int_{\mathbb{R}} \widehat{u}(\alpha + f_1 + f_2)\, \widehat{\omega}(\alpha, f_1, f_2)\, e^{i2\pi\alpha\tau_1} d\alpha,\end{aligned} \tag{15}$$

where $\widehat{\omega}(\alpha, f_1, f_2) = \widehat{\omega_2}(\alpha + f_1, f_2)\widehat{\omega_1}(\alpha, f_1)$, provided the integral exists. The composition $S_1 \circ S_2$ is an S-transform of three variables mapping from $\mathbb{R}$ into $\mathbb{R} \times (\mathbb{R} \setminus \{0\}) \times (\mathbb{R} \setminus \{0\})$ and analogous to the composition of the wavelet transform [8, pages 52-53].

Theorem 5. *If* $u(t) \in L^1(\mathbb{R})$, *then* $(\overset{*}{S}u)(\tau, f_1, f_2)$ *is continuous with respect to the variables* τ, f_1 *and* f_2.

3. Fractional Fourier Transform

The fractional Fourier transform (FRFT) has played an important role in signal processing. The ath order FRFT of a signal $u(t)$ is defined as [9]

$$F_u^a(f) = \int_{\mathbb{R}} u(t)\, K_a(t, f)\, dt, \tag{16}$$

where the transform kernel $K_a(t, f)$ is defined as

$$K_a(t, f) = \begin{cases} A_\theta e^{i\pi(f^2\cot\theta - 2ft\csc\theta + t^2\cot\theta)}, & \text{if } \theta \ne n\pi \\ \delta(t - f), & \text{if } \theta = 2n\pi \\ \delta(t + f), & \text{if } \theta + \pi = 2n\pi, \end{cases} \tag{17}$$

where $A_\theta = \sqrt{1 - i\cot\theta}$, $\theta = a\pi/2$, $a \in [0, 4)$, i is the complex unit, n is an integer, and f is the fractional Fourier frequency (FRFfr). The inverse FRFT of (16) is

$$u(t) = \int_{\mathbb{R}} F_u^a(f)\, \overline{K_a(t, f)}\, df. \tag{18}$$

We can write (16) as

$$\begin{aligned} F_u^a(f) &= \int_{\mathbb{R}} u(t)\, A_\theta e^{i\pi(f^2\cot\theta - 2ft\csc\theta + t^2\cot\theta)} dt \\ &= A_\theta e^{i\pi f^2\cot\theta} \int_{\mathbb{R}} u(t)\, e^{-i2\pi ft\csc\theta} e^{i\pi t^2\cot\theta} dt \\ &= A_\theta e^{i\pi f^2\cot\theta} \left[\widehat{e^{i\pi t^2\cot\theta} u(t)}\right](f\csc\theta). \end{aligned} \tag{19}$$

Replacing $u(t)$ by $e^{-i\pi t^2\cot\theta}\phi(t)$, we have

$$F_u^a(f) = A_\theta e^{i\pi f^2\cot\theta} \widehat{[\phi(t)]}(f\csc\theta). \tag{20}$$

Putting $f = \xi\sin\theta$, we have

$$\begin{aligned} F_u^a(\xi\sin\theta) &= A_\theta e^{i\pi\xi^2\sin 2\theta/2} \widehat{\phi}(\xi), \\ \widehat{\phi}(\xi) &= \frac{1}{A_\theta} e^{-i\pi\xi^2\sin 2\theta/2} F_u^a(\xi\sin\theta), \end{aligned} \tag{21}$$

where $u(t) = e^{-i\pi t^2\cot\theta}\phi(t)$.

4. The Fractional S-Transform

The fractional S-transform (FRST) is a generalization of the S-transform. The ath order continuous fractional S-transform of $u(t)$ is defined as [10]

$$\mathrm{FRST}_u^a(\tau, f) = \int_{\mathbb{R}} u(t)\, g(\tau - t, f)\, K_a(t, f)\, dt, \tag{22}$$

where the window g is

$$g(t, f) = \frac{|f\csc\theta|^p}{k\sqrt{2\pi}} e^{-t^2(f\csc\theta)^{2p}/2k^2}, \quad k, p > 0 \tag{23}$$

and satisfies the condition

$$\int_{\mathbb{R}} g(t, f)\, dt = 1 \quad \forall f \in \mathbb{R}\setminus\{0\}. \tag{24}$$

Inverse fractional S-transform is defined by

$$u(t) = \int_{\mathbb{R}} \left[\int_{\mathbb{R}} \mathrm{FRST}_u^a(\tau, f)\, d\tau\right] \overline{K_a(t, f)} df. \tag{25}$$

Note that the fractional S-transform depends on a parameter θ and can be interpreted as a rotation by an angle θ in the time-frequency plane. An FRST with $\theta = \pi/2$ corresponds to the S-transform, and an FRST with $\theta = 0$ corresponds to the zero operator. The parameters p and k can be used to adjust the window function space.

Let

$$h(t, \tau, f) = g(\tau - t, f)\, K_a(t, f), \tag{26}$$

$$\begin{aligned} H_a(\tau, f, f_1) &= \int_{\mathbb{R}} h(t, \tau, f)\, \overline{K_a(t, f_1)} \\ &= \int_{\mathbb{R}} g(\tau - t, f)\, K_a(t, f)\, \overline{K_a(t, f_1)} dt. \end{aligned} \tag{27}$$

Since

$$\begin{aligned} &K_a(t, f)\, \overline{K_a(t, f_1)} \\ &\quad = A_\theta \overline{A}_\theta e^{i\pi[(f^2 - f_1^2)\cot\theta - 2(f - f_1)t\csc\theta]}. \end{aligned} \tag{28}$$

By using (23) and (28) in (27) we obtain

$$\begin{aligned} H_a(\tau, f, f_1) &= \int_{\mathbb{R}} \frac{|f\csc\theta|^p}{k\sqrt{2\pi}} e^{-(\tau - t)^2(f\csc\theta)^{2p}/2k^2} \\ &\quad \times A_\theta \overline{A}_\theta e^{i\pi[(f^2 - f_1^2)\cot\theta - 2(f - f_1)t\csc\theta]} dt \\ &= \frac{|f\csc\theta|^p}{k\sqrt{2\pi}} A_\theta \overline{A}_\theta e^{i\pi[(f^2 - f_1^2)\cot\theta]} \\ &\quad \times \int_{\mathbb{R}} e^{-(\tau - t)^2(f\csc\theta)^{2p}/2k^2} e^{-i2\pi(f - f_1)t\csc\theta} dt. \end{aligned} \tag{29}$$

By using the technique of [10], we obtain

$$\begin{aligned} H_a(\tau, f, f_1) &= A_\theta \overline{A}_\theta e^{i\pi[(f^2 - f_1^2)\cot\theta]} e^{-i2\pi(f - f_1)\tau\csc\theta} \\ &\quad \times e^{-2\pi^2k^2(f - f_1)^2(\csc\theta)^2/(f\csc\theta)^{2p}}, \end{aligned} \tag{30}$$

and by using (28) we can write

$$\begin{aligned} &H_a(\tau, f, f_1) \\ &\quad = e^{-[2\pi^2k^2(f - f_1)^2(\csc\theta)^2/(f\csc\theta)^{2p}]} K_a(t, f)\, \overline{K_a(t, f_1)}. \end{aligned} \tag{31}$$

Also, the FRST can be also defined as operations on the fractional Fourier domain:

$$\begin{aligned} \mathrm{FRST}_u^a(\tau, f) &= \int_{\mathbb{R}} \left[\int_{\mathbb{R}} F_u^a(f)\, \overline{K_a(t, f)} df\right] \\ &\quad \times g(\tau - t, f)\, K_a(t, f)\, dt \\ &= \int_{\mathbb{R}} F_u^a(f_1)\, H_a(\tau, f, f_1)\, df_1. \end{aligned} \tag{32}$$

By using (19) and (30) we can write (32) as

$$\begin{aligned}
&\mathrm{FRST}_u^a(\tau, f) \\
&= \int_{\mathbb{R}} A_\theta e^{i\pi f_1^2 \cot\theta} \left[\widehat{e^{i\pi t^2 \cot\theta} u(t)}\right] \\
&\quad \times (f_1 \csc\theta) A_\theta \overline{A}_\theta e^{i\pi[(f^2 - f_1^2)\cot\theta]} \\
&\quad \times e^{-i2\pi(f-f_1)\tau\csc\theta} \\
&\quad \times e^{-2\pi^2 k^2 (f-f_1)^2 (\csc\theta)^2/(f\csc\theta)^{2p}} df_1 \\
&= A_\theta |A_\theta|^2 \int_{\mathbb{R}} e^{i\pi f_1^2 \cot\theta} \left[\widehat{e^{i\pi t^2 \cot\theta} u(t)}\right] \\
&\quad \times (f_1 \csc\theta) e^{i\pi[(f^2 - f_1^2)\cot\theta]} \qquad (33)\\
&\quad \times e^{-i2\pi(f-f_1)\tau\csc\theta} \\
&\quad \times e^{-2\pi^2 k^2 (f-f_1)^2 (\csc\theta)^2/(f\csc\theta)^{2p}} df_1 \\
&= A_\theta |A_\theta|^2 e^{i\pi f^2 \cot\theta} \\
&\quad \times \int_{\mathbb{R}} \left[\widehat{e^{i\pi t^2 \cot\theta} u(t)}\right] (f_1 \csc\theta) \\
&\quad \times e^{-i2\pi(f-f_1)\tau\csc\theta} \\
&\quad \times e^{-2\pi^2 k^2 (f-f_1)^2 (\csc\theta)^2/(f\csc\theta)^{2p}} df_1.
\end{aligned}$$

5. The Fractional S-Transform on Spaces of Type S

The spaces of type S play an important role in the theory of linear partial differential equations as intermediate spaces between the spaces of the C^∞ and the analytic functions. The Fourier transform has been studied on the spaces of type S by Friedman [11] and Gel'fand and Shilov [12]. Pathak and Singh [13] studied the wavelet transform on the spaces of type S. Also, Singh [6] studied the S-transform on the spaces of type S. In this section we study the fractional S-transform (22) on these spaces.

Let us recall the definitions of these spaces.

Definition 6. The space $S_\alpha(\alpha \geq 0)$ consists of all infinitely differentiable functions $\phi(x)$ $(-\infty < x < \infty)$, satisfying the inequalities

$$\begin{aligned}
\gamma_{k,q}(\phi) &:= \sup_{x\in\mathbb{R}} \left|x^k \phi^{(q)}(x)\right| \\
&\leq C_q A^k k^{k\alpha} \quad (k, q = 0, 1, 2, \ldots),
\end{aligned} \tag{34}$$

where the constants A and C_q depend on the function ϕ. For $k = 0$, the expression $k^{k\alpha}$ is considered to be equal to 1.

Definition 7. The space $S^\beta(\beta \geq 0)$ consists of all infinitely differentiable functions $\phi(x)$ $(-\infty < x < \infty)$, satisfying the inequalities

$$\begin{aligned}
\gamma_{k,q}(\phi) &:= \sup_{x\in\mathbb{R}} \left|x^k \phi^{(q)}(x)\right| \\
&\leq C_k B^q q^{q\beta} \quad (k, q = 0, 1, 2, \ldots),
\end{aligned} \tag{35}$$

where the constants B and C_k depend on the function ϕ.

Definition 8. The space $S_\alpha^\beta(\alpha \geq 0, \beta \geq 0)$ consists of all infinitely differentiable functions $\phi(x)$ $(-\infty < x < \infty)$, satisfying the inequalities

$$\begin{aligned}
\gamma_{k,q}(\phi) &:= \sup_{x\in\mathbb{R}} \left|x^k \phi^{(q)}(x)\right| \\
&\leq C A^k B^q k^{k\alpha} q^{q\beta} \quad (k, q = 0, 1, 2, \ldots),
\end{aligned} \tag{36}$$

where the constants A, B, and C depend on the function ϕ.

The spaces of type S are closely interrelated by means of the Fourier transformation; namely, the formulas

$$\widehat{S}_\alpha = S^\alpha, \qquad \widehat{S}^\beta = S_\beta, \qquad \widehat{S}_\alpha^\beta = S_\beta^\alpha \tag{37}$$

hold.

We will make use of the following inequalities in our investigation:

$$\frac{q!}{(q-k)!} = k! \binom{q}{k} \leq k! \sum_{k=0}^{q} \binom{q}{k} = k! 2^q, \tag{38}$$

$$(m+q)^{(m+q)\beta} \leq m^{m\beta} q^{q\beta} e^{m\beta} e^{q\beta} \quad [11, \text{ page } 265]. \tag{39}$$

In what follows we will also need the following similar test function space defined on $\mathbb{R} \times (\mathbb{R} \setminus \{0\})$, the space of type $\widetilde{S}$.

Definition 9. The space $\widetilde{S}_\alpha^\beta(\mathbb{R} \times (\mathbb{R} \setminus \{0\}))$, $\alpha, \beta \geq 0$, is defined to be the space of all functions $\phi \in C^\infty(\mathbb{R} \times (\mathbb{R} \setminus \{0\}))$ such that for all $s, t \in \mathbb{N}_0$,

$$\begin{aligned}
&\left|\frac{(f\csc\theta)^{2ps} \tau^s}{(1 + |f\csc\theta|)^t \left(2(f\csc\theta)^{2p} + |f\csc\theta| + 1\right)^s} \left(\frac{\partial}{\partial\tau}\right)^t \phi(\tau, f)\right| \\
&\quad \leq C_{\theta,s,t} A^s B^t s^{s\alpha} t^{t\beta},
\end{aligned} \tag{40}$$

where the constants A, B, and $C_{\theta,s,t}$ depend on ϕ.

We show that the fractional S-transform is a continuous linear map of the space of type S into the space of type $\widetilde{S}$ defined mentioned previously.

Theorem 10. *For $0 < p \leq 1$, the fractional S-transform is a continuous linear map of $S_\alpha(\mathbb{R})$ into $\widetilde{S}_{\widetilde{\alpha}}^{\widetilde{\beta}}(\mathbb{R} \times (\mathbb{R} \setminus \{0\}))$ where $\widetilde{\alpha} = (\alpha + 3)$ and $\widetilde{\beta} = 0$.*

Proof. Let $s, t \in \mathbb{N}$ and $I = \tau^s(\partial/\partial\tau)^t \mathrm{FRST}_u^a(\tau, f)$. Then, after differentiation and integration by parts, we obtain

$$
\begin{aligned}
I &= \tau^s\left(\frac{\partial}{\partial\tau}\right)^t A_\theta\left|A_\theta\right|^2 e^{i\pi f^2\cot\theta} \\
&\quad\times \int_{\mathbb{R}} e^{-i2\pi(f-f_1)\tau\csc\theta} \\
&\quad\times e^{-2\pi^2k^2(f-f_1)^2(\csc\theta)^2/(f\csc\theta)^{2p}} \\
&\quad\times \left[\widehat{e^{i\pi t^2\cot\theta}u(t)}\right](f_1\csc\theta)\,df_1 \\
&= \tau^s A_\theta\left|A_\theta\right|^2 e^{i\pi f^2\cot\theta} \\
&\quad\times \int_{\mathbb{R}} (-i2\pi(f-f_1)\csc\theta)^t e^{-i2\pi(f-f_1)\tau\csc\theta} \\
&\quad\times e^{-2\pi^2k^2(f-f_1)^2(\csc\theta)^2/(f\csc\theta)^{2p}} \\
&\quad\times \left[\widehat{e^{i\pi t^2\cot\theta}u(t)}\right](f_1\csc\theta)\,df_1 \\
&= A_\theta\left|A_\theta\right|^2 e^{i\pi f^2\cot\theta}(-i2\pi\csc\theta)^t \\
&\quad\times \int_{\mathbb{R}} \frac{1}{(-i2\pi\csc\theta)^s} e^{-i2\pi(f-f_1)\tau\csc\theta} \\
&\quad\times D_{f_1}^s (f-f_1)^t e^{-2\pi^2k^2(f-f_1)^2(\csc\theta)^2/(f\csc\theta)^{2p}} \\
&\quad\times \left[\widehat{e^{i\pi t^2\cot\theta}u(t)}\right](f_1\csc\theta)\,df_1 \\
&= A_\theta\left|A_\theta\right|^2 e^{i\pi f^2\cot\theta}(-i2\pi\csc\theta)^{t-s} \\
&\quad\times \int_{\mathbb{R}} e^{-i2\pi(f-f_1)\tau\csc\theta}\sum_{l=0}^{s}\binom{s}{l} D_{f_1}^{s-l}(f-f_1)^t \\
&\quad\times D_{f_1}^l\Big[e^{-2\pi^2k^2(f-f_1)^2(\csc\theta)^2/(f\csc\theta)^{2p}} \\
&\quad\times\left[\widehat{e^{i\pi t^2\cot\theta}u(t)}\right](f_1\csc\theta)\Big]\,df_1 \\
&= A_\theta\left|A_\theta\right|^2 e^{i\pi f^2\cot\theta}(-i2\pi\csc\theta)^{t-s} \\
&\quad\times \sum_{l=0}^{s}\binom{s}{l}\frac{t!(-1)^{s-l}}{(t-s+l)!} \\
&\quad\times \int_{\mathbb{R}} e^{-i2\pi(f-f_1)\tau\csc\theta}(f-f_1)^{t-s+l} \\
&\quad\times \sum_{m=0}^{l}\binom{l}{m} D_{f_1}^m\left[e^{-2[\pi k\csc\theta(f-f_1)/(f\csc\theta)^p]^2}\right] \\
&\quad\times D_{f_1}^{l-m}\left[\widehat{e^{i\pi t^2\cot\theta}u(t)}\right](f_1\csc\theta)\,df_1 \\
&= A_\theta\left|A_\theta\right|^2 e^{i\pi f^2\cot\theta}(-i2\pi\csc\theta)^{t-s} \\
&\quad\times \sum_{l=0}^{s}\binom{s}{l}\frac{t!(-1)^{s-l}}{(t-s+l)!}\sum_{m=0}^{l}\binom{l}{m} \\
&\quad\times \int_{\mathbb{R}} e^{-i2\pi(f-f_1)\tau\csc\theta}(f-f_1)^{t-s+l} \\
&\quad\times \left(-\frac{2\pi k\csc\theta}{(f\csc\theta)^p}\right)^m e^{-2[\pi k\csc\theta(f-f_1)/(f\csc\theta)^p]^2} \\
&\quad\times \sum_{n=0}^{[m/2]}\frac{(-1)^n m!(4\pi k\csc\theta)^{m-2n}}{n!\,(m-2n)!} \\
&\quad\times \left(\frac{f-f_1}{(f\csc\theta)^p}\right)^{(m-2n)} \\
&\quad\times D_{f_1}^{l-m}\left[\widehat{e^{i\pi t^2\cot\theta}u(t)}\right](f_1\csc\theta)\,df_1 \\
&= A_\theta\left|A_\theta\right|^2 e^{i\pi f^2\cot\theta}(-i2\pi\csc\theta)^{t-s} \\
&\quad\times \sum_{l=0}^{s}\binom{s}{l}\frac{t!(-1)^{s-l}}{(t-s+l)!}\sum_{m=0}^{l}\binom{l}{m} \\
&\quad\times \left(-\frac{2\pi k\csc\theta}{(f\csc\theta)^p}\right)^m \\
&\quad\times \sum_{n=0}^{[m/2]}\frac{(-1)^n m!(4\pi k\csc\theta)^{m-2n}}{n!\,(m-2n)!}\frac{1}{(f\csc\theta)^{p(m-2n)}} \\
&\quad\times \int_{\mathbb{R}} e^{-i2\pi(f-f_1)\tau\csc\theta}(f-f_1)^{t-s+l+m-2n} \\
&\quad\times e^{-2[\pi k\csc\theta(f-f_1)/(f\csc\theta)^p]^2} \\
&\quad\times D_{f_1}^{l-m}\left[\widehat{e^{i\pi t^2\cot\theta}u(t)}\right](f_1\csc\theta)\,df_1 \\
&= A_\theta\left|A_\theta\right|^2 e^{i\pi f^2\cot\theta}(-i2\pi\csc\theta)^{t-s} \\
&\quad\times \sum_{l=0}^{s}\binom{s}{l}\frac{t!(-1)^{s-l}}{(t-s+l)!}\sum_{m=0}^{l}\binom{l}{m} \\
&\quad\times \left(-\frac{2\pi k\csc\theta}{(f\csc\theta)^p}\right)^m \\
&\quad\times \sum_{n=0}^{[m/2]}\frac{(-1)^n m!(4\pi k\csc\theta)^{m-2n}}{n!\,(m-2n)!}\frac{1}{(f\csc\theta)^{p(m-2n)}} \\
&\quad\times \int_{\mathbb{R}} e^{-i2\pi(f-y\sin\theta)\tau\csc\theta}(f-y\sin\theta)^{t-s+l+m-2n} \\
&\quad\times e^{-2[\pi k\csc\theta(f-y\sin\theta)/(f\csc\theta)^p]^2} \\
&\quad\times (\sin\theta)^{-l+m+1} D_y^{l-m}\left[\widehat{e^{i\pi t^2\cot\theta}u(t)}\right](y)\,dy
\end{aligned}
$$

$$= A_\theta |A_\theta|^2 e^{i\pi f^2 \cot\theta} (-i2\pi\csc\theta)^{t-s}$$
$$\times \sum_{l=0}^{s} \binom{s}{l} \frac{t!(-1)^{s-l}}{(t-s+l)!} \sum_{m=0}^{l} \binom{l}{m}$$
$$\times \left(-\frac{2\pi k \csc\theta}{(f\csc\theta)^p}\right)^m$$
$$\times \sum_{n=0}^{[m/2]} \frac{(-1)^n m! (4\pi k \csc\theta)^{m-2n}}{n!\,(m-2n)!} \frac{1}{(f\csc\theta)^{p(m-2n)}}$$
$$\times (\sin\theta)^{t-s+2m-2n+1}$$
$$\times \int_{\mathbb{R}} e^{-i2\pi(f\csc\theta - y)\tau} (f\csc\theta - y)^{t-s+l+m-2n}$$
$$\times e^{-2[\pi k (f\csc\theta - y)/(f\csc\theta)^p]^2}$$
$$\times D_y^{l-m} \left[\widehat{e^{i\pi t^2 \cot\theta} u}(t) \right](y)\, dy$$
$$= A_\theta |A_\theta|^2 e^{i\pi f^2 \cot\theta} (-i2\pi\csc\theta)^{t-s}$$
$$\times \sum_{l=0}^{s} \binom{s}{l} \frac{t!}{(t-s+l)!} \sum_{m=0}^{l} \binom{l}{m}$$
$$\times \left(-\frac{2\pi k \csc\theta}{(f\csc\theta)^p}\right)^m$$
$$\times \sum_{n=0}^{[m/2]} \frac{(-1)^n m! (4\pi k \csc\theta)^{m-2n}}{n!\,(m-2n)!} \frac{1}{(f\csc\theta)^{p(m-2n)}}$$
$$\times (\sin\theta)^{t-s+2m-2n+1}$$
$$\times \int_{\mathbb{R}} e^{-i2\pi(f\csc\theta - y)\tau}$$
$$\times \sum_{r=0}^{t-s+l+m-2n} \binom{t-s+l+m-2n}{r}$$
$$\times (f\csc\theta)^{t-s+l+m-2n-r} (-y)^r$$
$$\times e^{-2[\pi k (f\csc\theta - y)/(f\csc\theta)^p]^2}$$
$$\times D_y^{l-m} \left[\widehat{e^{i\pi t^2 \cot\theta} u}(t) \right](y)\, dy. \tag{41}$$

By using the convergence factor $1+y^2$, we obtain

$$|I| \le |A_\theta|^3 (2\pi)^{t-s} \sum_{l=0}^{s} \binom{s}{l} \frac{t!}{(t-s+l)!} \sum_{m=0}^{l} \binom{l}{m}$$
$$\times \left| \frac{2\pi k}{(f\csc\theta)^p} \right|^m$$
$$\times \sum_{n=0}^{[m/2]} \frac{m!(4\pi k)^{m-2n}}{n!\,(m-2n)!} \frac{1}{|f\csc\theta|^{p(m-2n)}}$$
$$\times \sum_{r=0}^{t-s+l+m-2n} \binom{t-s+l+m-2n}{r}$$
$$\times |f\csc\theta|^{t-s+l+m-2n-r}$$
$$\times \int_{\mathbb{R}} \left| (1+y^2)\, y^r D_y^{l-m} \left[\widehat{e^{i\pi t^2 \cot\theta} u}(t) \right](y) \right|$$
$$\times \frac{1}{1+y^2} dy. \tag{42}$$

Using the property (38) we obtain

$$|I| \le C'_{\theta,s,t} \sum_{l=0}^{s} \binom{s}{l} (s-l)^{s-l} 2^t \sum_{m=0}^{l} \binom{l}{m}$$
$$\times \frac{1}{|f\csc\theta|^{pm}}$$
$$\times \sum_{n=0}^{[m/2]} \frac{(4\pi k)^{m-2n} (2n)^{2n} 2^m}{n!} \frac{1}{|f\csc\theta|^{p(m-2n)}}$$
$$\times \sum_{r=0}^{t-s+l+m-2n} \binom{t-s+l+m-2n}{r} \tag{43}$$
$$\times |f\csc\theta|^{t-s+l+m-2n-r}$$
$$\times \int_{\mathbb{R}} \left| (1+y^2)\, y^r D_y^{l-m} \right.$$
$$\left. \times \left[\widehat{e^{i\pi t^2 \cot\theta} u}(t) \right](y) \right| \frac{1}{1+y^2} dy.$$

Use the properties (35) and (39) to obtain

$$|I| \le C'_{\theta,s,t} \sum_{l=0}^{s} \binom{s}{l} s^s 2^t \sum_{m=0}^{l} \binom{l}{m}$$
$$\times \frac{1}{|f\csc\theta|^{pm}}$$
$$\times \sum_{n=0}^{[m/2]} \frac{(4\pi k)^{m-2n} (2n)^{2n} 2^m}{n!} \frac{1}{|f\csc\theta|^{p(m-2n)}}$$
$$\times \sum_{r=0}^{t-s+l+m-2n} \binom{t-s+l+m-2n}{r}$$
$$\times |f\csc\theta|^{t-s+l+m-2n-r}$$
$$\times C_r B^{l-m} (l-m)^{(l-m)\alpha}$$
$$\le C''_{\theta,s,t} s^s 2^t \sum_{l=0}^{s} \binom{s}{l} \sum_{m=0}^{l} \binom{l}{m}$$
$$\times \frac{1}{|f\csc\theta|^{pm}}$$

$$\begin{aligned}
&\times \sum_{n=0}^{[m/2]} \frac{(4\pi k)^{m-2n}(2n)^{2n}2^m}{n!} \frac{1}{|f\csc\theta|^{p(m-2n)}} \\
&\times \sum_{r=0}^{t-s+l+m-2n} \binom{t-s+l+m-2n}{r} \\
&\times |f\csc\theta|^{t-s+l+m-2n-r} B^l l^{l\alpha} \\
&\leq C''_{\theta,s,t} A_1^s B_1^t s^{(\alpha+3)s} \sum_{l=0}^{s} \binom{s}{l} \sum_{m=0}^{l} \binom{l}{m} \\
&\times \frac{1}{|f\csc\theta|^{pm}} \\
&\times \sum_{n=0}^{[m/2]} \frac{1}{|f\csc\theta|^{p(m-2n)}} (1+|f\csc\theta|)^{t-s+l+m-2n} \\
&= C''_{\theta,s,t} A_1^s B_1^t s^{(\alpha+3)s} \sum_{l=0}^{s} \binom{s}{l} \sum_{m=0}^{l} \binom{l}{m} \\
&\times \frac{1}{|f\csc\theta|^{pm}} \frac{(1+|f\csc\theta|)^{t-s+l+m}}{|f\csc\theta|^{pm}} \\
&\times \sum_{n=0}^{[m/2]} \left(\frac{|f\csc\theta|^{2p}}{(1+|f\csc\theta|)^2} \right)^n \\
&\leq C''_{\theta,s,t} A_1^s B_1^t s^{(\alpha+3)s} \sum_{l=0}^{s} \binom{s}{l} \\
&\times \sum_{m=0}^{l} \binom{l}{m} \frac{(1+|f\csc\theta|)^{t-s+l+m}}{|f\csc\theta|^{2pm}} \left[\frac{m}{2}\right] \\
&\leq C_{\theta,s,t} A_1^s B_1^t s^{(\alpha+3)s} (1+|f\csc\theta|)^t \\
&\times \sum_{l=0}^{s} \binom{s}{l} \sum_{m=0}^{l} \binom{l}{m} \frac{(1+|f\csc\theta|)^m}{|f\csc\theta|^{2pm}} \\
&= C_{\theta,s,t} A_1^s B_1^t s^{(\alpha+3)s} (1+|f\csc\theta|)^t \\
&\times \sum_{l=0}^{s} \binom{s}{l} \left(1 + \frac{1+|f\csc\theta|}{|f\csc\theta|^{2p}}\right)^l \\
&= C_{\theta,s,t} A_1^s B_1^t s^{(\alpha+3)s} (1+|f\csc\theta|)^t \\
&\times \left(2 + \frac{1+|f\csc\theta|}{|f\csc\theta|^{2p}}\right)^s \\
&= C_{\theta,s,t} A_1^s B_1^t s^{(\alpha+3)s} (1+|f\csc\theta|)^t \\
&\times \left(\frac{2|f\csc\theta|^{2p} + |f\csc\theta| + 1}{|f\csc\theta|^{2p}} \right)^s .
\end{aligned} \tag{44}$$

Hence, we obtain the estimate

$$\left| \frac{(f\csc\theta)^{2ps}\tau^s}{(1+|f\csc\theta|)^t \left(2(f\csc\theta)^{2p} + |f\csc\theta| + 1\right)^s} \times \left(\frac{\partial}{\partial\tau}\right)^t \mathrm{FRST}_u^a(\tau, f) \right| \leq C_{\theta,s,t} A_1^s B_1^t s^{(\alpha+3)s}, \tag{45}$$

where A_1 and B_1 are positive constants. □

Theorem 11. *For $0 < p \leq 1$, the fractional S-transform is a continuous linear map of $S^\alpha(\mathbb{R})$ into $\widetilde{S}_{\widetilde{\alpha}}^{\widetilde{\beta}}(\mathbb{R} \times (\mathbb{R} \setminus \{0\}))$ where $\widetilde{\alpha} = (\alpha + 3)$ and $\widetilde{\beta} = \alpha$.*

Theorem 12. *For $0 < p \leq 1$, the fractional S-transform is a continuous linear map of $S_\alpha^\beta(\mathbb{R})$ into $\widetilde{S}_{\widetilde{\alpha}}^{\widetilde{\beta}}(\mathbb{R} \times (\mathbb{R} \setminus \{0\}))$ where $\widetilde{\alpha} = (\alpha + 3)$ and $\widetilde{\beta} = \beta$.*

Proof. By inequality (43), we have

$$\begin{aligned}
|I| &\leq C'_{\theta,s,t} \sum_{l=0}^{s} \binom{s}{l} (s-l)^{s-l} 2^t \sum_{m=0}^{l} \binom{l}{m} \\
&\times \frac{1}{|f\csc\theta|^{pm}} \\
&\times \sum_{n=0}^{[m/2]} \frac{(4\pi k)^{m-2n}(2n)^{2n}2^m}{n!} \frac{1}{|f\csc\theta|^{p(m-2n)}} \\
&\times \sum_{r=0}^{t-s+l+m-2n} \binom{t-s+l+m-2n}{r} \\
&\times |f\csc\theta|^{t-s+l+m-2n-r} \\
&\times \int_{\mathbb{R}} \left| \left(1+y^2\right) y^r D_y^{l-m} \right. \\
&\times \left. \left[\widehat{e^{i\pi t^2 \cot\theta} u(t)} \right](y) \right| \frac{1}{1+y^2} dy.
\end{aligned} \tag{46}$$

Next, using the properties (36) and (39) we have

$$\begin{aligned}
|I| &\leq C'_{\theta,s,t} \sum_{l=0}^{s} \binom{s}{l} s^s 2^t \sum_{m=0}^{l} \binom{l}{m} \\
&\times \frac{1}{|f\csc\theta|^{pm}} \\
&\times \sum_{n=0}^{[m/2]} \frac{(4\pi k)^{m-2n}(2n)^{2n}2^m}{n!} \frac{1}{|f\csc\theta|^{p(m-2n)}} \\
&\times \sum_{r=0}^{t-s+l+m-2n} \binom{t-s+l+m-2n}{r} \\
&\times |f\csc\theta|^{t-s+l+m-2n-r}
\end{aligned}$$

$$
\begin{aligned}
&\times C\left(A_1^r r^{r\beta} + A_1^{(r+2)}(r+2)^{(r+2)\beta}\right)\\
&\times B_1^{(l-m)}(l-m)^{(l-m)\alpha}\\
\le{}& C'_{\theta,s,t}\sum_{l=0}^{s}\binom{s}{l}s^s 2^t\sum_{m=0}^{l}\binom{l}{m}\\
&\times\frac{1}{\left|f\csc\theta\right|^{pm}}\\
&\times\sum_{n=0}^{[m/2]}\frac{(4\pi k)^{m-2n}(2n)^{2n}2^m}{n!}\frac{1}{\left|f\csc\theta\right|^{p(m-2n)}}\\
&\times\sum_{r=0}^{t-s+l+m-2n}\binom{t-s+l+m-2n}{r}\\
&\qquad\times\left|f\csc\theta\right|^{t-s+l+m-2n-r}\\
&\times CA_2^r(r+2)^{(r+2)\beta}B_1^l l^{l\alpha}\\
\le{}& C'_{\theta,s,t}\sum_{l=0}^{s}\binom{s}{l}s^s 2^t\sum_{m=0}^{l}\binom{l}{m}\\
&\times\frac{1}{\left|f\csc\theta\right|^{pm}}\\
&\times\sum_{n=0}^{[m/2]}\frac{(4\pi k)^{m-2n}(2n)^{2n}2^m}{n!}\frac{1}{\left|f\csc\theta\right|^{p(m-2n)}}\\
&\times\sum_{r=0}^{t-s+l+m-2n}\binom{t-s+l+m-2n}{r}\\
&\qquad\times\left|f\csc\theta\right|^{t-s+l+m-2n-r}\\
&\qquad\times CA_2^r r^{r\beta}2^{2\beta}e^{r\beta}e^{2\beta}B_1^l l^{l\alpha}\\
\le{}& C_{\theta,s,t}A^sB^ts^{s(\alpha+3)}t^{t\beta}\\
&\times\sum_{l=0}^{s}\binom{s}{l}s^s 2^t\sum_{m=0}^{l}\binom{l}{m}\\
&\times\frac{1}{\left|f\csc\theta\right|^{pm}}\\
&\times\sum_{n=0}^{[m/2]}\frac{(4\pi k)^{m-2n}(2n)^{2n}2^m}{n!}\frac{1}{\left|f\csc\theta\right|^{p(m-2n)}}\\
&\times\sum_{r=0}^{t-s+l+m-2n}\binom{t-s+l+m-2n}{r}\\
&\qquad\times\left|f\csc\theta\right|^{t-s+l+m-2n-r}.
\end{aligned}
\tag{47}
$$

Next, by using the technique of Theorem 10, we obtain

$$
\begin{aligned}
|I| \le{}& C_{\theta,s,t}A^sB^ts^{s(\alpha+3)}t^{t\beta}\left(1+\left|f\csc\theta\right|\right)^t\\
&\times\left(\frac{2\left|f\csc\theta\right|^{2p}+\left|f\csc\theta\right|+1}{\left|f\csc\theta\right|^{2p}}\right)^s.
\end{aligned}
\tag{48}
$$

Hence, we obtain the estimate

$$
\begin{aligned}
&\left|\frac{(f\csc\theta)^{2ps}\tau^s}{\left(1+\left|f\csc\theta\right|\right)^t\left(2(f\csc\theta)^{2p}+\left|f\csc\theta\right|+1\right)^s}\right.\\
&\left.\times\left(\frac{\partial}{\partial\tau}\right)^t \mathrm{FRST}_u^a(\tau,f)\right| \le C_{\theta,s,t}A^sB^ts^{s(\alpha+3)}t^{t\beta},
\end{aligned}
\tag{49}
$$

where A and B are positive constants. □

6. Future Directions

The results of this paper will be useful in the investigation of time-frequency behaviour of test functions and distributions. We may use these results to obtain new results in modulation spaces, which may be defined by means of the fractional S-transform, by using the techniques developed in [14, Chapter 11–13].

Acknowledgment

The author would like to express their sincere thanks to Professor R. S. Pathak for his help and encouragement.

References

[1] R. G. Stockwell, L. Mansinha, and R. P. Lowe, "Localization of the complex spectrum: the S transform," *The IEEE Transactions on Signal Processing*, vol. 44, no. 4, pp. 998–1001, 1996.

[2] C. K. Chui, *An Introduction to Wavelets*, vol. 1, Academic Press, New York, NY, USA, 1992.

[3] L. Debnath, *Wavelet Transforms and Their Applications*, Birkhäuser, Boston, Mass, USA, 2002.

[4] S. Ventosa, C. Simon, M. Schimmel, J. J. Dañobeitia, and A. Mànuel, "The S-transform from a wavelet point of view," *IEEE Transactions on Signal Processing*, vol. 56, no. 7, pp. 2771–2780, 2008.

[5] A. H. Zemanian, *Generalized Integral Transformations*, Interscience Publishers, New York, NY, USA, 1996.

[6] S. K. Singh, "The S-transform on spaces of type S," *Integral Transforms and Special Functions*, vol. 23, no. 7, pp. 481–494, 2012.

[7] S. K. Singh, "The S-Transform on spaces of type W," *Integral Transforms and Special Functions*, vol. 23, no. 12, pp. 891–899, 2012.

[8] R. S. Pathak, *The Wavelet Transform*, vol. 4, Atlantis Press, Paris, France, 2009.

[9] L. B. Almeida, "The fractional Fourier transform and time-frequency representations," *The IEEE Transactions on Signal Processing*, vol. 42, no. 11, pp. 3084–3091, 1994.

[10] D. P. Xu and K. Guo, "Fractional *S* transform-part 1: theory," *Applied Geophysics*, vol. 9, no. 1, pp. 73–79, 2012.

[11] A. Friedman, *Generalized Functions and Partial Differential Equations*, Prentice-Hall, Englewood Cliffs, NJ, USA, 1963.

[12] I. M. Gel'fand and G. E. Shilov, *Generalized Functions*, vol. 2, Academic Press, New York, NY, USA, 1968.

[13] R. S. Pathak and S. K. Singh, "The wavelet transform on spaces of type *S*," *Proceedings of the Royal Society of Edinburgh A*, vol. 136, no. 4, pp. 837–850, 2006.

[14] K. Gröchenig, *Foundations of Time-Frequency Analysis*, Birkhäuser, Boston, Mass, USA, 2001.

2

On Solution Representation of Generalized Abel Integral Equation

Arsen Pskhu

Scientific Research Institute of Applied Mathematics and Automation, Kabardino-Balkar Scientific Centre, Russian Academy of Sciences, 89A Shortanov Street, Nalchik 360000, Russia

Correspondence should be addressed to Arsen Pskhu; pskhu@mail333.com

Academic Editor: Mike Tsionas

An explicit representation for solution of the generalized Abel integral equation $\sum_{k=0}^{m} \lambda_k D_{0x}^{-\alpha_k} u(x) = f(x)$, where $D_{0x}^{-\alpha_k}$ is the Riemann-Liouville fractional integral, in terms of the Wright function, is constructed.

1. Introduction

Consider the equation

$$\sum_{k=0}^{m} \lambda_k D_{0x}^{-\alpha_k} u(x) = f(x), \qquad (1)$$

where $\alpha_k \geq 0$, $\lambda_k \in \mathbb{R}$, $x \in (0, l)$, and D_{0x}^{β} is the Riemann-Liouville operator of fractional integrodifferentiation of order β, which is defined as follows:

$$D_{0x}^{\beta} \varphi(x) = \frac{1}{\Gamma(-\beta)} \int_0^x \varphi(t)(x-t)^{-\beta-1} dt \quad \text{for } \beta < 0,$$

$$D_{0x}^{0} \varphi(x) = \varphi(x),$$

$$D_{0x}^{\beta} \varphi(x) = \frac{d^n}{dx^n} D_{0x}^{\beta-n} \varphi(x) \quad \text{for } \beta \in (n-1, n], \; n \in \mathbb{N}. \qquad (2)$$

For suitable function $f(x)$, (1) with $m = 0$ and $\alpha_0 = \alpha > 0$, $\lambda_0 = 1$; that is, the Abel integral equation of the first kind (see [1])

$$D_{0x}^{-\alpha} u(x) = f(x) \qquad (3)$$

may be inverted by the formula

$$u(x) = D_{0x}^{\alpha} f(x) \qquad (4)$$

(see [2]). For $m = 1$, $\alpha_0 = 0$, $\alpha_1 = \alpha > 0$, $\lambda_0 = 1$, and $\lambda_1 = \lambda \neq 0$, (1) has the following form

$$u(x) + \lambda D_{0x}^{-\alpha} u(x) = f(x), \qquad (5)$$

which is the Abel equation of the second kind, and its solution may be given by the Hille-Tamarkin formula

$$u(x) = f(x) - \lambda \int_0^x (x-t)^{\alpha-1} E_{\alpha,\alpha}(-\lambda(x-t)^{\alpha}) f(t)\, dt \qquad (6)$$

(see [3]), which can be rewritten as

$$u(x) = \frac{d}{dx} \int_0^x E_{\alpha,1}(-\lambda(x-t)^{\alpha}) f(t)\, dt. \qquad (7)$$

Here,

$$E_{\alpha,\beta}(z) = \sum_{n=0}^{\infty} \frac{z^n}{\Gamma(\alpha n + \beta)} \qquad (8)$$

is the Mittag-Leffler function.

Equations of type (1) with rational parameters α_k were considered in [4, 5]. Generalized Abel integral equations of the second kind were investigated by operational method in [6]. For a more complete survey on results relating Abel type integral equations and fractional order integral and differential equations and their applications, we refer to [6–8].

In this paper, we obtain a representation for solution of (1). The results cover both cases, the solution of equation of the first kind ($\alpha_0 > 0$) and that of the second kind ($\alpha_0 = 0$). The solution is constructed in terms of the Wright function. It should be noted that (1) can be reduced to the generalized Abel integral equation of the second kind, and the method developed in [6] can be applied for (1). This provides an alternative approach to the equation under study.

2. Preliminaries

Now, we recall several properties of the operators of fractional integration and differentiation and special functions which are required for what follows.

The following three relations are valid:

$$D_{0x}^{\varepsilon} D_{0x}^{\delta} \varphi(x) = D_{0x}^{\varepsilon+\delta} \varphi(x) \tag{9}$$

for $\varepsilon \in \mathbb{R}$ and $\delta \le 0$ or $\varepsilon \in \mathbb{N}$ and $\delta \in \mathbb{R}$;

$$D_{0x}^{\varepsilon} D_{0x}^{\delta} \varphi(x) = D_{0x}^{\varepsilon+\delta} \varphi(x) - \sum_{j=1}^{n} \frac{x^{-\varepsilon-j}}{\Gamma(1-\varepsilon-j)} \left[D_{0x}^{\delta-j} \varphi(x) \right]_{x=0} \tag{10}$$

for $\varepsilon \in \mathbb{R}$ and $\delta \in (n-1, n]$, $n \in \mathbb{N}$;

$$D_{0x}^{\varepsilon} \frac{x^{\delta-1}}{\Gamma(\delta)} = \frac{x^{\delta-\varepsilon-1}}{\Gamma(\delta-\varepsilon)} \tag{11}$$

for $\varepsilon \in \mathbb{R}$ and $\delta > 0$ or $\varepsilon \in \mathbb{N}$ and $\delta \in \mathbb{R}$ (e.g., see [2] and [9, Section 2.1]).

By $(\varphi * \psi)(x)$, we denote the Laplace convolution of functions $\varphi(x)$ and $\psi(x)$:

$$(\varphi * \psi)(x) = \int_0^x \varphi(x-t)\, \psi(t)\, dt. \tag{12}$$

Operator D_{0x}^{β} may be written in the form of convolution

$$D_{0x}^{\beta} \varphi(x) = \varphi * \frac{x^{-\beta-1}}{\Gamma(-\beta)} \quad \text{for } \beta < 0,$$
$$D_{0x}^{\beta} \varphi(x) = \frac{d^n}{dx^n} \left(\varphi * \frac{x^{n-\beta-1}}{\Gamma(n-\beta)} \right) \tag{13}$$
$$\text{for } \beta \in (n-1, n], \quad n \in \mathbb{N}.$$

Using (11) and (13) and taking into account the associativity and commutativity of convolution, we get

$$\frac{1}{\prod_{i=1}^{m} \Gamma(\varepsilon_i)} \left(x^{\varepsilon_1 - 1} * \cdots * x^{\varepsilon_m - 1} \right) = \frac{x^{\varepsilon-1}}{\Gamma(\varepsilon)}, \tag{14}$$

where $\varepsilon = \sum_{j=1}^{m} \varepsilon_j$, $\varepsilon_j > 0$, and

$$\left(D_{0x}^{-\delta_1} \varphi_1 \right) * \cdots * \left(D_{0x}^{-\delta_m} \varphi_m \right) = D_{0x}^{-\delta} \left(\varphi_1 * \cdots * \varphi_m \right), \tag{15}$$

with $\delta = \sum_{j=1}^{m} \delta_j$ and $\delta_j \ge 0$. The relation

$$D_{0x}^{-\delta} (\varphi * \psi)(x) = \left(D_{0x}^{-\delta} \varphi * \psi \right)(x) = \left(\varphi * D_{0x}^{-\delta} \psi \right)(x) \tag{16}$$

is a particular case of (15).

Recall that the Wright function

$$\phi(\beta, \delta; z) = \sum_{n=0}^{\infty} \frac{z^n}{n! \Gamma(\beta n + \delta)} \tag{17}$$

satisfies the relations

$$\frac{d}{dz} \phi(\beta, \delta; z) = \phi(\beta, \delta + \beta; z) \tag{18}$$

(see [10]) and

$$D_{0x}^{\varepsilon} x^{\delta-1} \phi\left(\beta, \delta; z x^{\beta}\right) = x^{\delta-\varepsilon-1} \phi\left(\beta, \delta - \varepsilon; z x^{\beta}\right) \quad \text{for } \varepsilon \in \mathbb{R},\ \delta > 0. \tag{19}$$

For $\beta > 0$, relation (19) may be proved by termwise applying formula (11) to the representation of the Wright function as a series (17). Combining (18) and (19), we obtain

$$\frac{\partial}{\partial z} x^{\delta-1} \phi\left(\beta, \delta; \lambda z x^{\beta}\right) = \lambda D_{0x}^{-\beta} x^{\delta-1} \phi\left(\beta, \delta; \lambda z x^{\beta}\right) \quad (\lambda \in \mathbb{R}). \tag{20}$$

3. Auxiliary Results

Here, we investigate properties of functions in terms of which a solution of integral equation (1) will be constructed.

Consider the function

$$S_m^{\mu}(x; z_1, \ldots, z_m; \beta_1, \ldots, \beta_m) = (h_1 * h_2 * \cdots * h_m)(x) \tag{21}$$

(see [11]), where

$$h_k = h_k(x) = x^{\mu_k - 1} \phi\left(\beta_k, \mu_k; z_k x^{\beta_k}\right). \tag{22}$$

In what follows, the parameters and arguments of function (21) satisfy

$$x > 0, \qquad z_k \in \mathbb{R}, \qquad \beta_k > 0,$$
$$\mu_k > 0, \qquad \mu = \sum_{k=1}^{m} \mu_k. \tag{23}$$

Remark 1. The function $S_m^{\mu}(x; z_1, \ldots, z_m; \beta_1, \ldots, \beta_m)$ is independent of the distribution of parameters μ_k but depends only on their sum μ. Indeed, let S and $\widetilde{S}$ be the values of functions S_m^{μ} corresponding to the sets $\{\mu_1, \ldots, \mu_m\}$ and $\{\widetilde{\mu}_1, \ldots, \widetilde{\mu}_m\}$, with same sets of x, z_k, and β_k, and

$$\mu = \sum_{k=1}^{m} \mu_k = \sum_{k=1}^{m} \widetilde{\mu}_k. \tag{24}$$

Let

$$\nu = \min\left\{\mu_1, \ldots, \mu_m, \widetilde{\mu}_1, \ldots, \widetilde{\mu}_m\right\},$$
$$g_k = x^{\nu-1} \phi\left(\beta_k, \nu; z_k x^{\beta_k}\right). \tag{25}$$

Then, using (15) and (19), we obtain

$$S = \left(D_{0x}^{\nu-\mu_1} g_1\right) * \cdots * \left(D_{0x}^{\nu-\mu_m} g_m\right) = D_{0x}^{m\nu-\mu}\left(g_1 * \cdots * g_m\right) = \left(D_{0x}^{\nu-\tilde{\mu}_1} g_1\right) * \cdots * \left(D_{0x}^{\nu-\tilde{\mu}_m} g_m\right) = \widetilde{S}. \quad (26)$$

Lemma 2. *If*

$$Z = \max\left\{\left|z_1\right|, \ldots, \left|z_2\right|\right\}, \qquad \rho = \min\left\{\beta_1, \ldots, \beta_m\right\}, \qquad \tau = \max\left\{\beta_1, \ldots, \beta_m\right\}, \quad (27)$$

then

$$\left|S_m^{\mu}\left(x; z_1, \ldots, z_m; \beta_1, \ldots, \beta_m\right)\right| \leq C x^{\mu-1} \exp\left(\sigma Z^{1/(1+\rho)} x^{\tau/(1+\tau)}\right) \quad (28)$$

with positive constants C and σ independent of x and z_k.

Proof. Using the asymptotic formula of the Wright function (see [10]), we have

$$\left|\phi\left(\beta_k, \mu_k; z_k\right)\right| \leq C_k x^{\mu-1} \exp\left(\sigma_k \left|z_k\right|^{1/(1+\beta_k)} x_k^{\beta_k/(1+\beta_k)}\right), \quad (29)$$

where C_k and σ_k are positive constants independent of x and z_k. If we recall the definition (21), we get

$$\left|S_m^{\mu}\left(x; z_1, \ldots, z_m; \beta_1, \ldots, \beta_k\right)\right| \leq C\left(x^{\mu_1-1} * \cdots * x^{\mu_m-1}\right) \exp\left(\sigma Z^{1/(1+\rho)} x^{\tau/(1+\tau)}\right). \quad (30)$$

Using (14), we obtain (28). □

Lemma 3. *If $\mu > \nu$, then*

$$D_{0x}^{\nu} S_m^{\mu}\left(x; z_1, \ldots, z_m; \beta_1, \ldots, \beta_m\right) = S_m^{\mu-\nu}\left(x; z_1, \ldots, z_m; \beta_1, \ldots, \beta_m\right). \quad (31)$$

Proof. If $\nu < 0$, then combining (15), (19), and (21), we get

$$D_{0x}^{\nu} S_m^{\mu}\left(x; z_1, \ldots, z_m; \beta_1, \ldots, \beta_m\right) = \left(D_{0x}^{\nu} h_1\right) * \left(h_2 * \cdots * h_m\right)(x) = S_m^{\mu-\nu}\left(x; z_1, \ldots, z_m; \beta_1, \ldots, \beta_m\right). \quad (32)$$

Now, let $\nu \in (q-1, q]$, $q \in \mathbb{N}$. As follows from Remark 1, in (21), we can take a parameter μ_1 such that $\mu_1 > \nu$. Therefore, in view of the equality

$$\frac{d}{dx}(\phi * \psi)(x) = \left(\varphi' * \psi\right) - \varphi(0)\psi(x), \quad (33)$$

formula (19), and the relation

$$\lim_{x \to 0} D_{0x}^{\nu-j} h_1(x) = 0, \quad 1 \leq j \leq q, \quad (34)$$

we see that

$$D_{0x}^{\nu} S_m^{\mu}\left(x; z_1, \ldots, z_m; \beta_1, \ldots, \beta_m\right) = \frac{d^q}{dx^q}\left(D_{0x}^{\nu-q} h_1\right) * \left(h_2 * \cdots * h_m\right)(x) = \left(D_{0x}^{\nu} h_1\right) * \left(h_2 * \cdots * h_m\right)(x) = S_m^{\mu-\nu}\left(x; z_1, \ldots, z_m; \beta_1, \ldots, \beta_m\right). \quad (35)$$

This completes the proof. □

Now, we set

$$G_m^{\mu}\left(x; \gamma_1, \ldots, \gamma_m; \beta_1, \ldots, \beta_m\right) = \int_0^{\infty} e^{-t} S_m^{\mu}\left(x; \gamma_1 t, \ldots, \gamma_m t; \beta_1, \ldots, \beta_m\right) dt. \quad (36)$$

The convergence of the integral in (36) follows from (28).

Lemma 4. *The function $G_m^{\mu} = G_m^{\mu}(x; \gamma_1, \ldots, \gamma_m; \beta_1, \ldots, \beta_m)$ satisfies the relations*

$$G_m^{\mu} = O\left(x^{\mu-1}\right) \quad \text{as } x \longrightarrow 0, \quad (37)$$

$$D_{0x}^{\nu} G_m^{\mu} = G_m^{\mu-\nu} \quad \text{for } \mu > \nu. \quad (38)$$

Proof. From (28) and (36), we obtain

$$\left|G_m^{\mu}\right| \leq C x^{\mu-1} \int_0^{\infty} \exp\left(-t + \sigma(\gamma t)^{1/(1+\rho)} x^{\tau/(1+\tau)}\right) dt \leq C x^{\mu-1} \exp\left[\sup_{t>0}\left(-\frac{t}{2} + \sigma(\gamma t)^{1/(1+\rho)} x^{\tau/(1+\tau)}\right)\right] \times \int_0^{\infty} e^{-t/2} dt, \quad (39)$$

where

$$\rho = \min\left\{\beta_1, \ldots, \beta_m\right\}, \qquad \tau = \max\left\{\beta_1, \ldots, \beta_m\right\}, \qquad \gamma = \max\left\{\left|\gamma_1\right|, \ldots, \left|\gamma_m\right|\right\}, \quad (40)$$

and C and σ are positive constants independent of x, γ, and t. By the relation

$$\sup_{t>0}\left(-t + a t^{\varepsilon}\right) = a^{1/(1-\varepsilon)} \varepsilon^{\varepsilon/(1-\varepsilon)}(1-\varepsilon), \quad (41)$$

which holds for $a > 0$ and $\varepsilon \in (0, 1)$, (39) yields (37).

Relation (38) follows from (28), (31), and definition (36). □

Lemma 5. *Let $\lambda_0 \neq 0$, $0 \leq \alpha_0 < \alpha_k$ for $k = 1, 2, \ldots, m$, and $\mu > \alpha_0$. Denote that*

$$w_{\mu}(x) = \frac{1}{\lambda_0} G_m^{\mu-\alpha_0}\left(x; -\frac{\lambda_1}{\lambda_0}, \ldots, -\frac{\lambda_m}{\lambda_0}; \alpha_1 - \alpha_0, \ldots, \alpha_m - \alpha_0\right). \quad (42)$$

Then, the function $w_\mu(x)$ is a solution of the integral equation

$$\sum_{k=0}^{m} \lambda_k D_{0x}^{-\alpha_k} w_\mu(x) = \frac{x^{\mu-1}}{\Gamma(\mu)}. \tag{43}$$

Proof. Let

$$g_k = x^{\mu_k - 1}\phi\left(\alpha_k - \alpha_0, \mu_k; -\frac{\lambda_k t}{\lambda_0} x^{\alpha_k - \alpha_0}\right), \qquad \mu = \sum_{k=1}^{m} \mu_k. \tag{44}$$

Using (20), (21), and (31) and bearing in mind Remark 1, we get

$$\begin{aligned}
&\frac{\partial}{\partial t} S_m^\mu\left(x; -\frac{\lambda_1 t}{\lambda_0}, \dots, -\frac{\lambda_m t}{\lambda_0}; \alpha_1 - \alpha_0, \dots, \alpha_m - \alpha_0\right) \\
&= \sum_{k=1}^{m}\left(g_1 * \cdots * g_{k-1} * \frac{\partial g_k}{\partial t} * g_{k+1} * \cdots * g_m\right) \\
&= -\frac{1}{\lambda_0}\sum_{k=1}^{m} \lambda_k \left(g_1 * \cdots * g_{k-1} * D_{0x}^{\alpha_0 - \alpha_k} g_k * g_{k+1} * \cdots * g_m\right) \\
&= -\frac{1}{\lambda_0}\sum_{k=1}^{m} \lambda_k D_{0x}^{-\alpha_k} S_m^{\mu-\alpha_0}\left(x; -\frac{\lambda_1 t}{\lambda_0}, \dots, -\frac{\lambda_m t}{\lambda_0}; \alpha_1 - \alpha_0, \dots, \alpha_m - \alpha_0\right).
\end{aligned} \tag{45}$$

Hence, in view of (36) and (42),

$$\begin{aligned}
&\sum_{k=1}^{m} \lambda_k D_{0x}^{-\alpha_k} w_\mu(x) \\
&= -\int_0^\infty e^{-t}\frac{\partial}{\partial t} S_m^\mu\left(x; -\frac{\lambda_1 t}{\lambda_0}, \dots, -\frac{\lambda_m t}{\lambda_0}; \alpha_1 - \alpha_0, \dots, \alpha_m - \alpha_0\right) dt \\
&= S_m^\mu\left(x; -\frac{\lambda_1 t}{\lambda_0}, \dots, -\frac{\lambda_m t}{\lambda_0}; \alpha_1 - \alpha_0, \dots, \alpha_m - \alpha_0\right)\Bigg|_{t=0} - \lambda_0 D_{0x}^{-\alpha_0} w_\mu(x).
\end{aligned} \tag{46}$$

Taking account of the equality

$$\begin{aligned}
&S_m^\mu\left(x; -\frac{\lambda_1 t}{\lambda_0}, \dots, -\frac{\lambda_m t}{\lambda_0}; \alpha_1 - \alpha_0, \dots, \alpha_m - \alpha_0\right)\Bigg|_{t=0} \\
&= \frac{x^{\mu_1 - 1}}{\Gamma(\mu_1)} * \cdots * \frac{x^{\mu_m - 1}}{\Gamma(\mu_m)}
\end{aligned} \tag{47}$$

and (14), the relation (46) yields (43). □

4. Main Result

Theorem 6. *Let $\lambda_0 \neq 0$, $0 \leq \alpha_0 < \alpha_k$ for $k = 1, 2, \dots, m$, $f(x) \in L(0, l)$, and there exists a function $g(x) \in L(0, l)$ such that $f(x) = D_{0x}^{-\alpha_0} g(x)$. Then, (1) has a unique integrable solution $u(x)$, which for each $\mu > \alpha_0$ can be represented as follows:*

$$u(x) = D_{0x}^{\mu}\left(f * w_\mu\right)(x), \tag{48}$$

where $w_\mu = w_\mu(x)$ is defined by (42).

Proof. Let $u(x)$ be a solution of (1). Taking into account (16) and (43), we obtain

$$\begin{aligned}
f * w_\mu &= \left(\sum_{k=0}^{m} \lambda_k D_{0x}^{-\alpha_k} u\right) * w_\mu \\
&= u * \left(\sum_{k=0}^{m} \lambda_k D_{0x}^{-\alpha_k} w_\mu\right) \\
&= u * \frac{x^{\mu-1}}{\Gamma(\mu)},
\end{aligned} \tag{49}$$

or, in view of (13),

$$\left(f * w_\mu\right)(x) = D_{0x}^{-\mu} u(x). \tag{50}$$

Applying the operator D_{0x}^{μ} to both sides of this equality, bearing in mind (9), we obtain (48).

Now, we prove that a function $u(x)$ that is defined by (48) is a solution of (1). Let $\mu \in (q-1, q]$, $q \in \mathbb{N}$. Consider the function

$$v(x) = D_{0x}^{\mu-q}\left(f * w_\mu\right)(x). \tag{51}$$

Bearing in mind (16), (38), and (42), we get

$$v(x) = D_{0x}^{\mu-q}\left(D_{0x}^{-\alpha_0} g * w_\mu\right)(x) = \left(g * w_{q+\alpha_0}\right)(x). \tag{52}$$

Using (33), (37), and (38), we obtain

$$\begin{aligned}
D_{0x}^{\mu-j}\left(f * w_\mu\right)(x) &= \frac{d^{q-j}}{dx^{q-j}} v(x) \\
&= \left(g * w_{j+\alpha_0}\right)(x) \quad \text{for } j = 1, 2, \dots, q.
\end{aligned} \tag{53}$$

Taking this and relation (37) into account, we may conclude that the function $D_{0x}^{\mu-j}(f * w_\mu)(x)$ is absolutely continuous and

$$\lim_{x\to 0} D_{0x}^{\mu-j}\left(f * w_\mu\right)(x) = \frac{d^{q-j}}{dx^{q-j}} v(x)\Bigg|_{x=0} = 0 \tag{54}$$

for $j = 1, 2, \dots, q$, and

$$D_{0x}^{\mu}\left(f * w_\mu\right)(x) = \frac{d^q}{dx^q} v(x) \in L(0, l). \tag{55}$$

Using (10) and (54), we obtain

$$D_{0x}^{-\alpha_k} D_{0x}^{\mu} \left(f * w_\mu\right)(x) = D_{0x}^{\mu} D_{0x}^{-\alpha_k} \left(f * w_\mu\right)(x)$$
$$= D_{0x}^{\mu} \left(f * D_{0x}^{-\alpha_k} w_\mu\right)(x) \quad (56)$$
$$\text{for } k = 0, 1, \ldots, m.$$

Then, it follows that

$$\sum_{k=0}^{m} \lambda_k D_{0x}^{-\alpha_k} D_{0x}^{\mu} \left(f * w_\mu\right)(x) = D_{0x}^{\mu} \left(f * \sum_{k=0}^{m} \lambda_k D_{0x}^{-\alpha_k} w_\mu \right)(x)$$
$$= D_{0x}^{\mu} \left(f * \frac{x^{\mu-1}}{\Gamma(\mu)} \right)$$
$$= D_{0x}^{\mu} D_{0x}^{-\mu} f(x) = f(x). \quad (57)$$

This completes the proof. □

Remark 7. Let $\alpha_0 = \alpha$, $\lambda_0 = 1$, and $\lambda_k = 0$ for $k \geq 1$. In this case, (1) will appear like (3), and we get

$$w_\mu(x) = G_m^{\mu-\alpha}(x; 0, \ldots, 0; \alpha_1 - \alpha, \ldots, \alpha_m - \alpha)$$
$$= \frac{x^{\mu-\alpha-1}}{\Gamma(\mu-\alpha)}. \quad (58)$$

In addition, it follows from the equalities

$$G_1^{\mu}(x; \lambda; \beta) = \int_0^\infty e^{-t} x^{\mu-1} \phi\left(\beta, \mu; \lambda t x^\beta\right) dt$$
$$= \sum_{k=0}^{\infty} \frac{\lambda^k x^{\beta k + \mu - 1}}{k! \Gamma(\beta k + \mu)} \int_0^\infty e^{-t} t^k dt \quad (59)$$
$$= x^{\mu-1} E_{\beta,\mu}\left(\lambda x^\beta\right)$$

that if $m = 1$, $\alpha_0 = 0$, $\alpha_1 = \alpha > 0$, $\lambda_0 = 1$, and $\lambda_1 = \lambda \neq 0$, then

$$w_\mu(x) = G_1^{\mu}(x; -\lambda; \alpha) = x^{\mu-1} E_{\alpha,\mu}\left(-\lambda x^\alpha\right). \quad (60)$$

Thus, formula (48) coincides with (4) and (7) for the cases of (3) and (5) (i.e., for $m = 0$ and $m = 1$), respectively.

References

[1] N. H. Abel, "Auflösung einer mechanischen aufgabe," *Journal für Die Reine und Angewandte Mathematik*, vol. 1, pp. 153–157, 1826.

[2] J. D. Tamarkin, "On integrable solutions of Abel's integral equation," *Annals of Mathematics*, vol. 31, no. 2, pp. 219–229, 1930.

[3] E. Hille and J. D. Tamarkin, "On the theory of linear integral equations," *Annals of Mathematics*, vol. 31, no. 3, pp. 479–528, 1930.

[4] V. Kostitzin, "Sur une généralisation de l'équation intégrale d'Abel," *Comptes Rendus de l'Académie des Sciences*, vol. 224, pp. 885–887, 1947.

[5] P. Rieder, "Die Auflösung einer Klasse verallgemeinerter Abelscher Integralgleichungen," *Mathematische Zeitschrift*, vol. 109, pp. 29–52, 1969.

[6] R. Gorenflo and Y. Luchko, "Operational method for solving generalized Abel integral equation of second kind," *Integral Transforms and Special Functions*, vol. 5, no. 1-2, pp. 47–58, 1997.

[7] R. Gorenflo and S. Vessella, *Abel Integral Equations: Analysis and Applications*, vol. 1461, Springer, Berlin, Germany, 1991.

[8] S. G. Samko, A. A. Kilbas, and O. I. Marichev, *Fractional Integrals and Derivatives: Theory and Applications*, Gordon and Breach Science Publishers, Yverdon, Switzerland, 1993.

[9] A. A. Kilbas, H. M. Srivastava, and J. J. Trujillo, *Theory and Applications of Fractional Differential Equations*, vol. 204 of *North-Holland Mathematics Studies*, Elsevier, Amsterdam, The Netherlands, 2006.

[10] E. M. Wright, "On the coefficients of power series having exponential singularities," *The Journal of the London Mathematical Society*, vol. 8, no. 1, pp. 71–79, 1933.

[11] A. V. Pskhu, "Initial-value problem for a linear ordinary differential equation of noninteger order," *Sbornik: Mathematics*, vol. 202, no. 4, pp. 111–122, 2011.

3

The Periodic Solutions to Kawahara Equation by Means of the Auxiliary Equation with a Sixth-Degree Nonlinear Term

Zehra Pınar and Turgut Öziş

Department of Mathematics, Science Faculty, Ege University, Bornova, 35100 İzmir, Turkey

Correspondence should be addressed to Zehra Pınar; zehra.pnar@gmail.com

Academic Editor: Anjan Biswas

It is well known that different types of exact solutions of an auxiliary equation produce new types of exact travelling wave solutions to nonlinear equations. In this paper, by means of symbolic computation, the new solutions of original auxiliary equation of first-order nonlinear ordinary differential equation with a sixth-degree nonlinear term are presented to obtain novel exact solutions of the Kawahara equation. By the aid of the solutions of the original auxiliary equation, some other physically important nonlinear equations can be solved to construct novel exact solutions.

1. Introduction

Over the past decades a number of approximate methods for finding travelling wave solutions to nonlinear evolution equations have been proposed or developed and furthermore modified. The solutions to various evolution equations have been found by one or other of these methods. References [1–24] and references within cite some of this accomplishment. Among these methods, one of the current methods is so called auxiliary equation method [20–24]. The technique of this method consist of the solutions of the nonlinear evolution equations such that the target solutions of the nonlinear evolution equations can be expressed as a polynomial in a linearly independent elementary function which satisfies a particular ordinary differential equation which is named as auxiliary equation in general. Recently, to determine the solutions of nonlinear evolution equations, many exact solutions of various auxiliary equations have been utilized [20–32].

A concise observation shows that novel exact solutions of the auxiliary equations may fabricate original exact travelling wave solutions of nonlinear equations. Therefore, it is very important to find novel exact solutions of the auxiliary equations.

In this paper, we will examine the consequences of the choice of the auxiliary equation for determining the solutions of the nonlinear evolution equation in consideration and to seek more types of new exact solutions of nonlinear differential equations which satisfy a first-order nonlinear ordinary differential equation with a sixth-degree nonlinear term.

2. Auxiliary Equation Technique and Some Remarks

We assume that the given nonlinear partial differential equation for $u(x,t)$ to be in the form

$$P\left(u, u_x, u_t, u_{xx}, u_{xt}, u_{tt}, \ldots\right) = 0, \tag{1}$$

where P is a polynomial in its arguments, which include nonlinear terms and the highest-order derivatives.

Then using the transformation $u(x,t) = u(\xi), \xi = \mu(x - ct)$ one can reduce (1) to the ordinary differential equation

$$Q\left(u, u_\xi, u_{\xi\xi}, u_{\xi\xi\xi}, \ldots\right) = 0. \tag{2}$$

Now, we seek the exact travelling wave solutions of (2) by means of auxiliary equation method.

As it is well known, most exact travelling wave solutions [4–24] of nonlinear evolution equations were obtained on the assumption that the exact solution can be expressed as a finite expansion of linearly independent elementary functions. One common way of obtaining linearly independent elementary functions is to assume them as a solution of appropriate algebraic or differential equations which are exactly solvable.

Thus, it is natural to ask the following question: What type of finite expansion or what kind of function can be used to construct new travelling wave solutions?

Hence, for the solution of (2), let us assume that the exact solution can be expressed as simple expansion in the form of

$$u(\xi) = \sum_{i=0}^{N} a_i z^i(\xi), \tag{3}$$

where a_i, $(i = 0, 1, \ldots, N)$ are constants which will be determined later and function $z(\xi)$ is an appropriate function that yields new travelling wave solutions and is defined by the solution of the auxiliary equation considered.

Now, let us remember the process for finding the unknown coefficients a_i, $(i = 0, 1, \ldots, N)$, where $u(\xi) = \sum_{i=0}^{N} a_i z^i(\xi)$. Substituting the auxiliary equation into the given nonlinear equation and equating each coefficient of a power of $z(\xi)$ to zero yield an algebraic system. Hence, all coefficients a_i can be determined by solving the algebraic system and the parameter N is a positive integer and can be determined by balancing the highest order derivative terms and the highest power nonlinear terms in (3) in general.

In most auxiliary equation methods, the function $z(\xi)$ is defined as the solution of an auxiliary ordinary differential equation. Even if it is possible to construct more complex types of exact solutions for many auxiliary ordinary differential equations, one also encounters some difficulties. Hence, the following vital question arise: How do we find a variety of exact solutions of the auxiliary equations? Unfortunately, the answer of this question is not unique and various authors utilize different auxiliary ordinary differential equations which are exactly solvable.

For example, Nickel [25] has utilized the nonlinear differential equation

$$\left(\frac{dz}{d\xi}\right)^2 = \alpha z^4 + 4\beta z^3 + 6\gamma z^2 + 4\delta z + \varepsilon \equiv R(z) \tag{4}$$

as an auxiliary ordinary differential equation which transforms (2) into polynomial equation $P(z) = 0$, $z'Q(z) = 0$, and $P(z) + z'Q(z) = 0$, respectively. Vanishing coefficients in the polynomials $P(z) = 0$, $z'Q(z) = 0$ and $P(z) + z'Q(z) = 0$, respectively, lead to equations which partially determine the coefficients α, β, γ, δ, and ε in (4). We think that Nickel preferred (4) due to it is analytically solvability. Because, Whittaker and Watson [26] showed that the solution $z(\xi)$ of (4) can be expressed in terms of Weierstrass' elliptic function.

At the same time, Fan [27] developed a new direct algebraic method to seek additional types of new exact solutions of evolutionary nonlinear differential equations that can be expressed as polynomials in several elementary or special functions which satisfy a first-order nonlinear ordinary differential equation with a fourth-degree nonlinear term

$$\left(\frac{dz}{d\xi}\right)^2 = a_0 + a_1 z(\xi) + a_2 z^2(\xi) + a_3 z^3(\xi) + a_4 z^4(\xi), \tag{5}$$

where a_i, $(i = 0, 1, 3, 4)$ are constants. The method has been applied to find many exact traveling wave solutions of various types of nonlinear equations [27–29]. Sirendaorejii [21] and Abdou [30] utilized (5) for the case a_i, $(i = 2, 3, 4)$. Perhaps, it is significant to point out that the choice of the case, that is, a_i $(i = 2, 3, 4)$, permits analytical solution of the first-order nonlinear ordinary differential equation with a fourth-degree nonlinear term as Nickel's. This property of the nonlinear ordinary differential equation with a fourth-degree nonlinear term used in large in [20, 23], for example, for the case a_i $(i = 0, 2, 4)$ and in [21, 22] for the case a_i $(i = 2, 3, 4)$ for (5) to find many exact traveling wave solutions of various types of nonlinear equations.

In this paper, we seek for the solution of (2) in terms of ansatz (3) with $z(\xi)$ satisfying the following new auxiliary equation with a sixth-degree nonlinear term, that is,

$$\left(\frac{dz}{d\xi}\right)^2 = a_0 + a_2 z^2(\xi) + a_4 z^4(\xi) + a_6 z^6(\xi), \tag{6}$$

where a_i $(i = 0, 2, 4, 6)$ are real constants. Cases 1–8 report the new solutions of the auxiliary equation (6) with a sixth-degree nonlinear term under 8 distinct cases which include solutions including Bessel function and Lambert function and one can use these new solutions to seek exact travelling wave solutions for nonlinear equations.

Case 1. If

$$\left(\frac{dz}{d\xi}\right)^2 = a_0 + a_6 z^6(\xi) \tag{7}$$

then

$$z(\xi) = \text{Bessel}\, J\left(\frac{1}{6}, 2z^6\right) z\pi + 3\xi\Gamma\left(\frac{5}{6}\right). \tag{8}$$

Case 2. If

$$\left(\frac{dz}{d\xi}\right)^2 = a_4 z^4(\xi) + a_6 z^6(\xi) \tag{9}$$

then

$$z(\xi) = \frac{\sqrt{\left(-a_6 + a_4^2\xi^2 - 2a_4^2\xi c_1 + a_4^2 c_1^2\right) a_4}}{-a_6 + a_4^2\xi^2 - 2a_4^2\xi c_1 + a_4^2 c_1^2}. \tag{10}$$

Case 3. If

$$\left(\frac{dz}{d\xi}\right)^2 = a_0 + a_2 z^2(\xi) + a_6 z^6(\xi) \tag{11}$$

then

$$z(\xi) = \frac{\left(\left(-(3/8)\,\xi\left(\sqrt{-10a_6a_2}\,a_2/a_6\right)^{3/2} + (3/8)\left(\sqrt{-10a_6a_2}\,a_2/a_6\right)^{3/2}c_1 + a_0\sqrt{a_0\left(-8+9a_2\xi^2-18a_2\xi c_1+9a_2c_1^2\right)/a_2}/8a_6\right)a_2^2\right)^{1/3}}{a_2} + \frac{\sqrt{-10a_6a_2}\,a_2}{\left(2a_6\left(-(3/8)\,\xi\left(\sqrt{-10a_6a_2}\,a_2/a_6\right)^{3/2} + (3/8)\left(\sqrt{-10a_6a_2}\,a_2/a_6\right)^{3/2}c_1 + a_0\sqrt{a_0\left(-8+9a_2\xi^2-18a_2\xi c_1+9a_2c_1^2\right)/a_2}/8a_6\right)a_2^2\right)^{1/3}}. \tag{12}$$

Case 4. If

$$\left(\frac{dz}{d\xi}\right)^2 = a_2z^2(\xi) + a_4z^4(\xi) + a_6z^6(\xi) \tag{13}$$

then

$$z(\xi) = e^{\left(\sqrt{a_2}(-2\xi+c_1)\right)}\sqrt{2} \times 2\left(-\frac{a_6e^{\left(2\sqrt{a_2}(-2\xi+c_1)\right)}}{a_4\,\text{Lambert}\,W\left(-2a_6e^{\left(2\sqrt{a_2}(-2\xi+c_1)\right)}/3a_4\right)}\right)^{-1/2}. \tag{14}$$

Case 5. If

$$\left(\frac{dz}{d\xi}\right)^2 = a_0 + a_4z^4(\xi) + a_6z^6(\xi) \tag{15}$$

then

$$z(\xi) = \xi - \int^{z(\xi)} \frac{da}{\sqrt{a_0 + a_4a^4 + a_6a^6}} - c_1 = 0. \tag{16}$$

Case 6. If

$$\left(\frac{dz}{d\xi}\right)^2 = a_2z^2(\xi) + a_6z^6(\xi) \tag{17}$$

then

$$z(\xi) = e^{\left(\sqrt{a_2}(-\xi+c_1)\right)}2^{1/4} \times \left(\frac{a_6e^{\left(4\sqrt{a_2}(-\xi+c_1)\right)}}{a_2\,\text{Lambert}\,W\left(-a_6e^{\left(4\sqrt{a_2}(-\xi+c_1)\right)}/2a_2\right)}\right)^{-1/4}. \tag{18}$$

Case 7. If

$$\left(\frac{dz}{d\xi}\right)^2 = a_6z^6(\xi) \tag{19}$$

then

$$z(\xi) = \frac{1}{\sqrt{-2\xi\sqrt{a_6} + 2c_1\sqrt{a_6}}}. \tag{20}$$

Case 8. If

$$\left(\frac{dz}{d\xi}\right)^2 = a_0 + a_2z^2(\xi) + a_4z^4(\xi) + a_6z^6(\xi) \tag{21}$$

then

$$\begin{aligned} z(\xi) &= \Bigg(a_6 81^{2/3}\Bigg(\Bigg(\Bigg(-\frac{3}{8}\xi a_0^{3/2} + \frac{3}{8}a_0^{3/2}c_1 + \frac{3\sqrt{8}a_0}{8} \times \sqrt{\frac{a_0\left(-8a_6+9a_4^2\xi^2-18a_4^2\xi c_1+9a_4^2c_1^2\right)}{a_4^2}}\Bigg) \\ &\quad \times a_4^4\Bigg)\left(a_6^2\right)^{-1}\Bigg)^{1/3}\Bigg)\left(18a_4^2\right)^{-1} \\ &\quad + a_0 81^{1/3}\Bigg(2\Bigg(\Bigg(\Bigg(-\frac{3}{8}\xi a_0^{3/2} + \frac{3}{8}a_0^{3/2}c_1 + \frac{3\sqrt{8}a_0}{8} \times \sqrt{\frac{a_0\left(-8a_6+9a_4^2\xi^2-18a_4^2\xi c_1+9a_4^2c_1^2\right)}{a_4^2}}\Bigg) \\ &\quad \times a_4^4\Bigg)\left(a_6^2\right)^{-1}\Bigg)^{-1/3}\Bigg)^{-1}. \end{aligned} \tag{22}$$

As it is noticeable, by choosing the particular values of a_i $(i = 0, 2, 4, 6)$, (6) provides numerous types of special solutions. Due to limited space, we omit those results; instead we affirm the following remarks.

Hence, in ansatz (3), the parameter N is a positive integer and can be determined by balancing the higher-order derivative term and the highest power nonlinear terms in (2). The highest degree of $\partial^p u/\partial\xi^p$ can be calculated by

$$\mathcal{O}\left[\frac{\partial^p u}{\partial \xi^p}\right] = N + p, \quad p = 0, 1, 2, \ldots,$$
$$\mathcal{O}\left[u^q \frac{\partial^p u}{\partial \xi^p}\right] = qN + p, \quad q, p = 0, 1, 2, \ldots. \tag{23}$$

Consequently substituting (3) and (6) into (2) and equating the coefficients of all powers of $z(\xi)$ and $z^j(\xi)\sqrt{a_0 + a_2 z^2(\xi) + a_4 z^4(\xi) + a_6 z^6(\xi)}$ $(j = 0, 1, 2, \ldots)$ to zero in the resulting equation, several algebraic equations will be obtained. Then solving these algebraic equations by the symbolic computation system Maple, and combining (3) and the auxiliary equation (6), we can get the exact solutions for (1).

3. Exact Travelling Wave Solutions

In this section, we are going to combine the auxiliary equation method with the new solutions of original auxiliary equation (6) with a six-degree nonlinear term to consider the new solutions of the generalized nonlinear Klein-Gordon equation which is used as a test problem by a number of researchers [20, 21, 24].

The Kawahara equation is

$$u_t + uu_x + \alpha u_{xxx} - u_{xxxxx} = 0, \tag{24}$$

where α is nonzero arbitrary constant. Equation (24), proposed first by Kawahara in 1972, occurs in the theory of shallow water waves and plays an important role in the modeling of many physical phenomena such as plasma waves and magnetoacoustic waves. Making the transformation $u(x,t) = u(\xi)$, $\xi = x - ct$ and integrating once with respect to ξ, (24) becomes

$$-cu + \frac{1}{2}u^2 + \alpha u'' - u^{(4)} = 0. \tag{25}$$

From (23), we have $N = 4$. Therefore, the ansatz yields

$$u(\xi) = g_0 + g_1 z(\xi) + g_2 z^2(\xi) + g_3 z^3(\xi) + g_4 z^4(\xi) \tag{26}$$

and $z(\xi)$ may be determined from one of the Cases 1–8.

Now, for convenience, we give the calculation of typical two cases only for the practical purposes and the rest can be determined in a similar manner.

Example 1. We first consider the auxiliary equation with six-degree nonlinearity, that is,

$$\left(\frac{dz}{d\xi}\right)^2 = a_4 z^4(\xi) + a_6 z^6(\xi). \tag{27}$$

in Cases 1–8. The solution of (27)

$$z(\xi) = \frac{\sqrt{\left(-a_6 + a_4^2\xi^2 - 2a_4^2\xi c_1 + a_4^2 c_1^2\right) a_4}}{-a_6 + a_4^2\xi^2 - 2a_4^2\xi c_1 + a_4^2 c_1^2} \tag{28}$$

is, we belive, a new solution in the literature.

Hence, substituting (26) and (27) into (25) and letting each coefficient of $z^i(\xi)\sqrt{a_2 z^2(\xi) + a_6 z^6(\xi)}$, $(0 \le i \le 8)$ to be zero, we obtain

$$\begin{gathered}
-384 g_2 a_6 - 2688 g_4 a_4 = 0,\\
-105 g_1 a_6 - 1260 g_3 a_4 = 0,\\
24 a_6 \alpha g_4 + \frac{1}{2} g_4^3 - 840 g_4 a_4^2 - 480 a_6 a_4 g_2 = 0,\\
15 a_6 \alpha g_3 + g_3 g_4 - 360 g_3 a_4^2 - 120 a_6 a_4 g_1 = 0,\\
20 a_4 \alpha g_4 + \frac{1}{2} g_3^2 - 120 g_2 a_4^2 + 8 a_6 \alpha g_2 + a g_4 g_2 = 0,\\
3 a_6 \alpha g_1 + g_1 g_4 + g_2 g_3 - 24 g_1 a_4^2 + 12 \alpha a_4 g_3 = 0,\\
g_0 g_4 + g_1 g_3 - c g_4 + \frac{1}{2} g_2^2 + 6 \alpha a_4 g_2 = 0,\\
g_0 g_3 + g_1 g_2 - c g_3 + 2 \alpha a_4 g_1 = 0,\\
g_0 g_2 - c g_2 + \frac{1}{2} g_1^2 = 0,\\
g_0 - c = 0.
\end{gathered} \tag{29}$$

Solving the system (29) by the aid of Maple 13, we can determine the coefficients:

$$\begin{gathered}
g_0 = \frac{4c}{3}, \quad g_2 = g_1 = g_3 = 0, \quad \mu = \mu,\\
a_4 = a_4, \quad a_6 = a_6,\\
c = 8064 a_4 - 72 \alpha a_6 - \frac{3}{2} g_4 + 2520 a_4^2 - 60 \alpha a_4.
\end{gathered} \tag{30}$$

Substituting the above coefficients into ansatz (26) with the solution (28) of auxiliary equation, we obtain one of the new solutions of Kawahara equation:

$$\begin{aligned}
u(x,t) := {} & 10752 a_4 - 96 \alpha a_6 - 2 g_4 + 3360 a_4^2 \\
& - 80 a_4 \alpha + g_4 a_4^2 \\
& \times \Bigg(-a_6 + a_4^2 \Big(x + \Big(8064 a_4 - 72 \alpha a_6 \\
& - \frac{3}{2} g_4 + 2520 a_4^2 - 60 a_4 \alpha\Big) t\Big)^2 \\
& - 2_\mathrm{Cl} a_4^2 \Big(x + \Big(8064 a_4 - 72 \alpha a_6 - \frac{3}{2} g_4 \\
& + 2520 a_4^2 - 60 a_4 \alpha\Big) t\Big) + a_4^2 _\mathrm{Cl}^2\Bigg)^{-2}.
\end{aligned} \tag{31}$$

The graph of (31) is illustrated for selected values of coefficients given in Figure 1.

Example 2. We, next, consider the auiliary equation with six-degree nonlinearity, that is,

$$\left(\frac{dz}{d\xi}\right)^2 = a_2 z^2(\xi) + a_6 z^6(\xi) \tag{32}$$

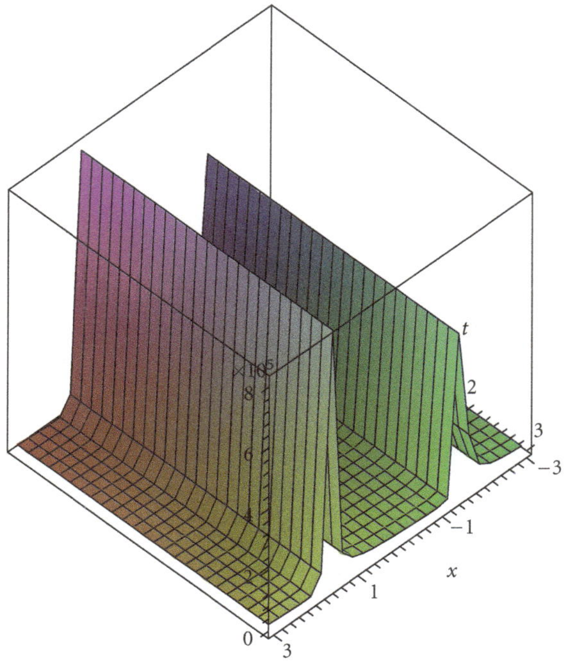

FIGURE 1: Graph of (31) for $g_4 = 1$, $\alpha = \mu = 1$, $a_4 = \tan x$, $a_6 = -1$.

in Cases 1–8. The solution of (32)

$$z(\xi) = e^{(\sqrt{a_2}(-\xi+c_1))} 2^{1/4} \times \left(a_6 e^{(4\sqrt{a_2}(-\xi+c_1))}\left(a_2 \text{LambertW}\left(-\frac{a_6 e^{(4\sqrt{a_2}(-\xi+c_1))}}{2a_2}\right)\right)^{-1}\right)^{-1/4} \quad (33)$$

is, we belive, a new solution in the literature.

Substituting (26) and (32) into (25) and letting each coefficient of $z^i(\xi)\sqrt{a_2 z^2(\xi) + a_6 z^6(\xi)}$, $(0 \le i \le 8)$ to be zero, we obtain

$$50\alpha - 3188 a_2 = 0,$$
$$30\alpha a_2 - 354 a_2^2 - 4c = 0. \quad (34)$$

Solving the system (34) by the aid of Maple 13, we can determine the coefficients:

$$c = \frac{487125\alpha^2}{5081672}, \quad g_0 = g_0, \quad g_1 = g_1, \quad g_2 = g_2,$$
$$g_3 = g_3, \quad g_4 = g_4, \quad \mu = \mu,$$
$$a_2 = \frac{25\alpha}{1594}, \quad a_6 = a_6. \quad (35)$$

Substituting the above coefficients into ansatz (26) with the solution (33) of auxiliary equation, we obtain another new solution of Kawahara equation:

$$u(x,t) := g_0 + g_1 \left(e^{\left(-\left(\frac{1}{4}\right) \text{lambert}W\left(-\frac{797}{25} \frac{a_6 e^{\left(\frac{2\sqrt{25}\sqrt{1594}\sqrt{\alpha}\left(-x - \frac{487125\alpha^2 t}{5081672} + \_Cl\right)}{797}\right)}}{\alpha} \right) - \frac{\sqrt{25}\sqrt{1594}\sqrt{\alpha}\left(x + \frac{487125\alpha^2 t}{5081672}\right)}{1594} + \frac{\sqrt{25}\sqrt{1594}\sqrt{\alpha}\_Cl}{1594} \right)} \right)$$

$$+g_2\left(e^{-\left(\frac{1}{4}\right)\text{lambertW}\left(-\frac{797}{25}\frac{a_6 e^{\left(\frac{2\sqrt{25}\sqrt{1594}\sqrt{\alpha}\left(-x-\frac{487125\alpha^2 t}{5081672}+\_C1\right)}{797}\right)}}{\alpha}\right)-\frac{\sqrt{25}\sqrt{1594}\sqrt{\alpha}\left(x+\frac{487125\alpha^2 t}{5081672}\right)}{1594}+\frac{\sqrt{25}\sqrt{1594}\sqrt{\alpha}\_C1}{1594}}\right)^2$$

$$+g_3\left(e^{-\left(\frac{1}{4}\right)\text{lambertW}\left(-\frac{797}{25}\frac{a_6 e^{\left(\frac{2\sqrt{25}\sqrt{1594}\sqrt{\alpha}\left(-x-\frac{487125\alpha^2 t}{5081672}+\_C1\right)}{797}\right)}}{\alpha}\right)-\frac{\sqrt{25}\sqrt{1594}\sqrt{\alpha}\left(x+\frac{487125\alpha^2 t}{5081672}\right)}{1594}+\frac{\sqrt{25}\sqrt{1594}\sqrt{\alpha}\_C1}{1594}}\right)^3$$

$$+g_4\left(e^{-\left(\frac{1}{4}\right)\text{lambertW}\left(-\frac{797}{25}\frac{a_6 e^{\left(\frac{2\sqrt{25}\sqrt{1594}\sqrt{\alpha}\left(-x-\frac{487125\alpha^2 t}{5081672}+\_C1\right)}{797}\right)}}{\alpha}\right)-\frac{\sqrt{25}\sqrt{1594}\sqrt{\alpha}\left(x+\frac{487125\alpha^2 t}{5081672}\right)}{1594}+\frac{\sqrt{25}\sqrt{1594}\sqrt{\alpha}\_C1}{1594}}\right)^4. \quad (36)$$

The graph of (36) is illustrated for selected values of coefficients given in Figure 2.

4. Conclusion

As is seen, the key idea of obtaining new travelling wave solutions for the nonlinear equations is constructing different types of solutions of the given auxiliary equation. In this paper, the exact solutions of the auxiliary equation with six-degree nonlinearity (6) is

$$\left(\frac{dz}{d\xi}\right)^2 = a_0 + a_2 z^2(\xi) + a_4 z^4(\xi) + a_6 z^6(\xi), \quad (37)$$

where a_i ($i = 0, 2, 4, 6$) are real constants, used to construct the solutions of Klein-Gordon equation [32]. Using these solutions, we have successfully obtained some new exact periodic solutions of the Kawahara equation. For comparative purposes, it is noticeable that Wu et al. in [20] and Lv et al. in [23] have utilized (5) for a_i ($i = 0, 2, 4$), and Sirendaoreji in [21] and Jang in [22] have applied the ansatz $(dz/d\xi)^2 = a_2 z^2(\xi) + a_3 z^3(\xi) + a_4 z^4(\xi)$ which only covers, at most, fourth-order nonlinear term in auxiliary ordinary differential equation, and therefore their solutions are limited in the quantity to the order considered.

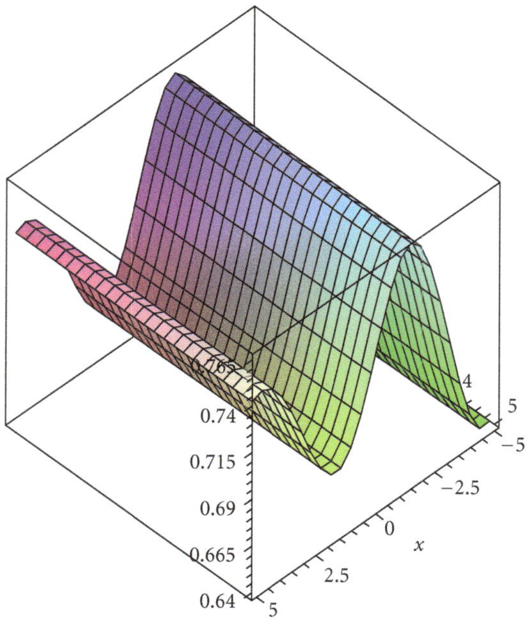

FIGURE 2: Graph of (36) for $\alpha = 1/2, a_6 = g_0 = g_2 = 1, g_1 = g_3 = -1, g_4 = \sin x$.

In this letter, we have obtained exact solutions of the auxiliary equation with sixth-degree nonlinearity (6) for distinct cases (see, Cases 1–8). Therefore, our approach gives additional new solutions beside the solutions obtained via fourth order auxiliary equation, and theoretically, some of our solutions may coincide with the solutions of Wu et al. in [20], Sirendaoreji in [21], Jang in [22], Lv et al. in [23], and some cases of Yomba [24] for certain choice of the parameters which we have left as an exercise.

However, it is well known that different types of auxiliary equations produce new travelling wave solutions for many nonlinear problems. Hence this also is our future work. The presented method could lead to finding new exact travelling wave solutions for other nonlinear problems. (33) of auxiliary equation, we obtain another new solution of Kawahara

References

[1] G. B. Whitham, "A general approach to linear and nonlinear waves using a Lagrangian," *Journal of Fluid Mechanics*, vol. 22, pp. 273–283, 1965.

[2] R. Knobel, *An Introduction to the Mathematical Theory of Waves*, American Mathematical Society, Providence, RI, USA, 2000.

[3] S. S. Shen, *A Course on Nonlinear Waves*, Kluwer, Dordrecht, The Netherlands, 1994.

[4] C. Yong, L. Biao, and Z. Hong-Qing, "Generalized Riccati equation expansion method and its application to the Bogoyavlenskii's generalized breaking soliton equation," *Chinese Physics*, vol. 12, no. 9, pp. 940–946, 2003.

[5] Y. Zhou, M. Wang, and Y. Wang, "Periodic wave solutions to a coupled KdV equations with variable coefficients," *Physics Letters A*, vol. 308, no. 1, pp. 31–36, 2003.

[6] G. Cai, Q. Wang, and J. Huang, "A modified F-expansion method for solving breaking soliton equation," *International Journal of Nonlinear Science*, vol. 2, pp. 122–128, 2006.

[7] X. Zeng and X. Yong, "A new mapping method and its applications to nonlinear partial differential equations," *Physics Letters A*, vol. 372, no. 44, pp. 6602–6607, 2008.

[8] X. Yong, X. Zeng, Z. Zhang, and Y. Chen, "Symbolic computation of Jacobi elliptic function solutions to nonlinear differential-difference equations," *Computers & Mathematics With Applications*, vol. 57, no. 7, pp. 1107–1114, 2009.

[9] T. Öziş and I. Aslan, "Exact and explicit solutions to the (3 + 1)-dimensional Jimbo-Miwa equation via the Exp-function method," *Physics Letters A*, vol. 372, no. 47, pp. 7011–7015, 2008.

[10] T. Öziş and C. Köroğlu, "A novel approach for solving the Fisher equation using Exp-function method," *Physics Letters A*, vol. 372, no. 21, pp. 3836–3840, 2008.

[11] A. Yıldırım and Z. Pınar, "Application of the Exp-function method for solving nonlinear reactiondiffusion equations arising in mathematical biology," *Computers and Mathematics with Applications*, vol. 60, no. 7, pp. 1873–1880, 2010.

[12] W. X. Ma, T. Huang, and Y. Zhang, "A multiple Exp-function method for nonlinear differential equations and its applications," *Physica Scripta*, vol. 82, Article ID 065003, 8 pages, 2010.

[13] T. Öziş and I. Aslan, "Application of the G'/G-expansion method to Kawahara type equations using symbolic computation," *Applied Mathematics and Computation*, vol. 216, no. 8, pp. 2360–2365, 2010.

[14] T. Öziş and I. Aslan, "Symbolic computation and construction of new exact traveling wave solutions to Fitzhugh-Nagumo and Klein-Gordon equations," *Zeitschrift für Naturforschung*, vol. 64, no. 1-2, pp. 15–20, 2009.

[15] I. Aslan and T. Öziş, "Analytic study on two nonlinear evolution equations by using the (G'/G)-expansion method," *Applied Mathematics and Computation*, vol. 209, no. 2, pp. 425–429, 2009.

[16] T. Öziş and M. Aslan, "Symbolic computations and exact and explicit solutions of some nonlinear evolution equations in mathematical physics," *Communications in Theoretical Physics*, vol. 51, no. 4, pp. 577–580, 2009.

[17] H. Zhang, "A note on some sub-equation methods and new types of exact travelling wave solutions for two nonlinear partial differential equations," *Acta Applicandae Mathematicae*, vol. 106, no. 2, pp. 241–249, 2009.

[18] B. Li, Y. Chen, and Y. Q. Li, "A generalized sub-equation expansion method and some analytical solutions to the inhomogeneous higher-order nonlinear Schrödinger equation," *Zeitschrift für Naturforschung*, vol. 63, no. 12, pp. 763–777, 2008.

[19] E. Yomba, "The modified extended Fan sub-equation method and its application to the (2+1)-dimensional Broer-Kaup-Kupershmidt equation," *Chaos, Solitons and Fractals*, vol. 27, no. 1, pp. 187–196, 2006.

[20] G. J. Wu, J. H. Han, W. L. Zhang, M. Zhang, and J. M. Wang, "New periodic wave solutions to generalized Klein-Gordon and Benjamin equations," *Communications in Theoretical Physics*, vol. 48, no. 5, pp. 815–818, 2007.

[21] Sirendaoreji, "Auxiliary equation method and new solutions of Klein-Gordon equations," *Chaos, Solitons and Fractals*, vol. 31, no. 4, pp. 943–950, 2007.

[22] B. Jang, "New exact travelling wave solutions of nonlinear Klein-Gordon equations," *Chaos, Solitons and Fractals*, vol. 41, no. 2, pp. 646–654, 2009.

[23] X. Lv, S. Lai, and Y. Wu, "An auxiliary equation technique and exact solutions for a nonlinear Klein-Gordon equation," *Chaos, Solitons and Fractals*, vol. 41, no. 1, pp. 82–90, 2009.

[24] E. Yomba, "A generalized auxiliary equation method and its application to nonlinear Klein-Gordon and generalized nonlinear Camassa-Holm equations," *Physics Letters A*, vol. 372, no. 7, pp. 1048–1060, 2008.

[25] J. Nickel, "Elliptic solutions to a generalized BBM equation," *Physics Letters A*, vol. 364, no. 3-4, pp. 221–226, 2007.

[26] E. T. Whittaker and G. N. Watson, *A Course of Modern Analysis*, Cambridge University Press, Cambridge, UK, 1927.

[27] E. Fan, "Multiple travelling wave solutions of nonlinear evolution equations using a unified algebraic method," *Journal of Physics A*, vol. 35, no. 32, pp. 6853–6872, 2002.

[28] E. Yomba, "The extended Fan's sub-equation method and its application to KdV-MKdV, BKK and variant Boussinesq equations," *Physics Letters A*, vol. 336, no. 6, pp. 463–476, 2005.

[29] Z. Yan, "An improved algebra method and its applications in nonlinear wave equations," *Chaos, Solitons and Fractals*, vol. 21, no. 4, pp. 1013–1021, 2004.

[30] M. A. Abdou, "A generalized auxiliary equation method and its applications," *Nonlinear Dynamics*, vol. 52, no. 1-2, pp. 95–102, 2008.

[31] D. J. Huang and H. Q. Zhang, "New exact travelling waves solutions to the combined KDV-MKDV and generalized Zakharov equations," *Reports on Mathematical Physics*, vol. 57, no. 2, pp. 257–269, 2006.

[32] Z. Pınar and T. Öziş, "An observation on the periodic solutions to nonlinear physical models by means of the auxiliary equation with a sixth-degree nonlinear term," *Communications in Nonlinear Science and Numerical Simulation*, 2013.

Categorical Abstract Algebraic Logic: Meet-Combination of Logical Systems

George Voutsadakis

School of Mathematics and Computer Science, Lake Superior State University, Sault Sainte Marie, MI 49783, USA

Correspondence should be addressed to George Voutsadakis; gvoutsad@yahoo.com

Academic Editor: Abdul Hamid Kara

The widespread and rapid proliferation of logical systems in several areas of computer science has led to a resurgence of interest in various methods for combining logical systems and in investigations into the properties inherited by the resulting combinations. One of the oldest such methods is *fibring*. In fibring the shared connectives of the combined logics inherit properties from *both* component logical systems, and this leads often to inconsistencies. To deal with such undesired effects, Sernadas et al. (2011, 2012) have recently introduced a novel way of combining logics, called *meet-combination*, in which the combined connectives share only the *common* logical properties they enjoy in the component systems. In their investigations they provide a sound and concretely complete calculus for the meet-combination based on available sound and complete calculi for the component systems. In this work, an effort is made to abstract those results to a categorical level amenable to *categorical abstract algebraic logic* techniques.

1. Introduction

The widespread and rapid proliferation of logical systems in several areas of computer science has led to a resurgence of interest in various methods for combining logical systems and in investigations into the properties inherited by the resulting combinations. One of the oldest methods for combining connectives is *fibring* [1]. In fibring one combines two logical systems by possibly imposing some sharing of common connectives or identification of connectives from the constituent logical systems. When such interaction occurs, the combined connectives inherit all properties of the components from both logical systems, and this leads often to inconsistencies. A typical example of this strong interaction is the combination of an intuitionistic negation from one logical system with a classical negation from another. The combined connective behaves like a classical negation, and this outcome defeats any intended purpose for the combination. Fibring has been studied substantially since its original introduction, and both its virtues and its vices are relatively well understood. For instance in [2], fibring was presented as a categorical construction (see also [3]), in [4] fibred logical systems were investigated from the point of view of preserving completeness, in [5] some work was carried out on the effect of fibring in logics belonging to specific classes of the classical abstract algebraic logic hierarchy [6–8], and more recently, in [9] fibring was employed to obtain some modal logics, first considered in [10], in a structured way and to draw some conclusions regarding their algebraic character.

To avoid some of the drawbacks and undesired effects involved in the application of fibring, Sernadas et al. [11, 12] introduced, recently, another way of combining logical systems, called *meet-combination*, in which the combined connectives, instead of inheriting all properties they enjoy in the component logical systems, inherit only those properties that are common to both connectives. A very illuminating example of the difference that this entails as contrasted to the fibring method consists of the result of combining two logics $\mathscr{L}_\wedge$ and $\mathscr{L}_\vee$, one including a classical conjunction $\wedge$ and one including a classical disjunction $\vee$, with the intention of obtaining a combined connective "identifying" these two connectives from the component logics. Roughly speaking, if fibring is used, then, since in the combination the combined connective $[\wedge\vee]$ has all properties that are enjoyed by each of the connectives in either logic, the derivation

$$\begin{array}{ll} \phi & \\ \phi\,[\wedge\vee]\,\psi & \text{(by the Property of Disjunction in } \mathscr{L}_2) \\ \psi & \text{(by the Property of Conjunction in } \mathscr{L}_1) \end{array} \quad (1)$$

shows that in the combined logic a single formula entails all other formulas; that is, there are only two possible theories, the empty theory and the entire set of formulas. On the other hand, this derivation would not be valid in the meet-combination of the two logics, since the afore-used Properties of Disjunction and Conjunction in $\mathscr{L}_2$ and $\mathscr{L}_1$, respectively, are not shared by $\wedge$ in $\mathscr{L}_1$ and by $\vee$ in $\mathscr{L}_2$, respectively. Commutativity, however, is a shared property, whence the derived rule $\phi[\wedge\vee]\psi/\psi[\wedge\vee]\phi$ is a derived rule of the meet-combination.

In [11] Sernadas et al. start from a given logical system $\mathscr{L}$ with a Hilbert style calculus and with a matrix semantics and define a new logic $\mathscr{L}^\times$ that incorporates all meet-combinations of connectives of $\mathscr{L}$ of the same arity. Moreover, this system includes in a canonical way the connectives of the original logical system. Roughly speaking, the Hilbert calculus of the combination consists of all old Hilbert rules plus two new rules that ensure that the combined connectives inherit the common properties of the component connectives and only those properties. The matrix semantics consists, also roughly speaking, of the direct squares of the matrices in the original matrix semantics. In the main results, [11, Theorems 3.9 and 3.13], it is shown that soundness and a special form of completeness, called concrete completeness, are inherited in $\mathscr{L}^\times$ from $\mathscr{L}$. Moreover, Sernadas et al. [11] investigate in some detail the case of classical propositional logic, which constitutes the main motivation and paradigmatic example behind their work. Based on classical propositional calculus, they present several interesting examples, which, in addition, serve as illustrations for various sensitive points of the general theory.

In the present paper, we adapt the framework of [11] to a categorical level, using notions and techniques of categorical abstract algebraic logic [13, 14]. Our main goal is providing a framework in which, starting from a π-institution whose closure system is axiomatized by a set of rules of inference, we may construct a new π-institution that includes, in a precise technical sense, natural transformations corresponding to meet-combinations of operations available in the original π-institution. The closure system of this new π-institution is created by essentially mimicking the process of [11] to create a new set of rules of inference, suitable for the new sentence functor, and by using this new set of rules to define the inferences in the newly created structure. Under conditions analogous to those imposed by Sernadas et al. in [11], we are also able to establish a form of soundness and a form of restricted completeness for the new system, with respect to a suitably constructed matrix system semantics, under the proviso that these properties are satisfied by the original system.

We close this section by providing an outline of the contents of the paper. In Section 2, we introduce the basic notions underlying the framework in which our work will be carried out. The inspiration comes from categorical abstract algebraic logic [13, 14] and, more specifically, uses the notion of a category of natural transformations on a given sentence functor and, implicitly, many aspects of the theory of N-rule based π-institutions, where N is a category of natural transformations on the sentence functor of the π-institution under consideration. A recent reference on this material is [15]. The reader should be aware that basic categorical notions are used rather heavily, but the elementary references to the subject [16–18] should be enough for necessary terminology and notation.

In Section 3 the basic constructions that take after corresponding constructions in [11] are presented. Here the meet-combination of logical systems refers to logical systems based on sentence functors, whose "signatures" are categories of natural transformations on the sentence functors and whose rules of inference and model classes are all categorical in nature. The goal is to work in a framework that would be amenable to categorical abstract algebraic logic methods and techniques so as to be able to consider aspects drawing from both theories.

In Sections 4 and 5, we show that a form of soundness and a form of restricted completeness are inherited by the meet-combination, subject to the condition that it is present in the components being combined. These results yield also results on conservativeness and on consistency, which are presented in Section 6.

Finally, based on the thorough work of [11], we present in Section 7 some examples showcasing various aspects of the general theory. These examples are relevant to both the theory developed in [11] and to its extension elaborated on in the present paper and, whenever appropriate, we draw attention to points where the two theories overlap and points where some differences occur.

2. Basic Framework

In the sequel we consider an arbitrary but fixed category Sign, called the category of signatures, and an arbitrary but fixed Set-valued functor SEN : Sign $\to$ Set, called the sentence functor. Also into the picture in a critical way will be an arbitrary but fixed category N of natural transformations on SEN, which we view as the clone of all algebraic operations on SEN. We remind the reader here of the precise definition of such a category, as presented, for example, in [15]. The clone of all natural transformations on SEN is defined to be the locally small category with collection of objects $\{\mathrm{SEN}^\alpha : \alpha \text{ an ordinal}\}$ and collection of morphisms $\tau : \mathrm{SEN}^\alpha \to \mathrm{SEN}^\beta$ β-sequences of natural transformations $\tau_i : \mathrm{SEN}^\alpha \to \mathrm{SEN}$. Composition

$$\mathrm{SEN}^\alpha \xrightarrow{\langle\tau_i:i<\beta\rangle} \mathrm{SEN}^\beta \xrightarrow{\langle\sigma_j:j<\gamma\rangle} \mathrm{SEN}^\gamma \quad (2)$$

is defined by

$$\langle\sigma_j : j < \gamma\rangle \circ \langle\tau_i : i < \beta\rangle = \langle\sigma_j(\langle\tau_i : i < \beta\rangle) : j < \gamma\rangle. \quad (3)$$

A subcategory N of this category containing *all* objects of the form SEN^k for $k < \omega$, and all projection morphisms $p^{k,i} : \mathrm{SEN}^k \to \mathrm{SEN}, i < k, k < \omega$, with $p^{k,i}_\Sigma : \mathrm{SEN}(\Sigma)^k \to \mathrm{SEN}(\Sigma)$ given by

$$p^{k,i}_\Sigma(\vec{\phi}) = \phi_i, \quad \forall \vec{\phi} \in \mathrm{SEN}(\Sigma)^k, \tag{4}$$

and such that, for every family $\{\tau_i : \mathrm{SEN}^k \to \mathrm{SEN} : i < l\}$ of natural transformations in N, the sequence $\langle \tau_i : i < l \rangle : \mathrm{SEN}^k \to \mathrm{SEN}^l$ is also in N, is referred to as a category of natural transformations on SEN.

A natural transformation $\sigma : \mathrm{SEN}^n \to \mathrm{SEN}$ in N is called a constant if, for all $\Sigma \in |\mathrm{Sign}|$ and all $\vec{\phi}, \vec{\psi} \in \mathrm{SEN}(\Sigma)^n$,

$$\sigma_\Sigma(\vec{\phi}) = \sigma_\Sigma(\vec{\psi}). \tag{5}$$

If $\sigma : \mathrm{SEN}^n \to \mathrm{SEN}$ is a constant, then we set $\sigma_\Sigma := \sigma_\Sigma(\vec{\phi})$, to denote the value of the constant in $\mathrm{SEN}(\Sigma)$, which is independent of $\vec{\phi} \in \mathrm{SEN}(\Sigma)^n$.

An N-rule of inference or simply an N-rule is a pair of the form $\langle\{\sigma^0, \ldots, \sigma^{n-1}\}, \tau\rangle$, sometimes written more legibly $\sigma^0, \ldots, \sigma^{n-1}/\tau$, where $\sigma^0, \ldots, \sigma^{n-1}$, τ are natural transformations in N. The elements σ^i, $i < n$, are called the premises and τ the conclusion of the rule.

An N-Hilbert calculus $\mathscr{R}$ is a set of N-rules. Using the N-rules in $\mathscr{R}$, one may define derivations of a natural transformation σ in N from a set Δ of natural transformations in N. Such a derivation is denoted by $\Delta \vdash^{\mathscr{R}} \sigma$. If the calculus $\mathscr{R}$ is fixed and clear in a particular context, we might simply write $\Delta \vdash \sigma$.

Given two functors $\mathrm{SEN} : \mathrm{Sign} \to \mathrm{Set}$ and $\mathrm{SEN}' : \mathrm{Sign}' \to \mathrm{Set}$, with categories of natural transformations N, N' on SEN, SEN', respectively, a pair $\langle F, \alpha\rangle$, where $F : \mathrm{Sign} \to \mathrm{Sign}'$ is a functor and $\alpha : \mathrm{SEN} \to \mathrm{SEN}' \circ F$ is a natural transformation, is called a translation from SEN to SEN'. Moreover, it is said to be (N, N')-epimorphic if there exists a correspondence $\sigma \mapsto \sigma'$ between the natural transformations in N and N' that preserves projections (and, thus, also arities), such that, for all $\sigma : \mathrm{SEN}^k \to \mathrm{SEN}$, all $\Sigma \in |\mathrm{Sign}|$ and all $\vec{\phi} \in \mathrm{SEN}(\Sigma)^k$,

$$\alpha_\Sigma\left(\sigma_\Sigma(\vec{\phi})\right) = \sigma'_{F(\Sigma)}\left(\alpha^k_\Sigma(\vec{\phi})\right). \tag{6}$$

An (N, N')-epimorphic translation from SEN to SEN' will be denoted by $\langle F, \alpha\rangle : \mathrm{SEN} \to \mathrm{SEN}'$, with the relevant categories N, N' of natural transformations on SEN, SEN', respectively, understood from context.

An N-algebraic system $\mathscr{A} = \langle \mathrm{SEN}', \langle F, \alpha\rangle\rangle$ consists of

(i) a functor $\mathrm{SEN}' : \mathrm{Sign}' \to \mathrm{Set}$, with a category N' of natural transformations on SEN';

(ii) an (N, N')-epimorphic translation $\langle F, \alpha\rangle : \mathrm{SEN} \to \mathrm{SEN}'$.

An N-matrix system or, simply, N-matrix $\mathfrak{A} = \langle \mathscr{A}, T\rangle$ is a pair consisting of

(i) an N-algebraic system $\mathscr{A} = \langle \mathrm{SEN}', \langle F, \alpha\rangle\rangle$;

(ii) an axiom family $T \in \mathrm{AxFam}(\mathrm{SEN}')$ on SEN', that is, a collection $T = \{T_\Sigma\}_{\Sigma \in |\mathrm{Sign}'|}$ of subsets $T_\Sigma \subseteq \mathrm{SEN}'(\Sigma), \Sigma \in |\mathrm{Sign}'|$.

We perceive of the elements of $\mathrm{SEN}'(F(\Sigma))$ as truth values for evaluating the natural transformations in N and those of $T_{F(\Sigma)}$ as being the designated ones. An N-matrix semantics $\mathscr{M}$ is a class of N-matrices. Given a natural transformation $\sigma : \mathrm{SEN}^k \to \mathrm{SEN}$ in N, we set

$$\sigma_{\Sigma'}\left(f(\vec{\phi})\right) := \sigma_{\Sigma'}\left(\mathrm{SEN}(f)^k(\vec{\phi})\right), \tag{7}$$

where $f \in \mathrm{Sign}(\Sigma, \Sigma')$ and $\vec{\phi} \in \mathrm{SEN}(\Sigma)^k$. The matrix $\mathfrak{A} = \langle \mathscr{A}, T\rangle$ satisfies σ at $\vec{\phi} \in \mathrm{SEN}(\Sigma)^k$ under $f \in \mathrm{Sign}(\Sigma, \Sigma')$, written $\mathfrak{A} \vDash_\Sigma \sigma[\vec{\phi}, f]$, if $\alpha_{\Sigma'}(\sigma_{\Sigma'}(f(\vec{\phi}))) \in T_{F(\Sigma')}$. An N-rule $\sigma^0, \ldots, \sigma^{n-1}/\tau$ is a rule of an N-matrix semantics $\mathscr{M}$, written

$$\sigma^0, \ldots, \sigma^{n-1} \vDash^{\mathscr{M}} \tau, \tag{8}$$

if $\mathfrak{A} \vDash_\Sigma \sigma^i[\vec{\phi}, f]$, for all $i < n$, implies $\mathfrak{A} \vDash_\Sigma \tau[\vec{\phi}, f]$, for every N-matrix $\mathfrak{A} \in \mathscr{M}$, all $\Sigma \in |\mathrm{Sign}|$, all Σ-assignments $\vec{\phi}$ in $\mathfrak{A}$, and all $f \in \mathrm{Sign}(\Sigma, \Sigma')$. If the semantics is clear from context, we simply write $\sigma^0, \ldots, \sigma^{n-1} \vDash \tau$.

In the remainder of this paper, by a logical system, or simply a logic, we understand a pentuple $\mathscr{L} = \langle \mathrm{Sign}, \mathrm{SEN}, N, \mathscr{R}, \mathscr{M}\rangle$, where

(i) Sign is a category;

(ii) SEN : Sign $\to$ Set is a sentence functor;

(iii) N is a category of natural transformations on SEN;

(iv) $\mathscr{R}$ is an N-Hilbert calculus;

(v) $\mathscr{M}$ is a N-matrix semantics.

3. Meet-Combinations

Let $\mathscr{L} = \langle \mathrm{Sign}, \mathrm{SEN}, N, \mathscr{R}, \mathscr{M}\rangle$ be a logical system. Define the product logical system or, simply, product logic $\mathscr{L}^\times = \langle \mathrm{Sign}, \mathrm{SEN}^\times, N^\times, \mathscr{R}^\times, \mathscr{M}^\times\rangle$ as follows:

the logic $\mathscr{L}^\times$ has the same signature category Sign as $\mathscr{L}$.

The sentence functor $\mathrm{SEN}^\times : \mathrm{Sign} \to \mathrm{Set}$ is defined by setting

$$\mathrm{SEN}^\times(\Sigma) = \mathrm{SEN}(\Sigma) \times \mathrm{SEN}(\Sigma), \tag{9}$$

for all $\Sigma \in |\mathrm{Sign}|$, and, similarly, for morphisms.

The category $N^\times$ of natural transformations on $\mathrm{SEN}^\times$ has the same objects as N and its morphisms $(\mathrm{SEN} \times \mathrm{SEN})^n \cong \mathrm{SEN}^n \times \mathrm{SEN}^n$ into $\mathrm{SEN} \times \mathrm{SEN}$ are pairs $\vec{\sigma} = \langle \sigma', \sigma''\rangle$ of natural transformations $\sigma', \sigma'' : \mathrm{SEN}^n \to \mathrm{SEN}$ in N. We call the members of $N^\times$ the combined natural transformations or combined operations or, following [11], but rather apologetic for abusing terminology, combined connectives.

Given $\sigma : \mathrm{SEN}^k \to \mathrm{SEN}$ in N, we set $\overline{\sigma} = \langle \sigma, \sigma\rangle$ in $N^\times$ and, accordingly, given $\vec{\sigma} = \langle \sigma', \sigma''\rangle$ in $N^\times$, we set

$$\overline{\sigma}' = \langle \sigma', \sigma'\rangle, \qquad \overline{\sigma}'' = \langle \sigma'', \sigma''\rangle. \tag{10}$$

Every N-rule $r = \sigma^0, \ldots, \sigma^{n-1}/\tau$ gives rise to an $N^\times$-rule

$$\overline{r} = \frac{\overline{\sigma}^0, \ldots, \overline{\sigma}^{n-1}}{\overline{\tau}}. \tag{11}$$

The calculus $\mathscr{R}^\times$ is an "enrichment" of $\mathscr{R}$ in the sense that it contains all rules of the form $\overline{r}$, for $r \in \mathscr{R}$, and some additional $N^\times$-rules devised for dealing with the combined operations:

(i) for each $\vec{\sigma} : (\mathrm{SEN}^\times)^k \to \mathrm{SEN}^\times$ in $N^\times$, the lifting rule (LFT)

$$\frac{\overline{\sigma}', \overline{\sigma}''}{\vec{\sigma}} \tag{12}$$

is included in $\mathscr{R}^\times$ to enforce inheritance by $\vec{\sigma}$ in $\mathscr{L}^\times$ of all the common properties of σ' and σ'' in $\mathscr{L}$;

(ii) for each constant $\vec{\sigma} : (\mathrm{SEN}^\times)^k \to \mathrm{SEN}^\times$ in $N^\times$, the special colifting rules (cLFT)

$$\frac{\vec{\sigma}}{\overline{\sigma}'}, \quad \frac{\vec{\sigma}}{\overline{\sigma}''} \tag{13}$$

are included in $\mathscr{R}^\times$ to enforce that $\vec{\sigma}$ should enjoy in $\mathscr{L}^\times$ only those properties that are common properties of σ' and σ'' in $\mathscr{L}$.

The reason for allowing only the special co-lifting rules (i.e., ones that admit only constants), rather than the (general) co-lifting rules, is that, unless this restriction is imposed, the rules are not in general sound. This will become apparent in the analysis to follow.

Before introducing the semantics $\mathscr{M}^\times$ of $\mathscr{L}^\times$, we show, following [11], that given constant natural transformations $\sigma', \sigma'' : (\mathrm{SEN}^\times)^k \to \mathrm{SEN}^\times$ in N, the two combined constructors $\langle \sigma', \sigma'' \rangle$ and $\langle \sigma'', \sigma' \rangle$ are closely related.

Theorem 1 (Sernadas, Sernadas, and Rasga). *Let $\mathscr{L} = \langle \mathrm{Sign}, \mathrm{SEN}, N, R, \mathscr{M} \rangle$ be a logical system. Consider a constant natural transformation $\vec{\sigma} = \langle \sigma', \sigma'' \rangle : (\mathrm{SEN}^\times)^k \to \mathrm{SEN}^\times$ in $N^\times$ and set $\overleftarrow{\sigma} = \langle \sigma'', \sigma' \rangle$. Then $\vec{\sigma}$ and $\overleftarrow{\sigma}$ are interderivable in $\mathscr{L}^\times$.*

Proof. Apply first cLFT twice and then LFT, in each direction. One gets the following proof:

$$\frac{\left(\vec{\sigma}/\overline{\sigma}''\right)\left(\vec{\sigma}/\overline{\sigma}'\right)}{\overleftarrow{\sigma}}. \tag{14}$$

□

Let $\mathscr{A}^\alpha = \langle \mathrm{SEN}', \langle F, \alpha \rangle \rangle$ and $\mathscr{A}^\beta = \langle \mathrm{SEN}', \langle F, \beta \rangle \rangle$ be N-algebraic systems with the same underlying sentence functors and the same signature functor component $F : \mathrm{Sign} \to \mathrm{Sign}'$. Let $\mathrm{SEN}'^\times = \mathrm{SEN}' \times \mathrm{SEN}' : \mathrm{Sign}' \to \mathrm{Set}$ be defined, for all $\Sigma \in |\mathrm{Sign}'|$, by

$$\mathrm{SEN}'^\times(\Sigma) = \mathrm{SEN}'(\Sigma) \times \mathrm{SEN}'(\Sigma), \tag{15}$$

and similarly for morphisms, and let $\langle F, \alpha \times \beta \rangle : \mathrm{SEN}^\times \to \mathrm{SEN}'^\times$ be given, for all $\Sigma \in |\mathrm{Sign}|$, by

$$(\alpha \times \beta)_\Sigma(\phi, \psi) = \langle \alpha_\Sigma(\phi), \beta_\Sigma(\psi) \rangle, \quad \forall \langle \phi, \psi \rangle \in \mathrm{SEN}^\times(\Sigma). \tag{16}$$

Denote by $\mathscr{A}^{\alpha\times\beta}$ the $N^\times$-algebraic system

$$\mathscr{A}^{\alpha\times\beta} = \left\langle \mathrm{SEN}'^\times, \langle F, \alpha \times \beta \rangle \right\rangle. \tag{17}$$

Moreover, given two N-matrix systems $\mathfrak{A}^\alpha = \langle \mathscr{A}^\alpha, T^\alpha \rangle$ and $\mathfrak{A}^\beta = \langle \mathscr{A}^\beta, T^\beta \rangle$, let

$$\mathfrak{A}^{\alpha\times\beta} = \langle \mathscr{A}^{\alpha\times\beta}, T^{\alpha\times\beta} \rangle, \tag{18}$$

where $T^{\alpha\times\beta} = \{T_\Sigma^{\alpha\times\beta}\}_{\Sigma \in |\mathrm{Sign}'|}$, such that, for all $\Sigma \in |\mathrm{Sign}'|$,

$$T_\Sigma^{\alpha\times\beta} = T_\Sigma^\alpha \times T_\Sigma^\beta. \tag{19}$$

The semantics $\mathscr{M}^\times$ is the class consisting of all $N^\times$-matrix systems of the form $\mathfrak{A}^{\alpha\times\beta}$, for $\mathfrak{A}^\alpha, \mathfrak{A}^\beta \in \mathscr{M}$, having underlying N-algebraic systems $\mathscr{A}^\alpha, \mathscr{A}^\beta$, respectively, with the same underlying sentence functors and the same signature functor components. The semantics $\mathscr{M}^\times$ will be called the product semantics, taking after [11].

Finally, we let $\vdash^\times$ and $\vDash^\times$ stand for satisfaction and entailment in the product logic $\mathscr{L}^\times = \langle \mathrm{Sign}, \mathrm{SEN}^\times, N^\times, \mathscr{R}^\times, \mathscr{M}^\times \rangle$.

4. Soundness

Recall that, given a natural transformation $\sigma : \mathrm{SEN}^k \to \mathrm{SEN}$ in N, we use the notation $\overline{\sigma}$ to denote the natural transformation $\overline{\sigma} = \langle \sigma, \sigma \rangle$ in $N^\times$.

Proposition 2. *Let $\mathscr{L} = \langle \mathrm{Sign}, \mathrm{SEN}, N, \mathscr{R}, \mathscr{M} \rangle$ be a logical system and consider the product system $\mathscr{L}^\times = \langle \mathrm{Sign}, \mathrm{SEN}^\times, N^\times, \mathscr{R}^\times, \mathscr{M}^\times \rangle$. Suppose that $\sigma : \mathrm{SEN}^k \to \mathrm{SEN}$ in N, $\Sigma \in |\mathrm{Sign}|$ and $\vec{\phi} \in \mathrm{SEN}^\times(\Sigma)^k$, where the ith component $\vec{\phi}_i$ of $\vec{\phi}$ is $\vec{\phi}_i = \langle \vec{\phi}_i', \vec{\phi}_i'' \rangle$, for all $i < k$. Then*

$$\overline{\sigma}_\Sigma(\vec{\phi}) = \left\langle \sigma_\Sigma(\vec{\phi}'), \sigma_\Sigma(\vec{\phi}'') \right\rangle. \tag{20}$$

Moreover, for all $\mathfrak{A}^\alpha = \langle \langle \mathrm{SEN}', \langle F, \alpha \rangle \rangle, T^\alpha \rangle$, $\mathfrak{A}^\beta = \langle \langle \mathrm{SEN}', \langle F, \beta \rangle \rangle, T^\beta \rangle \in \mathscr{M}$, all $\Sigma' \in |\mathrm{Sign}|$ and all $f \in \mathrm{Sign}(\Sigma, \Sigma')$,

$$\mathfrak{A}^{\alpha\times\beta} \vDash_\Sigma \overline{\sigma}\left[\vec{\phi}, f\right] \quad \text{iff } \mathfrak{A}^\alpha \vDash_\Sigma \sigma\left[\vec{\phi}', f\right], \mathfrak{A}^\beta \vDash_\Sigma \sigma\left[\vec{\phi}'', f\right]. \tag{21}$$

Proof. We have the following equivalences:

$$\mathfrak{A}^{\alpha\times\beta} \vDash_\Sigma \overline{\sigma}\left[\vec{\phi}, f\right] \tag{22}$$

$$\text{iff } (\alpha \times \beta)_{\Sigma'}(\overline{\sigma}_{\Sigma'}(f(\vec{\phi}))) \in T_{F(\Sigma')}^{\alpha\times\beta}$$

iff $\alpha_{\Sigma'}(\sigma_{\Sigma'}(f(\vec{\phi}'))) \in T^{\alpha}_{F(\Sigma')}$ and $\beta_{\Sigma'}(\sigma_{\Sigma'}(f(\vec{\phi}''))) \in T^{\beta}_{F(\Sigma')}$,

iff $\mathfrak{A}^{\alpha}\vDash_{\Sigma}\sigma[\vec{\phi}', f]$, and $\mathfrak{A}^{\beta}\vDash_{\Sigma}\sigma[\vec{\phi}'', f]$.

This proves the Proposition. □

Proposition 3. *Let $\mathscr{L} = \langle Sign, SEN, N, \mathscr{R}, \mathscr{M}\rangle$ be a logical system. If the N-rule $\langle\{\sigma^0,\ldots,\sigma^{n-1}\},\tau\rangle$ is sound in $\mathscr{L}$, then the $N^{\times}$-rule $\langle\{\overline{\sigma}^0,\ldots,\overline{\sigma}^{n-1}\},\overline{\tau}\rangle$ is sound in $\mathscr{L}^{\times}$.*

Proof. Suppose that $\mathfrak{A}^{\alpha} = \langle\langle \mathrm{SEN}', \langle F,\alpha\rangle\rangle, T^{\alpha}\rangle$ and $\mathfrak{A}^{\beta} = \langle\langle \mathrm{SEN}', \langle F,\beta\rangle\rangle, T^{\beta}\rangle$ are in $\mathscr{M}$ so that $\mathfrak{A}^{\alpha\times\beta} \in \mathscr{M}^{\times}$, $\Sigma,\Sigma' \in |\mathrm{Sign}|$, $\vec{\phi} \in \mathrm{SEN}^{\times}(\Sigma)^k$, and $f \in \mathrm{Sign}(\Sigma,\Sigma')$, such that $\mathfrak{A}^{\alpha\times\beta}\vDash_{\Sigma}\overline{\sigma}^i[\vec{\phi}, f]$, for all $i < n$. Then, by Proposition 2,

$$\mathfrak{A}^{\alpha}\vDash_{\Sigma}\sigma^i\left[\vec{\phi}', f\right], \qquad \mathfrak{A}^{\beta}\vDash_{\Sigma}\sigma^i\left[\vec{\phi}'', f\right], \tag{23}$$

for all $i < n$. Thus, by soundness of $\langle\{\sigma^0,\ldots,\sigma^{n-1}\},\tau\rangle$ in $\mathscr{L}$, we get that $\mathfrak{A}^{\alpha}\vDash_{\Sigma}\tau[\vec{\phi}', f]$ and $\mathfrak{A}^{\beta}\vDash_{\Sigma}\tau[\vec{\phi}'', f]$. Therefore, again by Proposition 2, $\mathfrak{A}^{\alpha\times\beta}\vDash_{\Sigma}\overline{\tau}[\vec{\phi}, f]$ and, hence, $\langle\{\overline{\sigma}^0,\ldots,\overline{\sigma}^{n-1}\},\overline{\tau}\rangle$ is sound for $\mathscr{L}^{\times}$. □

Let $\vec{\sigma} : (\mathrm{SEN}^{\times})^k \longrightarrow \mathrm{SEN}^{\times}$ be in $N^{\times}$ and suppose that $\Sigma,\Sigma' \in |\mathrm{Sign}|$, $\vec{\phi} \in \mathrm{SEN}^{\times}(\Sigma)^k$ and $f \in \mathrm{Sign}(\Sigma,\Sigma')$. Then, by the definition of $\vec{\sigma}$,

$$\begin{aligned}\vec{\sigma}_{\Sigma'}\left(f\left(\vec{\phi}\right)\right) &= \left\langle \sigma'_{\Sigma'}\left(f\left(\vec{\phi}'\right)\right), \sigma''_{\Sigma'}\left(f\left(\vec{\phi}''\right)\right)\right\rangle \\ &= \left\langle \overline{\sigma}'_{\Sigma'}\left(f\left(\vec{\phi}\right)\right)', \overline{\sigma}''_{\Sigma'}\left(f\left(\vec{\phi}\right)\right)''\right\rangle.\end{aligned} \tag{24}$$

Proposition 4. *Let $\mathscr{L} = \langle Sign, SEN, N, \mathscr{R}, \mathscr{M}\rangle$ be a logical system. The lifting rule LFT is sound in $\mathscr{L}^{\times} = \langle Sign, SEN^{\times}, N^{\times}, \mathscr{R}^{\times}, \mathscr{M}^{\times}\rangle$.*

Proof. Suppose that $\vec{\sigma}$ in $N^{\times}$, $\mathfrak{A}^{\alpha\times\beta} \in \mathscr{M}^{\times}$, $\Sigma \in |\mathrm{Sign}|$ and $\vec{\phi} \in \mathrm{SEN}^{\times}(\Sigma)$, such that, for some $\Sigma \in |\mathrm{Sign}'|$ and $f \in \mathrm{Sign}(\Sigma,\Sigma')$,

$$\mathfrak{A}^{\alpha\times\beta}\vDash_{\Sigma}\overline{\sigma}'\left[\vec{\phi}, f\right], \qquad \mathfrak{A}^{\alpha\times\beta}\vDash_{\Sigma}\overline{\sigma}''\left[\vec{\phi}, f\right]. \tag{25}$$

This implies that $(\alpha\times\beta)_{\Sigma'}(\overline{\sigma}'_{\Sigma'}(f(\vec{\phi}))) \in T^{\alpha\times\beta}_{F(\Sigma')}$ and $(\alpha\times\beta)_{\Sigma'}(\overline{\sigma}''_{\Sigma'}(f(\vec{\phi}))) \in T^{\alpha\times\beta}_{F(\Sigma')}$. These imply that $(\alpha\times\beta)_{\Sigma'}(\langle\overline{\sigma}'_{\Sigma'}(f(\vec{\phi}))', \overline{\sigma}''_{\Sigma'}(f(\vec{\phi}))''\rangle) \in T^{\alpha\times\beta}_{F(\Sigma')}$, whence, by (24),

$$(\alpha\times\beta)_{\Sigma'}\left(\vec{\sigma}_{\Sigma'}\left(f\left(\vec{\phi}\right)\right)\right) \in T^{\alpha\times\beta}_{F(\Sigma')}. \tag{26}$$

This proves the soundness of lifting. □

Let Sign be a category and SEN : Sign $\longrightarrow$ Set a sentence functor with N a category of natural transformations on SEN. Recall that a natural transformation $\sigma : \mathrm{SEN}^k \longrightarrow \mathrm{SEN}$ is called a *constant* if, for all $\Sigma \in |\mathrm{Sign}|$, all $\vec{\phi},\vec{\psi} \in \mathrm{SEN}(\Sigma)^k \sigma_{\Sigma}(\vec{\phi}) = \sigma_{\Sigma}(\vec{\psi})$ and that we use the notation $\sigma_{\Sigma} := \sigma_{\Sigma}(\vec{\phi})$, for this value, which is independent of $\vec{\phi} \in \mathrm{SEN}(\Sigma)^k$.

A class of N-matrix systems $\mathscr{M}$ is said to be a c-semantics if, for all $\mathfrak{A}^{\alpha} = \langle\langle \mathrm{SEN}', \langle F,\alpha\rangle\rangle, T^{\alpha}\rangle$ and $\mathfrak{A}^{\beta} = \langle\langle \mathrm{SEN}', \langle F,\beta\rangle\rangle, T^{\beta}\rangle$ in $\mathscr{M}$, every constant $\sigma : \mathrm{SEN}^k \longrightarrow \mathrm{SEN}$ in N and all $\Sigma \in |\mathrm{Sign}|$,

$$\alpha_{\Sigma}(\sigma_{\Sigma}) \in T^{\alpha}_{F(\Sigma)}, \quad \text{iff } \beta_{\Sigma}(\sigma_{\Sigma}) \in T^{\beta}_{F(\Sigma)}. \tag{27}$$

Intuitively, a semantics $\mathscr{M}$ is a c-semantics if and only if every constant is consistently interpreted as true or false under all matrix systems in the semantics, that is, under all combinations of interpretations and designated truth values included in the semantics.

Proposition 5. *Let $\mathscr{L} = \langle Sign, SEN, N, \mathscr{R}, \mathscr{M}\rangle$ be a logical system, where $\mathscr{M}$ is a c-semantics. For all constants $\vec{\sigma} : (SEN^{\times})^k \longrightarrow SEN^{\times}$ in $N^{\times}$, the special co-lifting rules*

$$\frac{\vec{\sigma}}{\overline{\sigma}'}, \qquad \frac{\vec{\sigma}}{\overline{\sigma}''} \tag{28}$$

are sound in $\mathscr{L}^{\times} = \langle Sign, SEN^{\times}, N^{\times}, \mathscr{R}^{\times}, \mathscr{M}^{\times}\rangle$.

Proof. Let $\mathfrak{A}^{\alpha\times\beta} \in \mathscr{M}^{\times}$, $\vec{\sigma}$ a constant in $N^{\times}$, $\Sigma \in |\mathrm{Sign}|$, and $\vec{\phi} \in \mathrm{SEN}^{\times}(\Sigma)^k$, such that, for some $\Sigma' \in |\mathrm{Sign}|$ and $f \in \mathrm{Sign}(\Sigma,\Sigma')$,

$$\mathfrak{A}^{\alpha\times\beta}\vDash_{\Sigma}\vec{\sigma}\left[\vec{\phi}, f\right]. \tag{29}$$

Then (recalling the notation for constants) $(\alpha\times\beta)_{\Sigma'}(\vec{\sigma}_{\Sigma'}) \in T^{\alpha\times\beta}_{F(\Sigma')}$, whence

$$\alpha_{\Sigma'}\left(\sigma'_{\Sigma'}\right) \in T^{\alpha}_{F(\Sigma')}, \qquad \beta_{\Sigma'}\left(\sigma''_{\Sigma'}\right) \in T^{\beta}_{F(\Sigma')}. \tag{30}$$

Since $\mathscr{M}$ is a c-semantics, we get that the four following relations hold:

$$\begin{aligned}&\alpha_{\Sigma'}\left(\sigma'_{\Sigma'}\right) \in T^{\alpha}_{F(\Sigma')}, \qquad \beta_{\Sigma'}\left(\sigma'_{\Sigma'}\right) \in T^{\beta}_{F(\Sigma')}, \\ &\alpha_{\Sigma'}\left(\sigma''_{\Sigma'}\right) \in T^{\alpha}_{F(\Sigma')}, \qquad \beta_{\Sigma'}\left(\sigma''_{\Sigma'}\right) \in T^{\beta}_{F(\Sigma')}.\end{aligned} \tag{31}$$

Therefore, we obtain that

$$\mathfrak{A}^{\alpha\times\beta}\vDash_{\Sigma}\overline{\sigma}'\left[\vec{\phi}, f\right], \qquad \mathfrak{A}^{\alpha\times\beta}\vDash_{\Sigma}\overline{\sigma}''\left[\vec{\phi}, f\right], \tag{32}$$

which show that the special co-lifting rules are sound in $\mathscr{L}^{\times}$. □

Theorem 6 (soundness). *Let $\mathscr{L} = \langle Sign, SEN, N, \mathscr{R}, \mathscr{M}\rangle$ be a logical system, where $\mathscr{M}$ is a c-semantics. If $\mathscr{L}$ is sound, then the product logic $\mathscr{L}^{\times} = \langle Sign, SEN^{\times}, N^{\times}, \mathscr{R}^{\times}, \mathscr{M}^{\times}\rangle$ is also sound.*

Proof. We have shown in Proposition 3 that all rules inherited by $\mathscr{L}$ are sound in $\mathscr{L}^{\times}$. By Proposition 4, the lifting rule is sound in $\mathscr{L}^{\times}$ and, since $\mathscr{M}$ is assumed to be a c-semantics, by Proposition 5, the special co-lifting rules are sound in $\mathscr{L}^{\times}$. Therefore the product logic $\mathscr{L}^{\times}$ is also sound. □

5. *c*-Completeness

A logic $\mathscr{L} = \langle \mathrm{Sign}, \mathrm{SEN}, N, \mathscr{R}, \mathscr{M}\rangle$ is c-complete if it is complete with respect to constant natural transformations.

More precisely, for all sets $\Delta \cup \{\sigma\}$ of constants in N, we have that

$$\Delta \vDash \sigma \text{ implies } \Delta \vdash \sigma. \tag{33}$$

Proposition 7. *If a logic $\mathscr{L} = \langle Sign, SEN, N, \mathscr{R}, \mathscr{M}\rangle$ is c-complete, then, for all sets of constants $\Delta \cup \{\sigma\}$ in N,*

$$\overline{\Delta}\nvdash^{\times}\overline{\sigma} \text{ implies } \overline{\Delta}\nvDash^{\times}\overline{\sigma}. \tag{34}$$

Proof. Suppose that $\overline{\Delta}\nvdash^{\times}\overline{\sigma}$. Then, since $\mathscr{R}^{\times}$ includes all $N^{\times}$-rules of the form $\overline{r}$, for all $r \in \mathscr{R}$, we get that $\Delta \nvdash \sigma$. Therefore, by the c-completeness of $\mathscr{L}$, we get that $\Delta \nvDash \sigma$. Thus, there exists a model $\mathfrak{A}^{\alpha} = \langle\langle \mathrm{SEN}', \langle F, \alpha\rangle\rangle, T^{\alpha}\rangle \in \mathscr{M}$, together with $\Sigma, \Sigma' \in |\mathrm{Sign}|$, $\vec{\phi} \in \mathrm{SEN}(\Sigma)^k$ and $f \in \mathrm{Sign}(\Sigma, \Sigma')$, such that $\mathfrak{A}^{\alpha}\vDash_{\Sigma}\Delta[\vec{\phi}, f]$ and $\mathfrak{A}^{\alpha}\nvDash_{\Sigma}\sigma[\vec{\phi}, f]$. Hence, the model $\mathfrak{A}^{\alpha\times\alpha} \in \mathscr{M}^{\times}$ is such that $\mathfrak{A}^{\alpha\times\alpha}\vDash_{\Sigma}^{\times}\overline{\Delta}[\langle\vec{\phi},\vec{\phi}\rangle, f]$ and $\mathfrak{A}^{\alpha\times\alpha}\nvDash_{\Sigma}^{\times}\overline{\sigma}[\langle\vec{\phi},\vec{\phi}\rangle, f]$. Therefore $\overline{\Delta}\nvDash^{\times}\overline{\sigma}$, showing that $\mathscr{L}^{\times}$ is also c-complete. □

Proposition 8. *Let $\mathscr{L} = \langle Sign, SEN, N, \mathscr{R}, \mathscr{M}\rangle$ be a logic and suppose that, for some $\Delta \cup \{\sigma, \tau\}$ in N,*

$$\left(\overline{\Delta}\nvdash^{\times}\overline{\sigma} \text{ implies } \overline{\Delta}\nvDash^{\times}\overline{\sigma}\right), \qquad \left(\overline{\Delta}\nvdash^{\times}\overline{\tau} \text{ implies } \overline{\Delta}\nvDash^{\times}\overline{\tau}\right). \tag{35}$$

Then it is also the case that

$$\overline{\Delta}\nvdash^{\times}\langle\sigma,\tau\rangle \text{ implies } \overline{\Delta}\nvDash^{\times}\langle\sigma,\tau\rangle. \tag{36}$$

Proof. Suppose that $\overline{\Delta}\nvdash^{\times}\langle\sigma,\tau\rangle$. By the lifting rule, we must have $\overline{\Delta}\nvdash^{\times}\overline{\sigma}$ or $\overline{\Delta}\nvdash^{\times}\overline{\tau}$. Therefore, by hypothesis, $\overline{\Delta}\nvDash^{\times}\overline{\sigma}$ or $\overline{\Delta}\nvDash^{\times}\overline{\tau}$. Suppose, without loss of generality, that the first holds. Thus, there exists a model $\mathfrak{A}^{\alpha\times\beta} \in \mathscr{M}^{\times}$, $\Sigma, \Sigma' \in |\mathrm{Sign}|$, $\vec{\phi} \in \mathrm{SEN}^{\times}(\Sigma)^k$ and $f \in \mathrm{Sign}(\Sigma, \Sigma')$, such that

$$\begin{aligned}&(\alpha\times\beta)_{\Sigma'}\left(\overline{\Delta}_{\Sigma'}\left(f\left(\vec{\phi}\right)\right)\right) \subseteq T^{\alpha\times\beta}_{F(\Sigma')},\\ &(\alpha\times\beta)_{\Sigma'}\left(\overline{\sigma}_{\Sigma'}\left(f\left(\vec{\phi}\right)\right)\right) \notin T^{\alpha\times\beta}_{F(\Sigma')}.\end{aligned} \tag{37}$$

Thus, we must have

$$\begin{aligned}&\alpha_{\Sigma'}\left(\Delta_{\Sigma'}\left(f\left(\vec{\phi}'\right)\right)\right) \subseteq T^{\alpha}_{F(\Sigma')},\\ &\alpha_{\Sigma'}\left(\sigma_{\Sigma'}\left(f\left(\vec{\phi}'\right)\right)\right) \notin T^{\alpha}_{F(\Sigma')}\end{aligned} \tag{38}$$

or

$$\begin{aligned}&\beta_{\Sigma'}\left(\Delta_{\Sigma'}\left(f\left(\vec{\phi}''\right)\right)\right) \subseteq T^{\beta}_{F(\Sigma')},\\ &\beta_{\Sigma'}\left(\sigma_{\Sigma'}\left(f\left(\vec{\phi}''\right)\right)\right) \notin T^{\beta}_{F(\Sigma')}.\end{aligned} \tag{39}$$

This implies that either $\mathfrak{A}^{\alpha\times\alpha}$ or $\mathfrak{A}^{\beta\times\beta}$ bears witness to $\overline{\Delta}\nvDash^{\times}\langle\sigma,\tau\rangle$ and concludes the proof. □

To formulate the following proposition we introduce a convenient notation: given a set Δ of natural transformations in $N^{\times}$, we write

$$\overline{\Delta}' = \left\{\overline{\sigma}' : \vec{\sigma} \in \Delta\right\}, \qquad \overline{\Delta}'' = \left\{\overline{\sigma}'' : \vec{\sigma} \in \Delta\right\}. \tag{40}$$

Proposition 9. *Let $\mathscr{L} = \langle Sign, SEN, N, \mathscr{R}, \mathscr{M}\rangle$ be a logic and suppose for some set of constants $\Delta \cup \{\vec{\sigma}\}$ in $N^{\times}$*

$$\overline{\Delta}' \cup \overline{\Delta}''\nvdash^{\times}\vec{\sigma} \text{ implies } \overline{\Delta}' \cup \overline{\Delta}''\nvDash^{\times}\vec{\sigma}. \tag{41}$$

Then it is also the case that

$$\Delta\nvdash^{\times}\vec{\sigma} \text{ implies } \Delta\nvDash^{\times}\vec{\sigma}. \tag{42}$$

Proof. If $\Delta\nvdash^{\times}\vec{\sigma}$, then, by the special co-lifting property, $\overline{\Delta}' \cup \overline{\Delta}''\nvdash^{\times}\vec{\sigma}$. Thus, by hypothesis, $\overline{\Delta}' \cup \overline{\Delta}''\nvDash^{\times}\vec{\sigma}$. Hence, there exists $\mathfrak{A}^{\alpha\times\beta} \in \mathscr{M}^{\times}$, $\Sigma, \Sigma' \in |\mathrm{Sign}|$, $\vec{\phi} \in \mathrm{SEN}^{\times}(\Sigma)^k$ and $f \in \mathrm{Sign}(\Sigma, \Sigma')$, such that

$$(\alpha\times\beta)_{\Sigma'}\left(\left(\overline{\Delta}' \cup \overline{\Delta}''\right)_{\Sigma'}\left(f\left(\vec{\phi}\right)\right)\right) \subseteq T^{\alpha\times\beta}_{F(\Sigma')}, \tag{43}$$

while, at the same time,

$$(\alpha\times\beta)_{\Sigma'}\left(\vec{\sigma}_{\Sigma'}\left(f\left(\vec{\phi}\right)\right)\right) \notin T^{\alpha\times\beta}_{F(\Sigma')}. \tag{44}$$

These relations imply that

$$\begin{aligned}&(\alpha\times\beta)_{\Sigma'}\left(\Delta_{\Sigma'}\left(f\left(\vec{\phi}\right)\right)\right) \subseteq T^{\alpha\times\beta}_{F(\Sigma')}\\ \text{but } &(\alpha\times\beta)_{\Sigma'}\left(\vec{\sigma}_{\Sigma'}\left(f\left(\vec{\phi}\right)\right)\right) \notin T^{\alpha\times\beta}_{F(\Sigma')},\end{aligned} \tag{45}$$

whence $\Delta\nvDash^{\times}\vec{\sigma}$. □

Theorem 10 (c-completeness). *If the logic $\mathscr{L} = \langle Sign, SEN, N, \mathscr{R}, \mathscr{M}\rangle$ is c-complete, then the product logic $\mathscr{L}^{\times} = \langle Sign, SEN^{\times}, N^{\times}, \mathscr{R}^{\times}, \mathscr{M}^{\times}\rangle$ is c-complete also.*

Proof. If $\mathscr{L}$ is c-complete, then, by Proposition 7, we get that, for all sets of constants $\Delta \cup \{\sigma\}$ in N,

$$\overline{\Delta}\nvdash^{\times}\overline{\sigma} \text{ implies } \overline{\Delta}\nvDash^{\times}\overline{\sigma}. \tag{46}$$

Thus, by Proposition 8, for all sets of constants $\Delta \cup \{\sigma, \tau\}$ in N,

$$\overline{\Delta}\nvdash^{\times}\langle\sigma,\tau\rangle \text{ implies } \overline{\Delta}\nvDash^{\times}\langle\sigma,\tau\rangle. \tag{47}$$

Finally, by Proposition 9, we get that, for all sets of constants $\Delta \cup \{\vec{\sigma}\}$ in $N^{\times}$,

$$\Delta\nvdash^{\times}\vec{\sigma} \text{ implies } \Delta\nvDash^{\times}\vec{\sigma}. \tag{48}$$

This proves that $\mathscr{L}^{\times}$ is c-complete. □

6. Conservativeness and Consistency

Theorem 11 (conservativeness). *Let $\mathscr{L} = \langle Sign, SEN, N, \mathscr{R}, \mathscr{M}\rangle$ be a logic. For every set of natural transformations $\Delta \cup \{\sigma\}$ in N,*

$$\overline{\Delta}\vDash^{\times}\overline{\sigma} \text{ implies } \Delta \vDash \sigma. \tag{49}$$

Proof. Suppose $\overline{\Delta}\vDash^{\times}\overline{\sigma}$. If $\mathfrak{A}^{\alpha} = \langle\langle \mathrm{SEN}', \langle F,\alpha\rangle\rangle, T^{\alpha}\rangle$ is such that, for some $\Sigma,\Sigma' \in |\mathrm{Sign}|$, $\vec{\phi}\in \mathrm{SEN}(\Sigma)^k$, $f \in \mathrm{Sign}(\Sigma,\Sigma')$, $\mathfrak{A}^\alpha \vDash_\Sigma \Delta[\vec\phi, f]$, then, we get that $\mathfrak{A}^{\alpha\times\alpha}\vDash^\times_\Sigma \overline{\Delta}[\vec{\phi}^2, f]$, whence, by the hypothesis, $\mathfrak{A}^{\alpha\times\alpha}\vDash^\times_\Sigma\overline{\sigma}[\vec\phi^2,f]$, and, therefore, $\mathfrak{A}^\alpha\vDash_\Sigma \sigma[\vec\phi,f]$. This shows that $\Delta\vDash\sigma$. □

Theorem 12 (consistency). *If the logic $\mathscr{L} = \langle \mathrm{Sign}, \mathrm{SEN}, N, \mathscr{R}, \mathscr{M}\rangle$ is consistent, then so is the product logic $\mathscr{L}^\times = \langle \mathrm{Sign}, \mathrm{SEN}^\times, N^\times, \mathscr{R}^\times, \mathscr{M}^\times\rangle$.*

Proof. This follows directly from conservativeness. □

7. Examples from Classical Propositional Logic

We present a simple example, essentially borrowed from [11], with the twofold goal of, first, seeing how the theory of [11] can be easily accommodated in the categorical framework (becoming actually a trivial case) and, second, showcasing the difference between the soundness of special co-lifting and the lack of soundness obtained by allowing the full power of the general co-lifting rule.

Suppose, first, that $\mathscr{L} = \langle \mathrm{Sign}, \mathrm{SEN}, N, \mathscr{R}, \mathscr{M}\rangle$ is a logic, such that N contains two binary natural transformations $\wedge, \vee : \mathrm{SEN}^2 \rightarrow \mathrm{SEN}$ and two constants T, F that obey the usual laws of conjunction, disjunction, truth, and falsity of classical propositional logic. Then, if $\mathsf{A}, \mathsf{B} \in \{\mathsf{T}, \mathsf{F}\}^2$, we have that

$$\langle \wedge, \vee\rangle (\mathsf{A}, \mathsf{B}) \vdash^\times \langle \wedge, \vee\rangle (\mathsf{B}, \mathsf{A}). \quad (50)$$

This can be shown by observing that the hypothesis yields, by special co-lifting, $\langle\wedge,\wedge\rangle(\overline{\mathsf{A}}', \overline{\mathsf{B}}')$ and $\langle\vee,\vee\rangle(\overline{\mathsf{A}}'', \overline{\mathsf{B}}'')$. These, by following usual derivations in $\mathscr{L}$, yield $\langle\wedge,\wedge\rangle(\overline{\mathsf{B}}', \overline{\mathsf{A}}')$ and $\langle\vee,\vee\rangle(\overline{\mathsf{B}}'', \overline{\mathsf{A}}'')$, whence, by lifting, we finally obtain the conclusion. In fact, if we arrange for $\mathscr{M}$ to consist, essentially, of Boolean algebras and evaluations together with Boolean filters, it is the case that

$$\langle \wedge, \vee\rangle \vDash^\times \langle \wedge, \vee\rangle \left(p^{1,1}, p^{1,0}\right), \quad (51)$$

where $p^{1,0}, p^{1,1} : (\mathrm{SEN}^\times)^2 \rightarrow \mathrm{SEN}^\times$ are the two projection natural transformations; that is, "commutativity" is valid in general, not just for constants. However, the derivation (50) cannot be inferred directly from this using c-completeness, since there are nonconstant natural transformations involved.

To illustrate, using the same example, that the general co-lifting rule fails, we may employ Boolean models to show that

$$\langle \wedge, \vee\rangle \nvDash^\times \langle \wedge, \wedge\rangle. \quad (52)$$

In fact, note that

$$\langle \wedge, \vee\rangle_\Sigma ((1,0),(1,1)) = (\wedge_\Sigma (1,1), \vee_\Sigma (0,1)) = (1,1), \quad (53)$$

whereas

$$\langle \wedge, \wedge\rangle_\Sigma ((1,0),(1,1)) = (\wedge_\Sigma (1,1), \wedge_\Sigma (0,1)) = (1,0), \quad (54)$$

the first belonging to the product filter of 2-element Boolean algebras, the second failing to do so.

Note, next, that

$$\langle \wedge, \vee\rangle (\langle \mathsf{T}, \mathsf{F}\rangle, \langle \mathsf{T}, \mathsf{T}\rangle) \nvDash^\times \langle \vee, \wedge\rangle (\langle \mathsf{T}, \mathsf{F}\rangle, \langle \mathsf{T}, \mathsf{T}\rangle). \quad (55)$$

A straightforward computation shows that in the direct product of 2-element Boolean algebras, the left-hand side evaluates to $(1,1)$, whereas the right-hand side to $(1,0)$. Even though this serves as a counterexample for an analog of Theorem 1 concerning the exchangeability of components in the context of [11], this problem does not arise in our context. In fact, our reformulation of [11, Theorem 2.1] in the form of Theorem 1 would only ensure that

$$\langle \wedge, \vee\rangle (\langle \mathsf{T}, \mathsf{F}\rangle, \langle \mathsf{T}, \mathsf{T}\rangle) \vDash^\times \langle \vee, \wedge\rangle (\langle \mathsf{F}, \mathsf{T}\rangle, \langle \mathsf{T}, \mathsf{T}\rangle). \quad (56)$$

Suppose now that in $\mathscr{L} = \langle \mathrm{Sign}, \mathrm{SEN}, N, \mathscr{R}, \mathscr{M}\rangle$, one has the, possibly derived, rule σ/τ, where σ, τ are both constants in N. Then it can be shown that

$$\langle \sigma, \tau\rangle \dashv\vdash^\times \overline{\sigma}. \quad (57)$$

In fact, $\langle\sigma,\tau\rangle \vdash^\times \overline{\sigma}$ follows from the special co-lifting, whereas lifting helps establish the opposite direction

$$\frac{(\overline{\sigma}/\overline{\sigma})\ (\overline{\sigma}/\overline{\tau})}{\langle \sigma, \tau\rangle}. \quad (58)$$

Finally, if one has available in $\mathscr{L}$ a disjunction $\vee$ and an implication $\rightarrow$, both behaving classically, then, since both derived rules

$$p^{1,1} \vdash \vee, \qquad p^{1,1} \vdash \rightarrow \quad (59)$$

are rules of $\mathscr{L}$, one obtains the rule $\overline{p^{1,1}} \vdash^\times \langle \vee, \rightarrow\rangle$ in $\mathscr{L}^\times$ by an application of lifting.

We close with a generally phrased (rather informally formulated) problem that would be of interest in the context developed in the present work from the point of view of *abstract algebraic logic*. For more details on the motivations and the state of the art in that theory, as well as the precise definitions and more insights on the notions employed in the phrasing of this problem, the reader is referred to [13–15] and further references therein.

Problem for Investigation. Suppose that we have some knowledge about the algebraic classification of the π-institution $\mathscr{I} = \langle \mathrm{Sign}, \mathrm{SEN}, C^{\mathscr{R}}\rangle$, where $\mathscr{L} = \langle \mathrm{Sign}, \mathrm{SEN}, N, \mathscr{R}, \mathscr{M}\rangle$ is a logic in the sense of the present paper, possibly satisfying some additional conditions. The closure system $C^{\mathscr{R}}$ is the system induced by the set $\mathscr{R}$ of N-rules, as detailed in, for example, [15]. What corresponding information may then be drawn about the π-institution $\mathscr{I} = \langle \mathrm{Sign}, \mathrm{SEN}^\times, C^{\mathscr{R}^\times}\rangle$, that corresponds, in a similar manner, to the product logic $\mathscr{L}^\times = \langle \mathrm{Sign}, \mathrm{SEN}^\times, N^\times, \mathscr{R}^\times, \mathscr{M}^\times\rangle$?

References

[1] D. M. Gabbay, "Fibred semantics and the weaving of logics. I. Modal and intuitionistic logics," *Journal of Symbolic Logic*, vol. 61, no. 4, pp. 1057–1120, 1996.

[2] A. Sernadas, C. Sernadas, and C. Caleiro, "Fibring of logics as a categorial construction," *Journal of Logic and Computation*, vol. 9, no. 2, pp. 149–179, 1999.

[3] C. Caleiro, W. Carnielli, J. Rasga, and C. Sernadas, "Fibring of logics as a universal construction," in *Handbook of Philosophical Logic*, vol. 13, pp. 123–187, 2nd edition, 2005.

[4] A. Zanardo, A. Sernadas, and C. Sernadas, "Fibring: completeness preservation," *Journal of Symbolic Logic*, vol. 66, no. 1, pp. 414–439, 2001.

[5] V. L. Fernández and M. E. Coniglio, "Fibring in the Leibniz hierarchy," *Logic Journal of the IGPL*, vol. 15, no. 5-6, pp. 475–501, 2007.

[6] W. J. Blok and D. Pigozzi, "Algebraizable logics," *Memoirs of the American Mathematical Society*, vol. 77, no. 396, 1989.

[7] J. Czelakowski, *Protoalgebraic Logics*, vol. 10 of *Trends in Logic-Studia Logica Library*, Kluwer Academic, Dodrecht, The Netherlands, 2001.

[8] J. M. Font, R. Jansana, and D. Pigozzi, "A survey of abstract algebraic logic," *Studia Logica*, vol. 74, no. 1-2, pp. 13–97, 2003.

[9] M. A. Martins and G. Voutsadakis, "Malinowski Modalization, Modalization through Fibring and the Leibniz Hierarchy," http://www.voutsadakis.com/RESEARCH/papers.html .

[10] J. Malinowski, "Modal equivalential logics," *The Journal of Non-Classical Logic*, vol. 3, no. 2, pp. 13–35, 1986.

[11] A. Sernadas, C. Sernadas, and J. Rasga, "On combined connectives," *Logica Universalis*, vol. 5, no. 2, pp. 205–224, 2011.

[12] A. Sernadas, C. Sernadas, and J. Rasga, "On meet-combination of logics," *Journal of Logic and Computation*, vol. 22, no. 6, pp. 1453–1470, 2012.

[13] G. Voutsadakis, "Categorical abstract algebraic logic: algebraizable institutions," *Applied Categorical Structures*, vol. 10, no. 6, pp. 531–568, 2002.

[14] G. Voutsadakis, "Categorical abstract algebraic logic: equivalent institutions," *Studia Logica*, vol. 74, no. 1-2, pp. 275–311, 2003.

[15] G. Voutsadakis, "Categorical abstract algebraic logic: algebraic semantics for pi-Institutions," *Mathematical Logic Quarterly*. In press.

[16] M. Barr and C. Wells, *Category Theory for Computing Science*, Les Publications CRM, Montreal, Canada, 3rd edition, 1999.

[17] F. Borceux, *Handbook of Categorical Algebra, Vol. I, Encyclopedia of Mathematics and Its Applications*, Cambridge University Press, Cambridge, UK, 1994.

[18] S. Mac Lane, *Categories for the Working Mathematician*, Springer, New York, NY, USA, 1971.

5

Summation of Multiple Fourier Series in Matrix Weighted L_p-Spaces

Morten Nielsen

Department of Mathematical Sciences, Aalborg University, Fredrik Bajers Vej 7G, 9220 Aalborg East, Denmark

Correspondence should be addressed to Morten Nielsen; mnielsen@math.aau.dk

Academic Editor: Baoding Liu

This paper is concerned with rectangular summation of multiple Fourier series in matrix weighted L_p-spaces. We introduce a product Muckenhoupt A_p condition for matrix weights W and prove that rectangular Fourier partial sums converge in the corresponding matrix weighted space $L_p(\mathbb{T}^d; W)$, $1 < p < \infty$, if and only if the weight satisfies the product Muckenhoupt A_p condition. The same result is shown to hold true for other summation methods such as Cesàro and summation with the Jackson kernel.

1. Introduction

Let $\mathcal{M}$ be the family of nonnegative-definite $m \times m$ complex-valued matrices. A (periodic) matrix weight is by definition an integrable map $W : \mathbb{T}^d \to \mathcal{M}$. For a measurable vector-valued function $\mathbf{f} = (f_1, \dots, f_m)^T : \mathbb{T}^d \to \mathbb{C}^m$, let

$$\|f\|_{L_p(\mathbb{T}^d;W)} := \left(\int_{\mathbb{T}^d} \left| W^{1/p}(t)\,\mathbf{f}(t) \right|^p dt \right)^{1/p}, \quad 1 \le p < \infty, \tag{1}$$

where $|\cdot|$ denotes the usual norm on $\mathbb{C}^m$. We let $L_p(\mathbb{T}^d; W)$ denote the family $\{\mathbf{f} : \mathbb{T}^d \to \mathbb{C}^m : \|\mathbf{f}\|_{L_p(\mathbb{T}^d;W)} < \infty\}$, and $L_p(\mathbb{T}^d; W)$ becomes a Banach space when we factorize over $\mathcal{N} = \{\mathbf{f} : \|\mathbf{f}\|_{L_p(\mathbb{T}^d;W)} = 0\}$.

In this paper we are interested in convergence properties of multiple trigonometric series in $L_p(\mathbb{T}^d; W)$ and how specific convergence properties of trigonometric series can be related to properties of the weight W. To be more specific, let $D_n(t) = \sum_{|k|\le n} e^{-2\pi i k t}$ denote the univariate Dirichlet kernel, and for $\mathbf{N} \in \mathbb{N}^d$ we define the *rectangular* kernel $D_{\mathbf{N}}(t) := \prod_{j=1}^{d} D_{N_j}(t_j)$. Then

$$S_{\mathbf{N}} f := f * D_{\mathbf{N}} := \int_{\mathbb{T}^d} f(t)\, D_{\mathbf{N}}(\cdot - t)\, dt, \quad f \in L_1(\mathbb{T}^d), \tag{2}$$

defines the rectangular partial sum operator for the trigonometric system. We define the action of $S_{\mathbf{N}}$ on vector-valued functions $\mathbf{f}$ by letting it act separately on each coordinate function; that is,

$$[S_{\mathbf{N}}(\mathbf{f})]_j = S_{\mathbf{N}}(f_j), \qquad j = 1, 2, \dots m. \tag{3}$$

It is well known (see, e.g., [1, Theorem 3.5.7]) that $\|f - S_{\mathbf{N}}(f)\|_{L_p(\mathbb{T}^d)} \to 0$, as $\min_j N_j \to +\infty$, for $1 < p < \infty$. An immediate corollary is that we have convergence of the partial sums $S_{\mathbf{N}}(\mathbf{f})$, in $L_p(\mathbb{T}^d; \mathrm{Id})$, for $1 < p < \infty$ in the vector-valued case. However, it is not obvious what can be said about convergence of $S_{\mathbf{N}}(\mathbf{f})$ in $L_p(\mathbb{T}^d; W)$ for a general matrix weight W. The main result of the preset paper completely characterizes the special class of weights that allow convergence; $S_{\mathbf{N}}(\mathbf{f})$ converges if and only if the weight W satisfies a certain matrix Muckenhoupt A_p product condition. Moreover, the characterization relies solely on certain localization properties of the Dirichlet kernels shared by many other summation kernels. So, in addition, we prove that the rectangular Cesàro means and approximation using the Jackson kernels converge in $L_p(\mathbb{T}^d; W)$ if and only if the weight W satisfies the mentioned matrix Muckenhoupt A_p product condition.

At a first glance, the study of vector-valued operators like $S_{\mathbf{N}}$ on $L_p(\mathbb{T}^d; W)$ may seem artificial, but let us mention one important application to finitely generated shift invariant spaces where such mappings appear naturally. For a finite set of functions $\{f_k\}_{k=1}^m$ in $L_2(\mathbb{R}^d)$, the associated shift invariant space S is given by

$$S := \overline{\text{Span}\{f_k(\cdot - l) : l \in \mathbb{Z}^d, k = 1, 2, \ldots, m\}} \subset L_2(\mathbb{R}^d). \tag{4}$$

A natural question to pose is whether $B := \text{Span}\{f_k(\cdot - l) : l \in \mathbb{Z}^d, k = 1, 2, \ldots, m\}$ forms some sort of "stable" generating system for S. Here stable can mean a Schauder basis, or even some weaker notion such as a block Schauder basis. Consider the vector valued system $\mathcal{T} := \{\mathbf{e}_j e^{2\pi i k\cdot t} : k \in \mathbb{Z}^d, j = 1, \ldots, m\}$, where $\{\mathbf{e}_j\}_{j=1}^m$ is the standard basis for $\mathbb{C}^m$, and let G the $m \times m$ Gram matrix be given by

$$G_{i,j} := \sum_{k\in\mathbb{Z}^d} \widehat{f}_i(\cdot - k)\overline{\widehat{f}_j(\cdot - k)}, \tag{5}$$

where the Fourier transform is given by $\widehat{f}(\xi) := \int_{\mathbb{R}^d} f(t)e^{-2\pi i t\cdot\xi}\, dt$. One can show (see [2]) that the map $U : L_2(\mathbb{T}^d; G) \to S$, given by

$$U(\tau) := \left(\sum_{\ell=1}^{N} \tau_\ell(\xi)\widehat{\psi_\ell}\right)^{\vee}, \tag{6}$$

is an isometric isomorphism between $L_2(\mathbb{T}^d; G)$ and S, satisfying $U(\mathbf{e}_j e^{-2\pi i k\cdot t}) = f_j(\cdot - k)$. Hence, the metric properties of B in S are equivalent to the properties of $\mathcal{T}$ in $L_2(\mathbb{T}^d; G)$. For example, U will map $\{S_\mathbf{N}\}$ to a corresponding rectangular partial sum operators for B on S. The same correspondence holds true for the other summation methods related to $\mathcal{T}$ such as rectangular Cesàro means and approximation using the Jackson kernels converge in $L_p(\mathbb{T}^d; W)$ that will be discussed below.

It was proved by the present author [2] that the rectangular partial sums $\{US_\mathbf{N}\}$ converge in S precisely when $\{S_\mathbf{N}\}$ are uniformly bounded on $L_2(\mathbb{T}; G)$ which happens exactly when G is a Muckenhoupt A_2 product matrix weight.

The structure of this paper is as follows. In Section 2 we introduce a product A_p condition for matrix weights. Then necessary and sufficient conditions for a convolution operator of product type (such as $S_\mathbf{N}$) to be bounded on $L_p(\mathbb{T}^d; W)$ are given. Section 3 contains applications of the results in Section 2 to convolution operators induced by rectangular Dirichlet, Fejér, and Jackson kernels.

2. The Muckenhoupt Condition and Operators on $L_p(\mathbb{T}^d; W)$

In this section we introduce a matrix Muckenhoupt A_p product condition suitable for dealing with convolution operators of product type such as the partial sum operator $S_\mathbf{N}$ defined by (3). A sufficient condition for convolution operators of product type bounded on $L_p(\mathbb{T}^d; W)$ is given in Proposition 5, while a converse type result is considered in Proposition 7. We prove that convolution operators with "nicely localized" kernels can only be uniformly bounded on $L_p(\mathbb{T}^d; W)$ when W satisfies the product A_p condition.

The scalar A_p condition was introduced by Muckenhoupt [3], and it was proved by Hunt et al. in their seminal paper [4] that the A_p condition on a weight w is necessary and sufficient for the Hilbert transform to be bounded on the weighted space $L_p(\mathbb{T}; w)$.

More recently, Hunt-Muckenhoupt-Wheeden type results for matrix weights have been considered. The matrix A_p condition was introduced by Nazarov et al. [5–7] and they showed that it is the right condition for "standard" singular integral operators to be bounded on $L_p(\mathbb{T}^d; W)$. The A_p condition ($1 < p < \infty$) for weights $W : \mathbb{R}^d \to \mathcal{M}$ was originally stated in terms of dual matrices and average, but it was shown by Roudenko [8] to be equivalent to

$$\sup_B \left(\int_B \left(\int_B \left\|W^{1/p}(x)W^{-1/p}(y)\right\|^{p'} \frac{dy}{|B|}\right)^{p/p'} \frac{dx}{|B|}\right)^{p'/p}, \tag{7}$$

where the sup is taken over all open balls in $\mathbb{R}^d$ and $p' = p/(p-1)$ is the conjugate exponent.

Since our goal is to study operators of product type related to rectangular trigonometric partial sums, the condition given by (7) is not the appropriate one. The periodic weights satisfying (7) are well behaved when it comes to the study of square or spherical partial sum operators for trigonometric series. Let us therefore introduce a new and slightly modified Muckenhoupt condition. Inspired by (7), we let $\mathcal{R}(d)$ denote the family of all rectangles in $\mathbb{R}^d$ of the form $R = I_1 \times I_2 \times \cdots \times I_d$, with I_j being a bounded open interval in $\mathbb{R}$. Then we consider the following more restrictive subclass of matrix weights.

Definition 1. Let $W : \mathbb{T}^d \to \mathcal{M}$ be a periodic matrix weight. For $1 < p < \infty$, let $p' = p/(p-1)$ denote the conjugate exponent to p. We say that W belongs to the matrix Muckenhoupt (product) class PA_p provided there exists a uniform constant c_W such that

$$A_W(R, p) := \int_R \left(\int_R \left\|W^{1/p}(x)W^{-1/p}(y)\right\|^{p'} \frac{dy}{|R|}\right)^{p/p'} \frac{dx}{|R|} \le c_W, \tag{8}$$

for any $R \in \mathcal{R}(d)$.

Remark 2. For $d = 1$, Definition 1 reduces to the standard matrix A_p condition on $\mathbb{T}$, which we denote by $A_p(\mathbb{T})$. In the scalar case (i.e., $m = 1$), Definition 1 reduces to the known product A_p condition for scalar weights, which has a long history; see [9] and references therein.

The similarity of conditions (8) and (7) implies that many results for matrix A_p weights have straightforward analogs in the product case; the proofs can be "translated verbatim." Let us state the following lemma which will be needed below.

Lemma 3. *Let $W : \mathbb{T}^d \to \mathcal{M}$ be a matrix weight. Then the following statements are equivalent for $1 < p < \infty$:*

(i) $W \in PA_p$;

(ii) $W^{-p'/p} \in PA_{p'}$;

(iii) $\int_R (\int_R \|W^{1/p}(x)W^{-1/p}(y)\|^p (dx/|R|))^{p'/p}(dy/|R|) \leq c_W$, *for all* $R \in \mathcal{R}(d)$.

We refer the reader to Roudenko [8] for the proof of Lemma 3 in the nonproduct case.

The following Lemma reveals why one can expect AP_p to be useful for product operators. A weight in AP_p is uniformly in $A_p(\mathbb{T})$ in each of its d variables.

Lemma 4. *Let $W : \mathbb{T}^d \to \mathcal{M}$ be a matrix weight, and let $1 < p < \infty$. Then the following holds.*

(a) *For any rectangles $R \subseteq \widetilde{R} \subset \mathcal{R}(d)$,*

$$A_W(R, p) \leq \left(\frac{|\widetilde{R}|}{|R|}\right)^p A_W(\widetilde{R}, p). \tag{9}$$

(b) *Suppose $W \in PA_p$; then the univariate weight $\xi_j \to W(\xi)$, obtained by fixing the variables ξ_k, $k \neq j$, is uniformly in $\mathbb{A}_p(\mathbb{T})$ for a.e. $(\xi_1, \ldots, \xi_{j-1}, \xi_{j+1}, \ldots, \xi_d) \in \mathbb{T}^{d-1}$.*

Proof. For (a), we notice that whenever $R \subseteq \widetilde{R} \subset \mathcal{R}(d)$,

$$\begin{aligned} &A_W(R, p) \\ &= \int_R \left(\int_R \left\|W^{1/p}(x) W^{-1/p}(t)\right\|^{p'} \frac{dt}{|R|}\right)^{p/p'} \frac{dx}{|R|} \\ &\leq \left(\frac{|\widetilde{R}|}{|R|}\right)^p \int_{\widetilde{R}} \left(\int_{\widetilde{R}} \left\|W^{1/p}(x) W^{-1/p}(t)\right\|^{p'} \frac{dt}{|\widetilde{R}|}\right)^{p/p'} \frac{dx}{|\widetilde{R}|} \\ &= \left(\frac{|\widetilde{R}|}{|R|}\right)^p A_W(\widetilde{R}, p). \end{aligned} \tag{10}$$

Now we turn to the proof of (b). It suffices to consider $\widetilde{W}(t) := W(t, \xi_2, \ldots, \xi_d)$ for $(\xi_2, \ldots, \xi_d) \in \mathbb{T}^{d-1}$ fixed. Given an interval $I \subset \mathbb{R}$, we form $R_\varepsilon = I_\varepsilon(\xi_2) \times \cdots \times I_\varepsilon(\xi_d)$, where $I_\varepsilon(\xi_j)$ is an interval of length 2ε centered at ξ_j. First suppose $p \leq p'$. Since $W \in PA_p$, there exists a constant c_W independent of $I \times R_\varepsilon$ such that

$$\begin{aligned} &\frac{1}{|R_\varepsilon|^2} \int_{R_\varepsilon} \int_{R_\varepsilon} \left[\int_I \left(\int_I \left\|W^{1/p}(t, \mathbf{u}) W^{-1/p}\right.\right.\right. \\ &\qquad \left.\left.\left. \times (w, \mathbf{v})\right\|^{p'} \frac{dw}{|I|}\right)^{p/p'} \frac{dt}{|I|}\right] d\mathbf{u}\, d\mathbf{v} \\ &\leq \int_{R_\varepsilon} \int_I \left(\int_{R_\varepsilon} \int_I \left\|W^{1/p}(t, \mathbf{u}) W^{-1/p}(w, \mathbf{v})\right\|^{p'} \right. \\ &\qquad \left. \times \frac{dw\, d\mathbf{v}}{|I| \cdot |R_\varepsilon|}\right)^{p/p'} \frac{dt\, d\mathbf{u}}{|I| \cdot |R_\varepsilon|} \\ &= A_W(R, p) \leq c_W, \end{aligned} \tag{11}$$

where we have used the continuous embedding $L_1(R_\varepsilon; d\mathbf{v}/|R_\varepsilon|) \hookrightarrow L_{p/p'}(R_\varepsilon; d\mathbf{v}/|R_\varepsilon|)$. Hence, by Lebesgue's differentiation theorem, for almost every $(\xi_2, \ldots, \xi_d) \in \mathbb{T}^{d-1}$,

$$\begin{aligned} c_W &\geq \lim_{\varepsilon \to 0^+} A_W(I \times R_\varepsilon, p) \\ &= \int_I \left(\int_I \left\|\widetilde{W}^{1/p}(t) \widetilde{W}^{-1/p}(w)\right\|^{p'} \frac{dw}{|I|}\right)^{p/p'} \frac{dt}{|I|} \\ &= A_{\widetilde{W}}(I, p), \end{aligned} \tag{12}$$

where the constant is independent of I and $(\xi_2, \ldots, \xi_d)$. It follows that $\widetilde{W}$ is uniformly in $\mathbb{A}_p(\mathbb{T})$ for a.e. $(\xi_2, \ldots, \xi_d) \in \mathbb{T}^{d-1}$. In the case $p' < p$, we use Lemma 3 to conclude that $W^{-p'/p} \in PA_{p'}$, which implies the following estimate:

$$\begin{aligned} &\int_{R_\varepsilon} \int_I \left(\int_{R_\varepsilon} \int_I \left\|W^{1/p}(t, \mathbf{u}) W^{-1/p}(w, \mathbf{v})\right\|^p \right. \\ &\qquad \left. \times \frac{dt d\mathbf{u}}{|I| \cdot |R_\varepsilon|}\right)^{p'/p} \frac{dw d\mathbf{v}}{|I| \cdot |R_\varepsilon|} \leq c_W. \end{aligned} \tag{13}$$

By repeating the argument from the $p \leq p'$ case, we conclude that $\widetilde{W}^{-p'/p}$ is uniformly in $A_{p'}(\mathbb{T})$ which again by Lemma 3 implies that $\widetilde{W}$ is uniformly in $A_p(\mathbb{T})$. □

We can now prove the following result that explains how to get from a bounded convolution operator on $L_p(\mathbb{T}; W)$ to a bounded convolution operator on $L_p(\mathbb{T}^d; W)$, for $W \in AP_p$, simply by forming the natural product kernel.

Proposition 5. *Suppose that $\{K_N\}_{N \geq 0}$ is a sequence of convolution kernels defined on $\mathbb{T}$ for which the corresponding operators*

$$T_N f := \int_{\mathbb{T}} f(t) K_N(\cdot - t)\, dt \tag{14}$$

are uniformly bounded on $L_p(\mathbb{T};W)$ *whenever* $W \in A_p(\mathbb{T})$. *Then the associated product convolution kernels*

$$K_{\mathbf{N}}(\xi) = \prod_{j=1}^{d} K_{N_j}(\xi_j), \qquad \mathbf{N} = (N_1,\dots,N_d) \in \mathbb{N}_0^d, \quad \xi \in \mathbb{T}^d, \tag{15}$$

induce a uniformly bounded family of operators on $L_p(\mathbb{T}^d;W)$ *for* $W \in PA_p$.

Proof. Suppose that $W \in PA_p$. In the case $d = 1$, there is nothing to prove. We focus on the case $d = 2$; the reader can easily verify that the argument below generalizes to any $d \geq 3$.

According to Lemma 4(b), $W_{\xi_1} := W(\xi_1,\cdot)$ and $W_{\xi_2} := W(\cdot,\xi_2)$ satisfy uniform Muckenhoupt A_p-conditions a.e. on $\mathbb{T}$. Pick any $f \in L_p(\mathbb{T}^2,W)$. By Fubini's theorem, $f_{\xi_1} := f(\xi_1,\cdot) \in L_p(\mathbb{T},W_{\xi_1})$ and $f_{\xi_2} := f(\cdot,\xi_2) \in L_p(\mathbb{T},W_{\xi_2})$ for a.e. $[\xi_1]$ and $[\xi_2]$, respectively.

We define

$$T_N^1 f := K_N * f_{\xi_2} := \int_{\mathbb{T}} f_{\xi_2}(t)\, K_N(\cdot - t)\, dt,$$
$$T_M^2 f := K_M * f_{\xi_1} := \int_{\mathbb{T}} f_{\xi_1}(t)\, K_M(\cdot - t)\, dt. \tag{16}$$

Notice that $T_{N,M} f = T_N^1 T_M^2 f$. By assumption,

$$\int_{\mathbb{T}} \left| W_{\xi_1}^{1/p}(\xi_2)\, T_M^2 f_{\xi_1}(\xi_2) \right|^p d\xi_2 \leq C \int_{\mathbb{T}} \left| W_{\xi_1}^{1/p}(\xi_2)\, f_{\xi_1}(\xi_2) \right|^p d\xi_2, \quad \text{a.e. } [\xi_1]. \tag{17}$$

An integration yields

$$\int_{\mathbb{T}}\int_{\mathbb{T}} \left| W^{1/p}(\xi_1,\xi_2)\, T_M^2 f(\xi_1,\xi_2) \right|^p d\xi_2 d\xi_1 \leq C \int_{\mathbb{T}}\int_{\mathbb{T}} \left| W^{1/p}(\xi_1,\xi_2)\, f(\xi_1,\xi_2) \right|^p d\xi_2 d\xi_1. \tag{18}$$

Similarly,

$$\begin{aligned} \|T_{N,M} f\|_{L_{p(\mathbb{T}^2,W)}}^p &= \int_{\mathbb{T}}\int_{\mathbb{T}} \left| W^{1/p}(\xi_1,\xi_2)\, T_N^1 T_M^2 f(\xi_1,\xi_2) \right|^p d\xi_1 d\xi_2 \\ &\leq C \int_{\mathbb{T}}\int_{\mathbb{T}} \left| W^{1/p}(\xi_1,\xi_2)\, T_M^2 f(\xi_1,\xi_2) \right|^p d\xi_1 d\xi_2 \\ &\leq C^2 \int_{\mathbb{T}}\int_{\mathbb{T}} \left| W^{1/p}(\xi_1,\xi_2)\, f \right|^p d\xi_1 d\xi_2. \end{aligned} \tag{19}$$

It follows that the family $\{T_{\mathbf{N}}\}_{\mathbf{N}\in\mathbb{N}_0^2}$ is uniformly bounded on $L_p(\mathbb{T}^2;W)$. □

We now turn to a converse type result to Proposition 5. Proposition 7 will show that well-localized trigonometric convolution kernels of product type can only be uniformly bounded on $L_p(\mathbb{T}^d;W)$ when $W \in AP_p$.

We need the following Lemma which gives an estimate of the norm of integral operators on $L_2(\mathbb{T}^d;W)$ with nice compactly supported kernels.

Lemma 6. *Suppose* $Sf(\xi) = \int_{\mathbb{T}^d} S(\xi,\eta) f(\eta) d\eta$ *is an integral operator with a scalar kernel* $S(\xi,\eta)$ *that satisfies* $|S(\xi,\eta)| \leq \alpha|R|^{-1}\chi_{R\times R}$ *for some bounded rectangle* $R \subset \mathbb{R}^d$. *For* $1 < p < \infty$, *there exists a constant* C_d *independent of the particular choice of* S *such that the norm of* S *on* $L_p(\mathbb{T}^d;W)$ *is at most* $C_d \cdot \alpha \cdot A_W(R,p)$, *with* $A_W(R,p)$ *given by* (8). *Moreover, the kernel* $\alpha|R|^{-1}\chi_R \times \chi_R$ *induces an operator with norm at least* $C_d^{-1} \cdot \alpha \cdot A_W(R,p)$ *on* $L_p(\mathbb{T}^d;W)$.

The proof of Lemma 6 for nonproduct A_p-weights can be found in Goldberg [10]. We leave the straightforward adaptation of the proof in [10] to the product case for the reader.

We can now give a proof of Proposition 7. For $K \in \mathbb{N}$, we let $\mathcal{P}_K = \text{span}\{e^{2\pi i k\cdot} : k \in \mathbb{Z}; |k| \leq K\}$.

Proposition 7. *Let* $W : \mathbb{T}^d \to \mathcal{M}$ *be a periodic matrix weight, and let* $\{K_n\}_{n\geq 1}$ *be a sequence of real-valued trigonometric convolution kernels defined on* $\mathbb{T}$. *Assume there exist constants* c, C *such that* $K_n \in \mathcal{P}_{c\cdot n}$, *with* $C^{-1}n \leq \|K_n\|_\infty = K_n(0) \leq Cn$, *for* $n \in \mathbb{N}$. *Suppose that the corresponding product kernels*

$$K_{\mathbf{N}}(\xi) = \prod_{j=1}^{d} K_{N_j}(\xi_j), \quad \mathbf{N} \in \mathbb{N}_0^d, \tag{20}$$

induce a uniformly bounded family $\{T_{K_{\mathbf{N}}}\}$ *of convolution operators on* $L_p(\mathbb{T}^d;W)$. *Then* $W \in PA_p$.

Proof. We have to estimate $A_W(R,p)$ for an arbitrary rectangle $R \in \mathcal{R}(d)$. The idea is to form a suitable product kernel $K_{\mathbf{N}}$ that is "large" on R in the sense that the corresponding operator can be well approximated by an integral operator of the type considered in Lemma 6.

By assumption, the kernel $K_n \in \mathcal{P}_{c\cdot n}$ is real and $\|K_n\|_\infty = K_n(0) \leq Cn$, so by Bernstein's inequality, $\|K_n'\|_\infty \leq cCn^2$. We can thus find an integer M (independent of n) such that for $t \in [-1/Mn, 1/Mn]$ we have $K_n(t) \geq (1 - (1/2C_d^2))^{1/d}\|K_n\|_\infty$, where C_d is the constant from Lemma 6.

Let a rectangle $R = I_1 \times I_2 \times \cdots \times I_d$ be given. For $j = 1,2,\dots,d$, with $|I_j| > 1/2M$, we define $N_j = 0$ and replace I_j with $[-1/2,1/2)$ and obtain a possibly larger rectangle $\widetilde{R}$. By Lemma 4(b), there exists a universal constant b such that $A_W(R,p) \leq bA_W(\widetilde{R},p)$ since $|\widetilde{R}| \leq (2M)^d|R|$. Next, for each $j = 1,2,\dots,d$ with $|I_j| \leq 1/2M$, we choose an integer $N_j \geq 1$ such that

$$\frac{1}{4M}\cdot\frac{1}{N_j} \leq |I_j| \leq \frac{1}{2M}\cdot\frac{1}{N_j}. \tag{21}$$

Notice that for $t, u \in I_j$, we have $t - u \in I_j - I_j \subset [-1/MN_j, 1/MN_j]$ so

$$K_{N_j}(t-u) \geq \left(1 - \frac{1}{2C_d^2}\right)^{1/d} \left\|K_{N_j}\right\|_\infty. \tag{22}$$

For notational convenience we put $K_0 := 1$ and form the product kernel

$$K_{\mathbf{N}}(\xi) = \prod_{j=1}^{d} K_{N_j}(\xi_j). \tag{23}$$

The plan of attack is to use the simple fact that $f \to \chi_{\widetilde{R}} T_{K_{\mathbf{N}}}(\chi_{\widetilde{R}} f)$ is uniformly bounded in both $\widetilde{R}$ and $\mathbf{N} \in \mathbb{N}_0^d$. We notice that $f \to \chi_{\widetilde{R}} T_{K_{\mathbf{N}}}(\chi_{\widetilde{R}} f)$ has integral kernel

$$S_2(\xi, \eta) := \chi_{\widetilde{R}}(\eta)\, \chi_{\widetilde{R}}(\xi)\, K_{\mathbf{N}}(\eta - \xi). \tag{24}$$

We wish to estimate the operator norm of S_2 from below. For that purpose we first consider the operator with kernel

$$\begin{aligned} S(\xi, \eta) &:= S_1(\xi, \eta) - S_2(\xi, \eta) \\ &:= \|K_{\mathbf{N}}\|_\infty \chi_{\widetilde{R}}(\xi)\, \chi_{\widetilde{R}}(\eta) - \chi_{\widetilde{R}}(\xi)\, \chi_{\widetilde{R}}(\eta)\, K_{\mathbf{N}}(\xi - \eta). \end{aligned} \tag{25}$$

Notice that estimate (22), together with the fact that $\|K_{\mathbf{N}}\|_\infty = \prod_{j=1}^{d} K_{N_j}(0)$, implies the following size estimate

$$\begin{aligned} &|S(\xi, \eta)| \\ &\quad = \left| \|K_{\mathbf{N}}\|_\infty \chi_{\widetilde{R}}(\xi)\, \chi_{\widetilde{R}}(\eta) - \chi_{\widetilde{R}}(\xi)\, \chi_{\widetilde{R}}(\eta)\, K_{\mathbf{N}}(\xi - \eta) \right| \\ &\quad \leq \frac{\|K_{\mathbf{N}}\|_\infty}{2C_d^2} \chi_{\widetilde{R}}(\xi)\, \chi_{\widetilde{R}}(\eta) \\ &\quad = \frac{\left|\widetilde{R}\right| \cdot \|K_{\mathbf{N}}\|_\infty}{2C_d^2} \left|\widetilde{R}\right|^{-1} \chi_{\widetilde{R}}(\xi)\, \chi_{\widetilde{R}}(\eta). \end{aligned} \tag{26}$$

According to Lemma 6, the kernel S induces an operator of norm at most $(1/2)C_d^{-1}|\widetilde{R}| \cdot \|K_{\mathbf{N}}\|_\infty A_W(\widetilde{R}, p)$ on $L_p(\mathbb{T}^d; W)$. At the same time, Lemma 6 shows that the operator with kernel $S_1(\xi, \eta) = |\widetilde{R}| \|K_{\mathbf{N}}\|_\infty \cdot |\widetilde{R}|^{-1} \chi_{\widetilde{R}}(\xi) \chi_{\widetilde{R}}(\eta)$ has norm at least $C_d^{-1}|\widetilde{R}| \cdot \|K_{\mathbf{N}}\|_\infty A_W(\widetilde{R}, p)$ on $L_p(\mathbb{T}^d; W)$. The triangle inequality for operator norms now implies that

$$\begin{aligned} &\frac{1}{2C_d} \left|\widetilde{R}\right| \cdot \|D_{\mathbf{N}}\|_\infty M\left(\widetilde{R}, W\right) \\ &\quad \geq \|S_1 - S_2\| \geq \left| \|S_1\| - \|S_2\| \right| \\ &\quad \geq C_d^{-1} \left|\widetilde{R}\right| \cdot \|D_{\mathbf{N}}\|_\infty M\left(\widetilde{R}, W\right) - \|S_2\|, \end{aligned} \tag{27}$$

so $\|S_2\| \geq (1/2)C_d^{-1}|\widetilde{R}| \cdot \|D_{\mathbf{N}}\|_\infty A_W(\widetilde{R}, p)$. Moreover, by (21), we see that $|\widetilde{R}| \cdot \|K_{\mathbf{N}}\|_\infty \geq C(4M)^{-d}$, so we may conclude that

$$\begin{aligned} A_W(R, p) &\leq bA_W\left(\widetilde{R}, p\right) \\ &\leq 2bCC_d (4M)^d \|S_2\| \\ &= C' \sup_{\|f\|_{L_p(\mathbb{T}^d; W)} = 1} \left\| \chi_{\widetilde{R}} T_{K_{\mathbf{N}}} \left(\chi_{\widetilde{R}} f\right) \right\|_{L_p(\mathbb{T}^d; W)} \\ &\leq C' \sup_{\|f\|_{L_p(\mathbb{T}^d; W)} = 1} \left\| T_{K_{\mathbf{N}}} f \right\|_{L_p(\mathbb{T}^d; W)} \leq C'' \end{aligned} \tag{28}$$

with constant C'' independent of R. We may finally conclude that $W \in PA_p(d)$. □

3. Summation of Multiple Trigonometric Series

This section contains applications of the results of Section 2 to convolution operators induced by rectangular Dirichlet, Fejér, and Jackson kernels. The Dirichlet kernels correspond to standard rectangular trigonometric summation while the Fejér kernels generate the corresponding Cesàro means. The Jackson kernels are (normalized) squares of the Fejér kernels, and they induce the well-known Jackson approximation by trigonometric polynomials.

We begin by studying the univariate Dirichlet kernel. The Hilbert transform H is defined on $L_p(\mathbb{T})$, $1 < p < \infty$, by

$$H(f)(x) := \text{p.v.} \int_{\mathbb{T}} f(t) \cot(\pi (x - t))\, dt. \tag{29}$$

We lift H to a linear operator on $L_p(\mathbb{T}; W)$, for any matrix weight $W : \mathbb{T} \to \mathcal{M}$, by letting it act coordinatewise.

Treil and Volberg completely characterized when the Hilbert transform H is bounded in the matrix case on $\mathbb{T}$ when $p = 2$; see [11]. Later, Nazarov and Treĭl introduced in a new "Bellman function" method [6] to extend the theory to $1 < p < \infty$. Volberg presented a different solution to the matrix weighted L_p boundedness of the Hilbert transform via Littlewood-Paley theory [7]. The fundamental result is the following.

Theorem 8 (see [6, 7, 11]). *Let $W : \mathbb{T} \to \mathcal{M}$ be a matrix weight. Suppose $1 < p < \infty$. Then the Hilbert transform is bounded on $L_p(\mathbb{T}; W)$ if and only if $W \in A_p(\mathbb{T})$.*

We recall that the univariate Dirichlet kernel D_N is given by

$$D_N(t) = \frac{\sin 2\pi (N + 1/2) t}{\sin \pi t}, \quad N \geq 1, \tag{30}$$

and for $f \in L_p(\mathbb{T})$ we define the associated partial sum operators,

$$S_N(f) := \sum_{k=-N}^{N} \widehat{f}(k) e^{2\pi i k \cdot} = f * D_N := \int_{\mathbb{T}} f(t) D_N(\cdot - t)\, dt. \quad (31)$$

We have the following lemma which follows easily from Theorem 8.

Lemma 9. *Let $W : \mathbb{T} \to \mathcal{M}$ be a matrix weight in $A_p(\mathbb{T})$. Then the partial sum operators $f \to f * D_N$ are uniformly bounded on $L_p(\mathbb{T}; W)$.*

Proof. We let $P_+ = (1/2)(I + iH + S_0)$ denote the Riesz projection onto H^p for $f \in L_p(\mathbb{T}; W)$, where $S_0 f := \int_{\mathbb{T}} f(y)dy$ is the 0-order partial sum operator. It follows that P_+ is bounded on $L_p(\mathbb{T}; W)$ since H is bounded according to Theorem 8, and S_0 is bounded according to [12, Lemma 1.5]. Notice that $f \to f e^{2\pi i M \cdot}$ is a norm preserving operator on $L_p(\mathbb{T}; W)$, just as in the scalar case. Then we observe that

$$f * D_N = e^{-2\pi i N \cdot} P_+ \left(e^{2\pi i N \cdot} f\right) - e^{2\pi i (N+1) \cdot} P_+ \left(e^{-2\pi i (N+1) \cdot} f\right), \quad (32)$$

and the result follows. □

For $\mathbf{N} = (N_1, \ldots, N_d) \in \mathbb{N}_0^d$, we form the product kernel $D_{\mathbf{N}}(\xi) := \prod_{j=1}^{d} D_{N_j}(\xi_j)$. One has

$$S_{\mathbf{N}}(\mathbf{f}) := \int_{\mathbb{T}^d} \mathbf{f}(t) D_{\mathbf{N}}(\cdot - t)\, dt. \quad (33)$$

We notice that $D_N \in \mathcal{P}_N$ and $D_N(0) = \|D_N\|_\infty = 2N + 1$, so the following corollary follows directly from Propositions 5 and 7 and Lemma 9.

Corollary 10. *Let $W : \mathbb{T}^d \to \mathcal{M}$ be a matrix weight. For $1 < p < \infty$, the operators $\{S_{\mathbf{N}} : \mathbf{N} \in \mathbb{N}^d\}$ are uniformly bounded on $L_p(\mathbb{T}^d; W)$ if and only if $W \in AP_p$.*

Remark 11. It is easy to verify that vectors of trigonometric polynomials are dense in $L_p(\mathbb{T}^d; W)$, $1 < p < \infty$, whenever W is a matrix weight (since $W \in L_1$ so each entry in W is in $L_1(\mathbb{T})$). It therefore follows by standard techniques that the family $\{S_{\mathbf{N}} : \mathbf{N} \in \mathbb{N}^d\}$ is uniformly bounded on $L_p(\mathbb{T}^d; W)$ if and only if $\|\mathbf{f} - S_{\mathbf{N}}(\mathbf{f})\|_{L_p(\mathbb{T};W)} \to 0$, as $\min_j N_j \to +\infty$, for all $\mathbf{f} \in L_p(\mathbb{T}; W)$.

Corollary 10 relies on basic localization properties of the Dirichlet kernel. However, many well-known summation kernels share the necessary properties needed to apply Propositions 5 and 7. Let us illustrate this fact by considering two specific examples.

The rectangular Cesàro summation is given by

$$\sigma_{\mathbf{N}}(\mathbf{f}) = \frac{1}{(N_1 + 1) \cdots (N_d + 1)} \sum_{k_1=0,\ldots,k_d=0}^{N_1,\ldots,N_d} S_{(k_1,k_2,\ldots,k_d)}(\mathbf{f}) = \int_{\mathbb{T}^d} \mathbf{f}(t) F_{\mathbf{N}}(\cdot - t)\, dt \quad (34)$$

with the product Féjer kernel given by $F_{\mathbf{N}}(\xi) = \prod_{j=1}^{d} F_{N_j}(\xi_j)$, where the scalar Féjer kernel is defined by

$$F_n(t) = \frac{1}{n}\left(\frac{\sin(n\pi t)}{\sin(\pi t)}\right)^2. \quad (35)$$

Notice that $F_b \in \mathcal{P}_n$ and $F_n(0) = \|F_n\|_\infty = n$. The scalar Jackson kernel is the normalized square of the Féjer kernel and given by

$$J_n(t) = \frac{3}{n(2n^2 + 1)}\left(\frac{\sin(n\pi t)}{\sin(\pi t)}\right)^4. \quad (36)$$

The corresponding product kernel is $J_{\mathbf{N}}(\xi) = \prod_{j=1}^{d} J_{N_j}(\xi_j)$, $\mathbf{N} \in \mathbb{N}_0^d$, and the rectangular Jackson summation operator is given by $\mathcal{J}_{\mathbf{N}}(f) := f * J_{\mathbf{N}}$. Notice that $J_n \in \mathcal{P}_{2n}$ and $J_n(0) = \|J_n\|_\infty = n$.

We now conclude by stating the main result, which summarizes the results obtained in the present paper. The theorem shows that uniform boundedness of the rectangular operators $S_{\mathbf{N}}$, $\sigma_{\mathbf{N}}$, and $\mathcal{J}_{\mathbf{N}}$ on $L_p(\mathbb{T}^d; w)$ is equivalent to the condition $W \in AP_p$.

Theorem 12. *Let $W : \mathbb{T}^d \to \mathcal{M}$ be a matrix weight. For $1 < p < \infty$, the following conditions are equivalent:*

(i) *$W \in AP_p$;*

(ii) *the operators $\{S_{\mathbf{N}} : \mathbf{N} \in \mathbb{N}_0^d\}$ are uniformly bounded on $L_p(\mathbb{T}^d; W)$;*

(iii) *the operators $\{\sigma_{\mathbf{N}} : \mathbf{N} \in \mathbb{N}_0^d\}$ are uniformly bounded on $L_p(\mathbb{T}^d; W)$;*

(iv) *the operators $\{\mathcal{J}_{\mathbf{N}} : \mathbf{N} \in \mathbb{N}_0^d\}$ are uniformly bounded on $L_p(\mathbb{T}^d; W)$.*

Proof. We first notice that each of the univariate kernels D_n, F_n, and J_n satisfies the hypothesis of Proposition 7, so (ii), (iii), and (iv) each implies that $W \in AP_p$. Now, suppose that $W \in AP_p$. Then (ii) holds by Corollary 10. To conclude, we just need to recall that Cesàro and Jackson summations are both regular summation methods, so (ii) implies both (iii) and (iv). □

References

[1] L. Grafakos, *Classical Fourier Analysis*, vol. 249 of *Graduate Texts in Mathematics*, Springer, New York, NY, USA, 2nd edition, 2008.

[2] M. Nielsen, "On stability of finitely generated shift-invariant systems," *The Journal of Fourier Analysis and Applications*, vol. 16, no. 6, pp. 901–920, 2010.

[3] B. Muckenhoupt, "Weighted norm inequalities for the Hardy maximal function," *Transactions of the American Mathematical Society*, vol. 165, pp. 207–226, 1972.

[4] R. Hunt, B. Muckenhoupt, and R. Wheeden, "Weighted norm inequalities for the conjugate function and Hilbert transform," *Transactions of the American Mathematical Society*, vol. 176, pp. 227–251, 1973.

[5] F. Nazarov, S. Treil, and A. Volberg, "The Bellman functions and two-weight inequalities for Haar multipliers," *Journal of the American Mathematical Society*, vol. 12, no. 4, pp. 909–928, 1999.

[6] F. L. Nazarov and S. R. Treĭl′, "The hunt for a Bellman function: applications to estimates for singular integral operators and to other classical problems of harmonic analysis," *Algebra i Analiz*, vol. 8, no. 5, pp. 32–162, 1996.

[7] A. Volberg, "Matrix A_p weights via S-functions," *Journal of the American Mathematical Society*, vol. 10, no. 2, pp. 445–466, 1997.

[8] S. Roudenko, "Matrix-weighted Besov spaces," *Transactions of the American Mathematical Society*, vol. 355, no. 1, pp. 273–314, 2003.

[9] S. Y.A. Chang and R. Fefferman, "Some recent developments in Fourier analysis and H^p-theory on product domains," *Bulletin of the American Mathematical Society*, vol. 12, no. 1, pp. 1–43, 1985.

[10] M. Goldberg, "Matrix A_p weights via maximal functions," *Pacific Journal of Mathematics*, vol. 211, no. 2, pp. 201–220, 2003.

[11] S. Treil and A. Volberg, "Wavelets and the angle between past and future," *Journal of Functional Analysis*, vol. 143, no. 2, pp. 269–308, 1997.

[12] S. Treil and A. Volberg, "Continuous frame decomposition and a vector Hunt-Muckenhoupt-Wheeden theorem," *Arkiv för Matematik*, vol. 35, no. 2, pp. 363–386, 1997.

6

Analytical Study of Some Important Generalized Nonlinear Partial Differential Equations

Marwan Alquran, Mahmoud Mohammad, and Ahmad Ababneh

Department of Mathematics and Statistics, Jordan University of Science and Technology, Irbid 22110, Jordan

Correspondence should be addressed to Marwan Alquran; marwan04@just.edu.jo

Academic Editor: Anjan Biswas

The aim of this paper is to extend the use of homotopy perturbation method (HPM) to study the solutions for some important generalized nonlinear partial differential equations (PDEs) such as Fisher equation with convection term, Sharma-Tasso-Olver (STO) equation, and Fitzhugh-Nagumo (FN) equation.

1. Introduction

The main purpose of this paper is to apply the homotopy perturbation method with the help of symbolic computation to obtain approximate solution of the following nonlinear equations.

Fisher equation with convection term serves as a basic model for population dynamics or chemical kinetics in mathematical biology. The solution shows exponential growth counteracted by nonlinear damping (where the damping rate is proportional to u). Fisher equation has the form

$$\frac{\partial u}{\partial t} + \alpha u \frac{\partial u}{\partial x} - \frac{\partial^2 u}{\partial x^2} - u(1-u) = 0, \quad (1)$$

subject to

$$u(x,0) = \frac{1}{2}\left(1 - \tanh\left(\frac{x}{4}\right)\right). \quad (2)$$

Fisher equation was studied by many authors using different methods such as the tanh method [1] and variational iteration method [2].

Sharma-Tasso-Olver (STO) equation is a good example to show the fusion of soliton solutions. It has the form

$$\frac{\partial u}{\partial t} + \alpha \frac{\partial}{\partial x}\left(u^3\right) + \frac{3}{2}\alpha \frac{\partial^2}{\partial x^2}\left(u^2\right) + \alpha \frac{\partial^3 u}{\partial x^3} = 0, \quad (3)$$

subject to

$$u(x,0) = \sqrt{\frac{1}{\alpha}} \tanh\left(\sqrt{\frac{1}{\alpha}} x\right). \quad (4)$$

Hirota's direct method [3] and extended tanh method [4] were used to solve the STO equation.

Fitzhugh-Nagumo (FN) equation is an important nonlinear reaction-diffusion equation and is usually used to model the transmission of nerve impulses; also it is used in circuit theory, biology, and the area of population genetics. FN equation has the form

$$\frac{\partial u}{\partial t} - \frac{\partial^2 u}{\partial x^2} + u(1-u)(\alpha - u) = 0, \quad (5)$$

subject to

$$u(x,0) = \frac{1}{2}\left(1 + \tanh\left(\frac{-\sqrt{2}}{4} x\right)\right). \quad (6)$$

FN equation was studied by means of improved sine-cosine method [5] and Exp-function method [6].

2. Survey of Homotopy Perturbation Method

To illustrate the basic ideas of this method, we consider the following nonlinear differential equation:

$$A(u) - f(r) = 0, \quad r \in \Omega, \quad (7)$$

with the boundary conditions

$$B\left(u, \frac{\partial u}{\partial \eta}\right) = 0, \quad r \in \Gamma, \tag{8}$$

where A is a general differential operator, B is a boundary operator, $f(r)$ is a known analytical function, and Γ is the boundary of the domain Ω. The operator A can be divided into two parts, which are L and N, where L is the linear, and N is the nonlinear operator. Therefore, (7) can be written as follows:

$$L(u) + N(u) - f(r) = 0. \tag{9}$$

Liao in [7] constructed a homotopy $v(r, p) : \Omega \times [0, 1] \rightarrow R$, which satisfies

$$H(v, p) = (1 - p)[L(v) - L(u_0)] + p[A(v) - f(r)] = 0, \tag{10}$$

which is equivalent to

$$H(v, p) = L(v) - L(u_0) + p[L(u_0) + N(v) - f(r)] = 0, \tag{11}$$

where $r \in \Omega$, $p \in [0, 1]$, that is, called embedding parameter, u_0 is an initial approximation solution of (7), which satisfies the boundary conditions. Obviously, from (10) and (11) we obtain

$$\begin{aligned} H(v, 0) &= L(v) - L(u_0) = 0, \\ H(v, 1) &= A(v) - f(r) = 0, \end{aligned} \tag{12}$$

and the changing process of p from 0 to 1, is just that of $H(v, p)$ from $L(v) - L(u_0)$ to $A(v) - f(r)$. In topology, this is called deformation, and $L(v) - L(u_0)$ and $A(v) - f(r)$ are called homotopic. Applying the perturbation technique, we can first use the embedding parameter p as a small parameter and assume that the solutions of (10) or (11) can be expressed as a power series in p as follows:

$$v = \sum_{i=0}^{\infty} p^i v_i. \tag{13}$$

Setting $p = 1$ gives the solution of (7)

$$u = \lim_{p \to 1} v = \sum_{i=0}^{\infty} v_i. \tag{14}$$

To study the convergence of the method, rewrite (11) in the following form:

$$L(v) = L(u_0) + p[f(r) - L(u_0) - N(v)], \tag{15}$$

applying the inverse operator L^{-1}, to both sides of (15), we obtain

$$v = u_0 + p[L^{-1}f(r) - L^{-1}N(v) - u_0], \tag{16}$$

substituting (13) into the right-hand side of (16), we get the following form:

$$v = u_0 + p\left[L^{-1}f(r) - (L^{-1}N)\left(\sum_{i=0}^{\infty} p^i v_i\right) - u_0\right]. \tag{17}$$

The exact solution may be obtained by using (14)

$$\begin{aligned} u &= \lim_{p \to 1} v \\ &= L^{-1}f(r) - (L^{-1}N)\left(\sum_{i=0}^{\infty} v_i\right) \\ &= L^{-1}f(r) - \sum_{i=0}^{\infty} (L^{-1}N)(v_i). \end{aligned} \tag{18}$$

The series (14) is convergent for most cases. However the following suggestions have been made by He [8], to find the convergence rate on nonlinear operator.

(1) The second derivative of $N(v)$ with respect to v must be small because the parameter may be relatively large, that is, $p \rightarrow 1$.

(2) The norm of $L^{-1}(\partial N/\partial v)$ must be smaller than one so that the series converges.

More about this method and illustrative examples, the reader may refer to the following articles [9–17].

3. Fisher Equation with Convection Term

Consider the Fisher equation with convection term as follows:

$$\frac{\partial u}{\partial t} + \alpha u \frac{\partial u}{\partial x} - \frac{\partial^2 u}{\partial x^2} - u(1 - u) = 0 \tag{19}$$

subject to

$$u(x, 0) = \frac{1}{2}\left(1 - \tanh\left(\frac{x}{4}\right)\right). \tag{20}$$

By means of homotopy perturbation technique, the homotopy construction of (19) is

$$\begin{aligned} &(1 - p)\left(\frac{\partial u}{\partial t} - \frac{\partial u_0}{\partial t}\right) \\ &\quad + p\left(\frac{\partial u}{\partial t} + \alpha u \frac{\partial u}{\partial x} - \frac{\partial^2 u}{\partial x^2} - u(1 - u)\right) = 0, \end{aligned} \tag{21}$$

or

$$\frac{\partial u}{\partial t} - \frac{\partial u_0}{\partial t} + p\left(\frac{\partial u_0}{\partial t} + \alpha u \frac{\partial u}{\partial x} - \frac{\partial^2 u}{\partial x^2} - u(1 - u)\right) = 0, \tag{22}$$

where the initial approximation $u_0 = u(x, 0)$. Suppose that the solution of (19) has the form

$$u = v_0 + p v_1 + p^2 v_2 + p^3 v_3 + \cdots. \tag{23}$$

Substituting (23) in (22) and comparing the coefficients of identical degrees of p, we obtain the following linear equations:

$$p^0: \frac{\partial v_0}{\partial t} - \frac{\partial u_0}{\partial t} = 0,$$

$$p^1: \frac{\partial v_1}{\partial t} + \frac{\partial u_0}{\partial t} + v_0\left(-1 + v_0 + \alpha\frac{\partial v_0}{\partial x}\right) - \frac{\partial^2 v_0}{\partial x^2} = 0,$$

$$p^2: \frac{\partial v_2}{\partial t} + v_1\left(-1 + 2v_0 + \alpha\frac{\partial v_0}{\partial x}\right) + \alpha v_0\frac{\partial v_1}{\partial x} - \frac{\partial^2 v_1}{\partial x^2} = 0,$$

$$\vdots \tag{24}$$

Solving this system, we obtain the following solutions for v_0, v_1, v_2, and so forth:

$$v_0 = u(x,0) = \frac{1}{2}\left(1 - \tanh\left(\frac{x}{4}\right)\right),$$

$$v_1 = -\int_0^t \left(\frac{\partial u_0}{\partial t} + v_0\left(-1 + v_0 + \alpha\frac{\partial v_0}{\partial x}\right) - \frac{\partial^2 v_0}{\partial x^2}\right)dt$$

$$= \frac{1}{16}\, t\, \operatorname{sech}^2\left(\frac{x}{4}\right)\left(4 + \alpha - (\alpha - 1)\tanh\left(\frac{x}{4}\right)\right),$$

$$v_2 = -\int_0^t \left(v_1\left(-1 + 2v_0 + \alpha\frac{\partial v_0}{\partial x}\right) + \alpha v_0\frac{\partial v_1}{\partial x} - \frac{\partial^2 v_1}{\partial x^2}\right)dt$$

$$= \frac{1}{256}\, t^2 \operatorname{sech}^4\left(\frac{x}{4}\right)\Bigg((17 + 2\alpha(3 + \alpha)) \times \sinh\left(\frac{x}{2}\right) - 2(\alpha - 1) \times \Bigg(-2(\alpha + 3) + (\alpha + 4) \times \cosh\left(\frac{x}{2}\right) + (-3 + 2\alpha) \times \tanh\left(\frac{x}{4}\right)\Bigg)\Bigg)$$

$$\vdots \tag{25}$$

Considering the first 11 terms of (23), then the approximate solution of (19) by setting $p = 1$ is

$$u_{\text{app}}(x,t) = \sum_{i=0}^{10} u_i = \frac{1}{2}\left(1 - \tanh\left(\frac{x}{4}\right)\right) + \frac{1}{16}\, t\, \operatorname{sech}^2\left(\frac{x}{4}\right)\left(4 + \alpha - (\alpha - 1)\tanh\left(\frac{x}{4}\right)\right)$$

$$+ \frac{1}{256}\, t^2 \operatorname{sech}^4\left(\frac{x}{4}\right)\Bigg((17 + 2\alpha(3 + \alpha)) \times \sinh\left(\frac{x}{2}\right) - 2(\alpha - 1) \times \Bigg(-2(\alpha + 3) + (\alpha + 4) \times \cosh\left(\frac{x}{2}\right) + (-3 + 2\alpha) \times \tanh\left(\frac{x}{4}\right)\Bigg)\Bigg) + \cdots. \tag{26}$$

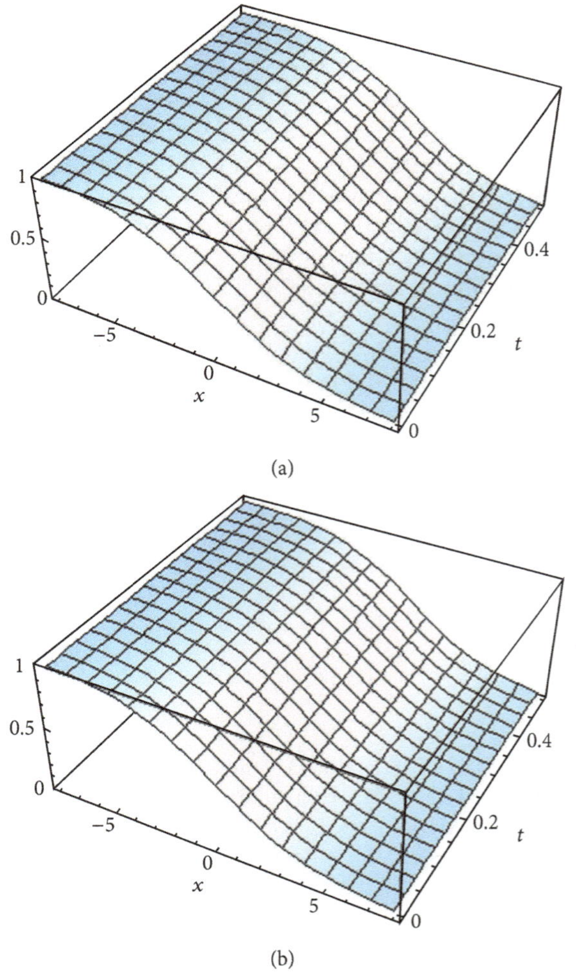

FIGURE 1: $u(x,t)$ and $u_{\text{app}}(x,t)$ are shown, respectively, from (b) to (a), when $-5 \le x \le 5$ and $0 \le t \le 0.5$, for Fisher equation, with $n = 11$, $\alpha = 1$.

The exact solution of (19) is

$$u(x,t) = \frac{1}{2}\left(1 - \tanh\left(\frac{\alpha}{4}\left(x - \left(\frac{\alpha^2 + 4}{2\alpha}\right)t\right)\right)\right). \tag{27}$$

The behaviors of $u(x,t)$ and $u_{\text{app}}(x,t)$ are shown in Figure 1 and the absolute error is shown in Table 1.

Figures 2 and 3 show that the best choice of α is 1.

TABLE 1: Absolute error $|u(x,t)-u_{app}(x,t)|$ for Fisher equation, with $n=11$.

x	t = 0.05	0.15	0.25	0.35
−6	0	4.4×10^{-16}	1.1×10^{-13}	4.5×10^{-12}
−3	0	3.9×10^{-15}	1.1×10^{-12}	4.5×10^{-11}
0	0	2.2×10^{-14}	5.9×10^{-12}	2.4×10^{-10}
3	2.8×10^{-17}	3.6×10^{-15}	9.5×10^{-13}	3.7×10^{-11}
6	1.4×10^{-17}	4×10^{-16}	1.2×10^{-13}	5×10^{-12}

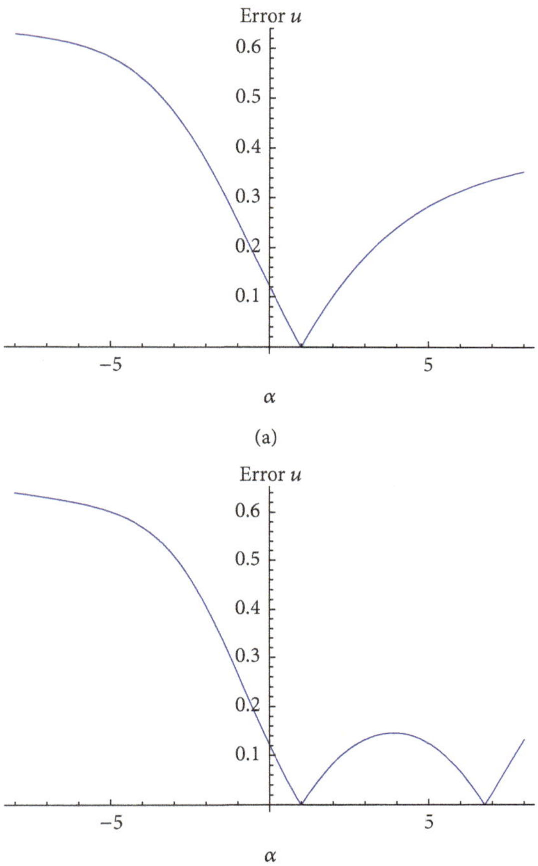

(b)

FIGURE 2: (a) Absolute error $|u(1,0.1)-u_{app}(1,0.1)|$, (b) absolute error $|u(1,0.3)-u_{app}(1,0.3)|$ with $n=11$, when $-8\le\alpha\le 8$.

4. Sharma-Tasso-Olver Equation

Consider the Sharma-Tasso-Olver equation as follows:

$$\frac{\partial u}{\partial t}+\alpha\frac{\partial}{\partial x}\left(u^3\right)+\frac{3}{2}\alpha\frac{\partial^2}{\partial x^2}\left(u^2\right)+\alpha\frac{\partial^3 u}{\partial x^3}=0 \quad (28)$$

subject to

$$u(x,0)=\sqrt{\frac{1}{\alpha}}\tanh\left(\sqrt{\frac{1}{\alpha}}\,x\right). \quad (29)$$

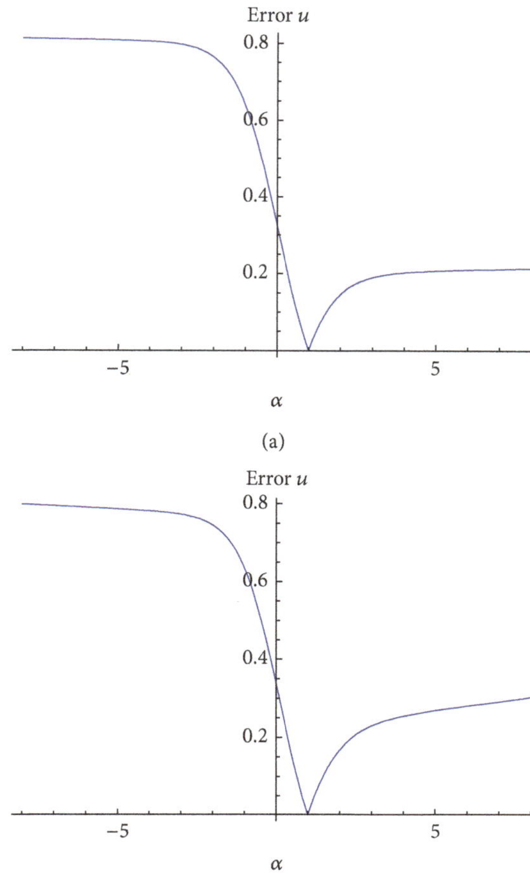

(b)

FIGURE 3: (a) Absolute error $|u(3,0.1)-u_{app}(3,0.1)|$, (b) absolute error $|u(3,0.3)-u_{app}(3,0.3)|$ with $n=11$, when $-8\le\alpha\le 8$.

By means of homotopy perturbation technique, the homotopy construction of (28) is

$$(1-p)\left(\frac{\partial u}{\partial t}-\frac{\partial u_0}{\partial t}\right)+p\left(\frac{\partial u}{\partial t}+\alpha\frac{\partial}{\partial x}\left(u^3\right)+\frac{3}{2}\alpha\frac{\partial^2}{\partial x^2}\left(u^2\right)+\alpha\frac{\partial^3 u}{\partial x^3}\right)=0, \quad (30)$$

or

$$\frac{\partial u}{\partial t}-\frac{\partial u_0}{\partial t}+p\left(\frac{\partial u_0}{\partial t}+\alpha\frac{\partial}{\partial x}\left(u^3\right)+\frac{3}{2}\alpha\frac{\partial^2}{\partial x^2}\left(u^2\right)+\alpha\frac{\partial^3 u}{\partial x^3}\right)=0, \quad (31)$$

where the initial approximation $u_0 = u(x,0)$. Suppose that the solution of (28) has the form

$$u=v_0+pv_1+p^2v_2+p^3v_3+\cdots. \quad (32)$$

Substituting (32) in (31) and comparing the coefficients of identical degrees of p, we obtain the following linear equations:

$$p^0: \frac{\partial v_0}{\partial t} - \frac{\partial u_0}{\partial t} = 0,$$

$$p^1: \frac{\partial v_1}{\partial t} + \frac{\partial u_0}{\partial t} + \alpha\left(3\frac{\partial v_0}{\partial x}\left(v_0^2 + \frac{\partial v_0}{\partial x}\right) + 3v_0\frac{\partial^2 v_0}{\partial x^2} + \frac{\partial^3 v_0}{\partial x^3}\right) = 0,$$

$$p^2: \frac{\partial v_2}{\partial t} + \alpha\left(3v_0^2\frac{\partial v_1}{\partial x} + 6\frac{\partial v_0}{\partial x}\frac{\partial v_1}{\partial x} + 3v_1\frac{\partial^2 v_0}{\partial x^2} + 3v_0\left(2v_1\frac{\partial v_0}{\partial x} + \frac{\partial^2 v_1}{\partial x^2}\right) + \frac{\partial^3 v_1}{\partial x^3}\right) = 0,$$

$$\vdots \tag{33}$$

Solving this system, we obtain the following solutions for v_0, v_1, v_2, and so forth:

$$v_0 = u(x,0) = \sqrt{\frac{1}{\alpha}}\tanh\left(\sqrt{\frac{1}{\alpha}}x\right),$$

$$v_1 = -\int_0^t\left(\frac{\partial u_0}{\partial t} + \alpha\left(3\frac{\partial v_0}{\partial x}\left(v_0^2 + \frac{\partial v_0}{\partial x}\right) + 3v_0\frac{\partial^2 v_0}{\partial x^2} + \frac{\partial^3 v_0}{\partial x^3}\right)\right)dt$$

$$= -\frac{1}{\alpha}t\,\mathrm{sech}^2\left(\sqrt{\frac{1}{\alpha}}x\right),$$

$$v_2 = -\int_0^t\left(\alpha\left(3v_0^2\frac{\partial v_1}{\partial x} + 6\frac{\partial v_0}{\partial x}\frac{\partial v_1}{\partial x} + 3v_1\frac{\partial^2 v_0}{\partial x^2} + 3v_0\left(2v_1\frac{\partial v_0}{\partial x} + \frac{\partial^2 v_1}{\partial x^2}\right) + \frac{\partial^3 v_1}{\partial x^3}\right)\right)dt$$

$$= -\alpha^{-3/2}t^2\,\mathrm{sech}^2\left(\sqrt{\frac{1}{\alpha}}x\right)\tanh\left(\sqrt{\frac{1}{\alpha}}x\right),$$

$$\vdots \tag{34}$$

Considering the first 8 terms of (32), then the approximate solution of (28) by setting $p = 1$ is

$$u_{\text{app}}(x,t) = \sum_{i=0}^{7} u_i = \sqrt{\frac{1}{\alpha}}\tanh\left(\sqrt{\frac{1}{\alpha}}x\right) - \frac{1}{\alpha}t\,\mathrm{sech}^2\left(\sqrt{\frac{1}{\alpha}}x\right) - \alpha^{-3/2}t^2\,\mathrm{sech}^2\left(\sqrt{\frac{1}{\alpha}}x\right)\tanh\left(\sqrt{\frac{1}{\alpha}}x\right) + \cdots. \tag{35}$$

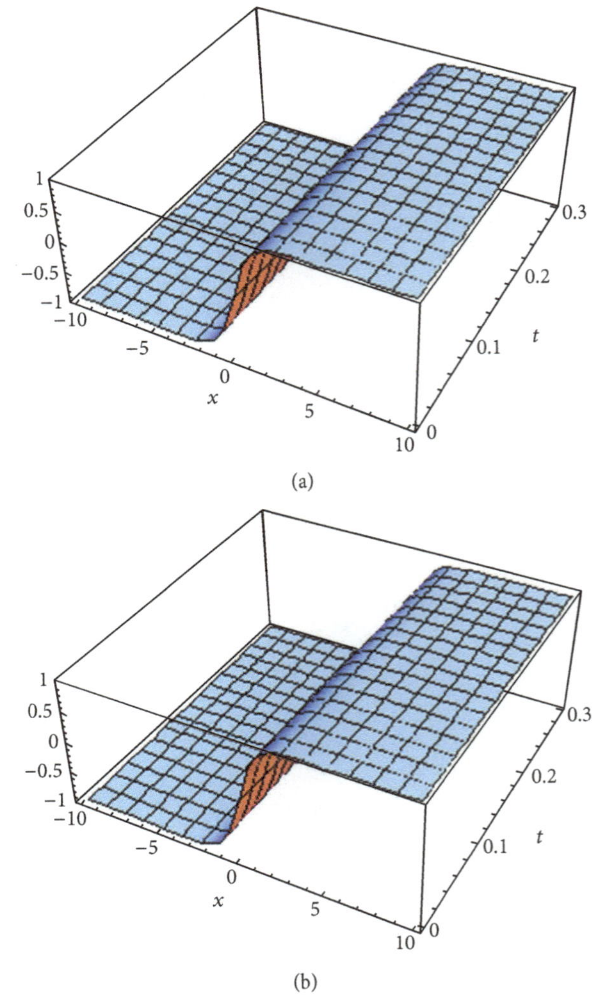

FIGURE 4: $u(x,t)$ and $u_{\text{app}}(x,t)$ are shown, respectively, from (b) to (a), when $-10 \le x \le 10$ and $0 \le t \le 0.3$, for STO equation with $n = 8$, $\alpha = 1$.

The exact solution of (28) is

$$u(x,t) = \sqrt{\frac{1}{\alpha}}\tanh\left(\sqrt{\frac{1}{\alpha}}(x-t)\right). \tag{36}$$

4.1. α=1. The behavior of $u(x,t)$ and $u_{\text{app}}(x,t)$ is shown in Figure 4 for the regions $-10 \le x \le 10$ and $0 \le t \le 0.3$. Absolute errors for differences between the exact solution and the 8th order approximate solution given by HPM for $\alpha = 1$ are given in Table 2.

4.2. α=5. In this part, we consider $\alpha = 5$ and compare it with the above case $\alpha = 1$. Figure 5 shows this comparison. One can see that for the case $\alpha = 5$ the solution converges rapidly than $\alpha = 1$.

TABLE 2: Absolute error $|u(x,t) - u_{app}(x,t)|$ for STO equation with $n = 8$, $\alpha = 1$.

x	t = 0.05	0.15	0.25	0.35
−6	1.1×10^{-16}	1.9×10^{-14}	1.1×10^{-12}	1.6×10^{-11}
−3	5.6×10^{-16}	3.3×10^{-12}	2×10^{-10}	2.9×10^{-9}
0	4.3×10^{-14}	8.3×10^{-10}	8.1×10^{-8}	1.6×10^{-6}
3	5.6×10^{-16}	3.2×10^{-12}	1.9×10^{-10}	2.7×10^{-9}
6	0	2.1×10^{-14}	1.3×10^{-12}	1.9×10^{-11}

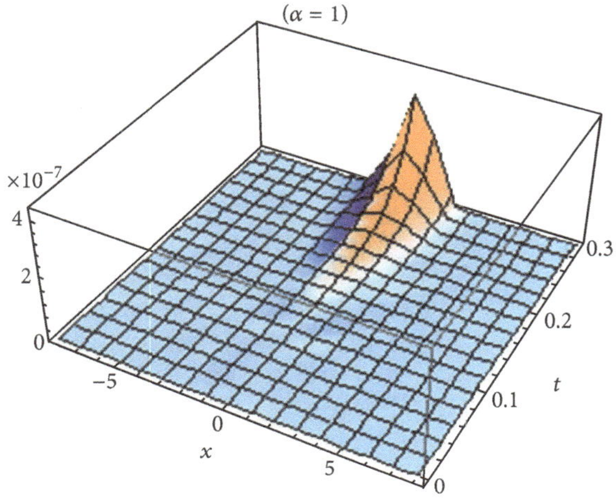

(a)

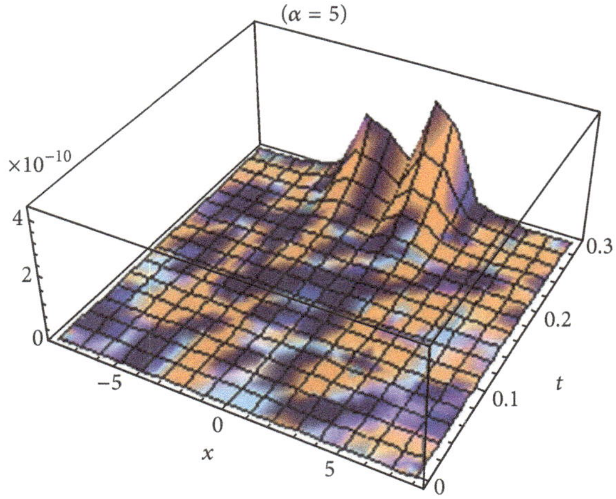

(b)

FIGURE 5: Absolute errors with respect to $u_{app}(x,t)$ are shown, respectively, for $\alpha = 1$ on (a) and for $\alpha = 5$ on (b) when $-8 \le x \le 8$ and $0 \le t \le 0.3$, for STO equation, with $n = 8$.

5. Fitzhugh-Nagumo Equation

Consider the Fitzhugh-Nagumo (FN) equation as follows:

$$\frac{\partial u}{\partial t} - \frac{\partial^2 u}{\partial x^2} + u(1-u)(\alpha - u) = 0, \quad (37)$$

subject to

$$u(x,0) = \frac{1}{2}\left(1 + \tanh\left(\frac{-\sqrt{2}}{4}x\right)\right). \quad (38)$$

By means of homotopy perturbation technique, the homotopy construction of (37) is

$$(1-p)\left(\frac{\partial u}{\partial t} - \frac{\partial u_0}{\partial t}\right) + p\left(\frac{\partial u}{\partial t} - \frac{\partial^2 u}{\partial x^2} + u(1-u)(\alpha - u)\right) = 0, \quad (39)$$

or

$$\frac{\partial u}{\partial t} - \frac{\partial u_0}{\partial t} + p\left(\frac{\partial u_0}{\partial t} - \frac{\partial^2 u}{\partial x^2} + u(1-u)(\alpha - u)\right) = 0, \quad (40)$$

where the initial approximation $u_0 = u(x,0)$. Suppose that the solution of (37) has the form

$$u = v_0 + pv_1 + p^2 v_2 + p^3 v_3 + \cdots. \quad (41)$$

Substituting (41) in (40) and comparing the coefficients of identical degrees of p, we obtain the following linear equations:

$$\begin{aligned} p^0 &: \frac{\partial v_0}{\partial t} - \frac{\partial u_0}{\partial t} = 0, \\ p^1 &: \frac{\partial v_1}{\partial t} + \frac{\partial u_0}{\partial t} + v_0\left(\alpha - (1+\alpha)v_0 + v_0^2\right) - \frac{\partial^2 v_0}{\partial x^2} = 0, \\ p^2 &: \frac{\partial v_2}{\partial t} + v_1\left(\alpha - 2(1+\alpha)v_0 + 3v_0^2\right) - \frac{\partial^2 v_1}{\partial x^2} = 0, \\ &\vdots \end{aligned} \quad (42)$$

Solving this system, we obtain the following solutions for v_0, v_1, v_2, and so forth:

$$\begin{aligned} v_0 &= u(x,0) = \frac{1}{2}\left(1 + \tanh\left(\frac{-\sqrt{2}}{4}x\right)\right), \\ v_1 &= -\int_0^t \left(\frac{\partial u_0}{\partial t} + v_0\left(\alpha - (1+\alpha)v_0 + v_0^2\right) - \frac{\partial^2 v_0}{\partial x^2}\right)dt \\ &= \frac{(1-2\alpha)\,t}{4\left(1 + \cosh\left(x/\sqrt{2}\right)\right)}, \\ v_2 &= -\int_0^t \left(v_1\left(\alpha - 2(1+\alpha)v_0 + 3v_0^2\right) - \frac{\partial^2 v_1}{\partial x^2}\right)dt \\ &= \frac{(1-2\alpha)^2}{4} t^2 \operatorname{csch}^3\left(\frac{x}{\sqrt{2}}\right) \sinh^4\left(\frac{x}{2\sqrt{2}}\right), \\ &\vdots \end{aligned} \quad (43)$$

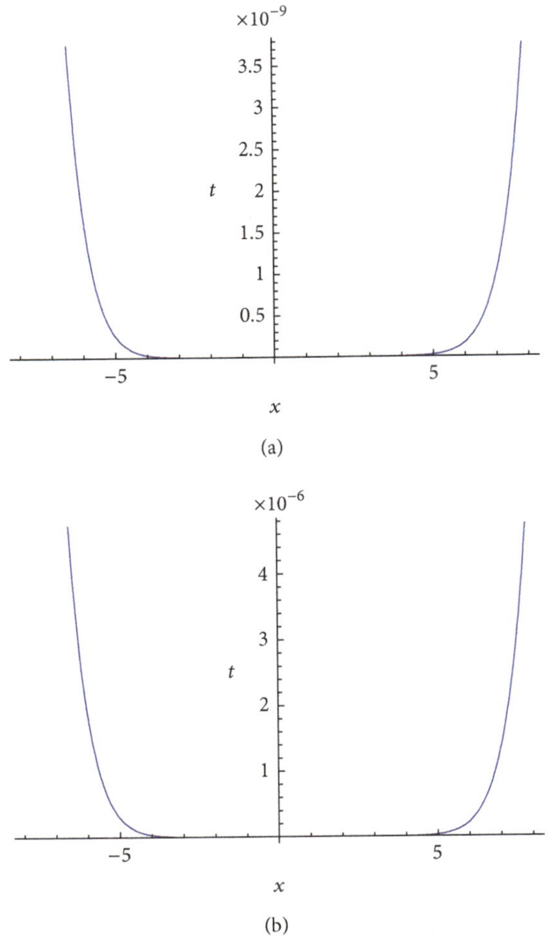

FIGURE 6: (a) Absolute error $|u(3, 0.1) - u_{\text{app}}(3, 0.1)|$, (b) absolute error $|u(7, 0.3) - u_{\text{app}}(7, 0.3)|$, for FN equation, with $n = 11$, when $-8 \leq \alpha \leq 8$.

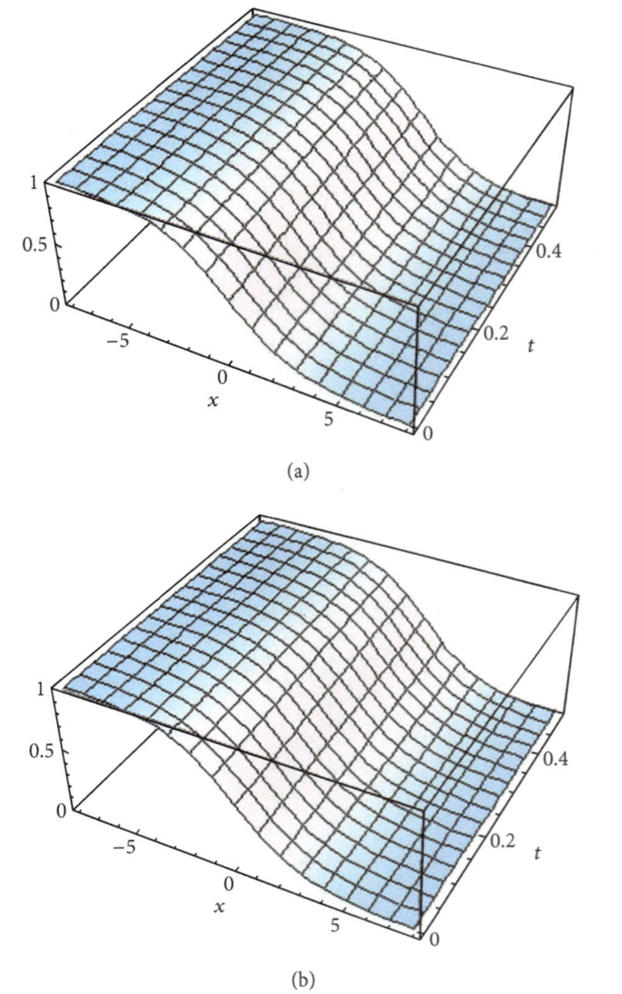

FIGURE 7: $u(x, t)$ and $u_{\text{app}}(x, t)$ are shown, respectively, from (b) to (a), when $-5 \leq x \leq 5$ and $0 \leq t \leq 0.5$, for NW equation, with $n = 11$.

Considering the first 11 terms of (41), then the approximate solution of (37) by setting $p = 1$ is

$$\begin{aligned} u_{\text{app}}(x, t) &= \sum_{i=0}^{10} u_i \\ &= \frac{1}{2}\left(1 + \tanh\left(\frac{-\sqrt{2}}{4} x\right)\right) \\ &\quad + \frac{(1 - 2\alpha) t}{4\left(1 + \cosh\left(x/\sqrt{2}\right)\right)} \\ &\quad + \frac{(1 - 2\alpha)^2}{4} t^2 \operatorname{csch}^3\left(\frac{x}{\sqrt{2}}\right) \sinh^4\left(\frac{x}{2\sqrt{2}}\right) \\ &\quad + \cdots . \end{aligned} \tag{44}$$

The exact solution of (37) is

$$u(x, t) = \frac{1}{2}\left(1 + \tanh\left(\frac{-\sqrt{2}}{4} x + \left(\frac{1}{4} - \frac{\alpha}{2}\right) t\right)\right). \tag{45}$$

Figure 6 shows that the best choice of α is $-5 \leq \alpha \leq 5$, which gives solution that converges rapidly than other choices for α. When $\alpha = -1$, the FN equation reduces to the Newell-Whitehead (NW) equation, which is an important nonlinear reaction-diffusion equation and is usually used to model the transmission of nerve impulses, also is used in circuit theory, biology, and the area of population genetics as mathematical models.

The behavior of $u(x, t)$ and $u_{\text{app}}(x, t)$ is shown in Figure 7 for the regions $-5 \leq x \leq 5$ and $0 \leq t \leq 0.5$, when $\alpha = -1$.

Absolute errors for differences between the exact solution and the 11th order approximate solution given by HPM for NW equation are given in Table 3.

TABLE 3: Absolute error $|u(x,t) - u_{app}(x,t)|$ for NW equation with $n = 11$.

x	t			
	0.05	0.15	0.25	0.35
−6	1.1×10^{-16}	0	2×10^{-14}	8.5×10^{-13}
−3	0	1×10^{-14}	2.7×10^{-12}	1×10^{-10}
3	4.2×10^{-17}	1.2×10^{-14}	3.6×10^{-12}	1.5×10^{-10}
6	1.2×10^{-17}	8×10^{-17}	1.5×10^{-14}	5.4×10^{-13}

6. Conclusion

In this paper, the homotopy perturbation method has been used for finding the approximate solutions of Fisher equation, Sharma-Tasso-Olver equation, and Fitzhugh-Nagumo equation. HPM provides highly accurate numerical solutions for our problems and they also do not require large computer memory and discretization of variable t. The approximations are not valid only for small parameters but also for larger ones and the initial approximation can be arbitrarily chosen with unknown constants.

The small size of computations in comparison with the computational size required in characteristics method and the rapid convergence show that the homotopy perturbation method is more reliable and introduces a significant improvement in solving nonlinear partial differential equations.

References

[1] W. Mafliet, *The tanh method in nonlinear wave theory [Habilitation thesis]*, Department of Physics, University of Antwerp, Antwerp, Belgium, 1994.

[2] M. Matinfar and M. Ghanbari, "Solving the Fisher's equation by means of variational iteration method," *International Journal of Contemporary Mathematical Sciences*, vol. 4, no. 5–8, pp. 343–348, 2009.

[3] S. Wang, X.-Y. Tang, and S.-Y. Lou, "Soliton fission and fusion: burgers equation and Sharma-Tasso-Olver equation," *Chaos, Solitons & Fractals*, vol. 21, no. 1, pp. 231–239, 2004.

[4] A.-M. Wazwaz, "New solitons and kinks solutions to the Sharma-Tasso-Olver equation," *Applied Mathematics and Computation*, vol. 188, no. 2, pp. 1205–1213, 2007.

[5] S. M. Sayed and G. M. Gharib, "Canonical reduction of self-dual Yang-Mills equations to Fitzhugh-Nagumo equation and exact solutions," *Chaos, Solitons and Fractals*, vol. 39, no. 2, pp. 492–498, 2009.

[6] A. Khajeh, A. Yousefi-Koma, M. Vahdat, and M. M. Kabir, "Exact travelling wave solutions for some nonlinear equations arising in biology and engineering," *World Applied Sciences Journal*, vol. 9, no. 12, pp. 1433–1442, 2010.

[7] S. J. Liao, "An approximate solution technique not depending on small parameters: a special example," *International Journal of Non-Linear Mechanics*, vol. 30, no. 3, pp. 371–380, 1995.

[8] J.-H. He, "Homotopy perturbation technique," *Computer Methods in Applied Mechanics and Engineering*, vol. 178, no. 3-4, pp. 257–262, 1999.

[9] J.-H. He, "A coupling method of a homotopy technique and a perturbation technique for non-linear problems," *International Journal of Non-Linear Mechanics*, vol. 35, no. 1, pp. 37–43, 2000.

[10] J.-H. He, "Homotopy perturbation method: a new nonlinear analytical technique," *Applied Mathematics and Computation*, vol. 135, no. 1, pp. 73–79, 2003.

[11] J.-H. He, "Recent development of the homotopy perturbation method," *Topological Methods in Nonlinear Analysis*, vol. 31, no. 2, pp. 205–209, 2008.

[12] J. H. He, "An elementary introduction to recently developed asymptotic methods and nanomechanics in textile engineering," *International Journal of Modern Physics B*, vol. 22, no. 21, pp. 3487–3578, 2008.

[13] J. Biazar, K. Hosseini, and P. Gholamin, "Homotopy perturbation method for solving KdV and Sawada-Kotera equations," *Journal of Applied Mathematics*, vol. 6, no. 21, pp. 23–29, 2009.

[14] M. A. Jafari and A. Aminataei, "Improved homotopy perturbation method," *International Mathematical Forum*, vol. 5, no. 29–32, pp. 1567–1579, 2010.

[15] E. Alibeiki and A. Neyrameh, "Application of homotopy perturbation method to nonlinear Drinfeld-Sokolov-Wilson equation," *Middle-East Journal of Scientific Research*, vol. 10, no. 4, pp. 440–443, 2011.

[16] M. Alquran and M. Mohammad, "Approximate solutions to system of nonlinear partial differential equations using homotopy perturbation method," *International Journal of Nonlinear Science*, vol. 12, no. 4, pp. 485–497, 2011.

[17] M. A. Abdou and A. A. Soliman, "Modified extended tanh-function method and its application on nonlinear physical equations," *Physics Letters A*, vol. 353, no. 6, pp. 487–492, 2006.

7

The Existence of Solutions for Impulsive Fractional Partial Neutral Differential Equations

Xiao-Bao Shu[1] and Fei Xu[2]

[1] *Department of Mathematics, Hunan University, Changsha 410082, China*
[2] *Department of Mathematics, Wilfrid Laurier University, Waterloo, ON, Canada N2L 3C5*

Correspondence should be addressed to Xiao-Bao Shu; sxb0221@163.com

Academic Editor: Beny Neta

This paper deals with the existence of mild solutions of a class of impulsive fractional partial neutral semilinear differential equations. A series of analytical results about the mild solutions are obtained by using fixed-point methods. Then, we present an example to further illustrate the applications of these results.

1. Introduction

Great progress in fractional differential equations has been achieved in recent years. Due to their broad applications in science and engineering such as physics, economics, biology, and mechanical engineering, fractional differential equations attract attention of researchers from different areas. Compared with ordinary differential equation systems and partial differential equation systems, fractional differential equation systems have the potential to model real-world problems with more accuracy. In order to further investigate these models, it is essential to study the fractional differential equations analytically. Though, mathematically, a fractional differential equation is closely related to its corresponding ordinary differential equation or partial differential equation, that is, the ordinary differential equation or the partial differential equation can be obtained by letting $\alpha = 1, 2, \ldots$ in its corresponding fractional differential equation, many mathematical methods which can be used in investigating ordinary differential equations or partial differential equations fail in analyzing fractional differential equations since fractional differential equations usually have more complicated structures and different properties. Thus, it is essential to develop novel methods to study fractional differential equations analytically. With the aim of analyzing fractional differential equations, extensive investigations had been carried out [1–15].

In applied mathematics, another two types of differential equations, impulsive differential equations and functional differential equations with different conditions have wide applications in modeling particular phenomena and dynamical processes in physics, automatics, robotics, biology, medicine, and so forth. For example, impulsive differential equations can be used to investigate natural phenomena or dynamical processes which are subject to great changes in a short period of time [16–18]. To further understand the mathematical models with impulsive differential equations or functional differential equations with different conditions, analytical investigation of these equations is required. During the last ten years, progress in studying these two types of differential equations has been made [19–25].

One of the important techniques for analyzing impulsive differential equations is the semigroup theory, which has been successfully used in investigating the existence, uniqueness, and continuous dependence of the solutions of impulsive differential equations. It also has wide applications in the study of periodic and almost periodic solutions of different kinds of differential equations. For example, Fan and Li studied the existence results for semilinear differential equations with nonlocal and impulsive conditions using semigroup theory [26]. The theory has also been used to investigate a mixed monotone iterative technique for a class of semilinear impulsive evolution equations in Banach spaces [27]. With the aid of the semigroup theory, a series of conclusions about the existence of the solutions of functional differential equations and functional integral equations have been obtained [19–23, 28–36].

Since the fractional derivative does not satisfy the Leibniz rule and the corresponding fractional operator equation does not satisfy the variation of constants formula [37], which implies that the corresponding operator does not satisfy the properties of semigroups (see [38]), $T_\alpha(t) \triangleq t^{\alpha-1}E_{\alpha,\alpha}(At^\alpha)$ does not satisfy the semigroup relations, that is, $T_\alpha(t+s) \neq T_\alpha(t)T_\alpha(s)$, indicating the properties of corresponding resolvent operator. It is difficult to define the mild solutions of fractional partial differential evolution equations. In particular, semigroup theory was used inappropriately to study the existence and uniqueness of mild solutions to impulsive fractional differential equations [13] and impulsive partial neutral functional differential equation [39]. The existence of mild solutions for a class of impulsive fractional partial semilinear differential equations was investigated in [15] with errors in [13] corrected.

A solution to this problem has been given in [15], in which the mild solutions of impulsive fractional differential equations are defined using piecewise functions, and a sufficient condition, which guarantees the existence and uniqueness of solutions to the equations, is obtained. The existence of solutions to a fractional neutral integrodifferential equation with unbounded delay was studied in [40].

In this paper, with the aim of investigating the existence theorem for the solutions of impulsive fractional neutral functional differential equations, we first define the mild solution of fractional neutral functional differential equation using Laplace transform. Then, by introducing the operator $S_{\alpha,k,l}(t)$, the mild solution of impulsive fractional differential equation is defined. Through analyzing operator $S_\alpha(t)$, $T_\alpha(t)$, and the corresponding semigroup $T(t)$, a sufficient condition, which guarantees the existence and uniqueness of solutions to the following system, is obtained:

$$\begin{aligned} &D_t^\alpha\left(x(t)+F(t,x_t)\right)+Ax(t)=G(t,x_t),\\ &\qquad t\in I=[0,T],\ t\neq t_k,\\ &x(0)=\varphi\in\mathscr{B},\Delta x|_{t=t_k}=I_k\left(x_{t_k^-}\right),\\ &\qquad k=1,\dots,m, \end{aligned} \tag{1}$$

where D_t^α, for $0<\alpha\le 1$, is the Caputo fractional derivative, $-A$ is the infinitesimal generator of an analytic semigroup $\{T(t)\}_{t\ge0}$, and $(X,\|\cdot\|)$ is a Banach space.

The rest of this paper is organized as follows. In Section 2, we present some notations, definitions, and theorems which will be used in the following sections. In Section 3, the definition of mild solution of the system (1) and the relation between analytic semigroup $T(t)$ and some solution operators are given. The main results of our paper are given in Section 4. In Section 5, application of the obtained results is presented.

2. Preliminaries

In this section, we will introduce some notations, definitions, and theorems, which will be used throughout this paper. Let $(X,\|\cdot\|_X)$ and $(Y,\|\cdot\|_Y)$ be Banach spaces, and let $\mathscr{L}(Y,X)$ be the Banach space of bounded linear operators from Y into X equipped with its natural topology. When $Y=X$, we use the notation $\mathscr{L}(X)$ to denote the Banach space. We say that a function $u(\cdot)$ is a normalized piecewise continuous function on $[\mu,\tau]$ if u is piecewise continuous, and left-continuous on $(\mu,\tau]$. The space formed by the normalized piecewise continuous functions from $[\mu,\tau]$ to X is denoted by $\mathscr{PC}([\mu,\tau];X)$, where the notation $\mathscr{PC}$ stands for the space formed by all functions $u\in\mathscr{PC}([0,\mu];X)$ such that $u(\cdot)$ is continuous at $t\neq t_i$, and $u(t_i^-)=u(t_i)$ and $u(t_i^+)$ exist for all $i=1,\dots,m$. In addition, we use $(\mathscr{PC},\|\cdot\|_{\mathscr{PC}})$ to denote the space $\mathscr{PC}$ endowed with the norm $\|x\|_{\mathscr{PC}}=\sup_{s\in I}\|x(s)\|$. Obviously, $(\mathscr{PC},\|\cdot\|_{\mathscr{PC}})$ is a Banach space. Moreover, an axiomatic definition for the phase space $\mathscr{B}$ is employed. Here, the definition of $\mathscr{B}$ is similar to that given in [41]. In particular, $\mathscr{B}$ is a linear space of functions mapping $(-\infty,0]$ into X endowed with a seminorm $\|\cdot\|_{\mathscr{B}}$ and the following axioms hold.

(I) If $x:(-\infty,\mu+b]\to X$, $b>0$, such that $x_\mu\in\mathscr{B}$ and $x|_{[\mu,\mu+b]}\in\mathscr{PC}([\mu,\mu+b]:X)$, then for every $t\in[\mu,\mu+b]$ the following conditions hold:

(i) x_t is in $\mathscr{B}$;
(ii) $\|x(t)\|\le H\|x_t\|_{\mathscr{B}}$;
(iii) $\|x_t\|_{\mathscr{B}}\le K(t-\mu)\sup\{\|x(s)\|:\mu\le s\le t\}+M(t-\mu)\|x_\mu\|_{\mathscr{B}}$,

where $H>0$ is a constant, $K,M:[0,\infty)\to[1,\infty)$, K is continuous, M is locally bounded, and H,K,M are independent of $x(\cdot)$.

(II) The space $\mathscr{B}$ is complete.

Next, we give an example to illustrate the above definitions.

Example 1 (the phase space $\mathscr{PC}_r\times L^2(g,X)$). Let $r>0$ and $g:(-\infty,-r)\to R$ be a nonnegative, locally Lebesgue integrable function. Assume that there is a non-negative measurable, locally bounded function $\eta(\cdot)$ on $(-\infty,0]$ such that $g(\xi+\theta)\le\eta(\xi)g(\theta)$ for all $\xi\in(-\infty,0]$ and $\theta\in(-\infty,-r]\setminus N_\xi$, where $N_\xi\subset(-\infty,-r)$ is a set with Lebesgue measure zero. $\mathscr{PC}_r\times L^2(g,X)$ denotes the set of all functions $\varphi:(-\infty,0]\to X$ such that $\varphi|_{[-r,0]}\in\mathscr{PC}([-r,0],X)$ and $\int_{-\infty}^{-r}g(\theta)\|\varphi(\theta)\|_X^2d\theta<\infty$. In $\mathscr{PC}_r\times L^2(g,X)$, we consider the seminorm defined by

$$\|\varphi\|_{\mathscr{B}}=\sup_{\theta\in[-r,0]}\|\varphi(\theta)\|_X+\left(\int_{-\infty}^{-r}g(\theta)\|\varphi(\theta)\|_X^2d\theta\right)^{1/2}. \tag{2}$$

The preceding conditions indicate that the space $\mathscr{PC}_r\times L^2(g,X)$ satisfies the axioms (I) and (II). Moreover, when $r=0$, we can take $H=1$, $K(t)=\left(\int_{-t}^0 g(\theta)\right)^{1/2}$, and $M(t)=\eta(-t)$ for $t\ge0$.

Definition 2. Let $a,\alpha \in \mathbb{R}$. A function $f : [a,\infty) \to X$ is said to be in the space $C_{a,\alpha}$ if there exist a real number $p > \alpha$ and a function $g \in C([a,\infty), X)$ such that $f(t) = t^p g(t)$. Moreover, f is said to be in the space $C^m_{a,\alpha}$ for some positive integer m if $f^{(m)} \in C_{a,\alpha}$.

Definition 3. If the function $f \in C^m_{a,-1}$, where $m \in \mathbb{N}^+$, the fractional derivative of order $\alpha > 0$ of f in the Caputo sense is defined as

$$D_t^\alpha f(t) = \frac{1}{\Gamma(m-\alpha)} \int_0^t (t-s)^{m-\alpha-1} f^{(m)}(s)\, ds, \quad m-1 < \alpha \le m. \tag{3}$$

Definition 4. Let $A : \mathscr{D} \subseteq X \to X$ be a closed linear operator. A is said to be sectorial if there exist $0 < \theta < \pi/2$, $M > 0$, and $\mu \in \mathbb{R}$ such that the resolvent of A exists outside the sector

$$\mu + S_\theta = \{\mu + \lambda : \lambda \in \mathbb{C}, |\arg(-\lambda)| < \theta\}, \qquad \left\|(\lambda I - A)^{-1}\right\| \le \frac{M}{|\lambda - \mu|}, \quad \lambda \notin \mu + S_\theta \tag{4}$$

(for short, we say that A is sectorial of type (M,θ,μ)).

Let $-A$ denote the infinitesimal generator of an analytic semigroup in a Banach space and $0 \in \rho(A)$, where $\rho(A)$ is the resolvent set of $\rho(A)$. We define the fractional power A^{-q} by

$$A^{-q} = \frac{1}{\Gamma(q)} \int_0^\infty (t)^{q-1} T(t)\, dt, \quad q > 0. \tag{5}$$

For $0 < q \le 1$, $A^q = (A^{-q})^{-1}$ is a closed linear operator whose domain $D(A^q) \supset D(A)$ is dense in X. For analytic semigroup $\{T(t)\}_{t\ge 0}$, we have the following theorem.

Theorem 5 ([42, Theorem 6.13]). *Let $-A$ be the infinitesimal generator of an analytic semigroup $\{T(t)\}_{t\ge0}$. If $0 \in \rho(A)$, then the following statements hold.*

(a) $T(t) : X \to D(A^\alpha)$ *for every $t > 0$ and $\alpha \ge 0$.*

(b) *For every $x \in D(A^\alpha)$, we have $T(t)A^\alpha x = A^\alpha T(t)x$.*

(c) *For every $t > 0$, the operator $A^\alpha T(t)$ is bounded and*

$$\|A^\alpha T(t)\| \le M_\alpha t^{-\alpha} e^{-\delta t}. \tag{6}$$

(d) $0 < \alpha \le 1$ *and $x \in D(A^\alpha)$, then*

$$\|T(t)x - x\| \le c_\alpha t^\alpha \|A^\alpha x\|. \tag{7}$$

3. Mild Solutions

In this section, we investigate the classical solution of the following equation:

$$D_t^\alpha (x(t) + F(t,x_t)) + Ax(t) = G(t,x_t), \quad t \in I, 0 < \alpha \le 1, \quad x(0) = \varphi \in \mathscr{B}. \tag{8}$$

Based on the classical solution, the mild solution of system (1) is defined. Then, the relationship between the analytic semigroup $T(t)$ and some solution operators is given.

Lemma 6. *Let $-A$ be the infinitesimal generator of an analytic semigroup $\{T(t)\}_{t\ge0}$. If G and F satisfy the uniform Hölder condition with exponent $\beta \in (0,1]$, then the solutions of the Cauchy problem (8) are fixed points of the operator equation*

$$\Psi x(t) = \begin{cases} \mathscr{S}_\alpha(t)(\varphi(0) + F(0,\varphi)) - F(t,x_t) \\ \quad - \int_0^t T_\alpha(t-s) AF(s,x_s)\, ds \\ \quad + \int_0^t T_\alpha(t-s) G(s,x_s)\, ds, \\ x_0 = \varphi \in \mathscr{B}, \end{cases} \tag{9}$$

where

$$\mathscr{S}_\alpha(t) = \frac{1}{2\pi i}\int_c e^{\lambda t}\lambda^{\alpha-1} R(\lambda^\alpha, -A)\, d\lambda, \qquad T_\alpha(t) = \frac{1}{2\pi i}\int_c e^{\lambda t} R(\lambda^\alpha, -A)\, d\lambda, \tag{10}$$

with c being a suitable path such that $\lambda^\alpha \notin \mu + S_\theta$ for $\lambda \in c$.

Proof. Applying the Laplace transform to (9), we obtain

$$\lambda^\alpha (\mathscr{L}x(t))(\lambda) + \lambda^\alpha (\mathscr{L}F(t,x_t))(\lambda) - \lambda^{\alpha-1}(\varphi(0) + F(0,\varphi)) + A(\mathscr{L}x)(\lambda) = (\mathscr{L}G(t,x_t))(\lambda). \tag{11}$$

It follows that

$$(\mathscr{L}x)(\lambda) = \lambda^{\alpha-1}(\lambda^\alpha I + A)^{-1}(\varphi(0) + F(0,\varphi)) - \lambda^\alpha(\lambda^\alpha + A)^{-1}(\mathscr{L}F(t,x_t))(\lambda) + (\lambda^\alpha I + A)^{-1}(\mathscr{L}G(t,x_t))(\lambda). \tag{12}$$

Noting that $\lambda^\alpha(\lambda^\alpha + A)^{-1} = I - A(\lambda^\alpha + A)^{-1}$, and applying the inverse Laplace transform, we have

$$x(t) = \mathscr{S}_\alpha(t)(\varphi(0) + F(0,\varphi)) - F(t,x_t) + \int_0^t T_\alpha(t-s) AF(s,x_s)\, ds + \int_0^t T_\alpha(t-s) G(s,x_s)\, ds. \tag{13}$$

Since F and G satisfy a uniform Hölder condition, with exponent $\beta \in (0,1)$, then the classical solutions of Cauchy problem (8) are fixed points of the following operator equations (see [43]):

$$\Psi x(t) = \mathscr{S}_\alpha(t)(\varphi(0) + F(0,\varphi)) - F(t,x_t) + \int_0^t T_\alpha(t-s) AF(s,x_s)\, ds + \int_0^t T_\alpha(t-s) G(s,x_s)\, ds. \tag{14}$$

□

In what follows, we use the notations $\mathcal{S}_{\alpha,k,l}(t)x \triangleq \mathcal{S}_\alpha(t-t_k)\circ\mathcal{S}_\alpha(t_k-t_{k-1})\circ\cdots\circ\mathcal{S}_\alpha(t_{l+1}-t_l)x$, $0\le l\le k-1$, $t_0=0$, for all $x\in X$, and $\overline{\mathcal{S}_{\alpha,k,l}}(t)x \triangleq \mathcal{S}_\alpha(t-t_k)\circ\mathcal{S}_\alpha(t_k-t_{k-1})\circ\cdots\circ\mathcal{S}_\alpha(t_{l+2}-t_{l+1})x$, $0\le l\le k-1$.

Definition 7. A function $u\in\mathscr{PC}(I,X)$ is called a mild solution of (1) if it satisfies the operator equation

$$\begin{aligned}
u(t) &= \mathcal{S}_{\alpha,k,0}(t)(\varphi(0)+F(0,\varphi)) - F(t,u_t) \\
&\quad + \sum_{0<t_{l+1}<t,0\le l\le k}\overline{\mathcal{S}_{\alpha,k,l}}(t) \\
&\quad \times \int_{t_l}^{t_{l+1}} AT_\alpha(t_{l+1}-s)F(s,u_s)\,ds \\
&\quad + \int_{t_k}^{t} AT_\alpha(t-s)F(s,u_s)\,ds \\
&\quad + \sum_{0<t_{l+1}<t,0\le l\le k}\overline{\mathcal{S}_{\alpha,k,l}}(t) \\
&\quad \times \int_{t_l}^{t_{l+1}} T_\alpha(t_{l+1}-s)G(s,u_s)\,ds \\
&\quad + \int_{t_k}^{t} T_\alpha(t-s)G(s,u_s)\,ds \\
&\quad + \sum_{0<t_{l+1}<t,0\le l\le k}\overline{\mathcal{S}_{\alpha,k,l}}(t) \\
&\quad \times [I_{l+1}(u_{t_{l+1}}) + \Delta F(t_{l+1},u_{t_{l+1}})], \quad t\in I.
\end{aligned} \tag{15}$$

Remark 8. It is easy to verify that a classical solution of (1) is a mild solution of the same system. Thus, Definition 7 is well defined (see [42, 43]).

Lemma 9. *If $-A$ is the infinitesimal generator of an analytic semigroup $\{T(t)\}_{t\ge0}$ and $0\in\rho(A)$, then we have*

$$\begin{aligned}
T_\alpha(t) &= \alpha\int_0^\infty \theta\phi_\alpha(\theta)t^{\alpha-1}T(t^\alpha\theta)\,d\theta, \\
\mathcal{S}_\alpha(t) &= \int_0^\infty \phi_\alpha(\theta)T(t^\alpha\theta)\,d\theta,
\end{aligned} \tag{16}$$

where $\phi_\alpha(\theta)$ is the probability density function defined on $(0,\infty)$ such that its Laplace transform is given by

$$\int_0^\infty e^{-\theta x}\phi_\alpha(\theta)\,d\theta = \sum_{j=0}^{\infty}\frac{(-x)^j}{\Gamma(1+\alpha j)}, \quad x>0, \tag{17}$$

and satisfies

$$\begin{aligned}
&\int_0^\infty \phi_\alpha(\theta)\,d\theta = 1, \\
&\int_0^\infty \theta^\eta\phi_\alpha(\theta)\,d\theta \le 1, \quad 0\le\eta\le1.
\end{aligned} \tag{18}$$

Proof. For all $x\in D(A)\subset X$, we have

$$(\lambda+A)^{-1}x = \int_0^\infty e^{-\lambda s}T(s)x\,ds. \tag{19}$$

Let

$$\int_0^\infty e^{-\lambda\theta}\psi_\alpha(\theta)\,d\theta = e^{-\lambda^\alpha}, \tag{20}$$

where $\alpha\in(0,1)$, $\psi_\alpha(\theta) = (1/\pi)\Sigma_{1\le n<\infty}(-1)^n\theta^{-\alpha n-1}(\Gamma(n\alpha+1)/n!)\sin(n\pi\alpha)$, and $\theta\in(0,\infty)$ (see [44]). Thus, we obtain

$$\begin{aligned}
&(\lambda^\alpha+A)^{-1}x \\
&= \int_0^\infty e^{-\lambda^\alpha s}T(s)x\,ds \\
&= \int_0^\infty\int_0^\infty \alpha t^{\alpha-1}e^{-(\lambda t)^\alpha}T(t^\alpha)x\,dt \\
&= \int_0^\infty\int_0^\infty \alpha t^{\alpha-1}e^{-\lambda t\theta}\psi_\alpha(\theta)T(t^\alpha)x\,d\theta\,dt \\
&= \int_0^\infty\int_0^\infty \alpha\psi_\alpha(\theta)e^{-\lambda t}T\left(\frac{t^\alpha}{\theta^\alpha}\right)\frac{t^{\alpha-1}}{\theta^\alpha}dt\,d\theta \\
&= \int_0^\infty e^{-\lambda t}\left(\alpha\int_0^\infty \theta\phi_\alpha(\theta)t^{\alpha-1}T(t^\alpha\theta)x\,d\theta\right)dt,
\end{aligned} \tag{21}$$

where $\phi_\alpha(\theta) = (1/\alpha)\theta^{-1-1/\alpha}\psi_\alpha(\theta^{-1/\alpha})$ is the probability density function defined on $(0,\infty)$ and satisfies $\int_0^\infty\phi_\alpha(\theta)d\theta = 1$ and $\int_0^\infty\theta^\eta\phi_\alpha(\theta)d\theta\le1$, $0\le\eta\le1$ (see [44]). Then, from (21), we get

$$\begin{aligned}
T_\alpha(t) &= \frac{1}{2\pi i}\int_\Gamma e^{\lambda t}(\lambda^\alpha+A)^{-1}d\lambda \\
&= \alpha\int_0^\infty \theta\phi_\alpha(\theta)t^{\alpha-1}T(t^\alpha\theta)\,d\theta.
\end{aligned} \tag{22}$$

On the other hand, for all $x\in D(A)\subset X$, we notice that

$$\begin{aligned}
&\lambda^{\alpha-1}(\lambda^\alpha+A)^{-1}x \\
&= \int_0^\infty \lambda^{\alpha-1}e^{-\lambda^\alpha s}T(s)x\,ds \\
&= \int_0^\infty \alpha(\lambda t)^{\alpha-1}e^{-(\lambda t)^\alpha}T(t^\alpha)x\,dt \\
&= \int_0^\infty -\frac{1}{\lambda}\frac{d}{dt}\left[e^{-(\lambda t)^\alpha}\right]T(t^\alpha)\,dt \\
&= \int_0^\infty\int_0^\infty \theta\psi_\alpha(\theta)e^{-\lambda t\theta}T(t^\alpha)x\,d\theta\,dt \\
&= \int_0^\infty e^{-\lambda t}\left[\int_0^\infty \psi_\alpha(\theta)T\left(\frac{t^\alpha}{\theta^\alpha}\right)x\,d\theta\right]dt \\
&= \int_0^\infty e^{-\lambda t}\left[\int_0^\infty \phi_\alpha(\theta)T(t^\alpha\theta)x\,d\theta\right]dt.
\end{aligned} \tag{23}$$

Thus,

$$\mathcal{S}_\alpha(t) = \int_0^\infty \phi_\alpha(\theta)T(t^\alpha\theta)\,d\theta. \tag{24}$$

□

Remark 10. If $M_\alpha = \sup_{0\le t\le T}\|T(t)\|$, then from Lemma (9), we have

$$\|\mathcal{S}_{\alpha,k,l}(t)\| \le M_\alpha^{k+1-l}, \qquad \overline{S_{\alpha,k,l}}(t) \le M_\alpha^{k-l}. \tag{25}$$

4. Main Results

In this section, we will present the main results of this paper. The mild solution will be understood in the sense of Definition 7. To investigate the uniqueness of the mild solution, we require the following assumptions.

(H_1) The functions $F, G : I \times \mathscr{B} \to X$ and $I_i : \mathscr{B} \to X$, $i = 1, \ldots, m$, are continuous and satisfy the following conditions

(i) For every $x : (-\infty, T] \to X$ such that $x_0 = \varphi$ and $x|_I \in \mathscr{PC}$, the function $t \to G(t, x_t)$ is strongly measurable and the function $t \to F(t, x_t)$ belongs to $\mathscr{PC}$.

(ii) There exist a positive constant $\beta \in (0,1)$, and a function $\mu_1 \in L^{+\infty}(I, R^+)$ as well as $L_F, L, i = 1, \ldots, m$, such that F is X_β-valued, $A^\beta F : I \times \mathscr{B} \to X$ is continuous and

$$\begin{aligned} &\left\|G(t,\psi_1) - G(t,\psi_2)\right\| \\ &\quad \le \mu_1(t)\left\|\psi_1 - \psi_2\right\|_{\mathscr{B}}, \quad \psi_1, \psi_2 \in \mathscr{B}, t \in I, \\ &\left\|A^\beta F(t,\psi_1) - A^\beta F(t,\psi_2)\right\| \\ &\quad \le L_F \left\|\psi_1 - \psi_2\right\|_{\mathscr{B}}, \quad \psi_1, \psi_2 \in \mathscr{B}, t \in I, \\ &\left\|I_i(\psi_1) - I_i(\psi_2)\right\| \le L \left\|\psi_1 - \psi_2\right\|_{\mathscr{B}}, \quad \psi_1, \psi_2 \in \mathscr{B}. \end{aligned} \tag{26}$$

(H_2) The function $\zeta(t) : I \to R^+$ is defined by

(i)

$$\begin{aligned} \zeta(t) &= K_\sigma \Bigg[L_F \left\|A^{(-\beta)}\right\| \left(1 + \frac{2M_\alpha\left(M_\alpha^{m+1} - 1\right)}{M_\alpha - 1}\right) \\ &\quad + \frac{M_\alpha\left(M_\alpha^{m+1} - 1\right)L}{M_\alpha - 1} + \left(1 + \frac{M_\alpha\left(M_\alpha^{m+1} - 1\right)}{M_\alpha - 1}\right) \\ &\quad \times \left(\frac{C_{1-\beta} L_F \Gamma(1+\beta)}{\beta\Gamma(1+\alpha\beta)} t^{\alpha\beta}\right) + \frac{M_\alpha \left\|\mu_1\right\|_{L^\infty(I,R^+)}}{\Gamma(\alpha+1)} t^\alpha \Bigg], \end{aligned} \tag{27}$$

when $M_\alpha \neq 1$, or

(ii)

$$\begin{aligned} \zeta(t) &= K_\sigma \Bigg[L_F \left\|A^{(-\beta)}\right\| (1 + 2m) + mL + (1+m) \\ &\quad \times \left(\frac{C_{1-\beta} L_F \Gamma(1+\beta)}{\beta\Gamma(1+\alpha\beta)} t^{\alpha\beta} + \frac{\left\|\mu_1\right\|_{L^\infty(I,R^+)}}{\Gamma(\alpha+1)} t^\alpha\right) \Bigg], \end{aligned} \tag{28}$$

when $M_\alpha = 1$. Here, $K_\sigma = \sup_{s \in I} K(s)$ and satisfies $0 < \zeta(t) \le \gamma < 1$, for all $t \in I$.

Remark 11. Let $x : (-\infty, T] \to X$ be such that $x_0 = \varphi$ and $x|_I \in \mathscr{PC}$, and assume that H_1 holds. We have the following estimations:

$$\begin{aligned} &\left\|AT_\alpha(t-s)F(s,x_s)\right\| \\ &= \left\|A^{1-\beta}T_\alpha(t-s)A^\beta F(s,x_s)\right\| \\ &\le \Bigg\|\left[\alpha \int_0^\infty \theta\phi_\alpha(\theta)(t-s)^{\alpha-1} A^{(1-\beta)} \right. \\ &\qquad \left. \times T\left((t-s)^\alpha\theta\right) d\theta\right] A^\beta F(s,x_s)\Bigg\| \\ &\le \alpha C_{1-\beta}(t-s)^{\alpha\beta-1} \\ &\qquad \times \int_0^\infty \theta^\beta \phi_\alpha(\theta)\, d\theta \left\|A^\beta F(s,x_s)\right\|. \end{aligned} \tag{29}$$

On the other hand, from $\int_0^\infty \theta^{-q}\psi_\alpha(\theta)d\theta = \Gamma(1+q/\alpha)/\Gamma(1+q)$, for all $q \in [0,1]$ (see [44]), we have

$$\begin{aligned} &\int_0^\infty \theta^\beta \phi_\alpha(\theta)\, d\theta \\ &\quad = \int_0^\infty \frac{1}{\theta^{\beta\alpha}} \psi_\alpha(\theta)\, d\theta \\ &\quad = \frac{\Gamma(1+\beta)}{\Gamma(1+\alpha\beta)}. \end{aligned} \tag{30}$$

Then, by (29) and (30), it is easy to see that

$$\begin{aligned} &\left\|AT_\alpha(t-s)F(s,x_s)\right\| \\ &\quad \le \frac{\alpha C_{1-\beta}\Gamma(1+\beta)}{\Gamma(1+\alpha\beta)(t-s)^{1-\alpha\beta}} \left\|A^\beta F(s,x_s)\right\|. \end{aligned} \tag{31}$$

It is obvious that the function $\theta \to AT_\alpha(t-\theta)F(\theta, x_\theta)$ is integrable on $[0,t)$ for every $t > 0$. Similarly, it can be verified that the function $\theta \to AT_\alpha(t-\theta)G(\theta, x_\theta)$ is integrable on $[0,t)$ for every $t > 0$.

Theorem 12. *Under the assumptions (H_1) and (H_2), the system (1) has a unique mild solution.*

Proof. On the metric space $\mathscr{BPC} = \{u : (-\infty, T] \to X,\ u_0 = \varphi,\ u|_I \in \mathscr{PC}\}$ endowed with the metric $d(u,v) = \|u - v\|_{\mathscr{PC}}$, we define the operator $\Gamma : \mathscr{BPC} \to \mathscr{BPC}$ by

$$\Gamma u(t) = \begin{cases} \varphi(t), & t \leq 0 \\ \mathcal{S}_{\alpha,k,0}(t)(\varphi(0) + F(0,\varphi)) - F(t,u_t) \\ \quad + \sum_{0<t_{l+1}<t, 0\leq l\leq k} \overline{\mathcal{S}_{\alpha,k,l}}(t) \times \int_{t_l}^{t_{l+1}} AT_\alpha(t_{l+1} - s) F(s,u_s)\, ds \\ \quad + \int_{t_k}^{t} AT_\alpha(t-s) F(s,u_s)\, ds \\ \quad + \sum_{0<t_{l+1}<t, 0\leq l\leq k} \overline{\mathcal{S}_{\alpha,k,l}}(t) \times \int_{t_l}^{t_{l+1}} T_\alpha(t_{l+1} - s) G(s,u_s)\, ds \\ \quad + \int_{t_k}^{t} T_\alpha(t-s) G(s,u_s)\, ds \\ \quad + \sum_{0<t_{l+1}<t, 0\leq l\leq k} \overline{\mathcal{S}_{\alpha,k,l}}(t) \times [I_{l+1}(u_{t_{l+1}}) + \Delta F(t_{l+1}, u_{t_{l+1}})] & t \in I. \end{cases} \tag{32}$$

From Remark 11, we know that $s \to T_\alpha(t-s)G(s,u_s)$ and $s \to T_\alpha(t-s)AF(s,u_s)$ are integrable on $[0,t)$ for every $t \in I$. Thus, Γ is well defined with values in $\mathscr{PC}$. Let $u, v \in \mathscr{BPC}$. Using $\|\Delta F(t_i,u_{t_i}) - \Delta F(t_i,v_{t_i})\| \leq 2L_F K_\sigma \|A^{(-\beta)}\| \|(u - v)|_I\|_{\mathscr{PC}}$, we get

$$\begin{aligned} &\|\Gamma u(t) - \Gamma v(t)\| \\ &\quad \leq \|F(t,u_t) - F(t,v_t)\| \\ &\qquad + \sum_{0<t_{l+1}<t, 0\leq l\leq k} \|\overline{\mathcal{S}_{\alpha,k,l}}(t)\| \\ &\qquad \times \int_{t_l}^{t_{l+1}} \|AT_\alpha(t_{l+1}-s)F(s,u_s) \\ &\qquad\qquad - AT_\alpha(t_{l+1}-s)F(s,v_s)\|\, ds \\ &\qquad + \int_{t_k}^{t} \|AT_\alpha(t-s)(F(s,u_s) - F(s,v_s))\|\, ds \\ &\qquad + \sum_{0<t_{l+1}<t, 0\leq l\leq k} \|\overline{\mathcal{S}_{\alpha,k,l}}(t)\| \\ &\qquad \times \int_{t_l}^{t_{l+1}} \|T_\alpha(t_{l+1}-s)(G(s,u_s) - G(s,v_s))\|\, ds \\ &\qquad + \int_{t_k}^{t} (t-s)^{(\alpha-1)} \\ &\qquad\qquad \times \|T_\alpha(t-s)(G(s,u_s) - G(s,v_s))\|\, ds \\ &\qquad + \sum_{0<t_{l+1}<t, 0\leq l\leq k} \|\overline{\mathcal{S}_{\alpha,k,l}}(t)\| \\ &\qquad \times [\|I_{l+1}(u_{t_{l+1}}) - I_{l+1}(v_{t_{l+1}})\| \\ &\qquad\qquad + \|\Delta F(t_{l+1},u_{t_{l+1}}) - \Delta F(t_{l+1},v_{t_{l+1}})\|] \\ &\quad \leq \|A^{(-\beta)}\| L_F \|u_t - v_t\|_{\mathscr{B}} \\ &\qquad + \sum_{0<t_{l+1}<t, 0\leq l\leq k} \|\overline{\mathcal{S}_{\alpha,k,l}}(t)\| \\ &\qquad \times \int_{t_l}^{t_{l+1}} (t_{l+1}-s)^{(\alpha\beta-1)} \\ &\qquad\qquad \times \frac{\alpha C_{1-\beta} L_F \Gamma(1+\beta)}{\Gamma(1+\alpha\beta)} \|u_s - v_s\|_{\mathscr{B}}\, ds \\ &\qquad + \int_{t_k}^{t} (t-s)^{\alpha\beta-1} \frac{\alpha C_{1-\beta} L_F \Gamma(1+\beta)}{\Gamma(1+\alpha\beta)} \|u_s - v_s\|_{\mathscr{B}}\, ds \\ &\qquad + \alpha M_\alpha \sum_{0<t_{l+1}<t, 0\leq l\leq k} \|\overline{\mathcal{S}_{\alpha,k,l}}(t)\| \int_{t_l}^{t_{l+1}} (t_{l+1}-s)^{\alpha-1} \\ &\qquad \times \mu_1(s) \|u_s - v_s\|_{\mathscr{B}} \left(\int_0^\infty \theta\phi_\alpha(\theta)\, d\theta\right) ds \\ &\qquad + \alpha M_\alpha \int_{t_k}^{t} (t-s)^{\alpha-1} \mu_1(s) \\ &\qquad\qquad \times \|u_s - v_s\|_{\mathscr{B}} \left(\int_0^\infty \theta\phi_\alpha(\theta)\, d\theta\right) ds \\ &\qquad + \sum_{0<t_{l+1}<t, 0\leq l\leq k} \|\overline{\mathcal{S}_{\alpha,k,l}}(t)\| \\ &\qquad \times \left(L\|u_{t_i} - v_{t_i}\|_{\mathscr{B}} + 2L_F K_\sigma \|A^{(-\beta)}\| \|(u-v)|_I\|_{\mathscr{PC}}\right) \\ &\quad \leq \|A^{(-\beta)}\| L_F \|u_t - v_t\|_{\mathscr{B}} \\ &\qquad + \left(1 + \frac{M_\alpha(M_\alpha^{m+1} - 1)}{M_\alpha - 1}\right) \frac{C_{1-\beta} L_F \Gamma(1+\beta) t^{\alpha\beta}}{\beta\Gamma(1+\alpha\beta)} \\ &\qquad \times \|u_s - v_s\|_{\mathscr{B}} \\ &\qquad + \frac{M_\alpha t^\alpha \|\mu_1\|_{L^\infty(I,R^+)}}{\Gamma(\alpha+1)} \left(1 + \frac{M_\alpha(M_\alpha^{m+1} - 1)}{M_\alpha - 1}\right) \\ &\qquad \times \|u_s - v_s\|_{\mathscr{B}} \\ &\qquad + \frac{M_\alpha(M_\alpha^{m+1} - 1)}{M_\alpha - 1} \\ &\qquad \times \left(L\|u_{t_i} - v_{t_i}\|_{\mathscr{B}} + 2L_F K_\sigma \|A^{(-\beta)}\| \|(u-v)|_I\|_{\mathscr{PC}}\right), \end{aligned} \tag{33}$$

for all $t \in [0, T]$. □

Remark 13. The above inequality makes use of

$$\int_0^{\infty} \theta \phi_\alpha (\theta)\, d\theta = \frac{1}{\Gamma(1+\alpha)}, \quad (34)$$

which can be obtained directly from (30).

Remark 14. If $M_\alpha = 1$ in the above inequality, we replace $(M_\alpha(M_\alpha^{m+1} - 1))/(M_\alpha - 1)$ by m.

Therefore, it can be deduced that

(i) when $M_\alpha \neq 1$,

$$\begin{aligned}
&\|\Gamma u(t) - \Gamma v(t)\| \\
&\leq K_\sigma \Bigg[L_F \left\| A^{(-\beta)} \right\| \left(1 + \frac{2M_\alpha (M_\alpha^{m+1} - 1) L_F}{M_\alpha - 1} \right) \\
&\quad + \frac{M_\alpha (M_\alpha^{m+1} - 1) L}{M_\alpha - 1} + \left(1 + \frac{M_\alpha (M_\alpha^{m+1} - 1)}{M_\alpha - 1} \right) \\
&\quad \times \left(\frac{C_{1-\beta} L_F \Gamma(1+\beta)}{\beta \Gamma(1+\alpha\beta)} t^{\alpha\beta} + \frac{M_\alpha \|\mu_1\|_{L^\infty(I,R^+)}}{\Gamma(\alpha+1)} t^\alpha \right) \Bigg] \\
&\quad \times \|(u-v)|_I\|_{\mathscr{PC}}, \text{ and}
\end{aligned} \quad (35)$$

(ii) when $M_\alpha = 1$,

$$\begin{aligned}
&\|\Gamma u(t) - \Gamma v(t)\| \\
&\leq K_\sigma \Bigg[L_F \left\| A^{(-\beta)} \right\| (1 + 2m) + mL + (1 + m) \\
&\quad \times \left(\frac{C_{1-\beta} L_F \Gamma(1+\beta)}{\beta \Gamma(1+\alpha\beta)} t^{\alpha\beta} + \frac{\|\mu_1\|_{L^\infty(I,R^+)}}{\Gamma(\alpha+1)} t^\alpha \right) \Bigg] \\
&\quad \times \|(u-v)|_I\|_{\mathscr{PC}}.
\end{aligned} \quad (36)$$

Thus,

$$\|\Gamma u(t) - \Gamma v(t)\| \leq \zeta(t) \|(u-v)|_I\|_{\mathscr{PC}}. \quad (37)$$

Therefore, in view of the contraction mapping principle, assumption (H_2) implies that Γ has a unique fixed point on $\mathscr{BPC}$. Thus, the system (1) has a unique mild solution.

5. An Example

In this section, we present an example to further illustrate the applications of the results given in Section 4.

Let $X = L^2([0,\pi])$ be the space of the square integrable functions and $\mathscr{B} = \mathscr{PC}_r \times L^2(g, X)$. We investigate the operator $-A : \mathscr{D}(-A) \subset X \to X$ defined by $(-A)x = x''$, where

$$\mathscr{D}(-A) = \left\{ x \in X : x'' \in X, x(0) = x(\pi) = 0 \right\}. \quad (38)$$

We note that $-A$ is the infinitesimal generator of an analytic compact semigroup $\{T(t)\}_{t\geq 0}$ on X. In particular, we have that $(T(t)_{t\geq 0})$ is a uniformly stable semigroup with $\|T(t)\| \leq e^{-t}$ for all $t \geq 0$. Moreover,

$$D(A^\beta) = \left\{ x : x \in X, \sum_{i=1}^{n} i^{2\beta} \langle x, z_i \rangle z_i \in X \right\}. \quad (39)$$

It follows from the above properties that analytic semigroup $(T(t))_{t\geq 0}$ satisfies

$$\left\| (-A)^{1/2} T(t) \right\|_X \leq \frac{e^{-t/2} t^{-1/2}}{\sqrt{2}}, \quad \left\| (-A)^{-1/2} \right\| = 1, \quad \text{for } t \geq 0. \quad (40)$$

Consider the impulsive fractional partial neutral differential equation system:

$$\begin{aligned}
&D_t^\alpha \left[\omega(t,\xi) + \int_{-\infty}^{t} \int_0^{\pi} b(s-t,\eta,\xi)\, \omega(\xi,s)\, d\eta\, ds \right] \\
&\quad - \frac{\partial^2}{\partial \xi^2} \omega(t,\xi) = \mathscr{N}(t, \omega(t,\xi)), \\
&\quad \omega(t,0) = \omega(t,\pi) = 0, \quad t \in I = [0,T], \\
&\quad \omega(\tau,\xi) = \varphi(\tau,\xi), \quad \tau \leq 0, \ \xi \in [0,\pi], \\
&\Delta\omega(t_i^+, \cdot) - \omega(t_i^-, \cdot) = \int_0^{\pi} p_i(\xi, \omega(t_i, s))\, ds, \\
&\qquad i = 1, \ldots, m,
\end{aligned} \quad (41)$$

where (t_i) is a strictly increasing sequence of positive real numbers. To treat this system, we assume that the following conditions hold.

$(a)^*$ The function $\mathscr{N} : R \times R \to R$ is continuous and there is a continuous integrable function $\mu_1 : R \to [0,\infty)$ such that

$$|\mathscr{N}(t,x)| \leq \mu_1(t)|x|, \quad t \in [0,T], \ x \in R. \quad (42)$$

$(b)^*$ The function $b(s,\eta,\xi), (\partial/\partial\xi) b(s,\eta,\xi)$ is integrable, $b(s,\eta,\pi) = b(s,\eta,0)$, and

$$\begin{aligned}
&L_F \\
&:= \sup \left\{ \int_0^{\pi} \int_{-\infty}^{0} \int_0^{\pi} \left(\frac{1}{g(s)} \frac{\partial^i b(s,\eta,\xi)}{\partial \xi^i} d\eta\, ds\, d\xi \right)^2 : i = 0, 1 \right\} \\
&< +\infty.
\end{aligned} \quad (43)$$

$(c)^*$ The function $p_i : [0,\pi] \times R \to R, (i = 1, \ldots, m)$ is continuous and there exists a positive constant L such that

$$\begin{aligned}
&[p_i(\xi,s) - p_i(\xi,\bar{s})] \leq L|s - \bar{s}|, \\
&\qquad \xi \in [0,\pi], \ s, \bar{s} \in R.
\end{aligned} \quad (44)$$

We now define the functions $F, G : [0,\alpha] \times \mathscr{B} \to X$ and $I_i : X \to X$ by

$$\begin{aligned} F(t,\phi)(\xi) &= \int_{-\infty}^{0}\int_{0}^{\pi} b(s,\eta,\xi)\,\phi(s,\eta)\,d\eta\,ds, \\ &\qquad t \in [0,T],\ \xi \in [0,\pi], \\ G(t,\phi)(\xi) &= \mathcal{N}(t,\phi(0,s)), \quad t \in [0,T],\ \xi \in [0,\pi], \\ I_i(\phi)(\xi) &= \int_0^{\pi} p_i(\xi,\phi(0,s))\,ds, \quad i \in N,\ \xi \in [0,\pi], \end{aligned} \tag{45}$$

respectively. It is easy to see that problem (41) can be modeled as the abstract impulsive Cauchy problem (1).

Theorem 15. *Assume that the conditions $(a)^*$, $(b)^*$, and $(c)^*$ hold and*

$$\left[\int_0^{\pi} |p_i(\xi,0)|^2 d\xi\right]^{1/2} < \infty. \tag{46}$$

If

$$\begin{aligned} K_\sigma \Bigg[L_F(1+2m) + \frac{(m+1)\,C_{1-\beta}L_F}{\beta\Gamma(1+\alpha\beta)} t^{\alpha\beta} \\ + \frac{(m+1)\,\|\mu_1\|_{L^\infty(I,R^+)}}{\Gamma(\alpha+1)} t^{\alpha} + \pi\sqrt{\pi}\, mL \Bigg] < 1, \end{aligned} \tag{47}$$

for all $t \in I$, then there is a mild solution of the impulsive system (41).

Proof. It follows from $(b)^*$ that F is $X_{1/2}$-valued and $F : I \times \mathscr{B} \to X_{1/2}$ is continuous. Moreover, $A^{1/2}F(t,\cdot)$ is a bounded linear operator, and as such $\|A^{1/2}F(t,\cdot)\| \le L_F$ for all $t \in I$. By $(a)^*$, one can easily conclude that the function $G(t,\cdot) : \mathscr{B} \to X$ is continuous for all $t \in [0,T]$, and $G(\cdot,\phi) : [0,\infty) \to X$ is a strongly measurable function for each $\phi \in \mathscr{B}$. In addition, for all $t \in [0,T]$ and $\phi \in \mathscr{B}$, we have

$$\begin{aligned} \|G(t,\phi)\|_X^2 &= \int_0^{\pi} |\mathcal{N}(t,\phi(0,\xi))|^2 d\xi \\ &\le \int_0^{\pi} \mu_1^2(t)\,|\phi(0,\xi)|^2 d\xi \\ &\le \mu_1^2(t)\,\|\phi\|_{\mathscr{B}}^2. \end{aligned} \tag{48}$$

It follows from condition $(c)^*$ that

$$\begin{aligned} &|I_i(\phi)(\xi) - I_i(\psi)(\xi)| \\ &\quad\le \int_0^{\pi} |p_i(\xi,\phi(0,s)) - p_i(\xi,\psi(0,s))|\,ds \\ &\quad\le L\int_0^{\pi} |\phi(0,s) - \psi(0,s)|\,ds \\ &\quad\le \pi L\|\phi - \psi\|_{\mathscr{B}}, \end{aligned} \tag{49}$$

for all $\varphi, \psi \in \mathscr{B}$. This implies that $\|I_i(\phi) - I_i(\psi)\|_X \le \pi\sqrt{\pi}L\|\phi - \psi\|_{\mathscr{B}}$. The proof is complete since the existence of a mild solution is a consequence of Theorem 15. □

Acknowledgments

This research was partially supported by the NNSF of China (11271115), the Postdoctoral Science Foundation of Hunan province (2006FJ4249), and the Doctoral Fund of Ministry of Education of China (200805321017).

References

[1] V. Lakshmikantham, "Theory of fractional functional differential equations," *Nonlinear Analysis: Theory, Methods and Applications*, vol. 69, no. 10, pp. 3337–3343, 2008.

[2] V. Lakshmikantham and A. S. Vatsala, "Basic theory of fractional differential equations," *Nonlinear Analysis: Theory, Methods and Applications*, vol. 69, no. 8, pp. 2677–2682, 2008.

[3] J. Henderson and A. Ouahab, "Fractional functional differential inclusions with finite delay," *Nonlinear Analysis: Theory, Methods and Applications*, vol. 70, no. 5, pp. 2091–2105, 2009.

[4] M. Benchohra, J. Henderson, S. K. Ntouyas, and A. Ouahab, "Existence results forfractional functional differential inclusions with infinite delay and applicationto control theory," *Fractional Calculus and Applied Analysis*, vol. 11, pp. 35–56, 2008.

[5] A. M. A. El-Sayed and A. G. Ibrahim, "Multivalued fractional differential equations," *Applied Mathematics and Computation*, vol. 68, no. 1, pp. 15–25, 1995.

[6] A. A. Kilbas, H. M. Srivastava, and J. J. Trujillo, *Theory and Applications of Fractional Differential Equations*, Elsevier, San Diego, Calif, USA, 2006.

[7] R. Gorenflo and F. Mainardi, *Fractional Calculus: Integral and Differential Equations of Fractional Order*, vol. 378 of *CISM Courses and lectures*, Springer, Berlin, Germany, 1997.

[8] R. P. Agarwal, M. Benchohra, and S. Hamani, "Boundary value problems for differential inclusions with fractional order," *Advanced Studies in Contemporary Mathematics*, vol. 16, no. 2, pp. 181–196, 2008.

[9] R. P. Agarwal, M. Benchohra, and B. A. Slimani, "Existence results for differentialequations with fractional order and impulses," *Memoirs on Differential Equations and Mathematical Physics*, vol. 44, pp. 1–21, 2008.

[10] M. Benchohra and B. A. Slimani, "Existence and uniqueness of solutions to impulsive fractional differential equations," *Electronic Journal of Differential Equations*, vol. 2009, pp. 1–11, 2009.

[11] B. Ahmad and S. Sivasundaram, "Existence of solutions for impulsive integral boundary value problems of fractional order," *Nonlinear Analysis: Hybrid Systems*, vol. 4, no. 1, pp. 134–141, 2010.

[12] B. Ahmad and S. Sivasundaram, "Existence results for nonlinear impulsive hybrid boundary value problems involving fractional differential equations," *Nonlinear Analysis: Hybrid Systems*, vol. 3, no. 3, pp. 251–258, 2009.

[13] G. M. Mophou, "Existence and uniqueness of mild solutions to impulsive fractional differential equations," *Nonlinear Analysis, Theory, Methods and Applications*, vol. 72, no. 3-4, pp. 1604–1615, 2010.

[14] Z. Tai and X. Wang, "Controllability of fractional-order impulsive neutral functional infinite delay integrodifferential systems in Banach spaces," *Applied Mathematics Letters*, vol. 22, no. 11, pp. 1760–1765, 2009.

[15] X. B. Shu, Y. Lai, and Y. Chen, "The existence of mild solutions for impulsive fractional partial differential equations," *Nonlinear Analysis: Theory, Methods and Applications*, vol. 74, no. 5, pp. 2003–2011, 2011.

[16] Z. Agur, L. Cojocaru, G. Mazor, R. M. Anderson, and Y. L. Danon, "Pulse mass measles vaccination across age cohorts," *Proceedings of the National Academy of Sciences of the United States of America*, vol. 90, no. 24, pp. 11698–11702, 1993.

[17] E. Kruger-Thiemr, "Formal theory of drug dosage regiments-I," *Journal of Theoretical Biology*, vol. 13, pp. 212–235, 1966.

[18] E. Kruger-Thiemr, "Formal theory of drug dosage regiments-II," *Journal of Theoretical Biology*, vol. 23, pp. 169–190, 1969.

[19] E. Hernández M., M. Rabello, and H. R. Henríquez, "Existence of solutions for impulsive partial neutral functional differential equations," *Journal of Mathematical Analysis and Applications*, vol. 331, no. 2, pp. 1135–1158, 2007.

[20] C. Cuevas, E. Hernández, and M. Rabelo, "The existence of solutions for impulsive neutral functional differential equations," *Computers and Mathematics with Applications*, vol. 58, no. 4, pp. 744–757, 2009.

[21] A. Anguraj and K. Karthikeyan, "Existence of solutions for impulsive neutral functional differential equations with nonlocal conditions," *Nonlinear Analysis: Theory, Methods and Applications*, vol. 70, no. 7, pp. 2717–2721, 2009.

[22] J. Y. Park, K. Balachandran, and N. Annapoorani, "Existence results for impulsive neutral functional integrodifferential equations with infinite delay," *Nonlinear Analysis: Theory, Methods and Applications*, vol. 71, no. 7-8, pp. 3152–3162, 2009.

[23] X. Fu and K. Mei, "Approximate controllability of semilinear partial functional differential systems," *Journal of Dynamical and Control Systems*, vol. 15, no. 3, pp. 425–443, 2009.

[24] J. P. Aubin, "Impulse differential inclusions and hybrid systems: a viability Approach," in *Lecture Notes*, Universit Paris-Dauphine, 2002.

[25] M. Benchohra, J. Henderson, and S. K. Ntouyas, *Impulsive Differential Equationsand Inclusions*, Hindawi Publishing Corporation, New York, NY, USA, 2007.

[26] Z. Fan and G. Li, "Existence results for semilinear differential equations with nonlocal and impulsive conditions," *Journal of Functional Analysis*, vol. 258, no. 5, pp. 1709–1727, 2010.

[27] P. Chen and Y. Li, "Mixed monotone iterative technique for a class of semilinear impulsive evolution equations in Banach spaces," *Nonlinear Analysis: Theory, Methods and Applications*, vol. 74, no. 11, pp. 3578–3588, 2011.

[28] B. Liu, "Controllability of impulsive neutral functional differential inclusions with infinite delay," *Nonlinear Analysis: Theory, Methods and Applications*, vol. 60, no. 8, pp. 1533–1552, 2005.

[29] N. Abada, M. Benchohra, and H. Hammouche, "Existence and controllability resultsfor impulsive partial functional differential inclusions," *Nonlinear Analysis: Theory, Methods and Applications* , vol. 69, pp. 2892–2909, 2008.

[30] K. Ezzinbi and X. Fu, "Existence and regularity of solutions for some neutral partial differential equations with nonlocal conditions," *Nonlinear Analysis: Theory, Methods and Applications*, vol. 57, no. 7-8, pp. 1029–1041, 2004.

[31] Y. K. Chang, A. Anguraj, and K. Karthikeyan, "Existence for impulsive neutral integrodifferential inclusions with nonlocal initial conditions via fractional operators," *Nonlinear Analysis: Theory, Methods and Applications*, vol. 71, no. 10, pp. 4377–4386, 2009.

[32] X. Fu and Y. Cao, "Existence for neutral impulsive differential inclusions with nonlocal conditions," *Nonlinear Analysis: Theory, Methods and Applications*, vol. 68, no. 12, pp. 3707–3718, 2008.

[33] R. Ye, "Existence of solutions for impulsive partial neutral functional differential equation with infinite delay," *Nonlinear Analysis: Theory, Methods and Applications*, vol. 73, no. 1, pp. 155–162, 2010.

[34] Y. K. Chang, A. Anguraj, and M. M. Arjunan, "Existence results for impulsive neutral functional differential equations with infinite delay," *Nonlinear Analysis: Hybrid Systems*, vol. 2, no. 1, pp. 209–218, 2008.

[35] J. Hu and X. Liu, "Existence results of impulsive partial neutral integrodifferential inclusions with infinity delay," *Nonlinear Analysis: Theory, Methods and Applications*, vol. 71, no. 12, pp. e1132–e1138, 2009.

[36] W. S. Li, Y. K. Chang, and J. J. Nieto, "Solvability of impulsive neutral evolution differential inclusions with state-dependent delay," *Mathematical and Computer Modelling*, vol. 49, no. 9-10, pp. 1920–1927, 2009.

[37] K. Diethelm, N. J. Ford, A. D. Freed, and Y. Luchko, "Algorithms for the fractional calculus: a selection of numerical methods," *Computer Methods in Applied Mechanics and Engineering*, vol. 194, no. 6-8, pp. 743–773, 2005.

[38] K. Li, J. Peng, and J. Gao, "On some properties of the of the α-Exponential function," *Integral Transforms and Special Function*. In press.

[39] D. N. Chalishajar, "Controllability of impulsive partial neutral functional differential equation with infinite delay," *International Journal of Mathematical Analysis*, vol. 5, no. 8, pp. 369–380, 2011.

[40] B. de Andrade and J. P. C. dos Santos, "Existence of solutions for a fractional neutral integro-differential equation with unbounded delay," *Electronic Journal of Differential Equations*, vol. 2012, no. 90, pp. 1–13, 2012.

[41] Y. Hino, S. Murakami, and T. Naito, "Functional-differential Equations with Infinite Delay," in *Lectures Notes in Mathematics*, vol. 1473, Spring, New York, NY, USA, 2003.

[42] A. Pazy, "Semigroups of linear operators and applications to partial dfifferential equations," in *Applied Mathematical Sciences*, vol. 44, Springer, New York, NY, USA, 1983.

[43] A. Lunardi, *Analytic Semigroups and Optimal Regularity in Parabolic Problems*, Birkhäuser, Basel, Switzerland, 1995.

[44] F. Mainardi, P. Paradisi, and R. Gorenflo, "Probability distributions generated by fractional diffusion equations," in *Econophysics: An Emerging Science*, J. Kertesz and I. Kondor, Eds., Kluwer Academic, Dordrecht, The Netherlands, 2000.

8

Some New Recurrence Relations Concerning Jacobi Functions

Hatem Mejjaoli[1] and Ahmedou Ould Ahmed Salem[2]

[1] *Department of Mathematics, College of Sciences, Taibah University, P.O. Box 30002, Al Madinah AL Munawara, Saudi Arabia*
[2] *Department of Mathematics, College of Sciences, King Khalid University, Mohayil, Saudi Arabia*

Correspondence should be addressed to Hatem Mejjaoli; hatem.mejjaoli@ipest.rnu.tn

Academic Editor: Baoding Liu

The aim of this paper is to obtain some new recurrence relations for the "modified" Jacobi functions $\Phi_{\lambda}^{\alpha,\beta}$. Based on an asymptotic relationship between the Jacobi function and the Bessel function, the expression of Bessel function in terms of elementary functions follows as particular cases.

1. Introduction

It is well known that the Jacobi function $t \to \varphi_{\lambda}^{\alpha,\beta}(t), \alpha, \beta, \lambda \in \mathbb{C}, \alpha \neq -1, -2, -3, \ldots$, is defined (cf. [1]) as the even C^{∞}-function on $\mathbb{R}$ which satisfies the differential equation

$$L_{\alpha,\beta}u(t) = -\left(\lambda^2 + (\alpha+\beta+1)^2\right)u(t), \qquad u(0) = 1, \qquad u'(0) = 0, \tag{1}$$

where $L_{\alpha,\beta}$ is the Jacobi operator given by

$$L_{\alpha,\beta} = \frac{d^2}{dt^2} + \left[(2\alpha+1)\coth t + (2\beta+1)\tanh t\right]\frac{d}{dt}. \tag{2}$$

In this paper some new summation formulas for the "modified" Jacobi functions

$$t \longmapsto \Phi_{\lambda}^{\alpha,\beta}(t) = \frac{\Gamma((\alpha+\beta+1+i\lambda)/2)\,\Gamma((\alpha+\beta+1-i\lambda)/2)}{2^{-\beta+1}\Gamma(1+\alpha)} \times (\sinh t)^{\alpha}\varphi_{\lambda}^{\alpha,\beta}(t) \tag{3}$$

are proved. We establish, under some conditions, that

$$\Phi_{\lambda}^{\alpha-1,\beta+1}(t) - \Phi_{\lambda}^{\alpha+1,\beta+1}(t) = \frac{2}{\cosh t}\frac{d}{dt}\left[\Phi_{\lambda}^{\alpha,\beta}\right](t),$$

$$\left[\frac{2}{\sinh(2t)}\frac{d}{dt}\right]^m\left[(\sinh t)^{\alpha}\Phi_{\lambda}^{\alpha,\beta}\right](t) = (\sinh t)^{\alpha-m}\Phi_{\lambda}^{\alpha-m,\beta+m}(t), \tag{4}$$

$$(\sinh t)^m\left[\Phi_{\lambda}^{\alpha-m,\beta+m}(t) - (-1)^m\Phi_{\lambda}^{\alpha+m,\beta+m}(t)\right] = \sum_{j=0}^{m-1}(-1)^j b_m^j(\alpha)(\sinh t)^j\Phi_{\lambda}^{\alpha+j,\beta+j}(t),$$

where $b_m^j(\alpha)$ stands for a constant that will be determined. A same of the last formula with respect to the dual variable is also shown, another formulas are proven and some examples are treated. We note that the subject for the generalization of recurrence relations concerning special functions was studied by many authors (cf. [2–5]).

The remaining part of the paper is organized as follows. In Section 2 we recall the main results about some necessary notions related to Jacobi functions, Bessel functions, and Macdonald's functions. Section 3 is devoted to establishing some new formulas concerning "modified" Jacobi functions and some examples are given.

2. Preliminaries

This section gives an introduction to the Jacobi function, Bessel function of the first kind, and Macdonald's function. Main references are [1, 6–11].

2.1. Jacobi Function. A Jacobi function $\varphi_\lambda^{\alpha,\beta}(\alpha, \beta, \lambda \in \mathbb{C}, \alpha \neq -1, -2, -3, \ldots)$ (cf. [1, 6]) can be expressed as a hypergeometric function under form

$$\varphi_\lambda^{\alpha,\beta}(t) = {}_2F_1\left(\frac{\rho+i\lambda}{2}, \frac{\rho-i\lambda}{2}; \alpha+1; -\sinh^2 t\right): \quad \rho = \alpha+\beta+1. \tag{5}$$

This function satisfies the following properties.

(i) For every $t \in [0, +\infty[$, $(\Gamma(\alpha+1))^{-1}\varphi_\lambda^{\alpha,\beta}(t)$ is an entire function of α, β, and λ.

(ii) For each $\alpha, \beta \in \mathbb{C}$ and for each nonnegative integer n there exists a positive constant k such that

$$\forall t \geq 0, \quad \forall \lambda \in \mathbb{C},$$

$$\left|(\Gamma(\alpha+1))^{-1}\left(\frac{d}{dt}\right)^n \varphi_\lambda^{\alpha,\beta}(t)\right| \leq k(1+|\lambda|)^{n+k}(1+x)e^{(|\operatorname{Im}\lambda|-\operatorname{Re}(\rho))t}, \tag{6}$$

where $k = 0$ if $\operatorname{Re}(\alpha) > -1/2$ and $k = [(1/2) - \operatorname{Re}(\alpha)]$ if $\operatorname{Re}(\alpha) \leq -1/2$.

(iii) The function $t \to \varphi_\lambda^{\alpha,\beta}(t)$ for $\operatorname{Re}(\alpha) > -1/2$ possesses the Laplace type integral representation

$$\forall t > 0,\ \forall \lambda \in \mathbb{C}, \quad \varphi_\lambda^{\alpha,\beta}(t) = \int_0^t \mathscr{K}(t, y)\cos(\lambda y)\,dy, \tag{7}$$

where $\mathscr{K}(t, \cdot)$ is explicitly given in [8] and it is a positive function on $(0, t)$ if $\alpha > -1/2$ and $|\beta| \leq \max\{1/2, \alpha\}$.

We note that the function $\varphi_\lambda^{\alpha,\beta}$ possesses also an integral representation with respect to the dual variable [12].

To finish this paragraph, we consider the "modified" Jacobi function

$$t \longmapsto \Phi_\lambda^{\alpha,\beta}(t) = \frac{\Gamma((\alpha+\beta+1+i\lambda)/2)\,\Gamma((\alpha+\beta+1-i\lambda/2))}{2^{-\beta+1}\Gamma(1+\alpha)} \times (\sinh t)^\alpha \varphi_\lambda^{\alpha,\beta}(t), \tag{8}$$

which can be written (cf. [7], page 693) as follows:

$$\forall t > 0, \quad \int_0^\infty x^\beta K_{i\lambda}(x)\,J_\alpha(x\sinh t)\,dx = \Phi_\lambda^{\alpha,\beta}(t), \quad |\operatorname{Im}\lambda| < \operatorname{Re}(\alpha+\beta+1), \tag{9}$$

where J_α and K_α denote, respectively, the Bessel function of the first kind and Macdonald's function.

Example 1. The functions $\varphi_\lambda^{1/2,1/2}$ and $\Phi_\lambda^{1/2,1/2}$ can be expressed in terms of elementary functions as

$$\forall t > 0, \quad \varphi_\lambda^{1/2,1/2}(t) = \frac{2\sin(\lambda t)}{\lambda\sinh(2t)}, \tag{10}$$

$$\forall t > 0, \quad \Phi_\lambda^{1/2,1/2}(t) = \sqrt{\frac{2}{\pi}}\,\Gamma\left(\frac{2+i\lambda}{2}\right)\Gamma\left(\frac{2-i\lambda}{2}\right) \times \frac{1}{\lambda}\frac{\sin(\lambda t)}{\sqrt{\sinh t}\cosh t}, \quad |\operatorname{Im}\lambda| < 2, \tag{11}$$

2.2. Bessel Function of the First Kind. We recall that the Bessel function of the first kind and order α denoted by J_α is defined as an analytic function on $\{z \in \mathbb{C} : |\arg z| < \pi\}$ by

$$J_\alpha(z) = \sum_{k=0}^\infty \frac{(-1)^k}{\Gamma(k+1)\,\Gamma(k+\alpha+1)}\left(\frac{z}{2}\right)^{\alpha+2k} \tag{12}$$

and possesses the following integral representation:

$$J_\alpha(z) = \frac{(z/2)^\alpha}{\sqrt{\pi}\,\Gamma(\alpha+1/2)}\int_{-1}^1 \left(1-t^2\right)^{\alpha-1/2}\cos(zt)\,dt, \quad \operatorname{Re}(\alpha) > -\frac{1}{2}. \tag{13}$$

It satisfies the following functional relations:

$$J_{\alpha-1}(z) - J_{\alpha+1}(z) = 2J'_\alpha(z), \tag{14}$$

$$\frac{1}{z}\frac{d}{dz}\left(z^\alpha J_\alpha(z)\right) = z^{\alpha-1}J_{\alpha-1}(z). \tag{15}$$

It also verifies [10]

$$z^m\left(J_{\alpha-m}(z) - (-1)^m J_{\alpha+m}(z)\right) = \sum_{j=0}^{m-1}(-1)^j b_m^j(\alpha)\,z^j J_{\alpha+j}(z), \tag{16}$$

with

$$b_m^j(\alpha) = \Gamma(\alpha+1)\frac{2^{m-j}C_m^j}{\Gamma(\alpha+j-m+1)}, \tag{17}$$

and satisfies

$$J_0(z) + 2\sum_{j=1}^\infty J_{2j}(z) = 1, \tag{18}$$

$$\sum_{j=0}^\infty (-1)^j J_{2j+1}(z) = \frac{\sin z}{2}, \tag{19}$$

$$J_0(z) + 2\sum_{j=1}^\infty (-1)^j J_{2j}(z) = \cos z, \tag{20}$$

$$\sum_{j=1}^{\infty}(-1)^{j+1}(2j)^2 J_{2j}(z) = \frac{z\sin z}{2}, \tag{21}$$

$$\sum_{j=0}^{\infty}(-1)^{j}(2j+1)^2 J_{2j+1}(z) = \frac{z\cos z}{2}. \tag{22}$$

It has, for $z \geq 0$ and $\alpha \geq 0$, the following asymptotic representation:

$$\begin{gathered} J_\alpha(z) \asymp \frac{z^\alpha}{2^\alpha \Gamma(\alpha+1)}, \quad z \longrightarrow 0, \\ J_\alpha(z) \asymp \sqrt{\frac{2}{\pi z}}\cos\left(z - \frac{1}{2}\alpha\pi - \frac{1}{4}\pi\right), \quad z \longrightarrow \infty. \end{gathered} \tag{23}$$

2.3. Modified Bessel Function of the First Kind and Macdonald's Function. We recall that the modified Bessel function of the first kind I_α of order $\alpha \in \mathbb{C}$ and Macdonald's function K_α of order $\alpha \in \mathbb{C}$ are defined as analytic functions on $\{z \in \mathbb{C} : |\arg z| < \pi\}$ by the formulas

$$\begin{gathered} I_\alpha(z) = \sum_{k=0}^{\infty}\frac{(z/2)^{\alpha+2k}}{\Gamma(k+1)\Gamma(k+\alpha+1)}, \\ K_\alpha(z) = \begin{cases} \dfrac{\pi}{2}\dfrac{[I_{-\alpha}(z) - I_\alpha(z)]}{\sin(\pi\alpha)}, & \text{if } \alpha \notin \mathbb{Z}, \\ \lim\limits_{\nu\to\alpha} K_\nu(z), & \text{if } \alpha \in \mathbb{Z}. \end{cases} \end{gathered} \tag{24}$$

The Macdonald's function satisfies

$$K_\alpha(z) = K_{-\alpha}(z), \tag{25}$$

$$K_{\alpha-1}(z) + K_{\alpha+1}(z) = -2K'_\alpha(z). \tag{26}$$

It also verifies (cf. [10])

$$\begin{aligned} z^m&\left(K_{\alpha+m}(z) - K_{\alpha-m}(z)\right) \\ &= \sum_{j=0}^{m-1}(-1)^{m-j-1} b_m^j(\alpha) z^j K_{\alpha+j}(z), \end{aligned} \tag{27}$$

where $b_m^j(\alpha)$ is the constant defined by (17).

For $\operatorname{Re} z > 0$ and $\alpha \in \mathbb{C}$, it can be written under the representation integral ([9], page 119)

$$K_\alpha(z) = \int_0^\infty e^{-z\cosh u}\cosh(\alpha u)\,du. \tag{28}$$

We recall also that the modified Bessel function of the first kind and Macdonald's function have, for $z > 0$ and $\alpha \geq 0$, the following asymptotic representations ([9], page 136):

$$I_\alpha(z) \asymp \frac{z^\alpha}{2^\alpha\Gamma(\alpha+1)}, \quad z \longrightarrow 0, \tag{29}$$

$$K_\alpha(z) \asymp \begin{cases} \log\dfrac{2}{z}, & \text{if } \alpha = 0 \\ 2^{\alpha-1}\dfrac{\Gamma(\alpha)}{z^\alpha}, & \text{if } \alpha > 0, \end{cases} \quad z \longrightarrow 0 \tag{30}$$

$$I_\alpha(z) \asymp \frac{e^z}{\sqrt{2\pi z}}, \quad z \longrightarrow \infty, \tag{31}$$

$$K_\alpha(z) \asymp \sqrt{\frac{\pi}{2z}}e^{-z}, \quad z \longrightarrow \infty. \tag{32}$$

Note that in the special case $\alpha = \pm 1/2$, Macdonald's function is given by

$$K_{\pm 1/2}(z) = \left(\frac{\pi}{2z}\right)^{1/2} e^{-z}. \tag{33}$$

To complete this paragraph, we recall the following (cf. [7], page 684 and page 747).

(i) If $|\operatorname{Im}\lambda| < 1 + \operatorname{Re}(\beta)$, then we have

$$\int_0^\infty x^\beta K_{i\lambda}(x)\,dx = 2^{\beta-1}\Gamma\left(\frac{1+\beta+i\lambda}{2}\right)\Gamma\left(\frac{1+\beta-i\lambda}{2}\right). \tag{34}$$

(ii) If $|\operatorname{Im}\lambda| < 1 + \operatorname{Re}(\beta)$ and $t > 0$, then we have

$$\int_0^\infty x^\beta K_{i\lambda}(x)\cos(x\sinh t)\,dx = \sqrt{\frac{\pi}{2}}\sinh t\,\Phi_\lambda^{-1/2,\beta+1/2}(t). \tag{35}$$

(iii) If $|\operatorname{Im}\lambda| < 2 + \operatorname{Re}(\beta)$ and $t > 0$, then we have

$$\int_0^\infty x^\beta K_{i\lambda}(x)\sin(x\sinh t)\,dx = \sqrt{\frac{\pi}{2}}\sinh t\,\Phi_\lambda^{1/2,\beta+1/2}(t). \tag{36}$$

3. Recurrence Relations for $\Phi_\lambda^{\alpha,\beta}$

3.1. Some Reccurrence Relations with Respect to Parameters α and β

Proposition 2. *Let $\alpha, \beta, \lambda \in \mathbb{C}$. If $|\operatorname{Im}\lambda| < \operatorname{Re}(\rho)$, with $\rho = \alpha + \beta + 1$ and $m \in \mathbb{N}$, then one has*

$$\begin{aligned} (\sinh t)^m&\left[\Phi_\lambda^{\alpha-m,\beta+m}(t) - (-1)^m\Phi_\lambda^{\alpha+m,\beta+m}(t)\right] \\ &= \sum_{j=0}^{m-1}(-1)^j b_m^j(\alpha)(\sinh t)^j\Phi_\lambda^{\alpha+j,\beta+j}(t), \end{aligned} \tag{37}$$

where $b_m^j(\alpha)$ stands for the constant given by (17).

Proof. Formula (16) asserts that

$$x^{\beta}(x\sinh t)^{m}\left(J_{\alpha-m}(x\sinh t)-(-1)^{m}J_{\alpha+m}(x\sinh t)\right) = \sum_{j=0}^{m-1}(-1)^{j}b_{m}(j)\,x^{\beta}(x\sinh t)^{j}J_{\alpha+j}(x\sinh t). \tag{38}$$

We multiply both sides of the last identity by $K_{i\lambda}(x)$ and integrate with respect to x from 0 to ∞. Then use formula (9) to obtain the required formula. □

Proposition 3. *For $\alpha,\beta,\lambda\in\mathbb{C}$ such that $|\operatorname{Im}\lambda|<\operatorname{Re}(\alpha+\beta+1)$, one has*

$$\Phi_{\lambda}^{\alpha-1,\beta+1}(t)-\Phi_{\lambda}^{\alpha+1,\beta+1}(t)=\frac{2}{\cosh t}\frac{d}{dt}\left[\Phi_{\lambda}^{\alpha,\beta}\right](t). \tag{39}$$

Proof. At first, we suppose that $\operatorname{Re}(\alpha)>-1/2$ and $|\operatorname{Im}\lambda|<\operatorname{Re}(\alpha+\beta+1)$. With the help of formulas (14) and (9), we obtain

$$\Phi_{\lambda}^{\alpha-1,\beta+1}(t)-\Phi_{\lambda}^{\alpha+1,\beta+1}(t) = \frac{2}{\cosh t}\int_{0}^{\infty}x^{\beta}K_{i\lambda}(x)\frac{d}{dt}\left[J_{\alpha}(x\sinh t)\right]dx. \tag{40}$$

On the other hand, by using formulas (14), (28), and (13), we can deduce that

$$\left|x^{\beta}K_{i\lambda}(x)\frac{d}{dt}\left[J_{\alpha}(x\sinh t)\right]\right| \le C(\alpha)\cosh t\cdot K_{|\operatorname{Im}(\lambda)|}(x)\,x^{\operatorname{Re}(\beta)+1} \times\left((x\sinh t)^{\operatorname{Re}(\alpha)-1}+(x\sinh t)^{\operatorname{Re}(\alpha)+1}\right), \tag{41}$$

where $C(\alpha)$ is a positive constant.

Taking account of the asymptotic formulas of Macdonald's function (30) and (32) (or formula (34)) and taking t in any compact of $(0,\infty)$, we can deduce the result in the case $\operatorname{Re}(\alpha)>-1/2$. According to principle of analytic continuation, the restriction $\operatorname{Re}(\alpha)>-1/2$ used can be dropped. □

Using Propositions 2 and 3, we obtain the following corollary.

Corollary 4. *If $|\operatorname{Im}\lambda|<\operatorname{Re}(\alpha+\beta+1)$, then one has*

$$\Phi_{\lambda}^{\alpha+1,\beta+1}(t)=\frac{\alpha}{\sinh t}\Phi_{\lambda}^{\alpha,\beta}(t)-\frac{1}{\cosh t}\frac{d}{dt}\left[\Phi_{\lambda}^{\alpha,\beta}\right](t), \tag{42}$$

$$\Phi_{\lambda}^{\alpha-1,\beta+1}(t)=\frac{\alpha}{\sinh t}\Phi_{\lambda}^{\alpha,\beta}(t)+\frac{1}{\cosh t}\frac{d}{dt}\left[\Phi_{\lambda}^{\alpha,\beta}\right](t). \tag{43}$$

Example 5. Using Example 1 and relation (42) for $\alpha=1/2$ and $\beta=1/2$, we obtain

$$\Phi_{\lambda}^{3/2,3/2}(t)=\frac{1}{\sqrt{2\pi}}\Gamma\left(\frac{2+i\lambda}{2}\right)\Gamma\left(\frac{2-i\lambda}{2}\right)\frac{1}{\lambda}\times\frac{2\sin(\lambda t)\cosh(2t)-\lambda\cos(\lambda t)\sinh(2t)}{(\sinh t)^{3/2}(\cosh t)^{3}},$$
$$\varphi_{\lambda}^{3/2,3/2}(t)=\frac{12}{\lambda(4+\lambda^{2})}\times\frac{2\sin(\lambda t)\cosh(2t)-\lambda\cos(\lambda t)\sinh(2t)}{(\sinh(2t))^{3}}. \tag{44}$$

Example 6. Using Example 1 and relation (43) for $\alpha=1/2$ and $\beta=1/2$, we obtain

$$\Phi_{\lambda}^{-1/2,3/2}(t)=\sqrt{\frac{2}{\pi}}\Gamma\left(\frac{2+i\lambda}{2}\right)\Gamma\left(\frac{2-i\lambda}{2}\right)\frac{1}{\lambda}\times\frac{\lambda\cos(\lambda t)\cosh t-\sin(\lambda t)\sinh t}{\sqrt{\sinh t}(\cosh t)^{3}}, \tag{45}$$
$$\varphi_{\lambda}^{-1/2,3/2}(t)=\frac{\lambda\cos(\lambda t)\cosh t-\sin(\lambda t)\sinh t}{\sinh t(\cosh t)^{3}}.$$

Example 7. Using Example 5 and relation (42) for $\alpha=1$ and $\beta=1$, we obtain

$$\Phi_{\lambda}^{5/2,5/2}(t)=\frac{1}{\sqrt{2\pi}}\Gamma\left(\frac{2+i\lambda}{2}\right)\Gamma\left(\frac{2-i\lambda}{2}\right)\frac{1}{4\lambda}\times\left(\left[16+\lambda^{2}-\left(\lambda^{2}-8\right)\cosh(4t)\right]\sin(\lambda t)-6\lambda\cos(\lambda t)\sinh(4t)\right)\times\left((\sinh t)^{5/2}(\cosh t)^{5}\right)^{-1},$$
$$\varphi_{\lambda}^{5/2,5/2}(t)=\frac{15}{8\lambda(16+\lambda^{2})(4+\lambda^{2})}\times\left(\left[16+\lambda^{2}-\left(\lambda^{2}-8\right)\cosh(4t)\right]\sin(\lambda t)-6\lambda\cos(\lambda t)\sinh(4t)\right)\times\left((\sinh t)^{5}(\cosh t)^{5}\right)^{-1}. \tag{46}$$

Example 8. Using Example 7 and relation (42) for $\alpha = 1$ and $\beta = 1$, we obtain

$$\begin{aligned}
&\Phi_\lambda^{7/2,7/2}(t) \\
&= \frac{1}{\sqrt{2\pi}}\Gamma\left(\frac{2+i\lambda}{2}\right)\Gamma\left(\frac{2-i\lambda}{2}\right)\frac{1}{8\lambda} \\
&\times\left(\frac{12\left[16+\lambda^2-\left(\lambda^2-4\right)\cosh(4t)\right]\cosh(2t)\sin(\lambda t)}{(\sinh t)^{7/2}(\cosh t)^7}\right. \\
&\left.+\frac{\lambda\left[-76-\lambda^2+\left(\lambda^2-44\right)\cosh(4t)\right]\sinh(2t)\cos(\lambda t)}{(\sinh t)^{7/2}(\cosh t)^7}\right), \\
&\varphi_\lambda^{7/2,7/2}(t) \\
&= \frac{105}{16\lambda(4+\lambda^2)(16+\lambda^2)(36+\lambda^2)} \\
&\times\left(\frac{12\left[16+\lambda^2-\left(\lambda^2-4\right)\cosh(4t)\right]\cosh(2t)\sin(\lambda t)}{(\sinh t)^{7}(\cosh t)^7}\right. \\
&\left.+\frac{\lambda\left[-76-\lambda^2+\left(\lambda^2-44\right)\cosh(4t)\right]\sinh(2t)\cos(\lambda t)}{(\sinh t)^{7}(\cosh t)^7}\right).
\end{aligned} \tag{47}$$

Example 9. Using Example 5 and relation (43) for $\alpha = 3/2$ and $\beta = 3/2$, we obtain

$$\begin{aligned}
\Phi_\lambda^{1/2,5/2}(t) &= \frac{1}{\sqrt{2\pi}}\Gamma\left(\frac{2+i\lambda}{2}\right)\Gamma\left(\frac{2-i\lambda}{2}\right)\frac{1}{\lambda} \\
&\times\left(\left[4+\lambda^2+\left(\lambda^2-2\right)\cosh(2t)\right]\sin(\lambda t)\right. \\
&\left.+3\lambda\cos(\lambda t)\sinh(2t)\right) \\
&\times\left((\sinh t)^{1/2}(\cosh t)^5\right)^{-1}.
\end{aligned} \tag{48}$$

Using the previous relation and relation (42) for $\alpha = 1/2$ and $\beta = 5/2$, we obtain

$$\begin{aligned}
&\Phi_\lambda^{3/2,7/2}(t) \\
&= \frac{1}{\sqrt{2\pi}}\Gamma\left(\frac{2+i\lambda}{2}\right)\Gamma\left(\frac{2-i\lambda}{2}\right)\frac{1}{4\lambda} \\
&\times\left(\frac{\left[\left(16+\lambda^2\right)(-3+4\cosh(2t))+\left(7\lambda^2-8\right)\cosh(4t)\right]\sin(\lambda t)}{(\sinh t)^{3/2}(\cosh t)^7}\right. \\
&\left.+\frac{2\lambda\left[16+\lambda^2+\left(\lambda^2-14\right)\cosh(2t)\right]\sinh(2t)\cos(\lambda t)}{(\sinh t)^{3/2}(\cosh t)^7}\right).
\end{aligned} \tag{49}$$

Proposition 10. *For $\alpha, \beta, \lambda \in \mathbb{C}$ such that $|\operatorname{Im}\lambda| < \operatorname{Re}(\alpha+\beta+1)$, one has*

$$\begin{aligned}
\forall t > 0,\quad &\frac{2}{\sinh(2t)}\frac{d}{dt}\left[(\sinh t)^\alpha\Phi_\lambda^{\alpha,\beta}\right](t) \\
&= (\sinh t)^{\alpha-1}\Phi_\lambda^{\alpha-1,\beta+1}(t).
\end{aligned} \tag{50}$$

Proof. According to formula (15), we obtain

$$\begin{aligned}
&\frac{1}{\sinh(t)\cosh(t)}\frac{d}{dt}\left[(\sinh t)^\alpha J_\alpha(x\sinh t)\right] \\
&= x(\sinh t)^{\alpha-1}J_{\alpha-1}(x\sinh t)
\end{aligned} \tag{51}$$

and consequently

$$\begin{aligned}
&(\sinh t)^{\alpha-1}\Phi_\lambda^{\alpha-1,\beta+1}(t) \\
&= \frac{2}{\sinh(2t)}\int_0^\infty x^\beta K_{i\lambda}(x)\frac{d}{dt}\left[(\sinh t)^\alpha J_\alpha(x\sinh t)\right]dx.
\end{aligned} \tag{52}$$

By using formulas (14), (28), (13), and (34), we deduce the result. □

Corollary 11. *For $\alpha, \beta, \lambda \in \mathbb{C}$ such that $|\operatorname{Im}\lambda| < \operatorname{Re}(\alpha+\beta+1)$ and $m \in \mathbb{N}$ one has*

$$\begin{aligned}
&\left[\frac{2}{\sinh(2t)}\frac{d}{dt}\right]^m\left[(\sinh t)^\alpha\Phi_\lambda^{\alpha,\beta}\right](t) \\
&= (\sinh t)^{\alpha-m}\Phi_\lambda^{\alpha-m,\beta+m}(t), \\
&\Phi_\lambda^{\alpha+m,\beta+m} \\
&= \frac{(-1)^m}{(\sinh t)^{\alpha-m}}\left[\frac{2}{\sinh(2t)}\frac{d}{dt}\right]^m\left[(\sinh t)^\alpha\Phi_\lambda^{\alpha,\beta}\right] \\
&+\frac{(-1)^{m+1}}{(\sinh t)^m}\sum_{j=0}^{m-1}(-1)^j b_m(j)(\sinh t)^j\Phi_\lambda^{\alpha+j,\beta+j}.
\end{aligned} \tag{53}$$

Proof. The result follows by induction argument. □

3.2. Some Reccurrence Relations with Respect to the Dual Variable. We give now some recurrence relations with respect to the dual variable λ in the next proposition.

Proposition 12. *If $|\operatorname{Im}\lambda| < \operatorname{Re}(\alpha+\beta+1)$ and $m \in \mathbb{N}$, then one has*

$$\begin{aligned}
&\forall t > 0, \\
&\Phi_{\lambda-im}^{\alpha,\beta+m}(t) - \Phi_{\lambda+im}^{\alpha,\beta+m}(t) \\
&= \sum_{k=0}^{m-1}(-1)^{m-k-1}b_m^k(i\lambda)\Phi_{\lambda-ik}^{\alpha,\beta+k}(t),
\end{aligned} \tag{54}$$

where $b_m^k(\alpha)$ is the constant given by (17). For all $t > 0$, we have

$$\begin{aligned}
&\Phi_{\lambda+i}^{\alpha,\beta+1}(t) + \Phi_{\lambda-i}^{\alpha,\beta+1}(t) \\
&= 2(\beta+1)\Phi_\lambda^{\alpha,\beta}(t) + 2\tanh t\frac{d}{dt}\Phi_\lambda^{\alpha,\beta}(t).
\end{aligned} \tag{55}$$

Proof. The first equality is an immediate consequence of formulas (27) and (9) while the second is proved by using (26) and the asymptotic representations of J_α and K_α. □

As a consequence of this proposition, we have the following corollary.

Corollary 13. *If* $|\operatorname{Im}\lambda| < \operatorname{Re}(\alpha+\beta+1)$ *and* $t > 0$*, one has*

$$\Phi_{\lambda-i}^{\alpha,\beta+1}(t) = (\beta+1+i\lambda)\,\Phi_{\lambda}^{\alpha,\beta}(t) + \tanh t\frac{d}{dt}\Phi_{\lambda}^{\alpha,\beta}(t),$$
$$\Phi_{\lambda+i}^{\alpha,\beta+1}(t) = (\beta+1-i\lambda)\,\Phi_{\lambda}^{\alpha,\beta}(t) + \tanh t\frac{d}{dt}\Phi_{\lambda}^{\alpha,\beta}(t). \tag{56}$$

Example 14. By using formulas (5.32) (in [1], page 44) and (10), we can see that

$$\varphi_{\lambda}^{1/2,-1/2}(t) = \frac{\sin(\lambda t)}{\lambda\sinh t}, \tag{57}$$

and therefore

$$\Phi_{\lambda}^{1/2,-1/2}(t) = \frac{\Gamma((1+i\lambda)/2)\,\Gamma((1-i\lambda)/2)}{\sqrt{2\pi}}\frac{\sin(\lambda t)}{\lambda\sqrt{\sinh t}}. \tag{58}$$

Using now the last corollary, we can get

$$\Phi_{\lambda+i}^{1/2,1/2}(t) = -i\frac{\Gamma((1+i\lambda)/2)\,\Gamma((1-i\lambda)/2)}{\sqrt{2\pi}}\frac{\sin((i+\lambda)t)}{\cosh t\sqrt{\sinh t}}, \tag{59}$$

which coincided with the value of $\Phi_{\Lambda}^{1/2,1/2}(t)$ already calculated at $\Lambda = i+\lambda$ and we have

$$\Phi_{\lambda+i}^{1/2,3/2}(t) = -\sqrt{\frac{2}{\pi}}\Gamma\left(\frac{2+i\lambda}{2}\right)\Gamma\left(\frac{2-i\lambda}{2}\right)\frac{i}{\lambda} \times \frac{(\lambda+i)\sin((\lambda+i)t)\cosh t + \cos((\lambda+i)t)\sinh t}{\sqrt{\sinh t}\cosh^3 t}. \tag{60}$$

Remark 15. By using (33) and the fact that (cf., e.g., [7], page 707)

$$\int_0^{\infty} e^{-x}J_{\alpha}(x\sinh t)\,dx = \frac{1}{\cosh t}\left(\frac{\cosh t-1}{\sinh t}\right)^{\alpha}, \quad t>0,\ \operatorname{Re}(\alpha)>-1, \tag{61}$$

we can get

$$\Phi_{i/2}^{\alpha,1/2}(t) = \sqrt{\frac{\pi}{2}}\frac{1}{\cosh t}\left(\frac{\cosh t-1}{\sinh t}\right)^{\alpha} = \sqrt{\frac{\pi}{2}}\frac{1}{\cosh t}\tanh^{\alpha}\left(\frac{t}{2}\right), \quad t>0,\ \operatorname{Re}(\alpha)>-1. \tag{62}$$

As an application of the last corollary, we obtain

$$\Phi_{3(i/2)}^{\alpha,3/2}(t) = \frac{\sqrt{\pi}}{4}\frac{1}{\cosh^3 t}\tanh^{\alpha}\left(\frac{t}{2}\right)(3+2\alpha\cosh t+\cosh(2t)),$$
$$\Phi_{-i/2}^{\alpha,3/2}(t) = \frac{\sqrt{\pi}}{2}\frac{1}{\cosh^3 t}\tanh^{\alpha}\left(\frac{t}{2}\right)(1+\alpha\cosh t). \tag{63}$$

3.3. Some Summations Intervening the "Modified" Jacobi Functions

Proposition 16. *If* $|\operatorname{Im}\lambda| < \operatorname{Re}(1+\beta)$ *and* $t \in (0, \arg\sinh 1)$*, then one has*

$$\Phi_{\lambda}^{0,\beta}(t) + 2\sum_{k=1}^{\infty}\Phi_{\lambda}^{2k,\beta}(t) = 2^{\beta-1}\Gamma\left(\frac{1+\beta+i\lambda}{2}\right)\Gamma\left(\frac{1+\beta-i\lambda}{2}\right). \tag{64}$$

$$\Phi_{\lambda}^{0,\beta}(t) + 2\sum_{k=1}^{\infty}(-1)^k\Phi_{\lambda}^{2k,\beta}(t) = \sqrt{\frac{\pi}{2}}\sinh t\,\Phi_{\lambda}^{-1/2,\beta+1/2}(t). \tag{65}$$

Proof. Using formulas (18), (34), and (9), we obtain

$$\Phi_{\lambda}^{0,\beta}(t) + 2\int_0^{\infty}\left(\sum_{k=1}^{\infty}J_{2k}(x\sinh t)\right)x^{\beta}K_{i\lambda}(x)\,dx = 2^{\beta-1}\Gamma\left(\frac{1+\beta+i\lambda}{2}\right)\Gamma\left(\frac{1+\beta-i\lambda}{2}\right). \tag{66}$$

With the help of formulas (13), (28), and (34), we can see that

$$\int_0^{\infty}\left|J_{2k}(x\sinh t)\,x^{\beta}K_{i\lambda}(x)\right|dx \le \int_0^{\infty}\frac{((x\sinh t)/2)^{2k}}{(2k)!}x^{\operatorname{Re}(\beta)}K_{|\operatorname{Im}(\lambda)|}(x)\,dx = \frac{((\sinh t)/2)^{2k}}{(2k)!}2^{2k+\operatorname{Re}(\beta)-1} \times\Gamma\left(\frac{1+2k+\operatorname{Re}(\beta)-|\operatorname{Im}(\lambda)|}{2}\right) \times\Gamma\left(\frac{1+2k+\operatorname{Re}(\beta)+|\operatorname{Im}(\lambda)|}{2}\right), \tag{67}$$

which permits to conclude the convergence of series

$$\sum_{k=1}^{\infty}\int_0^{\infty}\left|J_{2k}(x\sinh t)\,x^{\beta}K_{i\lambda}(x)\right|dx, \quad \forall t\in(0,\arg\sinh 1), \tag{68}$$

and consequently we obtain the first equality. The second identity is obtained by using formulas (20), (9), and (35). □

Remark 17. With the help of formulas (62) and (65), we deduce that

$$\Phi_{-i/2}^{-1/2,2}(t) = \frac{1-\sinh^2 t}{\sqrt{\sinh t}\cosh^4 t}. \tag{69}$$

Using now Corollary 13, we see that

$$\Phi_{(3/2)i}^{-1/2,2}(t) = \frac{2}{\sqrt{\sinh t}\cosh^4 t}. \tag{70}$$

Proposition 18. *If* $|\operatorname{Im}\lambda| < 2 + \operatorname{Re}(\beta)$ *then one has*

$$\sum_{k=1}^{\infty}(-1)^{k+1}(2k)^2\Phi_{\lambda}^{2k,\beta}(t) = \sqrt{\pi}(\sinh t)^{3/2}\Phi_{\lambda}^{1/2,\beta+3/2}(t), \quad \forall t\in(0,\arg\sinh 1),$$

$$\sum_{k=0}^{\infty}(-1)^{k}(2k+1)^2\Phi_{\lambda}^{2k+1,\beta}(t) = \sqrt{\frac{\pi}{2}}(\sinh t)^{3/2}\Phi_{\lambda}^{-1/2,\beta+3/2}(t), \quad \forall t\in(0,\arg\sinh 1). \tag{71}$$

Proof. The first identity deduced from (21), (9), and (36). The second follows from (22), (9), and (35). □

Remark 19. By using the fact

$$\sum_{k=1}^{\infty}(-1)^{k+1}(2k)^2\tanh^{2k}\left(\frac{t}{2}\right) = \frac{\sinh^2 t}{\cosh^3 t},$$

$$\sum_{k=0}^{\infty}(-1)^{k}(2k+1)^2\tanh^{2k+1}\left(\frac{t}{2}\right) = \frac{(3-\cosh(2t))}{4}\frac{\sinh t}{\cosh^3 t}, \tag{72}$$

formula (62), and the last proposition, we can see that

$$\Phi_{i/2}^{1/2,2}(t) = \frac{1}{\sqrt{2}}\frac{\sqrt{\sinh t}}{\cosh^4 t},$$

$$\Phi_{i/2}^{-1/2,2}(t) = \frac{1}{4}\frac{(3-\cosh(2t))}{\sqrt{\sinh t}\cosh^4 t}. \tag{73}$$

Remark 20. It is well known that the hypergeometric function ${}_2F_1(a,b;c;z)$ tends to the confluent hypergeometric function ${}_0F_1(c;Z)$ as $a,b\to\infty$ and $z\to\infty$ such a way that $abz\to Z$. Consequently, as $\varepsilon\to 0$,

$$\varphi_{\lambda/\varepsilon}^{\alpha,\alpha}(\varepsilon x) = {}_2F_1\left(\alpha+\frac{1}{2}+\frac{i\lambda}{2\varepsilon},\alpha+\frac{1}{2}-\frac{i\lambda}{2\varepsilon};\alpha+1;-\sinh^2(\varepsilon x)\right) \tag{74}$$

tends to the normalized Bessel function

$$j_\alpha(\lambda x) = {}_0F_1\left(\alpha+1;-\left(\frac{\lambda x}{2}\right)^2\right) = \Gamma(\alpha+1)\sum_{k=0}^{\infty}\frac{(-1)^k}{\Gamma(k+1)\Gamma(k+\alpha+1)}\left(\frac{\lambda x}{2}\right)^{2k}. \tag{75}$$

Using this remark, the expression of Bessel function in terms of elementary functions for some particular values of the parameters follows as particular cases of our findings; we get the results given in [7, 9, 11].

References

[1] T. H. Koornwinder, "Jacobi functions and analysis on noncompact semi simple Lie groups," in *Special Functions: Group Theoretical Aspects and ApplicaTions*, R. A. Askey, T. H. Koornwinder, and W. Schempp, Eds., D. Reidel, Dordrecht, The Netherlands, 1984.

[2] H. Exton, "Some new summation formulae for the generalised hypergeometric function of higher order," *Journal of Computational and Applied Mathematics*, vol. 79, no. 2, pp. 183–187, 1997.

[3] H. Exton, "Summation formulae involving the Laguerre polynomial," *Journal of Computational and Applied Mathematics*, vol. 100, no. 2, pp. 225–227, 1998.

[4] M. Neher, "A note on a sum associated with the generalized hypergeometric function," *Applied Mathematics and Computation*, vol. 187, no. 2, pp. 1527–1534, 2007.

[5] R. Tremblay and B. J. Fugère, "Products of two restricted hypergeometric functions," *Journal of Mathematical Analysis and Applications*, vol. 198, no. 3, pp. 844–852, 1996.

[6] M. Flensted-Jensen and T. H. Koornwinder, "Jacobi functions: the addition formula and the positivity of the dual convolution structure," *Arkiv för Matematik*, vol. 17, no. 1, pp. 139–151, 1979.

[7] I. S. Gradshteyn and I. M. Ryzhik, *Table of Integrals, Series, and Products*, Fourth edition prepared by Ju. V. Geronimus and M. Ju. Ceĭtlin. Translated from the Russian by Scripta Technica, Inc. Translation edited by Alan Jeffrey, Academic Press, New York, NY, USA, 1965.

[8] T. Koornwinder, "A new proof of a Paley-Wiener type theorem for the Jacobi transform," *Arkiv för Matematik*, vol. 13, pp. 145–159, 1975.

[9] N. N. Lebedev, *Special Functions and Their Applications*, Dover, New York, NY, USA, 1972.

[10] S. G. Samko, "On some index relations for Bessel functions," *Integral Transforms and Special Functions*, vol. 11, no. 4, pp. 397–402, 2001.

[11] G. N. Watson, *A Treatise on the Theory of Bessel Functions*, Merchant Books, 2008.

[12] N. Ben Salem and K. Trimèche, "Mehler integral transforms associated with Jacobi functions with respect to the dual variable," *Journal of Mathematical Analysis and Applications*, vol. 214, no. 2, pp. 691–720, 1997.

$f(R, \phi)$ and $f(R)$ Models: A Self-Similar Approach

José Antonio Belinchón

Departamento Física Atómica, Molecular y Nuclear, Universidad Complutense de Madrid, 28040 Madrid, Spain

Correspondence should be addressed to José Antonio Belinchón; abelcal@ciccp.es

Academic Editor: Jianming Zhan

We study the $f(R, \phi)$ models and their particular case, the so-called $f(R)$-models under the self-similarity hypothesis. In particular, we calculate the exact form that each quantity may take in order that field equations (FEs) admit self-similar solutions. The methods employed allow us to obtain general results that are valid not only for the FRW metric, but also for all the Bianchi types as well as for the Kantowski-Sachs model (under the self-similarity hypothesis and the power-law hypothesis for the scale factors).

1. Introduction

The physical and mathematical importance of the modified gravitational models has been recently pointed out for several authors (see, e.g., [1–4]), since this kind of theories explains in a better way the dynamics of the very early universe as well as its current acceleration. In a recent paper [5] (for a review see [6]), we have studied some particular cases of the so-called generalized scalar-tensor theories (STTs or $F(\phi)$ models). Obviously, although such class of theories is more general than the usual Jordan-Brans-Dicke models, they may be generalized in order to incorporate corrections to the Ricci scalar term as the $f(R)$ models (see, e.g., [7–9] and for a review [10]). Therefore, the purpose of this brief paper is to study, under the self-similarity hypothesis, the generalized $f(R, \phi)$ theories, where the $f(R)$ models (without ϕ) result as subclasses and the generalized SST with $f = F(\phi)R$ studied in [5]. In particular, we study the form that the different quantities may take in order that the field equations (FEs) admit self-similar solutions. To carry out our study, we employ two techniques: the Lie group method (LG) [11] and the matter collineation approach (MC) [12]. In this way, we shall be able to state and proof general theorems that are valid not only for the FRW symmetries, otherwise for all kind of Bianchi types as well as for the Kantowski-Sach models.

The self-similarity property of a space-time is characterized by the existence of a homothetic vector field $\mathscr{H}$ (HVF). For a metric, g, and for a vector field $\mathscr{H} \in \mathfrak{X}(\mathscr{M})$, $(\mathscr{H} = \mathscr{H}_i(t, x, y, z)\partial_{x_i})_{i=1}^4$ the homothetic equation reads: $\mathscr{L}_{\mathscr{H}} g = 2g$ [13]. We study self-similar solutions since, as it has been pointed out by Rosquist and Jantzen [14], they correspond to equilibrium points. In the same way, Coley [15] has stressed the fact that this kind of solutions describes asymptotic behaviour of the Bianchi models playing a dominant role in the dynamics of Bianchi cosmological models (see, e.g., [16] for a review of results in the context of the general relativity).

The paper is organized as follows. In Section 2 we study through the LG method the $f(R, \phi)$ models while Section 3 is devoted to study the $f(R)$ models. We summarize the main results in Section 4.

2. $f(R, \phi)$ Models

Our purpose is to describe a class of scalar-tensor theories of gravity represented by the action [17, 18]:

$$S = \int d^4x \frac{\sqrt{-g}}{\beta} \left[\frac{1}{2} f(\phi, R) - \frac{1}{2} Z(\phi) \phi^{;\mu} \phi_{;\mu} - U(\phi) + \beta L_m \right], \quad (1)$$

where R is the Ricci scalar, and ϕ stands for the scalar field, where we assume that the scalar field is homogeneous, that is, $\phi = \phi(t)$. We introduce the constant β in order to fix units, and L_m is a classical matter Lagrangian. The gravitational field equations derived from the action (1) are

$$FG_{\mu\nu} = \beta T^m_{\mu\nu} + Z\left(\phi_{,\mu}\phi_{,\nu} - \frac{1}{2}g_{\mu\nu}\phi_{,\sigma}\phi^{;\sigma}\right) + \left(F_{,\mu;\nu} - g_{\mu\nu}\Box F\right) - Ug_{\mu\nu} + \frac{1}{2}\left(f - FR\right)g_{\mu\nu}, \quad (2)$$

where $G_{\mu\nu}$ is the Einstein tensor, $T^m_{\mu\nu}$ is the energy-momentum tensor, $Z = Z(\phi)$, $U = U(\phi)$ and

$$f = f(R,\phi), \qquad F = \partial_R f(R,\phi). \quad (3)$$

The wave equation reads

$$2Z\Box\phi = f_{,\phi} - Z_\phi \phi'^2 - 2U_{,\phi}, \quad (4)$$

where a prime denotes derivation with respect to time.

We will work under the assumptions

$$f = K(R)\,g(\phi), \quad K = (R)^r, \; r \in \mathbb{R}, \quad (5)$$

that is, f may be split in separable variables and the function K takes this particular form, which is appropriate under the self-similarity hypothesis.

Theorem 1. *The scaling and in particular the self-similar solutions admitted for the FEs* (2) *and* (4) *under the assumptions* (5) *have the following form:*

$$\phi = \phi_0 t^n, \quad Z(\phi) = Z_0\phi^{-m}, \quad U(\phi) = U_0\phi^{-(mn-2n+2)}, \quad g(\phi) = C_1 + C_2\phi^A, \quad (6)$$

with $A = (2 - m - (2/n)(1 - r))$, *and therefore*

$$f = K(R)\cdot g(\phi) = f_0 t^{2n-mn-2}, \quad F = K_R(R)\cdot g(\phi) = F_0 t^{n(2-m)}, \quad (7)$$

with $K = R^r = (at^{-2})^r$, $a \in R$.

Proof. We are going to study (4) through the LG method, that is, we study the kind of functions $f = f(R,\phi)$, $Z(\phi)$, $U(\phi)$, and ϕ such that the FEs (2) and (4) admit power-law and in particular self-similar solutions.

We start by rewriting it in an appropriate way:

$$2Z\left(\phi'' + ht^{-1}\phi'\right) = -Kg_\phi + Z_\phi\phi'^2 + 2U_\phi, \quad (8)$$

since $H = ht^{-1}$, $h = \sum_{i=1}^{3} a_i \in \mathbb{R}^+$, being a_i the exponents of the scale factors, that is, $a(t) = a_0 t^{a_1}$, and so forth, since we are considering not only the FRW case, but also the Bianchi types as well. We also assume that the scalar field is homogeneous, that is, $\phi = \phi(t)$; therefore, a prime denotes a derivative with respect to time. Note that in the self-similar framework, the scalar curvature always behaves as $R \sim t^{-2}$ [14]; therefore, we may assume that $K(R) \sim K(t)$.

Roughly speaking, a symmetry, $X = \xi(x,y)(\partial/\partial x) + \eta(x,y)(\partial/\partial x)$, of a differential equation is an invertible transformation that leaves it form-invariant. By applying the standard Lie procedure (see, e.g., [11]), we need to solve the following overdetermined system of linear partial differential equations for η and ξ (from the extended infinitesimal or prolonged transformations), which allows us to determine the set of the symmetries admitted by (8):

$$Z_\phi\xi_\phi - 2Z\xi_{\phi\phi} = 0, \quad (9)$$

$$\left(ZZ_{\phi\phi} - Z_\phi^2\right)\eta + ZZ_\phi\eta_\phi + 4Z^2ht^{-1}\xi_\phi + 2Z^2\eta_{\phi\phi} - 4Z^2\xi_{t\phi} = 0, \quad (10)$$

$$3\left(2V_\phi - K(t)\,g_\phi\right)\xi_\phi + 2ht^{-2}Z\left(t\xi_t - \xi\right) + 4Z\eta_{t\phi} - 2Z\xi_{tt} + 2Z_\phi\eta_t = 0, \quad (11)$$

$$\left[K\left(Z_\phi g_\phi - Zg_{\phi\phi}\right) + 2\left(ZU_{\phi\phi} - Z_\phi U_\phi\right)\right]\eta + Z\left(Kg_\phi - 2U_\phi\right)\eta_\phi + 2Z\left(2U_\phi - Kg_\phi\right)\xi_t - K'Zg_\phi\xi + 2Z^2\left(ht^{-1}\eta_t + \eta_{tt}\right) = 0. \quad (12)$$

The knowledge of one symmetry X might suggests the form of a particular solution as an invariant of the operator X, that is, a solution of $dx/(\xi(x,y)) = dy/(\eta(x,y))$. This particular solution is known as an invariant solution (generalization of similarity solution). The symmetry, $\xi = t$, $\eta = n\phi$, which induces the invariant solution, $\phi = t^n$, leads us to obtain the following restrictions. From (10),

$$Z_{\phi\phi} = \frac{Z_\phi^2}{Z} - \frac{Z_\phi}{\phi} \Longrightarrow Z(\phi) = Z_0\phi^{-m}, \quad (13)$$

with, $m \in \mathbb{R}$. From (12), we get

$$2\left(ZU_{\phi\phi} - Z_\phi U_\phi\right)\eta - 2ZU_\phi\eta_\phi + 4ZU_\phi\xi_t = 0,$$
$$\left(KZ_\phi g_\phi - Zg_{\phi\phi}\right)\eta + KZg_\phi\eta_\phi - 2KZg_\phi\xi_t - K'Zg_\phi\xi = 0. \quad (14)$$

Therefore,

$$U_{\phi\phi} = \left(\frac{n-2}{n} - m\right)\frac{U_\phi}{\phi}, \quad (15)$$

where we have taken into account that $Z(\phi) = Z_0\phi^{-m}$. In this way,

$$U(\phi) = U_0\phi^{-(1/n)(mn-2n+2)}, \quad (16)$$

while

$$g_{\phi\phi} = \left(1 - m - \frac{1}{n}\left(2 + t\frac{K'}{K}\right)\right)\frac{g_\phi}{\phi}, \quad (17)$$

obtaining the following solution

$$g(\phi) = C_1 + C_2\phi^A, \qquad A = \left(2 - m - \frac{1}{n}\left(2 + t\frac{K'}{K}\right)\right). \tag{18}$$

Compare this result with the obtained one in [5]. Note that if $K = t^{-2}$ (in the self-similar framework the Ricci scalar always varies as t^{-2} [14]), that is, $r = 1$, then

$$2 + t\frac{K'}{K} = 0, \tag{19}$$

and therefore

$$g(\phi) = C_1 + C_2\phi^{2-m}. \tag{20}$$

In any case, when $K = R^r = t^{-2r}$, then

$$g(\phi) = C_1 + C_2\phi^{2-m-(2/n)(1-r)}, \tag{21}$$

thus

$$f = K(R)\cdot g(\phi) \approx R^r\left(t^{n(2-m-(2/n)(1-r))}\right) = f_0 t^{2n-mn-2},$$
$$F = K_R(R)\cdot g(\phi) \approx R^{r-1}\left(t^{n(2-m-(2/n)(1-r))}\right) = F_0 t^{n(2-m)}. \tag{22}$$

Therefore, the so-called $f(R,\phi)$ models admit power-law and in particular self-similar solutions. ☒

As we have shown, for different values of m and r we obtain different models. Note that the case $r = 1$ (the usual one) has been already studied in [5].

3. $f(R)$ Models

As a particular case of the above theories we may study the so-called $f(R)$ models. This kind of models is described by the action:

$$S = \int d^4x\sqrt{-g}\left[f(R) + L_m\right], \tag{23}$$

where R is the Ricci scalar, and L_m is a matter Lagrangian as in the above model. By defining $\equiv \partial f/\partial R(f = f(R))$, the gravitational field equations derived from the action (23) are [1]:

$$FG_{\mu\nu} = T^m_{\mu\nu} + F_{;\mu;\nu} - g_{\mu\nu}\Box F + \frac{1}{2}(f - FR)g_{\mu\nu}, \tag{24}$$

where R is the Ricci scalar, and $G_{\mu\nu}$ is the Einstein tensor. The trace of the field equations read

$$3\Box F + FR - 2f = T, \tag{25}$$

where T is the trace of the stress-energy tensor.

We deduce the form that can acquire the function $f(R)$ in order that the FE admit self-similar and power-law solutions (for the scale factors). For example, in the framework of FRW symmetries it is already known that $f(R) = R^n$ if the scale factor follow a power-law (see, e.g., [19]). In this section, we extend this result to all the Bianchi types as well as for the Kantowski-Sachs model through two methods. The first one, the matter collineation approach allows us to determine that the unique admitted form for the function $f(R)$ is precisely $f(R) = R^n$ compatible with the self-similar hypothesis. With the second one, by studying the trace equation with the Lie group method we proof that $f(R) = R^n$ is the unique possible form for the function $f(R)$ under the power-law hypothesis for the scale factors.

3.1. Matter Collineation Approach

Theorem 2. *Under the hypothesis of self-similarity, the unique form admitted for the function $f(R)$ is $f(R) = R^n$.*

Proof. We may define an effective stress-energy tensor as follows:

$$T^{\text{eff}}_{\mu\nu} = F^{-1}T^m_{\mu\nu} + F^{-1}\left(F_{;\mu;\nu} - g_{\mu\nu}\Box F\right) + \frac{1}{2}\left(fF^{-1} - R\right)g_{\mu\nu}, \tag{26}$$

which allows to use the standard Einstein equations by simply replacing the fluid quantities with the effective ones, so we have to calculate

$$\mathcal{L}_{\mathcal{H}}T^{\text{eff}}_{\mu\nu} = 0, \tag{27}$$

where $\mathcal{H}$ stands for a homothetic vector field. We calculate the Lie derivative of each component of the effective stress-energy tensor.

(1) Thus, $\mathcal{L}_{\mathcal{H}}(F^{-1}T^m_{\mu\nu}) = 0$, in components, $\mathcal{L}_{\mathcal{H}}\cdot(F^{-1}T^m_{11}) = 0$,

$$\frac{\rho'}{\rho} - \frac{F'}{F} = -\frac{2}{t} \Longleftrightarrow \rho F^{-1} = \rho_0 t^{-2}. \tag{28}$$

As it is noted, we are only able to obtain a relationship between the two quantities F and ρ, but we have not determine their particular form.

(2) $\mathcal{L}_{\mathcal{H}}\, T^f_{\mu\nu} = \mathcal{L}_{\mathcal{H}}\,((1/2)(fF^{(-1)} - R)g_{\mu\nu}) = 0$. The first term is $\mathcal{L}_{\mathcal{H}}(T^f_{11}) = 0$, that is,

$$t\left(f' - R'F - F'fF^{-1}\right) = -2(f - FR), \tag{29}$$

that we may rewrite in the following form:

$$t\left(FR' - FR' - f_{RR}R'fF^{-1}\right) = -2(f - FR), \tag{30}$$

that we may write as

$$t\left(-f_R^{-1}f_{RR}R'f\right) + 2(f - f_R R) = 0 \tag{31}$$

since $f' = (df/dR)R' = FR'$, and $F' = f_{RR}R'$. Now taking into account the fact that $tR' \approx -2R$, then the equation yields

$$f_{RR} = \frac{f_R^2}{f} - \frac{f}{R}, \tag{32}$$

whose solution is

$$f = C_1 R^n \quad n \in \mathbb{R}. \tag{33}$$

With the other components, $\mathscr{L}_{\mathscr{H}}(T^f_{1i}) = 0$, we only obtain restriction on the scale factors.

(3) $\mathscr{L}_{\mathscr{H}} T^F_{\mu\nu} = 0$, that is,

$$\mathscr{L}_{\mathscr{H}}\left(F^{-1}\left(F_{;\mu;\nu} - g_{\mu\nu}\Box F\right)\right) = 0. \tag{34}$$

We would like to emphasize that this term depends on the metric, that is, on the scale factors so, we mainly obtain restriction on the scale factors. Nevertheless, the first component is $\mathscr{L}_{\mathscr{H}}(T^F_{11}) = 0$,

$$t\left(F''H + F'H' - \frac{F'^2}{F}H\right) = -2HF', \tag{35}$$

now, taking into account that $H = ht^{-1}$, then we may rewrite it as follows:

$$F'' = \frac{F'^2}{F} - \frac{F'}{t}, \tag{36}$$

and whose solution is

$$F = F_0 t^{C_1}, \quad C_1 \in \mathbb{R}. \tag{37}$$

If, for example, we set $C_1 = 2 - 2n$, from the above result ($f(R) = R^n$), therefore, we arrive to the conclusion that

$$F = F_0 t^{2-2n} = nR^{n-1}. \tag{38}$$

Now if we take into account that $\rho = \rho_0 t^{-\alpha}$ (from the conservation equation for the matter field) being $\alpha = h(\gamma+1)$ and the result from (28), we arrive to the conclusion that $n = \alpha/2$.

This concludes the proof of the theorem as it is required. ☒

Since the hypothesis of self-similarity could be strong constraining the form of the function $f(R)$, in the next section we study if under the hypothesis of power-law solutions for the scale factors the function $f(R)$ could take a different form. For this purpose, we go next to study the trace equation through the Lie group method.

3.2. Lie Group Method

Theorem 3. *Under the hypothesis of power-law solutions for the scale factors, the unique form admitted for the function $f(R)$ is $f(R) = R^n$.*

Proof. By studying the trace equation (25), that we rewrite as follows:

$$F'' + ht^{-1}F' = \frac{r}{3}t^{-2}F - \left(\frac{3\gamma - 1}{3}\right)\rho_0 t^{-\alpha} - \frac{2}{3}f, \tag{39}$$

where we have set

$$H = ht^{-1}, \qquad R = rt^{-2}, \qquad T = (3\gamma - 1)\rho_0 t^{-\alpha}, \qquad f = f(t), \tag{40}$$

with $h, r \in \mathbb{R}$, that is, they are constants, $\alpha = h(\gamma+1)$.

The general solution of (39) is

$$F = C_2 t^{A_+} + C_1 t^{A_-} + D^{-1}t^{1/2} \times \left[t^{A_- - 1/2}\int\left(2t^{B_+}f + (3\gamma-1)\rho_0 t^{B_+ - \alpha}\right)dt - t^{A_+ - 1/2}\int\left(2t^{B_-}f + (3\gamma-1)\rho_0 t^{B_- - \alpha}\right)dt\right], \tag{41}$$

where

$$D = \sqrt{9(h-1)^2 + 12r}, \quad C_1, C_2 \in \mathbb{R}, \qquad A_\pm = \frac{1}{6}\left(3(1-h) \pm D\right), \qquad B_\pm = \frac{1}{6}\left(3(1+h) \pm D\right). \tag{42}$$

Note that if in (39) we set $F = a = \text{const.}$, then

$$f = \frac{3}{2}rat^{-2} - (3\gamma - 1)\rho_0 t^{-\alpha} \Longleftrightarrow f(R) \approx R \approx t^{-2}, \tag{43}$$

which implies that $\alpha = 2$ (as in the standard case).

We will determine the form of the unknown function f in order that the FEs (24)-(25) admit power-law solutions. By studying (39) through the Lie group method, then the standard Lie procedure brings us to get following system of PDEs:

$$2\xi_{FF}t^2 = 0, \tag{44}$$

$$2\xi_F ht + \eta_{FF}t^2 - 2\xi_{tF}t^2 = 0, \tag{45}$$

$$\xi_t ht - \xi_F\left(rF - Ct^{-\alpha+2} - 2ft^2\right) + 2\eta_{tF}t^2 - \xi_{tt}t^2 - \xi h = 0, \tag{46}$$

$$\frac{1}{3}\xi\left(2f't^2 + 2rFt^{-1} - C\alpha t^{-\alpha+1}\right) + ht\eta_t + \frac{2}{3}\xi_t\left(2t^2 f + Ct^{-\alpha+2} - rF\right) + t^2\eta_{tt} + \frac{1}{3}\eta_F\left(rF - 2ft^2 - Ct^{-\alpha+2}\right) - \frac{1}{3}r\eta = 0. \tag{47}$$

For example, the symmetry $= t/(2-2n)$, $\eta = F$, which induces the invariant solution $F = t^{2-2n}$, brings us to obtain the following restriction on the function f. From (47), we obtain

$$C(2n - \alpha) + 4nft^\alpha + 2tt^\alpha f' = 0, \tag{48}$$

whose solution is

$$f = \frac{C_1}{t^{2n}} - \frac{1}{2}\frac{C}{t^{\alpha}} \qquad f = t^{-\alpha} \Longleftrightarrow n = \frac{\alpha}{2}, \tag{49}$$

and which is consistent with the previous result $F^{-1}\rho = t^{-2}$.

To finish this proof, we would like to emphasize that if one tries to find a symmetry (ξ, η) whose invariant solution coincides with the solution $F = nR^{n-1} = n(rt^{-2})^{n-1}$, from the assumption $f = R^n = (rt^{-2})^n$, as for example, $\xi = 1/(-2n(n-1)rt^{-3}(rt^{-2})^{n-2})$, $\eta = 1$, then it is easily checked that it does not verify (46). ☒

We may come to the same conclusion, simply rewriting the trace equation in an appropriate way, by considering the relationship between the functions f and $F = f_R$ (i.e., $f_R = df/dR$). In this way, (39) yields

$$f_{RRR}R'^2 + f_{RR}R'' + f_{RR}R'H = \frac{1}{3}RF - \frac{1}{3}T - \frac{2}{3}f, \tag{50}$$

and taking into account that $R = rt^{-2}$, $r \in \mathbb{R}$, and $H = ht^{-1}$, then it may be rewritten in the following way:

$$\begin{aligned} &\frac{4}{r}f_{RRR}R^3 + \left(\frac{6}{r} - \frac{2h}{r}\right)f_{RR}R^2 \\ &= \frac{1}{3}RF - \frac{(1-3\gamma)\rho_0}{3r^{\alpha/2}}R^{\alpha/2} - \frac{2}{3}f, \end{aligned} \tag{51}$$

where we have identified $T = (1-3\gamma)\rho = (1-3\gamma)\rho_0 t^{-\alpha}$, and, therefore, we can write $T = (((1-3\gamma)\ \rho_0)/r^{(\alpha/2)})R^{\alpha/2} = CR^{\alpha/2}$. We have found that a particular solution of this third-order ODE is the next one:

$$f = f_0 R^{\alpha/2}, \tag{52}$$

induced by the symmetry $\xi = R$, $\eta = (\alpha/2)f$, which is a consistent solution with the previous one. Note that $f_0 = 6rC/(-4r + r\alpha - 3\alpha^2 + 9\alpha^2 - 6\alpha + 3\alpha h^2 - 6\alpha h)$.

4. Conclusions

We have studied under the self-similar hypothesis the modified gravitational $f(R,\phi)$ models as well as its particular case $f(R)$ models arriving at the conclusion that the $f(R,\phi)$ models admit power-law and in particular self-similar solutions if $f = K(R)g(\phi)$, and $K = (R)^r$ (see Theorem 1). In the same way, we have shown that the so-called $f(R)$-models, admit power-law and self-similar solutions (see Theorems 2 and 3) if $f(R) = R^n$. This result coincides with the previous one only obtained for FRW symmetries (studying the Friedmann equation). In this work, we have shown that this result is also valid for any Bianchi type under the self-similarity hypothesis as well as under the power-law solution for the scale factors.

Acknowledgments

The author would like to thank F. Navarro-Lérida for reading and criticizing this work. The author is also extremely grateful to all referees for their valuable comments and suggestions.

References

[1] S. Nojiri and S. D. Odintsov, "Unified cosmic history in modified gravity: from $f(R)$ theory to Lorentz non-invariant models," *Physics Reports*, vol. 505, no. 2–4, pp. 59–144, 2011.

[2] T. Clifton, P. G. Ferreira, A. Padilla, and C. Skordis, "Modified gravity and cosmology," *Physics Reports*, vol. 513, no. 1–3, pp. 1–189, 2012.

[3] L. Amendola and S. Tsujikawa, *Dark Energy. Theory and Observations*, CUP, Cambridge, UK, 2010.

[4] S. Capozziello and V. Faraoni, *Beyond Einstein Gravity: A Survey of Gravitational Theories for Cosmology and Astrophysics*, Springer, New York, NY, USA, 2011.

[5] J. A. Belinchón, "Generalized self-similar scalar-tensor theories," *The European Physical Journal*, vol. 72, p. 1866, 2012.

[6] V. Faraoni, *Cosmology in Scalar-Tensor Gravity*, Kluwer Academic, Dordrecht, The Netherlands, 2004.

[7] S. Nojiri and S. D. Odintsov, "Modified gravity and its reconstruction from the universe expansion history," *Journal of Physics: Conference Series*, vol. 66, Article ID 012005, 2007.

[8] S. Nojiri and S. D. Odintsov, "Modified gravity as an alternative for ΛCDM cosmology," *Journal of Physics A*, vol. 40, no. 25, p. 6725, 2007.

[9] G. Cognola, E. Elizalde, S. Nojiri, S. D. Odintsov, L. Sebastiani, and S. Zerbini, "Class of viable modified $f(R)$ gravities describing inflation and the onset of accelerated expansion," *Physical Review D*, vol. 77, Article ID 046009, 11 pages, 2008.

[10] A. De Felice and S. Tsujikawa, "$f(R)$ Theories," *Living Reviews in Relativity*, vol. 13, p. 3, 2010.

[11] G. W. Bluman and S. C. Anco, *Symmetry and Integration Methods for Differential Equations*, Springer, New York, NY, USA, 2002.

[12] G. Hall, *Symmetries and Curvature in General Relativity*, Lecture Notes in Physics, World Scientific, River Edge, NJ, USA, 2004.

[13] B. J. Carr and A. A. Coley, "Self-similarity in general relativity," *Classical and Quantum Gravity*, vol. 16, no. 7, p. R31, 1999.

[14] K. Rosquist and R. T. Jantzen, "Spacetimes with a transitive similarity group," *Classical and Quantum Gravity*, vol. 2, no. 6, pp. L129–L133, 1985.

[15] A. A. Coley, *Dynamical Systems and Cosmology*, Kluwer Academic, Dordrecht, The Netherlands, 2003.

[16] J. Wainwright and G. F. R. Ellis, *Dynamical Systems in Cosmology*, CUP, Cambridge, UK, 1997.

[17] J.-C. Hwang, "Unified analysis of cosmological perturbations in generalized gravity," *Physical Review D*, vol. 53, no. 2, pp. 762–765, 1996.

[18] F. Perrotta, C. Baccigalupi, and S. Matarrese, "Extended quintessence," *Physical Review D*, vol. 61, no. 2, Article ID 023507, 1999.

[19] N. Goheer, J. Larena, and P. K. S. Dunsby, "Power-law cosmic expansion in $f(R)$ gravity models," *Physical Review D*, vol. 80, no. 6, Article ID 061301, 4 pages, 2009.

10

Integral and Series Representations of Riemann's Zeta Function and Dirichlet's Eta Function and a Medley of Related Results

Michael S. Milgram

Consulting Physicist, Geometrics Unlimited, Ltd., P.O. Box 1484, Deep River, ON, Canada K0J 1P0

Correspondence should be addressed to Michael S. Milgram; mike@geometrics-unlimited.com

Academic Editor: Alfredo Peris

Contour integral representations of Riemann's Zeta function and Dirichlet's Eta (alternating Zeta) function are presented and investigated. These representations flow naturally from methods developed in the 1800s, but somehow they do not appear in the standard reference summaries, textbooks, or literature. Using these representations as a basis, alternate derivations of known series and integral representations for the Zeta and Eta function are obtained on a unified basis that differs from the textbook approach, and results are developed that appear to be new.

Dedicated to the memory of Leslie M. Saunders,
died September 8, 1972, age 29
(Physics Today obituary, issue of December 1972, page 63)

1. Introduction

Riemann's Zeta function $\zeta(s)$ and its sibling Dirichlet's (alternating zeta) function $\eta(s)$ play an important role in physics, complex analysis, and number theory and have been studied extensively for several centuries. In the same vein, the general importance of a contour integral representation of any function has been known for almost two centuries, so it is surprising that contour integral representations for both $\zeta(s)$ and $\eta(s)$ exist that cannot be found in any of the modern handbooks (NIST, [1, Section 25.5(iii)]; Abramowitz and Stegun, [2, Chapter 23]), textbooks (Apostol, [3, Chapter 12]; Olver, [4, Chapter 8.2]; Titchmarsh, [5, Chapter 4]; Whittaker and Watson, [6, Section (13.13)]), summaries (Edwards, [7], Ivić, [8, Chapter 4]; Patterson, [9]; Srivastava and Choi, [10, 11]), compendia (Erdélyi et al., [12, Section 1.12] and Chapter 17), tables (Gradshteyn and Ryzhik, [13], 9.512; Prudnikov et al, [14, Appendix II.7]), and websites [1, 15–19] that summarize what is known about these functions. One can only conclude that such representations, generally discussed in the literature of the late 1800s, have been long-buried and their significance has been overlooked by modern scholars.

It is the purpose of this work to disinter these representations and revisit and explore some of the consequences. In particular, it is possible to obtain series and integral representations of $\zeta(s)$ and $\eta(s)$ that unify and generalize well-known results in various limits and combinations and reproduce results that are only now being discovered and investigated by alternate means. Additionally, it is possible to explore the properties of $\zeta(s)$ in the complex plane and on the critical line and obtain results that appear to be new. Many of the results obtained are disparate and difficult to categorize in a unified manner but share the common theme that they are all somehow obtained from a study of the revived integral representations. That is the unifying theme of this work. To maintain a semblance of brevity, most of the derivations are only sketched, with citations sufficient enough to allow the reader to reproduce any result for her/himself.

2. Contour Integral Representations of $\zeta(s)$ and $\eta(s)$

Throughout, I will use $s = \sigma + i\rho$, where $s \in \mathfrak{C}$ (complex), $\sigma, \rho \in \mathfrak{R}$ (reals), and n and N are positive integers.

The Riemann Zeta function $\zeta(s)$ and the Dirichlet alternating zeta function $\eta(s)$ are well known and defined by (convergent) series representations:

$$\zeta(s) \equiv \sum_{k=1}^{\infty} k^{-s}, \quad \sigma > 1, \tag{1}$$

$$\eta(s) \equiv -\sum_{k=1}^{\infty} (-1)^k k^{-s}, \quad \sigma > 0, \tag{2}$$

with

$$\eta(s) = \left(1 - 2^{1-s}\right) \zeta(s). \tag{3}$$

It is easily found by an elementary application of the residue theorem that the following reproduces (1):

$$\zeta(s) = \pi \sum_{k=1}^{\infty} \text{residue} \left. \frac{t^{-s} e^{i\pi t}}{\sin(\pi t)} \right|_{t=k}, \quad \sigma > 1 \tag{4}$$

and the following reproduces (2):

$$\eta(s) = \pi \sum_{k=1}^{\infty} \text{residue} \left(t^{-s} \cot(\pi t) e^{i\pi t} \right) \Big|_{t=k}, \quad \sigma > 0. \tag{5}$$

Convert the result (4) into a contour integral enclosing the positive integers on the t-axis in a clockwise direction, and provided that $\sigma > 1$ so that contributions from infinity vanish, the contour may then be opened such that it stretches vertically in the complex t-plane, giving

$$\zeta(s) = \frac{i}{2} \int_{c-i\infty}^{c+i\infty} t^{-s} \cot(\pi t)\, dt, \quad 0 < c < 1,\ c \in \Re, \tag{6}$$

$$= -\frac{1}{2} \int_{-\infty}^{\infty} (c + it)^{-s} \cot(\pi (c + it))\, dt, \quad \sigma > 1, \tag{7}$$

a result that cannot be found in any of the modern reference works cited (in particular [20]). It should be noted that this procedure is hardly novel—it is often employed in an educational context (e.g., in the special case $s = 2$) to demonstrate the utility of complex integration to evaluate special sums ([10, Section 4.1]), ([21, Lemma 2.1]) and forms the basis for an entire branch of physics [22].

A similar application to (5), with some trivial trigonometric simplification, gives

$$\eta(s) = \frac{-i}{2} \int_{c-i\infty}^{c+i\infty} \frac{t^{-s}}{\sin(\pi t)}\, dt \tag{8}$$

$$= \frac{1}{2} \int_{-\infty}^{\infty} \frac{(c + it)^{-s}}{\sin(\pi (c + it))}\, dt. \tag{9}$$

Equation (9) reduces, with $c = 1/2$, to a Jensen result [23], presented (without proof) in 1895, and, to the best of my knowledge, reproduced, again, but only in the special case $c = 1/2$, only once in the modern reference literature [17]. In particular, although Lindelof's 1905 work [24] arrives at a result equivalent to that obtained here, attributing the technique to Cauchy (1827), it is performed in a general context, so that the specific results (7) and (9) never appear, particularly in that section of the work devoted to $\zeta(s)$. In modern notation, for a meromorphic function $f(z)$, Lindelof writes ([24, Equation III(4)]) *in general*

$$\sum_{\nu=-\infty}^{\infty} f(\nu) = -\frac{1}{2\pi i} \oint \pi \cot(\pi z)\, f(z)\, dz, \tag{10}$$

where the contour of integration encloses all the singularities of the integrand but never specifically identifies $f(z) = z^{-s}$. Much later, after the integral has been split into two by the invocation of identities for $\cot(\pi z)$, he does make this identification ([24, Equation 4.III (6)]), arriving (with (3)) at ((26)—see below) a now-well-known result, thereby bypassing the contour integral representation (7). In a similar vein, Olver ([4, page 290]) reproduces (10), employing it to evaluate remainder terms, the Abel-Plana formula and a Jensen result but again never writes (7) explicitly. Perhaps the omission of an explicit statement of (7) and (9) from Chapter IV of Lindelof [24] is why these forms seem to have vanished from the historical record, although citations to Jensen and Lindelof abound in modern summaries. (Recently, Srivastava and Choi [11, pages 169–172], have reproduced Lindelof's analysis, also failing to arrive explicitly at (7); although they do write a general form of (8) ([11, Section 2.3, Equation (38)]), it is studied only for the case $c = 1/2$.) (Note added in proof: after a draft copy of this paper was posted, I was made aware that Ruijsenaars has previously obtained a generalized form of (7) for the Hurwitz zeta function ([25, Equation (4.22)]). He also notes (private communication) that he could not find prior references to his result in the literature).

At this point, it is worthwhile to quote several related integral representations that define some of the relevant and important functions of complex analysis, starting with Riemann's original 1859 contour integral representation of the zeta function [26]:

$$\zeta(s) = \frac{\Gamma(1-s)}{2\pi i} \oint \frac{t^{s-1}}{e^{-t} - 1} dt, \tag{11}$$

where the contour of integration encloses the negative t-axis, looping from $t = -\infty - i0$ to $t = -\infty + i0$ enclosing the point $t = 0$.

The equivalence of (11) and (1) is well described in texts (e.g., [11]) and is obtained by reducing three components of the contour in (11). A different analysis is possible however (e.g., [7, Section 1.6])—open and translate the contour in (11) such that it lies vertically to the right of the origin in the complex t-plane (and thereby vanishes) with $\sigma < 0$ and evaluate the residues of each of the poles of the integrand lying at $t = \pm\, 2n\pi i$ to find

$$\zeta(s) = \frac{-i}{2\pi} \Gamma(1-s)(2\pi)^s \times \left(\exp\left(\frac{i\pi s}{2}\right) - \exp\left(-\frac{i\pi s}{2}\right) \right) \sum_{k=1}^{\infty} k^{s-1}. \tag{12}$$

This can be further reduced, giving the well-known functional equation for $\zeta(s)$:

$$\zeta(1-s) = \frac{2\cos(\pi s/2)\Gamma(s)}{(2\pi)^s}\zeta(s), \quad (13)$$

which is valid for all s by analytic continuation from $\sigma < 0$. This demonstrates that (11) and (6) are fundamentally different representations of $\zeta(s)$, since further deformations of (11) do not lead to (7).

Ephemerally, at one time it was noted in [17] that the following complex integral representation

$$\zeta(s) = \frac{\pi}{2(s-1)}\int_{-\infty}^{\infty}\frac{(1/2+it)^{1-s}}{\cosh^2(\pi t)}dt, \quad \sigma > 0 \quad (14)$$

can be obtained by the invocation of the Cauchy-Schlömilch transformation [27]. In fact, the representation (14) follows by the simple expedient of integrating (7) by parts in the special case $c = 1/2$ (see (35)), as also noted in [17]. Another result

$$\zeta(s) = \frac{2^{s-1}}{(1-2^{1-s})\Gamma(s+1)}\int_0^{\infty}\frac{t^s}{\cosh^2(t)}dt, \quad \sigma > -1 \quad (15)$$

obtained in [27] on the basis of the Cauchy-Schlömilch transformation does not follow from (14) as erroneously claimed at one time, but later corrected, in [17].

Based on (11), a number of other representations of $\zeta(s)$ are well known. From the tables cited, the following are worth noting here:

([12, Equation 1.12(6)])

$$\zeta(s) = \frac{1}{2\Gamma(s)(1-2^{-s})}\int_0^{\infty}\frac{t^{s-1}}{\sinh(t)}dt, \quad \sigma > 1, \quad (16)$$

([12, Equation 1.12(4)])

$$\zeta(s) = \frac{1}{\Gamma(s)}\int_0^{\infty}\frac{t^{s-1}}{\exp(t)-1}dt, \quad \sigma > 1, \quad (17)$$

and ([13, Equation 3.527(1)])

$$\zeta(s) = \frac{1}{4}\frac{(2a)^{(s+1)}}{\Gamma(s+1)}\int_0^{\infty}\frac{t^s}{\sinh^2(at)}dt, \quad \sigma > 1 \quad (18)$$

By way of comparison, Laplace's representation ([13, Equation 8.315(2)])

$$\frac{1}{\Gamma(s)} = \frac{1}{2\pi i}\int_{-\infty}^{\infty}\frac{e^{(c+it)}}{(c+it)^s}dt \quad (19)$$

defines the inverse Gamma function.

3. Simple Reductions

3.1. The Case $0 \le c \le 1$. The simplest reductions of (7) and (9) are obtained by uniting the two halves of their range corresponding to $t < 0$ and $t > 0$. With the use of

$$\left(-(c+it)^s + (c-it)^s\right)i = 2\left(c^2+t^2\right)^{(s/2)} \times \sin\left(s\arctan\left(\frac{t}{c}\right)\right), \quad (20)$$

$$(c+it)^s + (c-it)^s = 2\left(c^2+t^2\right)^{(s/2)}\cos\left(s\arctan\left(\frac{t}{c}\right)\right), \quad (21)$$

one finds (with $0 < c < 1$ and $0 \le c \le 1$, resp.)

$$\zeta(s) = -\int_0^{\infty}\left(\cos\left(s\arctan\left(\frac{t}{c}\right)\right)\cos(\pi c)\sin(\pi c) - \sinh(\pi t)\cosh(\pi t)\sin\left(s\arctan\left(\frac{t}{c}\right)\right)\right) \times\left(\left(c^2+t^2\right)^{s/2}\left(\cosh^2(\pi t)-\cos^2(\pi c)\right)\right)^{-1}dt, \quad \sigma > 1, \quad (22)$$

$$\eta(s) = \int_0^{\infty}\left(-\sin\left(s\arctan\left(\frac{t}{c}\right)\right)\cos(\pi c)\sinh(\pi t) + \cos\left(s\arctan\left(\frac{t}{c}\right)\right)\sin(\pi c)\cosh(\pi t)\right) \times\left(\left(c^2+t^2\right)^{s/2}\left(\cosh^2(\pi t)-\cos^2(\pi c)\right)\right)^{-1}dt, \quad (23)$$

the latter valid for all s. In the case that $c = 1/2$, (22) reduces to

$$\zeta(s) = 2^s\int_0^{\infty}\frac{\sinh(\pi t)\sin(s\arctan(2t))}{(1+4t^2)^{s/2}\cosh(\pi t)}dt, \quad \sigma > 1, \quad (24)$$

equivalent to the well-known result (attributed to Jensen by Lindelof, and a special case of Hermite's result for the Hurwitz zeta function ([6, Section 13.2])) (which also appears as an exercise in many textbooks—for example, [11, Equation 2.3(41)]):

$$\zeta(s) = \frac{2^{(s-1)}}{s-1} - 2^s\int_0^{\infty}\frac{\sin(s\arctan(2t))e^{-\pi t}}{(1+4t^2)^{s/2}\cosh(\pi t)}dt. \quad (25)$$

Similarly, (23) reduces to the known ([24, IV.III(6)]) result:

$$\eta(s) = 2^{s-1}\int_0^{\infty}\frac{\cos(s\arctan(t))}{(1+t^2)^{s/2}\cosh(\pi t/2)}dt. \quad (26)$$

Comparison of (24) and (25) immediately yields the known result ([13, Equation 3.975(2)]):

$$\zeta(s) = \frac{2^{(s-1)}}{1-s} + 2^s\int_0^{\infty}\frac{\sin(s\arctan(2t))e^{\pi t}}{(1+4t^2)^{s/2}\cosh(\pi t)}dt, \quad \sigma > 1. \quad (27)$$

Notice that (25) and (26) are valid for all s. Simple substitution of $c \to 0$ in (23) gives

$$\eta(s) = -\sin\left(\frac{s\pi}{2}\right)\int_0^{\infty}\frac{t^{-s}}{\sinh(\pi t)}dt, \quad \sigma < 0, \tag{28}$$

which, after taking (3) into account, leads to

$$\zeta(s) = \frac{\sin(s\pi/2)}{(2^{1-s}-1)}\int_0^{\infty}\frac{t^{-s}}{\sinh(\pi t)}dt, \quad \sigma < 0, \tag{29}$$

displaying the existence of the trivial zeros of $\zeta(s)$ at $s = -2N$, since the integral in (29) is clearly finite at each of these values of s. The result (29) is equivalent to the known result (16) after taking (13) into account. Other possibilities abound. For example, set $c = 1/3$ in (22) to obtain

$$\begin{aligned}\zeta(s) = 3^{s-1}\int_0^{\infty}&\left(2\sin(s\arctan(t))\sinh\left(\frac{2\pi t}{3}\right)\right.\\ &\left.-\sqrt{3}\cos(s\arctan(t))\right)\\ &\times\left(\left(1+t^2\right)^{s/2}\left(2\cosh\left(\frac{2\pi t}{3}\right)+1\right)\right)^{-1}dt,\\ &\sigma > 1.\end{aligned} \tag{30}$$

Further, a novel representation can be found by noting that (9) is invariant under the transformation of variables $t \to t + \rho$ together with the choice $c = \sigma$ along with the requirement that $0 \le \sigma \le 1$. This effectively allows one to choose $c = s$, yielding the following complex representation for $\eta(s)$, as well as the the corresponding representation for $\zeta(s)$ because of (3), which is valid in the critical strip $0 \le \sigma \le 1$:

$$\eta(s) = \frac{-i}{2}\int_{s-i\infty}^{s+i\infty}\frac{t^{-s}}{\sin(\pi t)}dt \tag{31}$$

$$= \frac{1}{2}\int_{-\infty}^{\infty}\frac{(s+it)^{-s}}{\sin(\pi(s+it))}dt. \tag{32}$$

Because (20) and (21) are valid for $c \in \mathfrak{C}$, it follows that (23) with $c = s$ is also a valid representation for $\eta(s)$, as well as the the corresponding representation for $\zeta(s)$ with recourse to (3), again only valid in the critical strip $0 \le \sigma \le 1$. Another interesting variation is the choice $c = 1/\rho$ in (23), valid for $\rho > 1$, again applicable to the corresponding representation for $\zeta(s)$ because of (3). Such a representation may find application in the analysis of $\zeta(\sigma + i\rho)$ in the asymptotic limit $\rho \to \infty$. Finally, the choice $c = \sigma$, $0 \le \sigma \le 1$ leads to an interesting variation that is discussed in Section 9.

3.2. Simple Recursion via Partial Differentiation. As a function of c, $\zeta(s)$ and $\eta(s)$ are constant, as are their integral representations over a range corresponding to $0 < N - c < 1$; however the integrals change discontinuously at each new value of c that corresponds to a new value of N. Thus, the operator $\partial/\partial c$ acting on (7) and (9) vanishes, except at nonnegative integers $c = N$, when it becomes indefinite. Therefore, it is of interest to investigate this operation in the neighbourhood of $c = N/2$ as well as the limit $c \to N$.

Applied to (9) (and equivalent to integration by parts), the requirement that

$$\frac{\partial\eta(s)}{\partial c} = 0, \quad c \ne N \tag{33}$$

and identification of one of the resulting terms yields

$$\eta(s+1) = -\frac{\pi}{2s}\int_{-\infty}^{\infty}\frac{(c+it)^{-s}\cos(\pi(c+it))}{\sin^2(\pi(c+it))}dt \qquad 0 \le c < 1, \tag{34}$$

a new result with improved convergence at infinity. Similarly operating with $\partial/\partial c$ on (7) yields

$$\zeta(s+1) = \frac{\pi}{2s}\int_{-\infty}^{\infty}\frac{(c+it)^{-s}}{\sin^2(\pi(c+it))}dt \quad 0 < c < 1. \tag{35}$$

(It should be noted that Srivastava et al. [20] study an incomplete form of (35) in order to obtain representations for sums both involving $\zeta(n)$, and $\zeta(n)$ by itself, without commenting on the significance of the contour integral with infinite bounds (35).) Using (20) and (21) to reduce the integration range of (34) gives a fairly lengthy result for general values of c. If $c = 1/2$, (34) becomes

$$\begin{aligned}\eta(s+1) = \frac{2^{s-1}\pi}{s}&\\ \times\int_0^{\infty}&\left(1+t^2\right)^{-s/2}\sinh\left(\frac{\pi t}{2}\right)\sin(s\arctan(t))\\ &\times\cosh^{-2}\left(\frac{\pi t}{2}\right)dt,\end{aligned} \tag{36}$$

which, by letting $s \to s - 1$, can be rewritten as

$$\begin{aligned}\eta(s+1) = \frac{2(s-1)\eta(s)}{s} + \frac{2^{s-1}\pi}{s}&\\ \times\int_0^{\infty}&\cos(s\arctan(t))\left(1+t^2\right)^{-s/2}\sinh\left(\frac{\pi t}{2}\right)\\ &\times\cosh^{-2}\left(\frac{\pi t}{2}\right)t\,dt,\end{aligned} \tag{37}$$

and, with regard to (3),

$$\begin{aligned}\zeta(s+1) = \frac{2(s-1)\left(1-2^{1-s}\right)\zeta(s)}{s\left(1-2^{-s}\right)} + \frac{2^{s-1}\pi}{s\left(1-2^{-s}\right)}&\\ \times\int_0^{\infty}&\cos(s\arctan(t))\left(1+t^2\right)^{-s/2}\sinh\left(\frac{\pi t}{2}\right)\\ &\times\cosh^{-2}\left(\frac{\pi t}{2}\right)t\,dt.\end{aligned} \tag{38}$$

Similarly, ((37) and (38) are the first two entries in a family of recursive relationships that will be discussed elsewhere (in preparation)). In the case that $c = 1/2$, (35) reduces to the well-known result ([24, Equation 4, III(5)] (attributed to Jensen [23]))

$$\zeta(s+1) = \frac{\pi 2^{s-1}}{s} \int_0^\infty \frac{\cos(s\arctan(t))}{(1+t^2)^{s/2}\cosh^2(\pi t/2)} dt. \quad (39)$$

If $c \to 0$ in (34) (for the case $c = 0$ in (35) see the following section), one finds a finite result (also equivalent to integrating (28) by parts):

$$\eta(s) = \frac{\pi\sin(s\pi/2)}{s-1} \int_0^\infty \frac{t^{1-s}\cosh(\pi t)}{\sinh^2(\pi t)} dt, \quad \sigma < 0, \quad (40)$$

and, when $c = 1$, one finds the unusual forms:

$$\begin{aligned} \eta(s+1) = {} & \{0,1\} + \frac{i\pi r}{2s} \\ & \times \int_0^\pi \frac{\left(1 \pm re^{i\theta}\right)^{-s}\cosh\left(\pi re^{i\theta}\right)e^{i\theta}}{\sinh^2\left(\pi re^{i\theta}\right)} d\theta \\ & - \frac{r\pi}{s}\int_1^\infty \left(1+r^2t^2\right)^{-s/2}\cos(s\arctan(rt)) \\ & \times \cosh(\pi rt)\sinh^{-2}(\pi rt)\, dt, \end{aligned} \quad (41)$$

which is valid for any $0 < r < 1$ and all s (particularly $r = 1/\pi$ or $r = 1/2$), after deformation of the contour of integration in (34) either to the left or to the right of the singularity at $t = 1$, respectively, belonging to a choice of $\pm$ where the symbol $\{0, 1\}$ means *zero* or *one* as the case may be.

4. Special Cases

4.1. The Case $c = 0$. With respect to (35), setting $c = 0$ together with some simplification gives

$$\zeta(s) = -\pi^{s-1}\frac{\sin(\pi s/2)}{s-1}\int_0^\infty \frac{t^{1-s}}{\sinh^2(t)} dt, \quad \sigma < 0, \quad (42)$$

reproducing (18), with $a = 1$, after taking (13) into account. This same case ($c = 0$) naturally suggests that the integral in (6) be broken into two parts:

$$\zeta(s) = \zeta_0(s) + \zeta_1(s), \quad (43)$$

where

$$\zeta_0(s) = \frac{i}{2}\oint_{-i}^{i} t^{-s}\cot(\pi t)\, dt, \quad (44)$$

$$\zeta_1(s) = \frac{i}{2}\int_{-i\infty}^{i} t^{-s}\cot(\pi t)\, dt + \frac{i}{2}\int_{i}^{i\infty} t^{-s}\cot(\pi t)\, dt \quad (45)$$

such that the contour in (44) bypasses the origin to the right. It is easy to show that the sum of the two integrals in (45) reduces to

$$\zeta_1(s) = \sin\left(\frac{\pi s}{2}\right)\int_1^\infty t^{-s}\coth(\pi t)\, dt, \quad \sigma > 1, \quad (46)$$

$$= \frac{\sin(\pi s/2)}{s-1} + \sin\left(\frac{\pi s}{2}\right)\int_1^\infty \frac{t^{-s}e^{(-\pi t)}}{\sinh(\pi t)} dt, \quad (47)$$

the latter being valid for all s. One, of several commonly used possibilities of reducing (44), is to apply ([28, Equation (54.3.2)])

$$t\cot(\pi t) = -\frac{2}{\pi}\sum_{k=0}^{\infty} t^{2k}\zeta(2k), \quad |t| < 1, \quad (48)$$

giving

$$\zeta_0(s) = -\frac{2}{\pi}\sin\left(\frac{\pi s}{2}\right)\sum_{k=0}^{\infty}(-1)^k\frac{\zeta(2k)}{2k-s}, \quad (49)$$

being convergent for all s. Such sums appear frequently in the literature (e.g., [10, 29] Section (3.2)) and have been studied extensively when $s = -n$ [10, 30]. In the case $s = 2n$, (49) reduces to $\zeta_0(s) = \zeta(s)$ and (47) vanishes.

Along with (42) the integral in (35) can be similarly split, using the convergent series representation:

$$\frac{1}{\sin^2(\pi v)} = \frac{2}{\pi^2}\sum_{k=0}^{\infty}(2k-1)\zeta(2k)\, v^{2k-2}, \quad |v| < 1, \quad (50)$$

where the Bernoulli numbers B_{2k} appearing in the original ([28, Equation (50.6.10)]) are related to $\zeta(2k)$ by

$$\zeta(2k) = -\frac{1}{2}\frac{(-1)^k(2\pi)^{2k}B_{2k}}{\Gamma(2k+1)}. \quad (51)$$

Thus the representation

$$\begin{aligned} \zeta(s) = {} & -\frac{1}{2}\frac{i\pi}{(s-1)}\int_{-i}^{i}\frac{t^{1-s}}{\sin^2(\pi t)} dt \\ & - \frac{\pi\sin(\pi s/2)}{s-1}\int_1^\infty \frac{t^{1-s}}{\sinh^2(\pi t)} dt \end{aligned} \quad (52)$$

becomes (formally)

$$\begin{aligned} \zeta(s) = {} & -\frac{\sin(\pi s/2)}{s-1} \\ & \times \left(\frac{2}{\pi}\sum_{k=0}^{\infty}\frac{(-1)^k(2k-1)\zeta(2k)}{2k-s} \right. \\ & \left. + \pi\int_1^\infty \frac{t^{1-s}}{\sinh^2(\pi t)} dt\right) \end{aligned} \quad (53)$$

after straightforward integration of (50) with $\sigma < 0$. In (53), the integral converges, and the formal sum conditionally converges or diverges according to several tests. Howver, further analysis is still possible. Following the method of [10, Section 3.3] applied to (49), together with [10, Equation 3.4(15)], we have the following general result for $|v| < 1$:

$$\sum_{k=0}^{\infty}\frac{\zeta(2k)\,v^{2k}}{2k-s} = \frac{1}{2s} + \int_0^v t^{-s}\sum_{k=1}^{\infty}\zeta(2k)\,t^{2k-1}dt \tag{54}$$

$$= \frac{1}{2s} + \frac{1}{2}v^s\int_0^v t^{-s}\left(\Psi(1+t) - \Psi(1-t)\right)dt, \quad \sigma < 2 \tag{55}$$

so that (formally), after premultiplying (55) with $1/v$, then operating with $v^2(\partial/\partial v)$,

$$\sum_{k=0}^{\infty}\frac{\zeta(2k)(2k-1)\,v^{2k}}{2k-s} = -\frac{1}{2s} + \frac{1}{2}v^s \times \int_0^v t^{1-s}\left(\Psi'(1+t) + \Psi'(1-t)\right)dt \tag{56}$$

$$= -\frac{1}{2s} + \frac{1}{2}v^s \times \int_0^v t^{1-s}\left(\pi^2\left(1+\cot^2(\pi t)\right) - \frac{1}{t^2}\right)dt, \quad \sigma < 2. \tag{57}$$

Notice that the right-hand side of (55) provides an analytic continuation of the left-hand side for $|v| > 1$ and the right-hand side of (57) gives a regularization of the left-hand side. Putting together (53) and (57), incorporating the first term of (57) into the second term of (53), setting $v = i$, and constraining $\sigma > 0$ eventually yields a new integral representation (compare with (121), [31, Equation (2.8)] and [25, Equation (4.80)]):

$$\zeta(s) = -\frac{\pi^{s-1}\sin(\pi s/2)}{s-1}\int_0^{\infty} t^{1-s}\left(\frac{1}{\sinh^2 t} - \frac{1}{t^2}\right)dt, \tag{58}$$

which is valid for $0 < \sigma < 2$, a nonoverlapping analytic continuation of (42) that spans the critical strip. Additionally, using (13) followed by replacing $s \to 1-s$ gives a second representation:

$$\zeta(s) = \frac{2^{s-1}}{\Gamma(1+s)}\int_0^{\infty} t^{s}\left(\frac{1}{\sinh^2 t} - \frac{1}{t^2}\right)dt \tag{59}$$

valid for $-1 < \sigma < 1$, a non-overlapping analytic continuation of (18). Over the critical strip $0 < \sigma < 1$ representations (58) and (59) are simultaneously true and augment the rather short list of other representations that share this property ([32], as cited in [12] (unnumbered equations immediately following 1.12(1), also reproduced in [11, Equations 2.3(44)–(46)].)). See Section 9 for further application of this result.

4.2. The Case $c \geq 1$. Interesting results can be obtained by translating the contour in (6) and (9) and adding the residues so-omitted. From (1) and (2) define the partial sums

$$\zeta_N(s) \equiv \sum_{k=1}^{N} k^{-s}$$
$$\eta_N(s) \equiv -\sum_{k=1}^{N}(-1)^k k^{-s}, \tag{60}$$

let $c \to c_N$, $N = 1, 2, 3, \ldots$ with

$$N \leq c_N < N+1, \tag{61}$$

and adjust all the above results accordingly by redefining $c \to c_N$. (If $c_N = N$, an extra half-residue is removed and the contour deformed to pass to the right of the point $t = N$.) This gives, for example, the generalization of (6):

$$\zeta(s) = \zeta_N(s) - \frac{1}{2N^s}\delta_{N,c_N} + \frac{i}{2}\int_{c_N-i\infty}^{c_N+i\infty} t^{-s}\cot(\pi t)\,dt, \quad \sigma > 1, \tag{62}$$

and (9):

$$\eta(s) = \eta_N(s) + \frac{(-1)^N}{2N^s}\delta_{N,c_N} - \frac{i}{2}\int_{c_N-i\infty}^{c_N+i\infty}\frac{t^{-s}}{\sin(\pi t)}dt, \tag{63}$$

so (22) and (23) with $c \to c_N$ become integral representations of the remainder of the partial sums $\zeta(s) - \zeta_N(s)$ and $\eta(s) - \eta_N(s)$, the former corresponding to a special case of the Hurwitz zeta function $\zeta(s, N)$.

In the case that $c_N = N = 1$, one finds

$$\zeta(s) = \frac{1}{2} + \int_0^{\infty}\frac{\cosh(\pi t)\sin(s\arctan(t))}{(1+t^2)^{s/2}\sinh(\pi t)}dt, \quad \sigma > 1, \tag{64}$$

$$= \frac{s+1}{2(s-1)} + \int_0^{\infty}\frac{e^{-\pi t}\sin(s\arctan(t))}{(1+t^2)^{s/2}\sinh(\pi t)}dt, \tag{65}$$

$$\eta(s) = \frac{1}{2} + \int_0^{\infty}\frac{\sin(s\arctan(t))}{\sinh(\pi t)(1+t^2)^{s/2}}dt, \tag{66}$$

with the latter two results being valid for all s. Neither (64) nor (66) appear in references; (65) corresponds to a limiting case of Hermite's representation of the Hurwitz zeta function ([12, Equation 1.10(7)]). In analogy to the developments leading to (27), equations (64), (66), and (3), together with the following modified form of [13, Equation (9.51.5)]

$$\zeta(s) = \frac{2^{s-1}s}{(2^s-1)(s-1)} + \frac{2^{s-1}}{2^s-1} \times \int_0^{\infty}\frac{\sin(s\arctan(t))\left(1+e^{-\pi t}\right)}{(1+t^2)^{s/2}\sinh(\pi t)}dt, \tag{67}$$

can be used to obtain the possibly new result:

$$\zeta(s) = \frac{s-3}{2(s-1)} + \int_0^{\infty}\frac{\sin(s\arctan(t))\,e^{\pi t}}{(1+t^2)^{s/2}\sinh(\pi t)}dt, \quad \sigma > 1. \tag{68}$$

For $c_N = 1/2 + N$ corresponding to general partial sums, two possibilities are apparent. Based on (20), (21), (22), and (23) (for comparison, see also [4, Section 8.3], and [7, Section (6.4)] and Chapter 7), we find

$$\zeta(s) - \sum_{k=1}^{N} \frac{1}{k^s} = \frac{1}{2}\left(\frac{1}{2}+N\right)^{1-s} \int_0^\infty \frac{\sinh(\pi t(1+2N))\left(1+t^2\right)^{-s/2} \sin(s\arctan(t))}{\cosh^2(\pi t(1/2+N))} dt \quad \sigma > 1 \tag{69}$$

$$= \frac{(1/2+N)^{1-s}}{s-1} - 2\left(\frac{1}{2}+N\right)^{1-s} \int_0^\infty \frac{\left(1+t^2\right)^{-s/2} \sin(s\arctan(t))}{1+e^{\pi t(1+2N)}} dt, \tag{70}$$

$$\eta(s) + \sum_{k=1}^{N} \frac{(-1)^k}{k^s} = (-1)^N \left(\frac{1}{2}+N\right)^{1-s} \int_0^\infty \frac{\left(1+t^2\right)^{-s/2} \cos(s\arctan(t))}{\cosh(\pi t(1/2+N))} dt \tag{71}$$

which together extend (24) and (26), respectively. Application of (34) and (35) with (20) and (21), also with $c_N = 1/2 + N$, gives the remainders

$$\zeta(s+1) - \sum_{k=1}^{N} \frac{1}{k^{s+1}} = \frac{\pi(1/2+N)^{1-s}}{s} \int_0^\infty \frac{\left(1+t^2\right)^{-s/2} \cos(s\arctan(t))}{\cosh^2(\pi t(1/2+N))} dt,$$
$$\eta(s+1) + \sum_{k=1}^{N} \frac{(-1)^k}{k^{s+1}} = \frac{\pi(-1)^N(1/2+N)^{1-s}}{s} \int_0^\infty \frac{\sinh(\pi t(1/2+N))\left(1+t^2\right)^{-s/2} \sin(s\arctan(t))}{\cosh^2(\pi t(1/2+N))} dt \tag{72}$$

which are valid for all s. For the case $c_N = N$ applying (62) and (63) gives the remainders

$$\zeta(s) - \sum_{k=1}^{N} \frac{1}{k^s} = -\frac{1}{2N^s} + N^{1-s} \int_0^\infty \frac{\cosh(\pi N t)\left(1+t^2\right)^{-s/2} \sin(s\arctan(t))}{\sinh(\pi N t)} dt, \quad \sigma > 1 \tag{73}$$

$$= -\frac{1}{2N^s} + \frac{N^{1-s}}{s-1} + 2N^{1-s} \int_0^\infty \frac{\left(1+t^2\right)^{-s/2} \sin(s\arctan(t))}{e^{2N\pi t}-1} dt, \tag{74}$$

$$\eta(s) + \sum_{k=1}^{N} \frac{(-1)^k}{k^s} = \frac{(-1)^N}{2N^s} + (-1)^{1+N} N^{1-s} \int_0^\infty \frac{\left(1+t^2\right)^{-s/2} \sin(s\arctan(t))}{\sinh(\pi N t)} dt, \tag{75}$$

which extend (64) and (66), the latter being valid for all s. It should be noted that (74), valid for $s \in \mathfrak{C}$, generalizes well-established results (e.g., [13, Equation 3.411(12)]) that are only valid in the limiting case $s = n$. (Although [13, Equation 3.411(21)] gives (with a missing minus sign) an integral representation for finite sums such as those appearing in (74), specified to be valid only for integer exponents (k^n), that particular result in [13] also appears to apply after the generalization $n \rightarrow s$.) Omitting the integral term from (70) gives an upper bound for the remainder; similarly, omitting the integral term in (74) gives a lower bound for the remainder, provided that $s \in \mathfrak{R}$ in both cases. In addition, many of the above results, in particular (70) and (74), appear to be new and should be compared to classical results such as [8, Theorem 1.8, page 23 and pages 99-100], and asymptotic approximations such as [33, Equation (8.2.1)].

5. Series Representations

5.1. By Parts. By simple manipulation of the integral representations given, it is possible to obtain new series representations. Starting from the integral representation (29), integrate by parts, giving

$$\zeta(s) = \frac{\sin(\pi s/2)\,\pi^2}{\left(2^{(1-s)}-1\right)(-s/2+1)} \times \int_0^\infty t^{2-s}\,{}_1F_2\left(-\frac{s}{2}+1;\frac{3}{2},2-\frac{s}{2};\frac{t^2\pi^2}{4}\right) \times \cosh(\pi t)\sinh^{-3}(\pi t)\,dt, \quad \sigma<0, \tag{76}$$

where the hypergeometric function arises because of [34, Equation 6.2.11(6)]. Write the hypergeometric function as a (uniformly convergent) series, thereby permitting the sum and integral to be interchanged, apply a second integration by parts, and find

$$\zeta(s) = \left(\frac{\pi}{2^{1-s}-1}\right)\sin\left(\frac{\pi s}{2}\right)\sum_{k=0}^{\infty}\frac{\pi^{2k}}{\Gamma(2k+2)} \times \int_0^\infty \frac{t^{1-s+2k}}{\sinh^2(\pi t)}\,dt, \quad \sigma<0. \tag{77}$$

Identify the integral in (77) using (18) giving

$$\zeta(s) = \frac{2^s\pi^{s-1}\sin(\pi s/2)}{2^{1-s}-1} \times \sum_{k=0}^{\infty}\frac{2^{-2k}\zeta(-s+1+2k)\,\Gamma(-s+2+2k)}{\Gamma(2k+2)}, \tag{78}$$

which is valid for all values of σ by the ratio test and the principle of analytic continuation. This result, which appears to be new, can be rewritten by extracting the $k=0$ term, and applying (13) to yield

$$\zeta(s) = \frac{2^s\pi^{s-1}\sin(\pi s/2)}{2^{1-s}-2+s} \times \sum_{k=1}^{\infty}\frac{2^{-2k}\,\zeta(-s+1+2k)\,\Gamma(-s+2+2k)}{\Gamma(2k+2)}, \tag{79}$$

a representation, valid for all values of s, that is reminiscent of the following similar sum recently obtained by Tyagi and Holm ([31, Equation (3.5)]):

$$\zeta(s) = \frac{\pi^{s-1}}{1-2^{1-s}}\sin\left(\frac{\pi s}{2}\right) \times \sum_{k=1}^{\infty}\frac{\left(2-2^{s-2k}\right)\zeta(-s+1+2k)\,\Gamma(-s+1+2k)}{\Gamma(2k+2)}, \quad \sigma>0 \tag{80}$$

(convergent by Gauss' test). Equation (79) can also be rewritten in the form

$$\zeta(s) = \frac{1}{\Gamma(s)\left(2^s-1-s\right)} \times \sum_{k=1}^{\infty}\frac{2^{-2k}\zeta(s+2k)\,\Gamma(1+s+2k)}{\Gamma(2k+2)}, \tag{81}$$

by replacing $s\to 1-s$ and applying (13).

With reference to (7), express the numerator *cosine* factor as a difference of exponentials each of which in turn is written as a convergent power series in $\pi(c+it)$, and recognize that the resulting series can be interchanged with the integral and each term identified using (9) to find

$$\zeta(s) = -\sqrt{\pi}\sum_{k=0}^{\infty}\frac{\left(-\pi^2/4\right)^k\eta(s-2k)}{\Gamma(k+1/2)\,\Gamma(k+1)}, \quad \sigma>1. \tag{82}$$

A more interesting variation of (82) can be obtained by applying (3) and explicitly identifying the $k=0$ term of the sum giving

$$\zeta(s) = -\frac{1}{2}\,\frac{\sum_{k=1}^{\infty}\left(-\pi^2\right)^k\left(1-2^{1-s+2k}\right)/\Gamma(2k+1)\,\zeta(s-2k)}{1-2^{-s}}, \tag{83}$$

a convergent representation if $\sigma>1$ (by Gauss' test). An immediate consequence is the identification

$$\zeta(2) = -\frac{\pi^2}{3}\zeta(0), \tag{84}$$

because only the $k=1$ term in the sum contributes when $s=2$. Additionally, when $s=2n$, (83) terminates at n terms, becoming a recursion for $\zeta(2n)$ in terms of $\zeta(2n-2k)$, $k=1,\ldots,n$—see Section 8 for further discussion. If the functional equation (13) is used to extend its region of convergence, (83) becomes

$$\zeta(s) = \frac{1}{2}\left(\pi^s\sum_{k=1}^{\infty}\frac{\zeta(1-s+2k)\left(2^{s-2k-1}-1\right)}{\Gamma(2k+1)\,\Gamma(s-2k)} \times \left(\left(2^{-s}-1\right)\cos\left(\frac{\pi s}{2}\right)\right)^{-1}\right), \quad \sigma>1, \tag{85}$$

which augments similar series appearing in [10, Sections 4.2 and 4.3], and corresponds to a combination of entries in [28, Equations (54.6.3) and (54.6.4)].

5.2. By Splitting

5.2.1. The Lower Range. After applying (13), the integral in (42) is frequently (e.g., [11]) and conveniently split in two, defining

$$\zeta_-(s) = \int_0^\omega \frac{t^s}{\sinh^2(t)}\,dt, \quad \sigma>1, \tag{86}$$

$$\zeta_+(s) = \int_\omega^\infty \frac{t^s}{\sinh^2(t)}\,dt \tag{87}$$

with

$$\zeta(s)\,\frac{\Gamma(s+1)}{2^{s-1}} = \zeta_-(s)+\zeta_+(s) \tag{88}$$

for arbitrary ω. Similar to the derivation of (53), after application of (13) we have

$$\zeta_{-}(s) = -2\sqrt{\pi}\sum_{k=0}^{\infty}\frac{B_{2k}\omega^{2k+s-1}}{\Gamma(k-1/2)\Gamma(k+1)(2k-1+s)}, \quad (89)$$

which can be rewritten using (51) as

$$\zeta_{-}(s) = 4\omega^{s-1}\sum_{k=0}^{\infty}\frac{\left(-\omega^2/\pi^2\right)^k\Gamma(k+1/2)\zeta(2k)}{\Gamma(k-1/2)(2k-1+s)}, \quad \omega\le\pi. \quad (90)$$

Substitute (1) in (90), interchange the order of summation, and identify the resulting series to obtain

$$\zeta_{-}(s) = \frac{2s\omega^{s+1}}{\pi^2(s+1)}\times\sum_{m=1}^{\infty}\frac{1}{m^2}\,{}_2F_1\left(1,\frac{1}{2}+\frac{s}{2};\frac{3}{2}+\frac{s}{2};-\frac{\omega^2}{\pi^2m^2}\right) - \omega^s\coth(\omega)+\frac{s\omega^{s-1}}{s-1}, \quad (91)$$

a convergent representation that is valid for all s.

Other representations are easily obtained by applying any of the standard hypergeometric linear transforms as in (118)—see Section 6. One such leads to

$$\zeta_{-}(s) = s\omega^{s-1}\Gamma\left(\frac{1}{2}+\frac{s}{2}\right)\times\sum_{k=0}^{\infty}\frac{\Gamma(k+1)}{\Gamma(k+3/2+s/2)}\sum_{m=1}^{\infty}\left(\frac{1}{\pi^2m^2/\omega^2+1}\right)^{k+1} - \omega^s\coth(\omega)+\frac{s\omega^{s-1}}{s-1}, \quad (92)$$

a rapidly converging representation for small values of ω, valid for all s. The inner sum of (92) can be written in several ways. By expanding the denominator as a series in $1/(t^2m^2)$ with $r<2p-1$ and interchanging the two series, we find the general form

$$\kappa(r,p,t) \equiv \sum_{m=1}^{\infty}\frac{m^r}{(1+t^2m^2)^p} = \sum_{m=0}^{\infty}\frac{\Gamma(p+m)(-1)^m t^{-2m-2p}\zeta(2m+2p-r)}{\Gamma(p)\Gamma(m+1)}, \quad (93)$$

where the right-hand side continues the left if $|t|>1$ and $r-2p\ge -1$. Alternatively, [28, Equation (6.1.32)] identifies

$$\sum_{m=0}^{\infty}\frac{1}{(m^2t^2+y^2)} = \frac{1}{2y^2}+\frac{\pi}{2ty}\coth\left(\frac{\pi y}{t}\right), \quad (94)$$

so for $p=k+1$ and $r=0$ in (93), each term of the inner series (92) can be obtained analytically by taking the kth derivative of the right-hand side with respect to the variable y^2 at $y=1$, leading to a rational polynomial in $\coth(\omega)$ and $\sinh(\omega)$. The first few terms are given in Table 1 for two choices of ω.

Application of (93) to (92) leads to a less rapidly converging representation that can be expressed in terms of more familiar functions:

$$\zeta_{-}(s) = s\omega^{s-1}\Gamma\left(\frac{1}{2}+\frac{s}{2}\right)\times\sum_{k=0}^{\infty}\sum_{m=0}^{\infty}\frac{(-1)^m\Gamma(m+k+1)\zeta(2m+2k+2)}{\Gamma(k+3/2+s/2)\Gamma(m+1)}\times\left(\frac{\omega^2}{\pi^2}\right)^{(m+k+1)} - \omega^s\coth(\omega)+\frac{s\omega^{s-1}}{s-1}. \quad (95)$$

A second transform of the intermediate hypergeometric function yields another useful result:

$$\zeta_{-}(s) = \frac{2s\omega^{s-1}}{\Gamma(1/2+s/2)}\sum_{k=0}^{\infty}\left(\frac{\omega^2}{\pi^2}\right)^{(k+1)}\times\sum_{m=0}^{\infty}\left(\frac{\omega^2}{\pi^2}\right)^m(-1)^m\Gamma\left(m+k+\frac{1}{2}+\frac{s}{2}\right)\times\zeta(2m+2k+2)\times\left(\Gamma(m+1)\Gamma(k+1)(2k+1+s)\right)^{-1} - \omega^s\coth(\omega)+\frac{s\omega^{s-1}}{s-1}. \quad (96)$$

Both (95) and (96) can be reordered along a diagonal of the double sum ($\sum_{k=0}^{\infty}\sum_{m=0}^{\infty}f(m,k) = \sum_{k=0}^{\infty}\sum_{m=0}^{k}f(m,k-m)$), allowing one of the resulting (terminating) series to be evaluated in closed form. In either case, both eventually give

$$\zeta_{-}(s) = -2s\omega^{s-1}\sum_{k=0}^{\infty}\left(\frac{-\omega^2}{\pi^2}\right)^k\frac{\zeta(2k)}{(2k-1+s)} - \omega^s\coth(\omega), \quad (97)$$

reducing to (49) in the case $\omega = \pi$ taking (13) into consideration. Series of the form (97) have been studied extensively [10, 11] when $s=n$ for special values of ω. These representations will find application in Section 8.

5.2.2. The Upper Range. Rewrite (87) in exponential form and expand the factor $(1-e^{-2t})^{-2}$ in a convergent power series, interchange the sum and integral, and recognize that the resulting integral is a generalized exponential integral $E_s(z)$ usually defined in terms of incomplete Gamma functions ([1, 35], Section 8.19) as follows:

$$E_s(z) = z^{s-1}\Gamma(1-s,z), \quad (98)$$

giving a rapidly converging series representation:

$$\zeta_{+}(s) = 4\omega^{s+1}\sum_{k=1}^{\infty}kE_{-s}(2\omega k). \quad (99)$$

TABLE 1: The first few values of $\kappa(0,\kappa+1,\pi^2/\omega^2)$ in (93).

k	Analytic	$\omega = 1$	$\omega = \pi$
0	$-\frac{1}{2}+\frac{\omega\coth(\omega)}{2}$	0.156518	1.076674
1	$-\frac{1}{2}+\frac{1}{4}\frac{\omega^2}{\sinh^2\omega}+\frac{1}{4}\omega\coth(\omega)$	0.927424×10^{-2}	0.306837
2	$\frac{3\omega}{16}\coth^3\omega-\frac{1}{2}\coth^2\omega+\frac{\omega\coth\omega(-3+2\omega^2)}{16\sinh^2\omega}+\frac{8+3\omega^2}{16\sinh^2\omega}$	0.795491×10^{-3}	0.134296

A useful variant of (99) is obtained by manipulating the well-known recurrence relation ([1, Equation 8.19.12]) for $E_s(z)$ to obtain

$$E_{-s}(z)=\frac{1}{z^2}\left(e^{-z}(s+z)+s(s-1)E_{2-s}(z)\right). \tag{100}$$

Apply (100) repeatedly N times to (99) and find

$$\begin{aligned}\zeta_+(s) &= -s\omega^{-1+s}\log\left(e^{2\omega}-1\right)+2s\omega^s-\frac{2\omega^s}{1-e^{2\omega}}\\ &\quad+2\omega^s\Gamma(s+1)\sum_{k=2}^{N+1}\frac{\mathrm{Li}_k\left(e^{-2\omega}\right)}{\Gamma(s-k+1)(2\omega)^k}\\ &\quad+\frac{2\omega^{s-N-1}\Gamma(s+1)}{\Gamma(s-N-1)2^{N+1}}\sum_{k=1}^{\infty}\frac{E_{N+2-s}(2k\omega)}{k^{N+1}}\end{aligned} \tag{101}$$

in terms of logarithms and polylogarithms ($\mathrm{Li}_k(z)$), a result that loses numerical accuracy as N increases, but which also, usefully, terminates if $s=n$, $N\ge n\ge 0$. This is discussed in more detail in Section 8.

Another useful representation emerges from (99) by writing the function $E_s(2k\omega)$ in terms of confluent hypergeometric functions. From [35, Equation (2.6b)] and [36, Equation (9.10.3)], we find

$$\begin{aligned}\zeta_+(s) &= 4\omega^{s+1}\sum_{k=1}^{\infty}kE_{-s}(2k\omega)\\ &= 2^{1-s}\Gamma(s+1)\zeta(s)-\frac{4\omega^{s+1}}{s+1}\\ &\quad\times\sum_{k=1}^{\infty}ke^{-2k\omega}{}_1F_1(1;2+s;2k\omega),\end{aligned} \tag{102}$$

where the series on the right-hand side converges absolutely if $\sigma>1$ independent of ω. Comparison of (99), (88), and (102) identifies

$$\zeta_-(s)=\frac{4\omega^{s+1}}{s+1}\sum_{k=1}^{\infty}ke^{-2k\omega}{}_1F_1(1;2+s;2k\omega) \tag{103}$$

which can be interpreted as a transformation of sums, or an analytic continuation, for the various guises of $\zeta_-(s)$ given in Section 5.2.1, if both, or either, of the two sides converge for some choice of the variables s and/or ω, respectively. The fact that a term containing $\zeta(s)$ naturally appears in the expression (102) for $\zeta_+(s)$ suggests that there will be a severe cancellation of digits between the two terms $\zeta_-(s)$ and $\zeta_+(s)$ if (88) is used to attempt a numerical evaluation of $\zeta(s)$, as has been reported elsewhere [37]. Essentially, (88) is numerically useless (with common choices of arithmetic precision) for any choices of $\rho>10$, even if ω is carefully chosen in an attempt to balance the cancellation of digits. However, with 20 digits of precision, using 20 terms in the series, it is possible to find the first zero of $\zeta(1/2+i\rho)$ correct to 10 digits, deteriorating rapidly thereafter as ρ increases. For any choice of ω, the convergence properties of the two sums representing $\zeta_-(s)$ and $\zeta_+(s)$ anti-correlate, as can be seen from Table 1. This can also be seen by rewriting (102) to yield the combined representation:

$$\begin{aligned}&2^{1-s}\Gamma(s+1)\zeta(s)\\ &\quad=4\omega^{s+1}\left(\sum_{k=1}^{\infty}k\left(E_{-s}(2k\omega)+\frac{{}_1F_1(s+1;2+s;-2k\omega)}{s+1}\right)\right).\end{aligned} \tag{104}$$

In (104), the first inner term ($E_{-s}(2k\omega)$) vanishes like $\exp(-2k\omega)/(2k\omega)$ whereas the second term (${}_1F_1/(s+1)$) vanishes like $1/k^{(1+s)}-\exp(-2k\omega)/(2k\omega)$, so for large values of k cancellation of digits is bound to occur. (In fact, the coefficients of both asymptotic series prefaced by the exponential terms cancel exactly term by term.)

It should be noted that (102) is representative of a family of related transformations. For example, although setting $s=0$ in (102) does not lead to a new result, operating first with $\partial/\partial s$ followed by setting $s=0$ leads to a transformation of a related sum:

$$\begin{aligned}\sum_{k=1}^{\infty}E_1(2k\omega) &= -\frac{\omega\,\mathrm{arctanh}\left(\sqrt{-\omega^2}/\pi\right)}{\pi\sqrt{-\omega^2}}\\ &\quad-\sum_{k=2}^{\infty}\frac{\omega\,\mathrm{arctanh}\left(\sqrt{-\omega^2}/\pi k\right)}{\pi k\sqrt{-\omega^2}}\\ &\quad+\frac{1}{2}\log\left(\frac{\omega}{\pi}\right)+\frac{1}{2\omega}+\frac{\gamma}{2},\end{aligned} \tag{105}$$

where the first term corresponding to $k=1$ has been isolated, since it contains the singularity that occurs due to divergence of the left-hand sum if $\omega=i\pi$. (In (86) and (87), $\omega\in\mathbb{C}$ is permitted by a fundamental theorem of integral calculus.) Similarly, operating with $\partial/\partial s$ on (102) at $s=1$ leads to the identification

$$\sum_{k=1}^{\infty}\log\left(1+\frac{1}{\pi^2k^2}\right)=-1-\log(2)+\log\left(e^2-1\right) \tag{106}$$

(which can alternatively be obtained by writing the infinite sum as a product). A second application of $\partial/\partial s$ on (102), again at $s = 1$ using (106), leads to the following transformation of sums:

$$\frac{1}{2}\sum_{k=1}^{\infty}\frac{\zeta(2k)(-1)^k}{\pi^{2k}k^2} = \frac{1}{2}\ln(2)^2 - \frac{\gamma^2}{2} + \frac{\pi^2}{12} - 1 - \gamma(1) - \sum_{k=1}^{\infty}\frac{E_1(2k)}{k}. \quad (107)$$

This transformation can be further reduced by applying an integral representation given in [1, Section (6.2)], leading to

$$\sum_{k=1}^{\infty}\frac{E_1(2k)}{k} = -\int_1^{\infty}\frac{\log\left(1-e^{-2t}\right)}{t}dt. \quad (108)$$

Correcting a misprinted expression given in ([38, Equation (4.8)], [10, 11], Equation 3.3(43)) which should read

$$\sum_{k=1}^{\infty}\frac{t^k}{k^2}\zeta(2k) = \int_0^t \log\left(\pi\sqrt{v}\csc\left(\pi\sqrt{v}\right)\right)\frac{dv}{v}, \quad (109)$$

followed by a change of variables, eventually yields the representation

$$\sum_{k=1}^{\infty}\frac{\zeta(2k)(-1)^k}{\pi^{2k}k^2} = -2\int_0^1\frac{1}{t}\left(\log\left(\frac{1}{2}e^t - \frac{1}{2}e^{-t}\right) - \log(t)\right)dt. \quad (110)$$

Substituting (108) and (110) into (107) followed by some simplification finally gives

$$\int_0^1 \log\left(\frac{\left(1-e^{-2/v}\right)\left(e^{2v}-1\right)}{2v}\right)\frac{dv}{v} = -\frac{\pi^2}{12} + \frac{\gamma^2}{2} + \gamma(1) - \frac{1}{2}\log^2(2) + 2, \quad (111)$$

a result that appears to be new.

5.3. *Another Interpretation.* These expressions for the split representation can be interpreted differently. For the moment, let $\sigma < -3$, and insert the contour integral representation [35, Equation (2.6a)], for $E_{-s}(z)$ into the converging series representation (99), noting that contributions from infinity vanish, giving

$$\zeta_+(s) = -4\omega^{s+1}\sum_{k=1}^{\infty}\frac{1}{2\pi i}\int_{c-i\infty}^{c+i\infty}\frac{\Gamma(-t)(2k\omega)^t}{s+1+t}dt, \quad (112)$$

where the contour is defined by the real parameter $c < -2$. With this caveat, the series in (112) converges, allowing the integration and summation to be interchanged, leading to the identification

$$\zeta_+(s) = -\frac{4\omega^{s+1}}{2\pi i}\int_{c-i\infty}^{c+i\infty}\frac{\Gamma(-t)(2\omega)^t\zeta(-1-t)}{s+1+t}dt. \quad (113)$$

The contour may now be deformed to enclose the singularity at $t = -s - 1$, translated to the right to pick up residues of $\Gamma(-t)$ at $t = n$, $n = 0, 1, \ldots$ and the residue of $\zeta(-1-t)$ at $t = -2$. After evaluation, each of the residues so obtained cancels one of the terms in the series expression (90) for $\zeta_-(s)$ leaving (after a change of integration variables)

$$\frac{\Gamma(s+1)\zeta(s)}{2^{s-1}\omega^{s+1}} = -\frac{1}{2\pi i}\oint\frac{\Gamma(t+1)\zeta(t)}{(s-t)2^{t-1}\omega^{t+1}}dt, \quad (114)$$

the standard Cauchy representation of the left-hand side, valid for all s, where the contour encloses the simple pole at $t = s$.

6. Integration by Parts

Integration by parts applied to previous results yields other representations.

From (17), we obtain (known to the Maple computer code, but only in the form of a Mellin transform)

$$\zeta(s) = -\frac{1}{\Gamma(s-1)}\int_0^{\infty}t^{s-2}\log(1-\exp(-t))\,dt, \quad \sigma > 1, \quad (115)$$

and from (64), we find

$$\zeta(s) = \frac{1}{2} - \frac{s}{\pi} \times \int_0^{\infty}\frac{\log(\sinh(\pi t))\cos((s+1)\arctan(t))}{\left(1+t^2\right)^{(1+s)/2}}dt, \quad \sigma > 1 \quad (116)$$

as well as the following intriguing result (discovered by the Maple computer code):

$$\begin{aligned}\zeta(s) = {} & \frac{1}{2} + \frac{1}{2}\frac{\sqrt{\pi}\sin(\pi s/2)\Gamma(s/2-1/2)}{\Gamma(s/2)} \\ & + \pi\int_0^{\infty}\frac{t\sin(s\arctan(t))}{\sinh^2(\pi t)} \times {}_2F_1\left(\frac{1}{2},\frac{s}{2};\frac{3}{2};-t^2\right)dt \\ & - s\int_0^{\infty}\frac{t\cosh(\pi t)\cos(s\arctan(t))}{\sinh(\pi t)\left(1+t^2\right)} \times {}_2F_1\left(\frac{1}{2},\frac{s}{2};\frac{3}{2};-t^2\right)dt, \quad \sigma > 0.\end{aligned} \quad (117)$$

This result has the interesting property that it is the only integral representation of which I am aware in which the independent variable s does not appear as an exponent. Many variations of (117) can be obtained through the use of any of the transformations of the hypergeometric function, for example,

$$ {}_2F_1\left(\frac{1}{2},\frac{s}{2};\frac{3}{2};-t^2\right)=\left(1+t^2\right)^{-s/2}\times{}_2F_1\left(1,\frac{s}{2};\frac{3}{2};\frac{t^2}{(1+t^2)}\right), \tag{118}$$

which choice happens to restore the variable s to its usual role as an exponent.

7. Conversion to Integral Representations

From (79), identify the product $\zeta(\cdots)\Gamma(\cdots)$ in the form of an integral representation from one of the primary definitions of $\zeta(s)$ ([13, Equation (9.511)]) and interchange the series and sum. The sums can now be explicitly evaluated, leaving an integral representation valid over an important range of s:

$$\zeta(s)=\frac{\pi^{s-1}4^s\sin(\pi s/2)}{2-2^{1+s}+s2^s}\times\int_0^\infty\frac{t^{-s}\left(s-(2s\sinh(t/2)/t)-1+\cosh(t/2)\right)}{e^t-1}dt,\quad \sigma<2. \tag{119}$$

Each of the terms in (119) corresponds to known integrals ([13, Equations (3.552.1) and (8.312.2)]), and if the identification is taken to completion, (119) reduces to the identity $\zeta(s)=\zeta(s)$ with the help of (13). Other useful representations are immediately available by evaluating the integrals in (119) corresponding to pairs of terms in the numerator, giving

$$\zeta(s)=\frac{\pi^s2^{s-2}}{(1-2^s)\cos(\pi s/2)}\times\left(\frac{\sin(\pi s)2^s}{\pi}\int_0^\infty\frac{t^{-s}(-1+\cosh(t/2))}{e^t-1}dt+\frac{1}{\Gamma(s)}\right),\quad\sigma<2, \tag{120}$$

$$\zeta(s)=\pi^{s-1}\sin\left(\frac{\pi s}{2}\right)\times\left(\int_0^\infty v^{-s-1}\left(\frac{v}{\sinh(v)}-1\right)e^{-v}dv-\frac{\pi}{\sin(\pi s)\Gamma(1+s)}\right),\quad\sigma<2. \tag{121}$$

8. The Case $s=n$

Several of the results obtained in previous sections reduce to helpful representations when $s=n$.

8.1. Finite Series Representations. From (79), we find, for even values of the argument,

$$\zeta(2n)=-\frac{1}{2}\left(\sum_{k=1}^{n}\left(\frac{(-1)^k\pi^{2k}(2n-2k-1)\zeta(2n-2k)}{\Gamma(2k+2)}\right)\times\left(2^{-2n}+n-1\right)^{-1}\right), \tag{122}$$

from (83), we find

$$\zeta(2n)=\frac{1}{2(2^{-2n}-1)}\times\sum_{k=1}^{n}\frac{\left(-\pi^2\right)^k\left(1-2^{1-2n+2k}\right)\zeta(2n-2k)}{\Gamma(2k+1)}, \tag{123}$$

and, based on (80), we find [39] a similar, but independent, result:

$$\zeta(2n)=2\frac{2^{-2n}}{1-2^{1-2n}}\times\sum_{k=1}^{n}\left(\frac{\left(1-2^{2n-2k-1}\right)(2\pi)^{2k}\zeta(2n-2k)(-1)^k}{\Gamma(2k+2)}\right), \tag{124}$$

three of an infinite number of such recursive relationships [40]. The latter two may be identified with known results by reversing the sums, giving

$$\zeta(2n)=-\frac{\left(-\pi^2\right)^n}{2^{1-2n}+2n-2}\times\left(\sum_{k=0}^{n-1}\frac{(2k-1)\zeta(2k)}{\left(-\pi^2\right)^k\Gamma(2n-2k+2)}\right), \tag{125}$$

thought to be new,

$$\zeta(2n)=\frac{\left(-\pi^2\right)^n}{2^{1-2n}-2}\left(\sum_{k=0}^{n-1}\frac{\left(1-2^{1-2k}\right)\zeta(2k)}{\left(-\pi^2\right)^k\Gamma(2n-2k+1)}\right) \tag{126}$$

equivalent to [11, Equation 4.4(11)], and

$$\zeta(2n)=\frac{2\left(-\pi^2\right)^n}{1-2^{1-2n}}\left(\sum_{k=0}^{n-1}\frac{\left(1-2^{2k-1}\right)\zeta(2k)(-1)^k}{(2\pi)^{2k}\Gamma(2n-2k+2)}\right), \tag{127}$$

equivalent to [11, Equation 4.4(9)].

8.2. Novel Series Representations. By adding and subtracting the first $n-1$ terms in the series representation of the hypergeometric function in (91), we find ([36, Chapter 9]), for the case $s = 2n + 2$ $(n \geq 0)$,

$$\frac{{}_2F_1\left(1, n+1/2;\ n+3/2,\ -\omega^2/\pi^2 m^2\right)}{m^2} = \left(n+\frac{1}{2}\right)\left(\frac{-\omega^2}{\pi^2 m^2}\right)^{-n} \times\left(\frac{2\pi}{m\omega}\arctan\left(\frac{\omega}{\pi m}\right) - \frac{1}{m^2}\sum_{k=0}^{n-1}\frac{\left(-\omega^2/\pi^2 m^2\right)^k}{1/2+k}\right), \tag{128}$$

and for the case $s = 2n + 3$

$$\frac{{}_2F_1\left(1, n+1;\ n+2;\ -\omega^2/\pi^2 m^2\right)}{m^2} = (n+1)\left(-\omega^2/\pi^2 m^2\right)^{-n} \times\left(\frac{\pi^2}{\omega^2}\ln\left(1+\frac{\omega^2}{\pi^2 m^2}\right) - \frac{1}{m^2}\sum_{k=0}^{n-1}\frac{\left(-\omega^2/\pi^2 m^2\right)^k}{k+1}\right). \tag{129}$$

Substituting (128) or (129) in (88), (91), and (101) gives series representations for $\zeta(2n)$ and $\zeta(2n+1)$, respectively, for which I find no record in the literature, with particular reference to works involving Barnes double Gamma function (e.g., [30]). For example, employing (128) in the case $n = 0$, we obtain what appears to be a new representation (or the evaluation of a new sum):

$$\zeta(2) = -4\omega\sum_{m=1}^{\infty}\left(\frac{\pi m}{\omega}\arctan\left(\frac{\omega}{\pi m}\right) - 1\right) + \mathrm{Li}_2\left(e^{-2\omega}\right) - 2\omega\log\left(-1+e^{2\omega}\right) + 3\omega^2 + 2\omega. \tag{130}$$

The series (130) (compare with ([28, Equation (42.1.1)])) is absolutely convergent by Gauss' test for arbitrary values of ω and reduces to an identity in the limiting case $\omega \to 0$.

8.3. Connection with Multiple Gamma Function. From Section 5.2 with $\omega = 1$, we have, after some simplification,

$$\zeta(2) = 5 - 4\left(\sum_{m=1}^{\infty}\frac{(-1)^m \pi^{(-2m)}\zeta(2m)}{2m+1}\right) + \mathrm{Li}_2\left(e^{-2}\right) - 2\log\left(e^2-1\right). \tag{131}$$

Comparison of the previous infinite series with [10, Equation 3.4(464)] using, $z = i/\pi$ identifies

$$\log\left(\frac{G(1+i/\pi)}{G(1-i/\pi)}\right) = \frac{i}{4\pi}\left(6 + 4\log(2\pi) + 2\mathrm{Li}_2\left(e^{-2}\right) - 4\log\left(e^2-1\right) - \frac{\pi^2}{3}\right) = 0.232662175652953i, \tag{132}$$

where G is Barnes double Gamma function. From [10, Section 1.3], unnumbered equation following 26, we also find

$$\log\left(\frac{G(1+i/\pi)}{G(1-i/\pi)}\right) = \frac{i\log(2\pi)}{\pi} - \frac{i}{\pi} + \left(\sum_{k=1}^{\infty}\left(-\frac{2i}{\pi} + k\ln\left(\frac{k+i/\pi}{k-i/\pi}\right)\right)\right) \tag{133}$$

thereby recovering (130) after representing the arctan function in (130) as a logarithm (with $\omega = 1$).

9. The Critical Strip

In (58) and (59) with $0 \leq \sigma \leq 1$ let

$$f(s) = -\frac{\pi^{s-1}\sin(\pi s/2)}{s-1}, \tag{134}$$

$$g(s) = \frac{2^{s-1}}{\Gamma(s+1)}, \tag{135}$$

$$h(t) = \sqrt{t}\left(\frac{1}{\sinh^2 t} - \frac{1}{t^2}\right). \tag{136}$$

Then, after adding and subtracting, we find a composition of symmetric and antisymmetric integrals about the critical line $\sigma = 1/2$:

$$\zeta(s) = \frac{2f(s)g(s)}{g(s)-f(s)} \times \int_0^{\infty} h(t)\cos(\rho\ln(t))\sinh\left(\left(\frac{1}{2}-\sigma\right)\ln(t)\right)dt + \frac{2if(s)g(s)}{f(s)-g(s)} \times \int_0^{\infty} h(t)\sin(\rho\ln(t))\cosh\left(\left(\frac{1}{2}-\sigma\right)\ln(t)\right)dt, \tag{137}$$

and simultaneously

$$\zeta(s) = \frac{2f(s)g(s)}{f(s)+g(s)} \times \int_0^\infty h(t)\cos(\rho\ln(t))\cosh\left(\left(\frac{1}{2}-\sigma\right)\ln(t)\right)dt - \frac{2if(s)g(s)}{f(s)+g(s)} \times \int_0^\infty h(t)\sin(\rho\ln(t))\sinh\left(\left(\frac{1}{2}-\sigma\right)\ln(t)\right)dt. \quad (138)$$

Let

$$C(\rho) = \int_0^\infty h(t)\cos(\rho\ln(t))\,dt, \qquad S(\rho) = \int_0^\infty h(t)\sin(\rho\ln(t))\,dt. \quad (139)$$

For $\zeta(s) = 0$ on the critical line $\sigma = 1/2$, both (137) and (138) can be simultaneously satisfied when both $C(\rho) \neq 0$ and $S(\rho) \neq 0$ only if

$$\left|f\left(\frac{1}{2}+i\rho\right)\right|^2 = \left|g\left(\frac{1}{2}+i\rho\right)\right|^2, \quad (140)$$

a true relationship that can be verified by direct computation. Set $\sigma = 1/2$ in (137) and (138) and find

$$\zeta\left(\frac{1}{2}+i\rho\right) = \frac{2f(1/2+i\rho)g(1/2+i\rho)}{f(1/2+i\rho)+g(1/2+i\rho)}C(\rho) = \frac{2if(1/2+i\rho)g(1/2+i\rho)}{f(1/2+i\rho)-g(1/2+i\rho)}S(\rho). \quad (141)$$

Equation (141) can be used to obtain a relationship between the real and imaginary components of $\zeta(1/2+i\rho)$. Let

$$\zeta\left(\frac{1}{2}+i\rho\right) = \zeta_R\left(\frac{1}{2}+i\rho\right) + i\zeta_I\left(\frac{1}{2}+i\rho\right). \quad (142)$$

Then, by breaking $f(1/2+i\rho)$ and $g(1/2+i\rho)$ into their real and imaginary parts, (141) reduces to

$$\frac{\zeta_R(1/2+i\rho)}{\zeta_I(1/2+i\rho)} = \frac{C_m(\rho)\cos(\rho\ln(2\pi)) - C_p(\rho)\sin(\rho\ln(2\pi))}{C_p(\rho)\cos(\rho\ln(2\pi)) + C_m(\rho)\sin(\rho\ln(2\pi)) - \sqrt{\pi}}, \quad (143)$$

where

$$C_p(\rho) = \Gamma_R\left(\frac{1}{2}+i\rho\right)\cosh\left(\frac{\pi\rho}{2}\right) + \Gamma_I\left(\frac{1}{2}+i\rho\right)\sinh\left(\frac{\pi\rho}{2}\right), \quad (144)$$

$$C_m(\rho) = \Gamma_I\left(\frac{1}{2}+i\rho\right)\cosh\left(\frac{\pi\rho}{2}\right) - \Gamma_R\left(\frac{1}{2}+i\rho\right)\sinh\left(\frac{\pi\rho}{2}\right), \quad (145)$$

in terms of the real and imaginary parts of $\Gamma(1/2+i\rho)$. This is a known result [41] that depends only on the functional equation (13) (See the Appendix for further discussion on this point). In [41], it was shown that the numerator and denominator on the right-hand side of (11) can never vanish simultaneously, so this result defines the half-zeros (only one of $\zeta_R(1/2+i\rho)$ or $\zeta_I(1/2+i\rho)$ vanishes) (and usually brackets the full-zeros) of $\zeta(1/2+i\rho)$ according to whether the numerator or denominator vanishes. This was studied in [41]. The full-zeros ($\zeta_R(1/2+i\rho_0) = \zeta_I(1/2+i\rho_0) = 0$) of $\zeta(1/2+i\rho)$ occur when the following condition

$$S(\rho_0) = 0, \qquad C(\rho_0) = 0 \quad (146)$$

is simultaneously satisfied for some value(s) of $\rho = \rho_0$. In that case, the defining equation (143) becomes

$$\frac{\zeta_R(1/2+i\rho)'}{\zeta_I(1/2+i\rho)'} = -\frac{\zeta'(1/2+i\rho)_I}{\zeta'(1/2+i\rho)_R} = \left(C_m(\rho)\cos(\rho\ln(2\pi)) - C_p(\rho)\sin(\rho\ln(2\pi))\right) \times \left(C_p(\rho)\cos(\rho\ln(2\pi)) + C_m(\rho)\sin(\rho\ln(2\pi)) - \sqrt{\pi}\right)^{-1}, \quad (147)$$

where the derivatives are taken with respect to the variable ρ, and (147) is evaluated at $\rho = \rho_0$. Many solutions to (146) are known [42], so there is no doubt that these two functions can simultaneously vanish at certain values of ρ. (A proof that this cannot occur in (137) and (138) when $\sigma \neq 1/2$ is left as an exercise for the reader.)

A further interesting representation can be obtained inside the critical strip by setting $c = \sigma$ in (9) and separating $\eta(s)$ into its real and imaginary components:

$$\eta_R(\sigma+\rho i) = \frac{1}{2}\sigma^{(1-\sigma)} \int_{-\infty}^{\infty} \frac{\left(1+t^2\right)^{-\sigma/2} e^{\rho\arctan(t)} \left(\cos(\Omega(t))\sin(\pi\sigma)\cosh(\pi\sigma t) - \sin(\Omega(t))\cos(\pi\sigma)\sinh(\pi\sigma t)\right)}{\cosh^2(\pi\sigma t) - \cos^2(\pi\sigma)} dt, \tag{148}$$

$$\eta_I(\sigma+\rho i) = -\frac{1}{2}\sigma^{(1-\sigma)} \int_{-\infty}^{\infty} \frac{\left(1+t^2\right)^{-\sigma/2} e^{\rho\arctan(t)} \left(\sin(\Omega(t))\sin(\pi\sigma)\cosh(\pi\sigma t) + \cos(\Omega(t))\cos(\pi\sigma)\sinh(\pi\sigma t)\right)}{\cosh^2(\pi\sigma t) - \cos^2(\pi\sigma)} dt \tag{149}$$

where

$$\Omega(t) = \frac{1}{2}\rho\log\left(\sigma^2\left(1+t^2\right)\right) + \sigma\arctan(t). \tag{150}$$

In analogy to (137) and (138), the representations (148) and (149) include a term $\cos(\pi\sigma)$ that vanishes on the critical line $\sigma = 1/2$, both reducing to the equivalent of (26) in that case. (I am speculating here that such forms contain information about the distribution and/or existence of zeros of $\eta(s)$ and hence $\zeta(s)$ inside the critical strip.) The integrands in both representations vanish as $\exp(-\pi\sigma t)$ as $t \to \infty$ and are finite at $t = 0$; most of the numerical contribution to the integrals comes from the region $0 < t \cong \rho$. Numerical evaluation in the range $0 \le t$ can be acclerated by writing

$$e^{\rho\arctan(t)} \longrightarrow e^{\rho(\arctan(t)-\pi/2)}, \tag{151}$$

and extracting a factor $e^{\rho\pi/2}$ outside the integral in that range. It remains to be shown how the remaining integral conspires to cancel this factor in the asymptotic range $\rho \to \infty$. Finally, (148) is invariant and (149) changes sign under the transformations $\rho \to -\rho$ and $t \to -t$, demonstrating that $\eta(s)$ is self-conjugate.

10. Comments

A number of related results that are not easily categorized and which either do not appear in the reference literature or appear incorrectly should be noted.

Comparison of (49) and (54) suggests the identification $v = i$. Such sums and integrals have been studied [30, 43] for the case $s = -n$ in terms of Barnes double Gamma function $G(z)$ and Clausen's function defined by

$$\mathrm{Cl}_2(z) = \sum_{k=1}^{\infty} \frac{\sin(zk)}{k^2}, \tag{152}$$

with the (tacit) restriction that $z \in \Re$. That is not the case here. Thus straightforward application of that analysis will sometimes lead to incorrect results. For example [30, Equation (4.3)], gives (with $n = 1$)

$$\pi\int_0^t v^n \cot(\pi v)\,dv = t\log(2\pi) + \log\left(\frac{G(1-t)}{G(1+t)}\right), \tag{153}$$

further ([43, Equation (13)]) identifying

$$\log\left(\frac{G(1-t)}{G(1+t)}\right) = t\log\left(\frac{\sin(\pi t)}{\pi}\right) + \frac{1}{2\pi}\mathrm{Cl}_2(2\pi t), \quad 0 < t < 1. \tag{154}$$

To allow for the possibility that $t \in \mathfrak{C}$, this should instead read

$$\log\left(\frac{G(1-t)}{G(1+t)}\right) = t\log\left(\frac{\sin(\pi t)}{\pi}\right) + \frac{i\pi}{2}B_2(t) + \frac{e^{-i\pi/2}}{2\pi}\mathrm{Li}_2\left(e^{2\pi i t}\right), \tag{155}$$

which is valid for all t in terms of Bernoulli polynomials $B_n(t)$ and polylogarithm functions. However, similar results are given in the equivalent, general form in [20, Equation (5.12)]. Interestingly, both the Maple and Mathematica computer codes somehow (neither [44] nor references contained therein, both of which study such integrals, list the computer results, which are likely based on [20, Equation (5.12)]) manage to evaluate the integral in (153) in terms similar to (155) for a variety of values of n provided that the user restricts $|t| < 1$, although the right-hand side provides the analytic continuation of the left when that restriction is lifted. This has application in the evaluation of $\zeta(3)$ according to the method of Section 8.3, where, using the methods of [10, Chapter 3], we identify

$$\sum_{k=1}^{\infty} \frac{\zeta(2k)\,t^{(2k)}}{k+1} = \frac{1}{2} - \frac{\pi}{t^2}\int_0^t v^2\cot(\pi v)\,dv, \tag{156}$$

a generalization of integrals studied in [44]. From the computer evaluation of the integral, (156) eventually yields

$$\sum_{k=1}^{\infty} \frac{\zeta(2k)\,t^{2k}}{k+1} = \frac{1}{3}i\pi t + \frac{1}{2} - \log\left(1 - e^{2i\pi t}\right) + i\frac{\mathrm{Li}_2\left(e^{2i\pi t}\right)}{\pi t} - \frac{\mathrm{Li}_3\left(e^{2i\pi t}\right) - \zeta(3)}{2\pi^2 t^2}, \tag{157}$$

as a new addition to the lists found in [10, Section (3.4)], and [11, Sections (3.3) and (3.4)], and a generalization of (97) when $s = 3$.

Appendix

In [41], an attempt was made to show that the functional equation (13) and a minimal number of analytic properties of $\zeta(s)$ might suffice to show that $\zeta(s) \neq 0$ when $\sigma \neq 1/2$. This was motivated by a comment attributed to H. Hamburger ([8, page 50]) to the effect that the functional equation for $\zeta(s)$ perhaps characterizes it completely. (An unequivocal equivalent claim can also be found in [15]. As demonstrated by the counterexample given here, and in [45], this claim is false.) The purported demonstration contradicts a result of Gauthier and Zeron [45], who show that infinitely many functions exist, arbitrarily close to $\zeta(s)$, that satisfy (13) and the same analytic properties that are invoked in [41], but which vanish when $\sigma \neq 1/2$. (To quote Prof. Gauthier: "We prove the existence of functions approximating the Riemann zeta-function, satisfying the Riemann functional equation and violating the analogue of the Riemann hypothesis." (private communication).)

In fact, an example of such a function is easily found. Let

$$\begin{aligned} \nu(s) &= C \sin(\pi(s - s_0)) \sin(\pi(s + s_0)) \\ &\quad \times \sin(\pi(s - s_0^*)) \sin(\pi(s + s_0^*)), \end{aligned} \tag{A.1}$$

where s_0^* is the complex conjugate of s_0, $s_0 = \sigma_0 + i\rho_0$ is arbitrary complex, and C, given by

$$C = \frac{1}{\left(\cosh^2(\pi\rho_0) - \cos^2(\pi\sigma_0)\right)^2}, \tag{A.2}$$

is a constant chosen such that $\nu(1) = 1$. Then

$$\zeta_\nu(s) \equiv \nu(s)\zeta(s) \tag{A.3}$$

satisfies analyticity conditions (1–3) of [45] (which include the functional equation (13)). Additionally, $\zeta_\nu(s)$ has zeros at the zeros of $\zeta(s)$ as well as complex zeros at $s = \pm(n + s_0)$ and $s = \pm(n + s_0^*)$ and a pole at $s = 1$ with residue equal to unity, where $n = 0, \pm1, \pm2, \ldots$.

Thus the last few paragraphs of Section 2 of [41] beginning with the words "There are two cases where (2.9) fails..." cannot be valid. However, the remainder of that paper, particularly those sections dealing with the properties of $\zeta(1/2 + i\rho)$ as referenced here in Section 9, are valid and correct. I am grateful to Professor Gauthier for discussions on this topic.

Acknowledgments

The author wishes to thank Vinicius Anghel and Larry Glasser for commenting on a draft copy of this paper. Leslie M. Saunders obituary may be found in the journal Physics Today, issue of Dec. 1972, page 63, http://www.physicstoday.org/resource/1/phtoad/v25/i12/p63_s2?bypassSSO=1.

References

[1] F. W. J. Olver, D. W. Lozier, R. F. Boisvert, and C. W. Clark, *NIST Handbook of Mathematical Functions*, Cambridge University Press, Cambridge, UK, 2010, http://dlmf.nist.gov/25.5.

[2] M. Abramowitz and I. Stegun, *Handbook of Mathematical Functions*, Dover, New York, NY, USA, 1964.

[3] T. M. Apostol, *Introduction to Analytic Number Theory*, Springer, New York, NY, USA, 2010.

[4] F. W. J. Olver, *Asymptotics and Special Functions*, Academic Press, New York, NY, USA, 1974, Computer Science and Applied Mathematics.

[5] E. C. Titchmarsh, *The Theory of Functions*, Oxford University Press, New York, NY, USA, 2nd edition, 1949.

[6] E. T. Whittaker and G. N. Watson, *A Course of Modern Analysis*, Cambridge Mathematical Library, Cambridge University Press, Cambridge, UK, 1963.

[7] H. M. Edwards, *Riemann's Zeta Function*, Academic Press, New York, NY, USA, 1974.

[8] A. Ivić, *The Riemann Zeta-Function: Theory and Applications*, Dover, New York, NY, USA, 1985.

[9] S. J. Patterson, *An introduction to the Theory of the Riemann Zeta-Function*, vol. 14 of *Cambridge Studies in Advanced Mathematics*, Cambridge University Press, Cambridge, UK, 1988.

[10] H. M. Srivastava and J. Choi, *Series Associated with the Zeta and Related Functions*, Kluwer Academic, Dordrecht, The Netherlands, 2001.

[11] H. M. Srivastava and J. Choi, *Zeta and q-Zeta Functions and Associated Series and Integrals*, Elsevier, New York, NY, USA, 2012.

[12] A. Erdélyi, W. Magnus, F. Oberhettinger, and F. G. Tricomi, Eds., *Higher Transcendental Functions*, vol. 1–3, McGraw-Hill, New York, NY, USA, 1953.

[13] I. S. Gradshteyn and I. M. Ryzhik, *Tables of Integrals, Series and Products*, Academic Press, New York, NY, USA, 1980.

[14] A. P. Prudnikov, Yu. A. Brychkov, and O. I. Marichev, *Integrals and Series: Volume 3: More Special Functions*, Gordon and Breach Science, New York, NY, USA, 1990.

[15] Encyclopedia of mathematics, http://www.encyclopediaofmath.org/index.php/Zeta-function.

[16] http://en.wikipedia.org/wiki/Riemann_zeta_function.

[17] http://en.wikipedia.org/wiki/Dirichlet_eta_function.

[18] Mathworld, http://mathworld.wolfram.com/HankelFunction.html.

[19] Mathworld, http://mathworld.wolfram.com/RiemannZeta-Function.html.

[20] H. M. Srivastava, M. L. Glasser, and V. S. Adamchik, "Some definite integrals associated with the Riemann zeta function," *Zeitschrift fur Analysis und ihre Anwendung*, vol. 19, no. 3, pp. 831–846, 2000.

[21] P. Flajolet and B. Salvy, "Euler sums and contour integral representations," *Experimental Mathematics*, vol. 7, no. 1, pp. 15–35, 1998.

[22] R. G. Newton, *The Complex J-Plane*, The Mathematical Physics Monograph Series, W. A. Benjamin, New York, NY, USA, 1964.

[23] J. L. W. V. Jensen, *L'Intermédiaire des mathématiciens*, vol. 2, 1895.

[24] E. Lindelof, *Le Calcul Des Residus et ses Applications a La Theorie Des Fonctions*, Gauthier-Villars, Paris, France, 1905.

[25] S. N. M. Ruijsenaars, "Special functions defined by analytic difference equations," in *Special Functions 2000: Current Perspective and Future Directions (Tempe, AZ)*, J. Bustoz, M. Ismail, and S. Suslov, Eds., vol. 30 of *NATO Science Series*, pp. 281–333, Kluwer Academic, Dordrecht, The Netherlands, 2001.

[26] B. Riemann, "Über die Anzahl der Primzahlen unter einer gegebenen Grösse," 1859.

[27] T. Amdeberhan, M. L. Glasser, M. C. Jones, V. Moll, R. Posey, and D. Varela, "The Cauchy-Schlömilch transformation," http://arxiv.org/abs/1004.2445.

[28] E. R. Hansen, *Table of Series and Products*, Prentice Hall, New York, NY, USA, 1975.

[29] R. E. Crandall, *Topics in Advanced Scientific Computation*, Springer, New York, NY, USA, 1996.

[30] J. Choi, H. M. Srivastava, and V. S. Adamchick, "Multiple gamma and related functions," *Applied Mathematics and Computation*, vol. 134, no. 2-3, pp. 515–533, 2003.

[31] S. Tyagi and C. Holm, "A new integral representation for the Riemann zeta function," http://arxiv.org/abs/math-ph/0703073.

[32] N. G. de Bruijn, "Mathematica," *Zutphen*, vol. B5, pp. 170–180, 1937.

[33] R. B. Paris and D. Kaminski, *Asymptotics and Mellin-Barnes Integrals*, vol. 85 of *Encyclopedia of Mathematics and Its Applications*, Cambridge University Press, Cambridge, UK, 2001.

[34] Y. L. Luke, *The Special Functions and Their Approximations*, vol. 1, Academic Press, New York, NY, USA, 1969.

[35] M. S. Milgram, "The generalized integro-exponential function," *Mathematics of Computation*, vol. 44, no. 170, pp. 443–458, 1985.

[36] N. N. Lebedev, *Special Functions and Their Applications*, Prentice Hall, New York, NY, USA, 1965.

[37] T. M. Dunster, "Uniform asymptotic approximations for incomplete Riemann zeta functions," *Journal of Computational and Applied Mathematics*, vol. 190, no. 1-2, pp. 339–353, 2006.

[38] V. S. Adamchik and H. M. Srivastava, "Some series of the zeta and related functions," *Analysis*, vol. 18, no. 2, pp. 131–144, 1998.

[39] M. S. Milgram, "Notes on a paper of Tyagi and Holm: 'a new integral representation for the Riemann Zeta function'," http://arxiv.org/abs/0710.0037v1.

[40] E. Y. Deeba and D. M. Rodriguez, "Stirling's series and Bernoulli numbers," *American Mathematical Monthly*, vol. 98, no. 5, pp. 423–426, 1991.

[41] M. S. Milgram, "Notes on the zeros of Riemann's zeta Function," http://arxiv.org/abs/0911.1332.

[42] A. M. Odlyzko, "The 10^{22}-nd zero of the Riemann zeta function," in *Dynamical, Spectral, and Arithmetic Zeta Functions (San Antonio, TX, 1999)*, vol. 290 of *Contemporary Mathematics Series*, pp. 139–144, American Mathematical Society, Providence, RI, USA, 2001.

[43] V. S. Adamchik, "Contributions to the theory of the barnes function, computer physics communication," http://arxiv.org/abs/math/0308086.

[44] D. Cvijović and H. M. Srivastava, "Evaluations of some classes of the trigonometric moment integrals," *Journal of Mathematical Analysis and Applications*, vol. 351, no. 1, pp. 244–256, 2009.

[45] P. M. Gauthier and E. S. Zeron, "Small perturbations of the Riemann zeta function and their zeros," *Computational Methods and Function Theory*, vol. 4, no. 1, pp. 143–150, 2004, http://www.heldermann-verlag.de/cmf/cmf04/cmf0411.pdf.

11

New Generalized Hyperbolic Functions to Find New Exact Solutions of the Nonlinear Partial Differential Equations

Yusuf Pandir and Halime Ulusoy

Department of Mathematics, Faculty of Science, Bozok University, 66100 Yozgat, Turkey

Correspondence should be addressed to Yusuf Pandir; yusufpandir@gmail.com

Academic Editor: Anjan Biswas

We firstly give some new functions called generalized hyperbolic functions. By the using of the generalized hyperbolic functions, new kinds of transformations are defined to discover the exact approximate solutions of nonlinear partial differential equations. Based on the generalized hyperbolic function transformation of the generalized KdV equation and the coupled equal width wave equations (CEWE), we find new exact solutions of two equations and analyze the properties of them by taking different parameter values of the generalized hyperbolic functions. We think that these solutions are very important to explain some physical phenomena.

1. Introduction

To find exact solutions to nonlinear evolution equations, some new techniques have been introduced in recent years such as the Hirota method, the extended Jacobi elliptic function expansion method, the exponential function method, (G'/G)-expansion method, the simplest equation method, the trial equation method, Kudryashov's method, and solitary ansatz method [1–19]. Also some authors define new functions which are named generalized hyperbolic functions for constructing new solutions [20–22]. There are a lot of nonlinear evolution equations that are integrated using the various mathematical methods. Soliton solutions, compactons, singular solitons, and other solutions have been found by using these approaches. These types of solutions are very important and appear in various areas of applied mathematics.

In Section 2, we give the definition and properties of generalized hyperbolic functions. In Section 3, as applications, we obtain exact solutions of the generalized KdV equation and the coupled equal width wave equations.

2. The Definition and Properties of the Symmetrical Hyperbolic Fibonacci and Lucas Functions

In this section, we will define new functions which are named the symmetrical hyperbolic Fibonacci and Lucas functions for constructing new exact solutions of NPDEs and then study the properties of these functions.

Definition 1. Suppose that ξ is an independent variable; p, q, and k are all constants. The generalized hyperbolic sine function is

$$\sinh_a(\xi) = \frac{pa^{k\xi} - qa^{-k\xi}}{2}, \tag{1}$$

generalized hyperbolic cosine function is

$$\cosh_a(\xi) = \frac{pa^{k\xi} + qa^{-k\xi}}{2}, \tag{2}$$

generalized hyperbolic tangent function is

$$\tanh_a(\xi) = \frac{pa^{k\xi} - qa^{-k\xi}}{pa^{k\xi} + qa^{-k\xi}}, \tag{3}$$

generalized hyperbolic cotangent function is

$$\coth_a(\xi) = \frac{pa^{k\xi} + qa^{-k\xi}}{pa^{k\xi} - qa^{-k\xi}}, \quad (4)$$

generalized hyperbolic secant function is

$$\mathrm{sech}_a(\xi) = \frac{2}{pa^{k\xi} + qa^{-k\xi}}, \quad (5)$$

generalized hyperbolic cosecant function is

$$\mathrm{cosech}_a(\xi) = \frac{2}{pa^{k\xi} - qa^{-k\xi}}, \quad (6)$$

and the above six kinds of functions are said to be generalized new hyperbolic functions. Thus we can prove the following theory of generalized hyperbolic functions on the basis of Definition 1.

Theorem 2. *The generalized hyperbolic functions satisfy the following relations:*

$$\begin{aligned}
\cosh_a^2(\xi) - \sinh_a^2(\xi) &= pq, \\
1 - \tanh_a^2(\xi) &= pq\,\mathrm{sech}_a^2(\xi), \\
1 - \coth_a^2(\xi) &= -pq\,\mathrm{cosech}_a^2(\xi), \\
\mathrm{sech}_a(\xi) &= \frac{1}{\cosh_a(\xi)}, \\
\mathrm{cosech}_a(\xi) &= \frac{1}{\sinh_a(\xi)}, \\
\tanh_a(\xi) &= \frac{\sinh_a(\xi)}{\cosh_a(\xi)}, \\
\coth_a(\xi) &= \frac{\cosh_a(\xi)}{\sinh_a(\xi)}.
\end{aligned} \quad (7)$$

The following just part of them are proved here for simplification.

Theorem 3. *The derivative formulae of generalized hyperbolic functions are as follows:*

$$\begin{aligned}
\frac{d(\sinh_a(\xi))}{d\xi} &= k\ln a\cosh_a(\xi), \\
\frac{d(\cosh_a(\xi))}{d\xi} &= k\ln a\sinh_a(\xi), \\
\frac{d(\tanh_a(\xi))}{d\xi} &= kpq\ln a\,\mathrm{sech}_a^2(\xi), \\
\frac{d(\coth_a(\xi))}{d\xi} &= -kpq\ln a\,\mathrm{cosech}_a^2(\xi), \\
\frac{d(\mathrm{sech}_a(\xi))}{d\xi} &= -k\ln a\,\mathrm{sech}_a(\xi)\tanh_a(\xi), \\
\frac{d(\mathrm{cosech}_a(\xi))}{d\xi} &= -k\ln a\,\mathrm{cosech}_a(\xi)\coth_a(\xi).
\end{aligned} \quad (8)$$

Proof of (8). According to (7) and (8), we can get

$$\begin{aligned}
\frac{d(\tanh_a(\xi))}{d\xi} &= \left(\frac{\sinh_a(\xi)}{\cosh_a(\xi)}\right)' \\
&= \frac{(\sinh_a(\xi))'\cosh_a(\xi) - (\cosh_a(\xi))'\sinh_a(\xi)}{\cosh_a^2(\xi)} \\
&= \frac{k\ln a\cosh_a^2(\xi) - k\ln a\sinh_a^2(\xi)}{\cosh_a^2(\xi)} \\
&= kpq\ln a\,\mathrm{sech}_a^2(\xi).
\end{aligned} \quad (9)$$

Similarly, we can prove other differential coefficient formulae in Theorem 3. □

Remark 4. We see that when $p = 1$, $q = 1$, $k = 1$, and $a = e$ in (1)–(6), new generalized hyperbolic functions $\sinh_a(\xi)$, $\cosh_a(\xi)$, $\tanh_a(\xi)$, $\coth_a(\xi)$, $\mathrm{sech}_a(\xi)$, and $\mathrm{cosech}_a(\xi)$ degenerate as hyperbolic functions $\sinh(\xi)$, $\cosh(\xi)$, $\tanh(\xi)$, $\coth(\xi)$, $\mathrm{sech}(\xi)$, and $\mathrm{cosech}(\xi)$, respectively. In addition, when $p = 0$ or $q = 0$ in (1)–(6), $\sinh_a(\xi)$, $\cosh_a(\xi)$, $\tanh_a(\xi)$, $\coth_a(\xi)$, $\mathrm{sech}_a(\xi)$, and $\mathrm{cosech}_a(\xi)$ degenerate as exponential function $(1/2)pa^{k(\xi)}$, $\pm(1/2)qa^{-k(\xi)}$, $2pa^{-k(\xi)}$, $\pm 2qa^{k(\xi)}$, and ± 1, respectively.

3. Applications

Example 5. Application to the generalized form of KdV equation is presented.

The generalized KdV equation that will be studied in this paper are given by [23]

$$\left(u^l\right)_t + \alpha u^n u_x + \beta(u^n u_{xx})_x + \gamma u(u^n)_{xxx} = 0, \quad (10)$$

where the first term represents the generalized evolution. The special case with $l = 1$ is the regular evolution term. The coefficients of α are the nonlinear terms while the coefficients of β and γ are the nonlinear dispersion terms. This equation with $l = 1$ has been already studied by [18], where, in addition to soliton solution, compactons and periodic solutions were also obtained. The hypothesis for solving this equation is

$$u(x,t) = \frac{A}{\cosh_a^s(\eta)}, \quad (11)$$

where

$$\eta = B(x - vt). \quad (12)$$

Here, in (11) and (12), A represents the amplitude of the soliton while B is the inverse width of the soliton and v is the velocity of the soliton. The exponent s is unknown at this point and will be evaluated during the course of the derivation of the solutions to (10). From (11), it is possible to obtain

$$\left(u^{l}\right)_{t} = \frac{klsvA^{l}B\ln a\tanh_{a}(\eta)}{\cosh_{a}^{sl}(\eta)},$$

$$u^{n}u_{x} = \frac{-ksA^{n+1}B\ln a\tanh_{a}(\eta)}{\cosh_{a}^{s(n+1)}(\eta)},$$

$$\left(u^{n}u_{xx}\right)_{x} = \frac{k^{3}sA^{n+1}B^{3}pq(\ln a)^{3}(s+1)(sn+s+2)\tanh_{a}(\eta)}{\cosh_{a}^{s(n+1)+2}(\eta)} - \frac{k^{3}s^{3}(n+1)A^{n+1}B^{3}(\ln a)^{3}\tanh_{a}(\eta)}{\cosh_{a}^{s(n+1)}(\eta)},$$

$$u\left(u^{n}\right)_{xxx} = \frac{-k^{3}s^{3}n^{3}A^{n+1}B^{3}(\ln a)^{3}\tanh_{a}(\eta)}{\cosh_{a}^{s(n+1)}(\eta)} + \frac{k^{3}snpqA^{n+1}B^{3}(\ln a)^{3}(sn+1)(sn+2)\tanh_{a}(\eta)}{\cosh_{a}^{s(n+1)+2}(\eta)}. \tag{13}$$

These results will now be substituted in (10) to obtain the 1-soliton solution of the generalized KdV equation. Equation (10) by virtue of (13) reduces to

$$\begin{aligned}&\frac{klsvA^{l}B\ln a\tanh_{a}(\eta)}{\cosh_{a}^{sl}(\eta)} - \frac{\alpha ksA^{n+1}B\ln a\tanh_{a}(\eta)}{\cosh_{a}^{s(n+1)}(\eta)} \\ &+ \frac{\beta k^{3}spqA^{n+1}B^{3}(\ln a)^{3}(s+1)(sn+s+2)\tanh_{a}(\eta)}{\cosh_{a}^{s(n+1)+2}(\eta)} \\ &- \frac{\beta k^{3}s^{3}(n+1)A^{n+1}B^{3}(\ln a)^{3}\tanh_{a}(\eta)}{\cosh_{a}^{s(n+1)}(\eta)} \\ &+ \frac{\gamma k^{3}snpqA^{n+1}B^{3}(\ln a)^{3}(sn+1)(sn+2)\tanh_{a}(\eta)}{\cosh_{a}^{s(n+1)+2}(\eta)} \\ &- \frac{\gamma k^{3}s^{3}n^{3}A^{n+1}B^{3}(\ln a)^{3}\tanh_{a}(\eta)}{\cosh_{a}^{s(n+1)}(\eta)} = 0.\end{aligned} \tag{14}$$

From (14), equating the exponents sl and $sn+s+2$ gives

$$sl = sn + s + 2, \tag{15}$$

that leads to

$$s = \frac{2}{l-n-1}. \tag{16}$$

Now from (14), the two linearly independent functions are $1/\cosh^{ns+s+j}$ for $j = 0, 2$. Thus setting their coefficients to zero gives

$$v = \frac{\alpha pq\left(\beta l(l+1-n) + n\gamma(l-1)(l+n-1)\right)}{2lA^{l-n-1}\left(\beta(n+1)+n^{3}\gamma\right)}, \tag{17}$$

$$B = \frac{\sqrt{-\alpha}(n+1-l)}{2k\ln a\sqrt{\beta(n+1)+n^{3}\gamma}}. \tag{18}$$

Thus the 1-soliton solution of the generalized KdV equation with generalized evolution is given by

$$u(x,t) = \frac{A}{\cosh_{a}^{2/(l-n-1)}[B(x-vt)]}. \tag{19}$$

This shows that the restriction on the nonlinear exponents of (10) must be

$$l > n + 1. \tag{20}$$

Remark 6. If we take the corresponding values for some parameters, the solution (19) can be reduced to the solution (17) mentioned in [23].

Example 7. Application to the coupled equal width wave equations is presented.

We consider the coupled equal width wave equations [24]

$$\begin{aligned}u_{t} + uu_{x} - u_{xxt} + rr_{x} &= 0,\\ r_{t} + rr_{x} - r_{xxt} &= 0.\end{aligned} \tag{21}$$

The hypothesis for solving these equations is

$$\begin{aligned}r(x,t) &= \frac{A_{1}}{\cosh_{a}^{s_{1}}(\eta)},\\ u(x,t) &= \frac{A_{2}}{\cosh_{a}^{s_{2}}(\eta)},\end{aligned} \tag{22}$$

where

$$\eta = B(x - vt). \tag{23}$$

Here in (22)–(23) A_1 and A_2 are the amplitudes of the r-soliton and u-soliton, respectively, while v_1 and v_2 are the velocity of the soliton and B_1 and B_2 are the inverse widths of the solitons. The exponents s_1 and s_2 are unknown at this point and their values will fall out in the process of deriving the solution of this equation. From (22), it is possible to obtain

$$u_t = \frac{ks_2 v A_2 B \ln a \tanh_a(\eta)}{\cosh_a^{s_2}(\eta)}, \quad r_t = \frac{ks_1 v A_1 B \ln a \tanh_a(\eta)}{\cosh_a^{s_1}(\eta)}, \tag{24}$$

$$u_x = \frac{-ks_2 A_2 B \ln a \tanh_a(\eta)}{\cosh_a^{s_2}(\eta)}, \quad r_x = \frac{-ks_1 A_1 B \ln a \tanh_a(\eta)}{\cosh_a^{s_1}(\eta)}, \tag{25}$$

$$u_{xxt} = \frac{k^3 s_2^3 v A_2 B^3 (\ln a)^3 \tanh_a(\eta)}{\cosh_a^{s_2}(\eta)} - \frac{k^3 s_2 pq (s_2+1) v A_2 B^3 (\ln a)^3 (s_2+2) \tanh_a(\eta)}{\cosh_a^{s_2+2}(\eta)}, \tag{26}$$

$$r_{xxt} = \frac{k^3 s_1^3 v A_1 B^3 (\ln a)^3 \tanh_a(\eta)}{\cosh_a^{s_1}(\eta)} - \frac{k^3 s_1 pq (s_1+1) v A_1 B^3 (\ln a)^3 (s_1+2) \tanh_a(\eta)}{\cosh_a^{s_1+2}(\eta)}. \tag{27}$$

Substituting these into (21) yields to

$$\frac{ks_2 v A_2 B \ln a \tanh_a(\eta)}{\cosh_a^{s_2}(\eta)} - \frac{ks_2 A_2^2 B \ln a \tanh_a(\eta)}{\cosh_a^{2s_2}(\eta)} - \frac{k^3 s_2^3 v A_2 B^3 (\ln a)^3 \tanh_a(\eta)}{\cosh_a^{s_2}(\eta)} + \frac{k^3 s_2 (s_2+1)(s_2+2) pqv A_2 B^3 (\ln a)^3 \tanh_a(\eta)}{\cosh_a^{s_2+2}(\eta)} - \frac{ks_1 A_1^2 B \ln a \tanh_a(\eta)}{\cosh_a^{2s_1}(\eta)} = 0, \tag{28}$$

$$\frac{ks_1 v A_1 B \ln a \tanh_a(\eta)}{\cosh_a^{s_1}(\eta)} - \frac{ks_1 A_1^2 B \ln a \tanh_a(\eta)}{\cosh_a^{2s_1}(\eta)} - \frac{k^3 s_1^3 v A_1 B^3 (\ln a)^3 \tanh_a(\eta)}{\cosh_a^{s_1}(\eta)} + \frac{k^3 s_1 (s_1+1)(s_1+2) vpq A_1 B^3 (\ln a)^3 \tanh_a(\eta)}{\cosh_a^{s_1+2}(\eta)} = 0. \tag{29}$$

Now from (28) and (29), equating the exponents $2s_1$ and s_1+2 and also $2s_2$ and s_2+2 gives, respectively,

$$2s_1 = s_1 + 2, \quad 2s_2 = s_2 + 2, \tag{30}$$

that leads to

$$s_1 = s_2 = 2. \tag{31}$$

So from (28) and (29), the four linearly independent functions are $1/\cosh^{s_1+j}$ and $1/\cosh^{s_2+j}$ for $j = 0, 2$. Therefore, setting their respective coefficients to zero we obtain A_2, B, and v as follow.

Case 1. One has

$$A_2 = \frac{A_1}{2}\left(1 \pm i\sqrt{3}\right), \quad B = -\frac{1}{2k \ln a}, \quad v = \frac{A_1}{3pq}, \tag{32}$$

where $A_1, k, p,$ and q are arbitrary constants. Substituting (32) into (22), we obtain new exact solution to (21),

$$r(x,t) = \frac{A_1}{\cosh_a^2(\eta)}, \quad u(x,t) = \frac{A_2}{\cosh_a^2(\eta)}, \tag{33}$$

where $\eta = -(1/2k \ln a)(x - (A_1/3pq)t)$.

Case 2. One has

$$A_2 = \frac{A_1}{2}\left(1 \pm i\sqrt{3}\right), \quad B = \frac{1}{2k \ln a}, \quad v = \frac{A_1}{3pq}, \tag{34}$$

where $A_1, k, p,$ and q are arbitrary constants. Substituting (34) into (22), we obtain new exact solution to (21),

$$r(x,t) = \frac{A_1}{\cosh_a^2(\eta)}, \quad u(x,t) = \frac{A_2}{\cosh_a^2(\eta)}, \tag{35}$$

where $\eta = (1/2k \ln a)(x - (A_1/3pq)t)$.

Remark 8. If we search the corresponding values for some parameters and make some transformations, in particular the solution (35) can be reduced to the solution (32) mentioned in [24].

4. Conclusions and Remarks

We consider generalized hyperbolic functions and new kinds of generalized hyperbolic function transformation to construct new exact solutions of nonlinear partial differential equations. This paper obtains 1-soliton solution to the generalized KdV equation and the coupled equal width wave equations. Our methods also can be applied to construct new exact solutions of other nonlinear partial differential equations.

References

[1] R. Hirota, "Exact N-soliton solutions of the wave equation of long waves in shallow-water and in nonlinear lattices," *Journal of Mathematical Physics*, vol. 14, no. 7, pp. 810–814, 1973.

[2] Y. Gurefe, Y. Pandir, and E. Misirli, "New exact solutions of stochastic KdV equation," *Applied Mathematical Sciences*, vol. 6, no. 65-68, pp. 3225–3234, 2012.

[3] X. Wen, "Extended Jacobi elliptic function expansion solutions of variant Boussinesq equations," *Applied Mathematics and Computation*, vol. 217, no. 6, pp. 2808–2820, 2010.

[4] Y. Gurefe and E. Misirli, "Exp-function method for solving nonlinear evolution equations with higher order nonlinearity," *Computers and Mathematics with Applications*, vol. 61, no. 8, pp. 2025–2030, 2011.

[5] G. Ebadi and A. Biswas, "The (G'/G) method and 1-soliton solution of the Davey-Stewartson equation," *Mathematical and Computer Modelling*, vol. 53, no. 5-6, pp. 694–698, 2011.

[6] Y. Gurefe and E. Misirli, "New variable separation solutions of two-dimensional Burgers system," *Applied Mathematics and Computation*, vol. 217, no. 22, pp. 9189–9197, 2011.

[7] N. A. Kudryashov, "Simplest equation method to look for exact solutions of nonlinear differential equations," *Chaos, Solitons and Fractals*, vol. 24, no. 5, pp. 1217–1231, 2005.

[8] N. K. Vitanov, "Modified method of simplest equation: Powerful tool for obtaining exact and approximate traveling-wave solutions of nonlinear PDEs," *Communications in Nonlinear Science and Numerical Simulation*, vol. 16, no. 3, pp. 1176–1185, 2011.

[9] C. S. Liu, "Applications of complete discrimination system for polynomial for classifications of traveling wave solutions to nonlinear differential equations," *Computer Physics Communications*, vol. 181, no. 2, pp. 317–324, 2010.

[10] Y. Gurefe, A. Sonmezoglu, and E. Misirli, "Application of an irrational trial equation method to high-dimensional nonlinear evolution equations," *Journal of Advanced Mathematical Studies*, vol. 5, no. 1, pp. 41–47, 2012.

[11] Y. Pandir, Y. Gurefe, U. Kadak, and E. Misirli, "Classifications of exact solutions for some nonlinear partial differential equations with generalized evolution," *Abstract and Applied Analysis*, vol. 2012, Article ID 478531, 16 pages, 2012.

[12] Y. Xie and J. Tang, "The superposition method in seeking the solitary wave solutions to the KdV-Burgers equation," *Pramana*, vol. 66, no. 3, pp. 479–483, 2006.

[13] N. A. Kudryashov, "One method for finding exact solutions of nonlinear differential equations," *Communications in Nonlinear Science and Numerical Simulation*, vol. 17, no. 6, pp. 2248–2253, 2012.

[14] Y. Pandir, Y. Gurefe, and E. Misirli, "A new approach to Kudryashovs method for solving some nonlinear physical models," *International Journal of Physical Sciences*, vol. 7, pp. 2860–2866, 2012.

[15] N. A. Kudryashov and S. G. Prilipko, "Exact solutions of the generalized K(m,m) equations," *Communications in Nonlinear Science and Numerical Simulation*, vol. 16, no. 3, pp. 1107–1113, 2011.

[16] G. Ebadi and A. Biswas, "Application of the (G'/G)-expansion method for nonlinear diffusion equations with nonlinear source," *Journal of the Franklin Institute*, vol. 347, no. 7, pp. 1391–1398, 2010.

[17] G. Ebadi and A. Biswas, "The (G'/G) method and topological soliton solution of the K(m,n) equation," *Communications in Nonlinear Science and Numerical Simulation*, vol. 16, no. 6, pp. 2377–2382, 2011.

[18] G. Ebadi, A. H. Kara, M. D. Petkovic, A. Yildirim, and A. Biswas, "1-Soliton solution of the generalized KdV equation with generalized evolution," *Proceedings of the Romanian Academy A*, vol. 13, pp. 215–224, 2012.

[19] A. Biswas, M. D. Petković, and D. Milović, "Topological exact soliton solution of the power law KdV equation," *Applied Mathematics and Computation*, vol. 217, no. 4, pp. 1780–1784, 2010.

[20] Y. J. Ren and H. Q. Zhang, "New generalized hyperbolic functions and auto-Bäcklund transformation to find new exact solutions of the (2+1)-dimensional NNV equation," *Physics Letters A*, vol. 357, no. 6, pp. 438–448, 2006.

[21] E. Yomba, "Generalized hyperbolic functions to find soliton-like solutions for a system of coupled nonlinear Schrödinger equations," *Physics Letters A*, vol. 372, no. 10, pp. 1612–1618, 2008.

[22] Z. I. A. Al-Muhiameed and E. A.-B. Abdel-Salam, "Generalized hyperbolic function solution to a class of nonlinear Schrödinger-type equations," *Journal of Applied Mathematics*, vol. 2012, Article ID 265348, 15 pages, 2012.

[23] M. S. Ismail and A. Biswas, "1-Soliton solution of the generalized KdV equation with generalized evolution," *Applied Mathematics and Computation*, vol. 216, no. 5, pp. 1673–1679, 2010.

[24] A. H. A. Ali, A. A. Soliman, and K. R. Raslan, "Soliton solution for nonlinear partial differential equations by cosine-function method," *Physics Letters A*, vol. 368, no. 3-4, pp. 299–304, 2007.

12

Existence Results for Vector Mixed Quasi-Complementarity Problems

Suhel Ahmad Khan[1] and Naeem Ahmad[2]

[1] *Department of Mathematics, BITS-Pilani, Dubai Campus, P.O. Box 345055, Dubai, UAE*
[2] *Department of Mathematics, Al-Jouf University, P.O. Box 2014, Skaka, Saudi Arabia*

Correspondence should be addressed to Suhel Ahmad Khan; khan.math@gmail.com

Academic Editor: Kaleem R. Kazmi

We introduce strong vector mixed quasi-complementarity problems and the corresponding strong vector mixed quasi-variational inequality problems. We establish equivalence between strong mixed quasi-complementarity problems and strong mixed quasi-variational inequality problem in Banach spaces. Further, using KKM-Fan lemma, we prove the existence of solutions of these problems, under pseudomonotonicity assumption. The results presented in this paper are extensions and improvements of some earlier and recent results in the literature.

1. Introduction

In 1980, Giannessi [1] introduced vector variational inequalities in a finite-dimensional Euclidean space. Motivated by Giannessi [1], Chen and Cheng [2] studied vector variational inequalities in infinite-dimensional Euclidean space and applied them to the vector optimization problems. Since then, vector variational inequalities and their generalizations have been studied and applied to vector optimization problems, vector complementarity problems, game theory, and so forth; see, for example, [1–20] and references therein. It is well known that the complementarity problems are closely related to variational inequality problems. Complementarity theory is introduced by Lemke [21] and Cottle and Dantzig [5]. It has emerged as an active and interesting field for researcher with wide range of applications in pure and applied sciences. Complementarity problems have been extended and generalized in various directions to study a large class of problems arising in industry, finance, optimization, physical, mathematical and engineering sciences; see, for example [4–12, 14, 15, 20]. Recently, vector complementarity problems and their relations with vector variational inequality problems have been investigated under pseudomonotone-type conditions and positiveness-type conditions; see, for example [6, 8–10, 20]. However, to the best of our knowledge, only a few existence results on the strong version of the vector variational inequality and vector complementarity problems were established.

Recently, Huang et al. [12] discussed equivalence results among a vector complementarity problem, a vector variational inequality problem, a vector optimization problem, and weak minimal element problem, under some monotonicity conditions and some inclusive-type conditions in ordered Banach spaces. In 2005, Huang and Fang [9] introduced several classes of strong vector F-complementarity problems and give some existence results for these problems in Banach spaces and discussed the least element problems of feasible sets and presented their relations with the strong vector F-complementarity problems.

Very recently, Khan [22] introduced and studied a generalized vector implicit Quasi-Complementarity problem and generalized vector implicit quasi variational inequality problem. He investigated the nonemptiness and closedness of solution sets of these problems and proved that solution sets of both the problems are equivalent to each other under some suitable conditions.

Inspired and motivated by the work going in this direction, in this paper we introduce and analyze a new class of strong vector Quasi-Complementarity problem and the corresponding strong vector mixed quasi variational inequality problem in the setting of Banach space and establish equivalence results between them. By using the KKM-Fan lemma, we derive the existence of solutions of strong vector mixed quasi variational inequalities under pseudomonotonicity assumption and show that the solution of the strong vector mixed quasi variational inequality is equivalent to the solution of strong vector mixed Quasi-Complementarity problems under suitable conditions. The results presented in this paper are the generalization and improvement of existing works of [6, 7, 9, 11, 15].

2. Preliminaries

Throughout this paper unless otherwise stated let X and Y be two real Banach spaces. Let K be a nonempty, closed, convex subset of a real Banach space X. A nonempty subset $P \subset Y$ is called convex, pointed, connected, and reproduced cone, respectively, if it satisfies the following conditions: (i) $\lambda P \subseteq P$, for all $\lambda > 0$ and $P + P \subseteq P$; (ii) $P \cap -P = \{0\}$; (iii) $P \cup -P = X$; (iv) $P - P = X$.

Given P in Y, we can define the relations "$\leq_P$" and "$\nleq_P$" as follows:

$$x \leq_P y \Longleftrightarrow y - x \in P,$$
$$x \nleq_P y \Longleftrightarrow y - x \notin P, \quad \forall x, y \in Y. \tag{1}$$

If "$\leq_P$" is a partial order, then $(Y, \leq_P)$ is called a Banach space ordered by P. Let $L(X, Y)$ denote the space of all continuous linear mappings from X into Y.

Now, we recall the following concepts and results needed in this paper.

Definition 1. A mapping $f : K \times K \to Y$ is said to be *P-convex* in first argument, if

$$f(tx + (1-t)y, z) \leq_P t f(x, z) + (1-t) f(y, z), \quad \forall x, y, z \in K, \quad t \in [0, 1]. \tag{2}$$

Definition 2. Let $T : K \to L(X, Y)$ and $F : K \times K \to Y$ be the two nonlinear mappings. T is said to be *monotone with respect to F* if

$$\langle Tx - Ty, x - y \rangle + F(y, x) - F(x, x) \geq_P 0, \quad \forall x, y \in K. \tag{3}$$

Definition 3. Let $T : K \to L(X, Y)$ and $F : K \times K \to Y$ be the two nonlinear mappings. T is said to be *pseudomonotone with respect to F* if, for any given $x, y \in K$,

$$\langle Tx, y - x \rangle + F(y, x) - F(x, x) \nleq_{P \setminus \{0\}} 0 \Longrightarrow \langle Ty, y - x \rangle + F(y, x) - F(x, x) \geq_P 0. \tag{4}$$

Remark 4. Every monotone with respect to F is pseudomonotone with respect to F but converse does not hold in general. Definition 3 is vector version of θ-pseudomonotonicity studied by Kazmi et al. in [23, 24].

Example 5. Let $X = \mathbb{R}$, $K = \mathbb{R}_+$, $Y = \mathbb{R}^2$, $P = \mathbb{R}^2_+$, and

$$T(x) = \begin{pmatrix} 0 \\ \sin x \cos x \end{pmatrix}, \quad F(y, x) = \begin{pmatrix} y + x \\ y + x \end{pmatrix}, \quad \forall x, y \in K. \tag{5}$$

Now,

$$\langle Tx, y - x \rangle + F(y, x) - F(x, x) = \begin{pmatrix} y - x \\ (1 + \sin x \cos x)(y - x) \end{pmatrix} \nleq_{P \setminus \{0\}} 0. \tag{6}$$

We have $y \geq x$. It follows that

$$\langle Ty, y - x \rangle + F(y, x) - F(x, x) = \begin{pmatrix} y - x \\ (1 + \sin y \cos y)(y - x) \end{pmatrix} \geq_P 0. \tag{7}$$

So, T is pseudomonotone with respect to F. However, for $x = \pi$ and $y = \pi/2$, it follows that

$$\langle Tx - Ty, x - y \rangle + F(y, x) - F(x, x) = \begin{pmatrix} -\frac{\pi}{2} \\ -\frac{\pi}{2} \end{pmatrix} \ngeq_P 0. \tag{8}$$

This shows that T is not a monotone with respect to F.

Definition 6. A mapping $T : K \to L(X, Y)$ is said to be *hemicontinuous* if, for any $x, y \in K$, the mapping $t \mapsto \langle T(x + t(y - x)), y - x \rangle$ is continuous at 0^+.

Definition 7. A mapping $F : K \times K \to Y$ is said to be *positively homogeneous* in first argument, if $F(tx, y) = tF(x, y)$ for all $x, y \in K$ and $t \geq 0$.

Definition 8. Let K be a nonempty subset of a topological vector space X. A set-valued map $T : K \to 2^X$ is said to be a *KKM mapping* if, for each nonempty finite subset $\{x_1, \dots, x_n\} \subset K$, $\operatorname{co}\{x_1, \dots, x_n\} \subset \bigcup_{i=1}^{n} T(x_i)$, where co denotes the convex hull.

Lemma 9 (KKM-Fan Lemma (see [25])). *Let K be a nonempty subset of Hausdorff topological vector space X. Let $T : K \to 2^X$ be a KKM-mapping such that for each $x \in K$, $T(x)$ is closed and for at least one $x \in K$, $T(x)$ is compact, then*

$$\bigcap_{x \in K} T(x) \neq \emptyset. \tag{9}$$

3. Strong Vector Mixed Quasi-Complementarity Problems

Throughout this section, let X be a real Banach space and let $K \subseteq X$ be a nonempty, closed, and convex subset of X. Let $(Y, \leq_P)$ be an ordered Banach space induced by a pointed, closed, convex cone P with nonempty interior. Let

$T : K \to L(X, Y)$ and $F : K \times K \to Y$ be the two nonlinear mappings. In this paper, we consider the following strong vector mixed Quasi-Complementarity problems:

(i) Strong vector mixed Quasi-Complementarity problem $(SVMQCP)_1$:

Find $x \in K$ such that $\langle Tx, x\rangle + F(x, x) \not\geq_{P\setminus\{0\}} 0$, $\langle Tx, y\rangle + F(y, x) \not\leq_{P\setminus\{0\}} 0$, for all $y \in K$.

(ii) Strong vector mixed Quasi-Complementarity problem $(SVMQCP)_2$:

Find $x \in K$ such that $\langle Tx, x\rangle + F(x, x) = 0$, $\langle Tx, y\rangle + F(y, x) \not\leq_{P\setminus\{0\}} 0$, for all $y \in K$.

Closely related to $(SVMQCP)_1$ and $(SVMQCP)_2$ problems, we consider the following strong vector mixed quasi variational inequality problem:

Strong vector mixed quasi variational inequality problem (SVMQVIP):

Find $x \in K$ such that $\langle Tx, y - x\rangle + F(y, x) - F(x, x) \not\leq_{P\setminus\{0\}} 0$, for all $y \in K$.

The strong vector mixed quasi variational inequality problem (SVMQVIP) is the generalization and extension of many previously known vectors as well as scalar mixed quasi variational inequalities. For the formulation, numerical results, existence results, sensitivity analysis, and dynamical aspects of the mixed quasi variational inequalities, see [3, 5, 7, 16, 17] and the references therein.

Remark 10. (1) If $Y = \mathbb{R}$ and $P = \mathbb{R}^+$, then $(SVMQCP)_1$ and $(SVMQCP)_2$ and (SVMQVIP) reduce, respectively, to the mixed Quasi-Complementarity problem (MQCP):

(MQCP) Find $x \in K$ such that $\langle Tx, y-x\rangle + F(x, x) = 0$, $\langle Tx, y\rangle + F(y, x) \geq 0$, for all $y \in K$

and mixed quasi variational inequality problem (MQVIP):

(MQVIP) Find $x \in K$ such that $\langle Tx, y-x\rangle + F(y, x) - F(x, x) \geq 0$, for all $y \in K$,

which were introduced and studied by Farazjadeh et al. [7].

(2) If $F = 0$, then $(SVMQCP)_1$ and $(SVMQCP)_2$ reduce to the following strong vector complementarity problems (SVCP):

$(SVCP)_1$ Find $x \in K$ such that $\langle Tx, x\rangle \not\geq_{P\setminus\{0\}} 0$, $\langle Tx, y\rangle \not\leq_{P\setminus\{0\}} 0$, for all $y \in K$,

$(SVCP)_2$ Find $x \in K$ such that $\langle Tx, x\rangle = 0$, $\langle Tx, y\rangle \not\leq_{P\setminus\{0\}} 0$, for all $y \in K$

and (SVMQVIP) reduces to the following strong vector variational inequality problem (SVVIP):

(SVVIP) Find $x \in K$ such that $\langle Tx, y - x\rangle \not\leq_{P\setminus\{0\}} 0$, for all $y \in K$.

First, we will investigate the equivalences among $(SVMQCP)_1$ and $(SVMQCP)_2$ and (SVMQVIP), under some suitable assumptions.

Theorem 11. (i) *Suppose that $\langle Tz, z\rangle + F(z, z) \in P \cup (-P)$, for all $z \in K$. If x solves $(SVMQCP)_1$ then x solves (SVMQVIP).*

(ii) *Let $F : K \times K \to Y$ satisfy $F(2x, y) = 2F(x, y)$, for all $x, y \in K$ and $\langle Tz, z\rangle + F(z, z) \in P \cup (-P)$, for all $z \in K$. If x solves (SVMQVIP) then x also solves $(SVMQCP)_1$.*

Proof. (i) Let $x \in K$ be the solution of $(SVMQCP)_1$. Then $x \in K$ such that

$$\langle Tx, x\rangle + F(x, x) \not\geq_{P\setminus\{0\}} 0, \tag{10}$$

$$\langle Tx, y\rangle + F(y, x) \not\leq_{P\setminus\{0\}} 0, \quad \forall y \in K. \tag{11}$$

Substituting $y = x$ in Inclusion (11), we get

$$\langle Tx, x\rangle + F(x, x) \not\leq_{P\setminus\{0\}} 0. \tag{12}$$

Since $\langle Tz, z\rangle + F(z, z) \in P \cup (-P)$, for all $z \in K$, we have

$$\langle Tx, x\rangle + F(x, x) \geq_P 0 \quad \text{or} \quad \langle Tx, x\rangle + F(x, x) \leq_P 0. \tag{13}$$

From Inclusions (10), (12), and (13), we have

$$\langle Tx, x\rangle + F(x, x) = 0. \tag{14}$$

From (11) and (14), we have

$$\begin{aligned} &\langle Tx, y - x\rangle + F(y, x) - F(x, x) \\ &= \langle Tx, y\rangle + F(y, x) - \langle Tx, x\rangle - F(x, x) \\ &= \langle Tx, y\rangle + F(y, x) \\ &\not\leq_{P\setminus\{0\}} 0, \end{aligned} \tag{15}$$

for all $y \in K$. Thus, $x \in K$ is the solution of (SVMQVIP).

(ii) Now, let $x \in K$ be the solution of (SVMQVIP), then

$$\langle Tx, y - x\rangle + F(y, x) - F(x, x) \not\leq_{P\setminus\{0\}} 0, \quad \forall y \in K. \tag{16}$$

Since $F(2x, y) = 2F(x, y)$, for all $x, y \in K$, therefore it follows that $F(0, y) = 0$, for all $y \in K$. By substituting $y = 2x$ and $y = 0$, respectively, in (16), we get

$$\begin{aligned} &\langle Tx, x\rangle + F(x, x) \not\leq_{P\setminus\{0\}} 0, \\ &\langle Tx, x\rangle + F(x, x) \not\geq_{P\setminus\{0\}} 0. \end{aligned} \tag{17}$$

Since $\langle Tz, z\rangle + F(z, z) \in P \cup (-P)$, for all $z \in K$, we have

$$\langle Tx, x\rangle + F(x, x) \geq_P 0 \quad \text{or} \quad \langle Tx, x\rangle + F(x, x) \leq_P 0. \tag{18}$$

From (17) and (18), we have

$$\langle Tx, x\rangle + F(x, x) = 0. \tag{19}$$

By using Inclusions (16) and (19), we have

$$\begin{aligned} &\langle Tx, y\rangle + F(y, x) \\ &= \langle Tx, y - x\rangle + F(y, x) - F(x, x) \\ &\quad + \langle Tx, x\rangle + F(x, x) \\ &= \langle Tx, y - x\rangle + F(y, x) - F(x, x) \\ &\not\leq_{P\setminus\{0\}} 0, \end{aligned} \tag{20}$$

for all $y \in K$, which implies that x solves $(\mathrm{SVMQCP})_1$. □

Remark 12. The condition $F(2x, y) = 2F(x, y)$, for all $x, y \in K$ holds if F is positively homogeneous; that is, $F(tx, y) = tF(x, y)$ for all $t \geq 0$. Hence, Theorem 11 generalizes and improves the theorems in [6, 9, 11, 14, 15].

Here we give an example of a function F, which satisfies the condition $F(2x, y) = 2F(x, y)$, for all $x, y \in K$ but not a positively homogeneous, which implies that previously known results in [6, 9, 11, 14, 15] cannot be applied.

Example 13. Let $F : \mathbb{R} \times \mathbb{R} \to \mathbb{R}$, defined by

$$F(x, y) = \begin{cases} 2x, & \text{if } x \text{ rational}, \\ 0, & \text{if } x \text{ irrational}. \end{cases} \tag{21}$$

Then F satisfies $F(2x, y) = 2F(x, y)$ but it is not positively homogeneous.

Theorem 14. (a) *If x solves Problem $(\mathrm{SVMQCP})_2$ then x solves (SVMQVIP).*

(b) *Let $F : K \times K \to Y$ satisfy $F(2x, y) = 2F(x, y)$, for all $x, y \in K$ and $\langle Tz, z\rangle + F(z, z) \in P \cup (-P)$, for all $z \in K$. If x solves (SVMQVIP) then x solves $(\mathrm{SVMQCP})_2$.*

Proof. (a) Let $x \in K$ be the solution of $(\mathrm{SVMQCP})_2$. Then $x \in K$ such that

$$\begin{gathered} \langle Tx, x\rangle + F(x, x) = 0, \\ \langle Tx, y\rangle + F(y, x) \nleq_{P\setminus\{0\}} 0, \quad \forall y \in K. \end{gathered} \tag{22}$$

Now,

$$\begin{aligned} &\langle Tx, y - x\rangle + F(y, x) - F(x, x) \\ &\quad = \langle Tx, y\rangle + F(y, x) - \langle Tx, x\rangle - F(x, x) \\ &\quad = \langle Tx, y\rangle + F(y, x) \\ &\quad \nleq_{P\setminus\{0\}} 0, \end{aligned} \tag{23}$$

for all $y \in K$. Thus, $x \in K$ is the solution of (SVMQVIP).

(b) Now, let $x \in K$ be the solution of (SVMQVIP), then

$$\langle Tx, y - x\rangle + F(y, x) - F(x, x) \nleq_{P\setminus\{0\}} 0, \quad \forall y \in K. \tag{24}$$

Since $F(2x, y) = 2F(x, y)$, for all $x, y \in K$, therefore it follows that $F(0, y) = 0$, for all $y \in K$. By substituting $y = 2x$ and $y = 0$, respectively, in (24), we get

$$\begin{gathered} \langle Tx, x\rangle + F(x, x) \nleq_{P\setminus\{0\}} 0, \\ \langle Tx, x\rangle + F(x, x) \ngeq_{P\setminus\{0\}} 0. \end{gathered} \tag{25}$$

Since $\langle Tz, z\rangle + F(z, z) \in P \cup (-P)$, for all $z \in K$, we have

$$\langle Tx, x\rangle + F(x, x) \geq_P 0 \quad \text{or} \quad \langle Tx, x\rangle + F(x, x) \leq_P 0. \tag{26}$$

From (25) and (26), we have

$$\langle Tx, x\rangle + F(x, x) = 0. \tag{27}$$

By using (27), we have

$$\begin{aligned} &\langle Tx, y\rangle + F(y, x) \\ &\quad = \langle Tx, y - x\rangle + F(y, x) - F(x, x) \\ &\qquad + \langle Tx, x\rangle + F(x, x) \\ &\quad = \langle Tx, y - x\rangle + F(y, x) - F(x, x) \\ &\quad \nleq_{P\setminus\{0\}} 0, \end{aligned} \tag{28}$$

for all $y \in K$. Then (27) and (28) imply that x solves $(\mathrm{SVMQCP})_2$. □

4. Existence Results

First, we prove following Minty-type lemma with the help of pseudomonotone mapping with respect to F.

Lemma 15. *Let $F : K \times K \to Y$ be P-convex in first argument and let $T : K \to L(X, Y)$ be a hemicontinuous mapping and pseudomonotone with respect to F. Then the following two problems are equivalent:*

(A)

$$x \in K, \ \langle Tx, y - x\rangle + F(y, x) - F(x, x) \nleq_{P\setminus\{0\}}, \quad \forall y \in K, \tag{29}$$

(B)

$$x \in K, \ \langle Ty, y - x\rangle + F(y, x) - F(x, x) \geq_P 0, \quad \forall y \in K. \tag{30}$$

Proof. (29) ⇒ (30). The result directly follows from pseudomonotonicity with respect to F.

Now, (30) ⇒ (29). For any given $y \in K$, we know that $y_t = ty + (1 - t)x \in K$, for all $t \in (0, 1)$, as K is convex. Since $x \in K$ is a solution of problem (30), so for each $x \in K$, it follows that

$$\langle Ty_t, y_t - x\rangle + F(y_t, x) - F(x, x) \geq_P 0. \tag{31}$$

Now, we have

$$\begin{aligned} &t\langle Ty_t, y - x\rangle + t(F(y, x) - F(x, x)) \\ &\quad \geq \langle Ty_t, y_t - x\rangle + F(y_t, x) - F(x, x) \\ &\quad \geq_P 0. \end{aligned} \tag{32}$$

For $t > 0$, we get

$$\langle Ty_t, y - x\rangle + F(y, x) - F(x, x) \geq_P 0. \tag{33}$$

Since T is hemicontinuous and P is closed, letting $t \to 0^+$ in inclusion (33), we get

$$\langle Tx, y - x\rangle + F(y, x) - F(x, x) \geq_P 0, \quad \forall y \in K. \tag{34}$$

Hence,

$$\langle Tx, y - x\rangle + F(y, x) - F(x, x) \nleq_{P\setminus\{0\}}, \quad \forall y \in K. \tag{35}$$

Therefore, $x \in K$ is solution of problem (29). This completes the proof. □

Now, with the help of Lemma 15, we have following existence theorem for (SVMQVIP).

Theorem 16. *Let X be real reflexive Banach space and let Y be a Banach space. Let $K \subset X$ be a nonempty, bounded, closed, and convex subset of X. Let $F : K \times K \to Y$ be P-convex and upper semicontinuous in first and second arguments, respectively. Let $T : K \to L(X,Y)$ be hemicontinuous and pseudomonotone with respect to F. Then (SVMQVIP) has solution.*

Proof. Define two set-valued mappings $F, G : K \to 2^K$ as follows:

$$\begin{aligned} G(y) &= \{x \in K : \langle Tx, y - x\rangle \\ &\quad + F(y,x) - F(x,x) \nleq_{P\setminus\{0\}}\}, \quad \forall y \in K, \\ H(y) &= \{x \in K : \langle Ty, y - x\rangle \\ &\quad + F(y,x) - F(x,x) \geq_P 0\}, \quad \forall y \in K. \end{aligned} \tag{36}$$

$G(y)$ and $H(y)$ are nonempty, since $y \in G(y) \cap H(y)$. We claim that G is a KKM mapping. If this is not true, then there exists a finite set $\{y_1, \ldots, y_n\} \subset K$ and $t_i \geq 0, i = 1, \ldots, n$ with $\sum_{i=1}^n t_i = 1$ such that $y = \sum_{i=1}^n t_i y_i \notin \bigcup_{i=1}^n G(y_i)$. Now, by the definition of G, we have

$$\langle Ty, y_i - y\rangle + F(y_i, y) - F(y,y) \leq_{P\setminus\{0\}}, \quad i = 1, \ldots, n. \tag{37}$$

Now, we have

$$\begin{aligned} 0 &= \langle Ty, y - y\rangle + F(y,y) - F(y,y) \\ &= \left\langle Ty, \sum_{i=1}^n t_i y_i - y \right\rangle + F\left(\sum_{i=1}^n t_i y_i, y\right) - F(y,y) \\ &= \sum_{i=1}^n t_i \left[\langle Ty, y_i - y\rangle + F(y_i, y) - F(y,y)\right] \\ &\leq_{P\setminus\{0\}}, \end{aligned} \tag{38}$$

which is not possible. Thus, our claim is verified. So G is a KKM mapping.

Now, since T is pseudomonotone with respect to F, therefore $G(y) \subset H(y)$ for every $y \in K$ and so H is also a KKM mapping. Now we claim that for each $y \in K, H(y) \subset K$ is closed in the weak topology of X.

Indeed, suppose $\bar{x} \in \overline{H(y)}^w$, the weak closure of $H(y)$. Since X is reflexive, there is a sequence $\{x_n\}$ in $H(y)$ such that $\{x_n\}$ converges weakly to $\bar{x} \in K$. Then

$$\langle Ty, y - x_n\rangle + F(y, x_n) - F(x, x_n) \geq_P 0. \tag{39}$$

Since $F(y,\cdot)$ is upper semicontinuous and P is closed, therefore,

$$\langle Ty, y - \bar{x}\rangle + F(y, \bar{x}) - F(x, \bar{x}) \geq_P 0 \tag{40}$$

and so $\bar{x} \in H(y)$. This shows that $H(y)$ is weakly closed, for each $y \in K$. Our claim is then verified. Since X is reflexive and $K \subset X$ is nonempty, bounded, closed and convex, K is a weakly compact subset of X and so $H(y)$ is also weakly compact. According to Lemma 9 (KKM-Fan Lemma),

$$\bigcap_{y\in K} H(y) \neq \emptyset. \tag{41}$$

This implies that there exists $x \in K$ such that

$$\langle Ty, y - x\rangle + f(y,x) - F(x,x) \geq_P 0, \quad \forall y \in K. \tag{42}$$

Therefore by Lemma 15, we conclude that there exists $x \in K$ such that

$$\langle Tx, y - x\rangle + F(y,x) - F(x,x) \nleq_{P\setminus\{0\}}, \quad \forall y \in K. \tag{43}$$

This completes the proof. □

Theorem 17. *Let $F : K \times K \to Y$ satisfy $F(2x, y) = 2F(x, y)$, for all $x, y \in K$. If all the assumptions of Theorem 16 hold, then $(SVMQCP)_1$ is solvable. In addition, if $\langle Tz, z\rangle + F(z,z) \in P \cup (-P)$, for all $z \in K$, then $(SVMQCP)_2$ is solvable.*

Proof. The conclusion follows directly from Theorems 11, 14, and 16. □

References

[1] F. Giannessi, "Theorems of alternative, quadratic programs and complementarity problems," in *Variational Inequalities and Complementarity Problems*, R. W. Cottle, F. Gianneesi, and J. L. Lions, Eds., pp. 151–186, Wiley, Chichester, UK, 1980.

[2] G. Y. Chen and G. M. Cheng, "Vector variational inequalities and vector optimization," in *Lecture Notes in Economics and Mathematical Systems*, vol. 285, pp. 408–416, Springer, Berlin, Germany, 1987.

[3] A. Bnouhachem and M. A. Noor, "Numerical method for general mixed quasi-variational inequalities," *Applied Mathematics and Computation*, vol. 204, no. 1, pp. 27–36, 2008.

[4] G. Y. Chen and X. Q. Yang, "The vector complementary problem and its equivalences with the weak minimal element in ordered spaces," *Journal of Mathematical Analysis and Applications*, vol. 153, no. 1, pp. 136–158, 1990.

[5] R. W. Cottle and G. B. Dantzig, "Complementary pivot theory of mathematical programming," *Linear Algebra and its Applications*, vol. 1, pp. 163–185, 1968.

[6] Y.-P. Fang and N.-J. Huang, "The vector F-complementary problems with demipseudomonotone mappings in Banach spaces," *Applied Mathematics Letters*, vol. 16, no. 7, pp. 1019–1024, 2003.

[7] A. P. Farajzadeh, M. Aslam Noor, and S. Zainab, "Mixed quasi complementarity problems in topological vector spaces," *Journal of Global Optimization*, vol. 45, no. 2, pp. 229–235, 2009.

[8] G. Isac, "A special variational inequality and the implicit complementarity problem," *Journal of the Faculty of Science. University of Tokyo. Section IA. Mathematics*, vol. 37, no. 1, pp. 109–127, 1990.

[9] N.-j. Huang and Y.-p. Fang, "Strong vector F-complementary problem and least element problem of feasible set," *Nonlinear Analysis: Theory, Methods & Applications*, vol. 61, no. 6, pp. 901–918, 2005.

[10] N.-J. Huang and C.-J. Gao, "Some generalized vector variational inequalities and complementarity problems for multivalued mappings," *Applied Mathematics Letters*, vol. 16, no. 7, pp. 1003–1010, 2003.

[11] N.-j. Huang and J. Li, "F-implicit complementarity problems in Banach spaces," *Zeitschrift für Analysis und ihre Anwendungen*, vol. 23, no. 2, pp. 293–302, 2004.

[12] N. J. Huang, X. Q. Yang, and W. K. Chan, "Vector complementarity problems with a variable ordering relation," *European Journal of Operational Research*, vol. 176, no. 1, pp. 15–26, 2007.

[13] K. R. Kazmi and S. A. Khan, "Existence of solutions to a generalized system," *Journal of Optimization Theory and Applications*, vol. 142, no. 2, pp. 355–361, 2009.

[14] B. S. Lee, M. F. Khan, and Salahuddin, "Vector F-implicit complementarity problems with corresponding variational inequality problems," *Applied Mathematics Letters*, vol. 20, no. 4, pp. 433–438, 2007.

[15] J. Li and N.-j. Huang, "Vector F-implicit complementarity problems in Banach spaces," *Applied Mathematics Letters*, vol. 19, no. 5, pp. 464–471, 2006.

[16] U. Mosco, "Implicit variational problems and quasi variational inequalities," in *Nonlinear Operators and the Calculus of Variations*, vol. 543 of *Lecture Notes in Mathematics*, pp. 83–156, Springer, Berlin, Germany, 1976.

[17] M. A. Noor, "Mixed quasi variational inequalities," *Applied Mathematics and Computation*, vol. 146, no. 2-3, pp. 553–578, 2003.

[18] F. Usman and S. A. Khan, "A generalized mixed vector variational-like inequality problem," *Nonlinear Analysis: Theory, Methods & Applications*, vol. 71, no. 11, pp. 5354–5362, 2009.

[19] R. U. Verma, "On generalized variational inequalities involving relaxed Lipschitz and relaxed monotone operators," *Journal of Mathematical Analysis and Applications*, vol. 213, no. 1, pp. 387–392, 1997.

[20] X. Q. Yang, "Vector complementarity and minimal element problems," *Journal of Optimization Theory and Applications*, vol. 77, no. 3, pp. 483–495, 1993.

[21] C. E. Lemke, "Bimatrix equilibrium points and mathematical programming," *Management Science*, vol. 11, pp. 681–689, 1965.

[22] S. A. Khan, "Generalized vector implicit quasi complementarity problems," *Journal of Global Optimization*, vol. 49, no. 4, pp. 695–705, 2011.

[23] K. R. Kazmi, A. Khaliq, and A. Raouf, "Iterative approximation of solution of generalized mixed set-valued variational inequality problem," *Mathematical Inequalities & Applications*, vol. 10, no. 3, pp. 677–691, 2007.

[24] K. R. Kazmi, F. A. Khan, and M. Shahza, "Two step algorithm for solving regularized generalized mixed variational inequality problem," *Bulletin of the Korean Mathematical Society*, vol. 47, no. 4, pp. 675–685, 2010.

[25] K. Fan, "A generalization of Tychonoff's fixed point theorem," *Mathematische Annalen*, vol. 142, pp. 305–310, 1961.

13

Adjusted Empirical Likelihood for Varying Coefficient Partially Linear Models with Censored Data

Peixin Zhao

Department of Mathematics, Hechi University, Yizhou, Guangxi 546300, China

Correspondence should be addressed to Peixin Zhao; zpx81@163.com

Academic Editor: Xuewen Lu

By constructing an adjusted auxiliary vector ingeniously, we propose an adjusted empirical likelihood ratio function for the parametric components of varying coefficient partially linear models with censored data. It is shown that its limiting distribution is standard central chi-squared. Then the confidence intervals for the parametric components are constructed. A simulation study and a real data analysis are undertaken to assess the finite sample performance of the proposed method.

1. Introduction

Outcome censoring often occurs in many disciplines such as econometrics, biostatistics, and bioinformatics. There have been many recent researches in the area of statistical inference for censored data. Wang and Zheng [1] and Wang and Li [2] considered the estimation problem for partly linear models based on different methods. Yang et al. [3] proposed an empirical likelihood method for a partially linear single-index model with right censored data. More results on the statistical inferences of censored data can be seen in [4–6]. In this paper, we consider the empirical likelihood inferences for a varying coefficient partially linear model with right censored data. Specifically, let $\{Y_i, i = 1, \dots, n\}$ be the response variable and let $\{(X_i, Z_i, U_i), i = 1, \dots, n\}$ be associated covariates. We consider the following varying coefficient partially linear model:

$$Y_i = X_i^T \theta(U_i) + Z_i^T \beta + \epsilon_i, \qquad (1)$$

where β is a $q \times 1$ vector of unknown parameters, $\theta(\cdot)$ is a $p \times 1$ vector of unknown function, and ϵ_i is the model error such as $E(\epsilon_i \mid X_i, Z_i, U_i) = 0$. Due to the curse of dimensionality, as in [7], we assume that U_i is univariate. In this paper, we assume that the available data $\{(Y_i^*, \delta_i, X_i, Z_i, U_i), i = 1, \dots, n\}$, are independent and identically distributed, where $Y_i^* = \min\{Y_i, C_i\}$, $\delta_i = I(Y_i \leq C_i)$ and C_i is a censoring variable. We also assume that $\{C_i, i = 1, \dots, n\}$ are *i.i.d.* variables with a distribution function $G(\cdot)$ and C_i is independent of the prognostic variables (X_i, Z_i, U_i) and the response Y_i.

Model (1) has proved to be very useful as it combines the flexibility of nonparametric models and the interpretation of linear models. Recently, a variety of methods have been proposed for estimating the parameters and nonparametric functions in model (1) based on the uncensored data (see [7, 8]). In this paper, we recommend using the empirical likelihood method to make inference for β in the model (1) under right censored data. We define an empirical log-likelihood function based on the synthetic data and show that its asymptotic distribution is a mixture of central chi-squared distribution. Hence, the confidence regions for parametric components can be constructed if the unknown weights are estimated. In addition, by constructing an adjusted auxiliary vector ingeniously, we also propose an adjusted empirical likelihood ratio function for the parametric components and show that its limiting distribution is a standard central chi-squared distribution. Then the confidence intervals for parametric components are constructed and the estimation for the unknown weights is avoided. To demonstrate the performance of the proposed method, based on some numerical experiments such as the simulation study and the real data analysis, we conducted a comparison with the normal approximation method. Compared with the normal approximation method, the confidence intervals based on the empirical likelihood method perform fairly well.

The confidence intervals obtained by the empirical likelihood method possess several attractive features compared to the conventional Wald-type confidence intervals, such as circumvention of asymptotic variance estimation to compute the standard error, flexible shapes of the confidence intervals determined by data and range-preserving property (see [9]). Wang and Li [2] considered the adjusted empirical likelihood inferences for a class of partially linear models with right censored data. Yang et al. [3] considered the adjusted empirical likelihood inferences for a partially linear single-index model with right censored data. This paper also contributes to the rapidly growing literature on the adjustment-based empirical likelihood method and provides additional positive results for the Wilks phenomena in the parametric components of varying coefficient partially linear models with censored data, which extends the application literature of the adjusted empirical likelihood method. The more works on empirical likelihood methods can be found in [10, 11].

2. Methodology and Results

As in [3], let $Y_{iG}^* = Y_i^*\delta_i/(1-G(Y_i^*))$; then we have $E(Y_{iG}^* \mid X_i, Z_i, U_i) = E(Y_i \mid X_i, Z_i, U_i)$. This implies that

$$Y_{iG}^* = X_i^T\theta(U_i) + Z_i^T\beta + \epsilon_i^*, \qquad (2)$$

where $E(\epsilon_i^* \mid X_i, Z_i, U_i) = 0$. When $G(\cdot)$ is known as well, model (2) is a standard semiparametric varying coefficient partially linear model. For given β, using the same arguments as Fan and Huang [7], we can get the weighted local least-squares estimator of $\theta(u)$ by minimizing

$$\sum_{i=1}^{n}\left\{Y_{iG}^* - Z_i^T\beta - \sum_{k=1}^{p}\left[a_k + b_k(U_i-u)\right]X_{ik}\right\}^2 K_h(U_i-u), \qquad (3)$$

where $K_h(\cdot) = h^{-1}K(\cdot/h)$, $K(\cdot)$ is a kernel function, h is a bandwidth, and X_{ik} denotes the kth component of X_i.

Let $Z = (Z_1,\dots,Z_n)^T$, let $Y_G^* = (Y_{1G}^*,\dots,Y_{nG}^*)^T$, let I_p be $p\times p$ identity matrix, let 0_p be $p\times p$ zero matrix, let $\Omega_u = \text{diag}(K_h(u-U_1),\dots,$ let $K_h(u-U_n))$ be $n\times n$ diagonal matrix. Then, the solution to minimize (3) can be given by

$$\tilde{\theta}(u) = (I_p, 0_p)\left(D_u^T\Omega_u D_u\right)^{-1}D_u^T\Omega_u\left(Y_G^* - Z\beta\right), \qquad (4)$$

where $D_u = \begin{pmatrix} X_1 & \cdots & X_n \\ h^{-1}(U_1-u)X_1 & \cdots & h^{-1}(U_n-u)X_n\end{pmatrix}^T$.

Let $(I_p,0_p)(D_u^T\Omega_u D_u)^{-1}D_u^T\Omega_u \equiv (S_1(u),\dots,S_n(u))$, let $\tilde{\mu}(u) = \sum_{k=1}^n S_k(u)Z_k^T$, and let $\tilde{g}(u) = \sum_{k=1}^n S_k(u)Y_{kG}^*$. Then we have $\tilde{\theta}(u) = \tilde{g}(u) - \tilde{\mu}(u)\beta$. Substituting this into (2) and by a simple calculation we have

$$\breve{Y}_{iG}^* = \breve{Z}_i^T\beta + \epsilon_i^*, \qquad (5)$$

where $\breve{Z}_i = Z_i - \tilde{\mu}(U_i)^TX_i$, $\breve{Y}_{iG}^* = Y_{iG}^* - \tilde{g}(U_i)^TX_i$. To construct the empirical likelihood ratio function for β, we introduce the auxiliary random vector

$$\breve{\eta}_i(\beta) = \breve{Z}_i\left(\breve{Y}_{iG}^* - \breve{Z}_i^T\beta\right). \qquad (6)$$

By (5) we have that β is true if and only if $E\{\breve{\eta}_i(\beta)\} = o(1)$. Using such information, we can define a profile empirical log-likelihood ratio function for β. However, $G(\cdot)$ is usually unknown, and $\breve{\eta}_i(\beta)$ cannot be used directly to make inference for β. To solve this problem, we replace $G(\cdot)$ in $\breve{\eta}_i(\beta)$ by its estimator. In this paper, we employ the Kaplan-Meier estimator

$$G_n(y) = 1 - \prod_{i=1}^{n}\left(\frac{N^+(Y_i^*)}{1+N^+(Y_i^*)}\right)^{I(Y_i^*\le y,\delta_i=0)}, \qquad (7)$$

where $N^+(y) = \sum_{j=1}^n I(Y_j^* > y)$. Hence, an estimator of $\breve{\eta}_i(\beta)$ can be defined as

$$\hat{\eta}_i(\beta) = \breve{Z}_i\left(\breve{Y}_{iG_n}^* - \breve{Z}_i^T\beta\right), \qquad (8)$$

where $\breve{Y}_{iG_n}^* = Y_{iG_n}^* - \tilde{g}(U_i)^TX_i$ and $\tilde{g}(u) = \sum_{k=1}^n S_k(u)Y_{kG_n}^*$. Then, we can define the profile empirical log-likelihood ratio function for β as

$$\hat{R}(\beta) = -2\max\left\{\sum_{i=1}^n \log(np_i) \mid p_i \ge 0, \sum_{i=1}^n p_i = 1, \sum_{i=1}^n p_i\hat{\eta}_i(\beta) = 0\right\}, \qquad (9)$$

where $p_i = p_i(\beta)$, $i = 1,\dots,n$. A unique value for $\hat{R}(\beta)$ exists, provided that 0 is inside the convex hull of the points $(\hat{\eta}_1(\beta),\dots,\hat{\eta}_n(\beta))$. Based on the method of Lagrange multipliers to find the optimal p_i, then the empirical log-likelihood ratio function can be represented as

$$\hat{R}(\beta) = 2\sum_{i=1}^n \log\left(1+\lambda^T\hat{\eta}_i(\beta)\right), \qquad (10)$$

where $\lambda = \lambda(\beta)$ is a $q\times 1$ vector, which satisfies

$$\sum_{i=1}^n \frac{\hat{\eta}_i(\beta)}{\left(1+\lambda^T\hat{\eta}_i(\beta)\right)} = 0. \qquad (11)$$

Let

$$\Psi(u) = E\left(XX^T \mid U=u\right), \qquad \Phi(u) = E\left(XZ^T \mid U=u\right),$$

$$H(s) = \frac{E\{[Z-\mu(U)^TX]\,Y_G^*I(s<Y^*)\}}{(1-G(s))(1-F(s-))},$$

$$\mu(u) = \Psi(u)^{-1}\Phi(u), \qquad \Lambda^G(u) = \int_{-\infty}^{u}\frac{1}{1-G(s-)}dG(s),$$

$$\Sigma_0(\beta) = E\left\{\left[Z-\mu(U)^TX\right]\left[Z-\mu(U)^TX\right]^T \times\left[Y_G^* - Z^T\beta - X^T\theta(U)\right]^2\right\},$$

$$\Sigma_1 = \int_{-\infty}^{\infty} H(s)H(s)^T(1-F(s-))\left(1-\Delta\Lambda^G(s)\right)dG(s). \qquad (12)$$

The following theorem gives the asymptotic distribution of $\hat{R}(\beta)$.

Theorem 1. *Suppose that conditions (C1)–(C5) in the Appendix hold. If β_0 is the true parameter, then*

$$\widehat{R}(\beta_0) \xrightarrow{\mathscr{L}} w_1\chi_{1,1}^2 + w_2\chi_{1,2}^2 + \cdots + w_q\chi_{1,q}^2, \tag{13}$$

where $w_1, \ldots, w_q$ are the eigenvalues of $\Sigma_0^{-1}(\beta_0)\Sigma(\beta_0)$, $\Sigma(\beta_0) = \Sigma_0(\beta_0) - \Sigma_1$, and $\chi_{1,1}^2, \ldots, \chi_{1,q}^2$ are independent standard chi-square random variables with 1 degree of freedom.

From Theorem 1, we can see that the asymptotic distribution of $\widehat{R}(\beta)$ is a mixture of central chi-square distributions. Hence, the confidence regions for the parametric components can be constructed if the unknown weights are estimated. Next, we give an adjusted empirical log-likelihood ratio function that has an asymptotic standard chi-square distribution. Then the confidence regions for the parametric components are constructed, and the estimation for the unknown weights is avoided. Let $F_n(\cdot)$ be the Kapaln-Meier estimator for the distribution function of Y, and

$$\widehat{\Sigma}_0(\beta) = \frac{1}{n}\sum_{i=1}^{n}\widehat{\eta}_i(\beta)\,\widehat{\eta}_i(\beta)^T,$$

$$\Lambda_n^{G_n}(u) = \int_{-\infty}^{u}\frac{1}{1-G_n(s-)}\,dG_n(s),$$

$$H_n(s) = \frac{n^{-1}\sum_{i=1}^{n}\left[Z_i - \widetilde{\mu}(U_i)^T X_i\right] Y_{iG_n}^{*} I(s < Y_i^{*})\}}{(1-G_n(s))(1-F_n(s-))},$$

$$\widehat{\Sigma}_1 = \frac{1}{n}\sum_{i=1}^{n}(1-\delta_i)\,H_n(Y_i^{*})\,H_n(Y_i^{*})^T\left(1-\Delta\Lambda_n^{G_n}(Y_i^{*}-)\right)$$

$$\widehat{\Sigma}(\beta) = \widehat{\Sigma}_0(\beta) - \widehat{\Sigma}_1, \qquad S_n(\beta) = \sum_{i=1}^{n}\widehat{\eta}_i(\beta)\sum_{i=1}^{n}\widehat{\eta}_i(\beta)^T. \tag{14}$$

Then, as in [3], an adjusted empirical log-likelihood ratio function can be defined by

$$\widehat{R}_{\text{ad}}(\beta) = \gamma_n(\beta)\,\widehat{R}(\beta), \tag{15}$$

where $\gamma_n(\beta) = \text{tr}\{\widehat{\Sigma}^{-1}(\beta)S_n(\beta)\}/\,\text{tr}\{\widehat{\Sigma}_0^{-1}(\beta)S_n(\beta)\}$. The following theorem shows that the asymptotic distribution of $\widehat{R}_{\text{ad}}(\beta)$ can be approximated by standard chi-square distribution.

Theorem 2. *Suppose that conditions (C1)–(C5) in the Appendix hold. If β_0 is the true parameter, then*

$$\widehat{R}_{\text{ad}}(\beta_0) \xrightarrow{\mathscr{L}} \chi_q^2. \tag{16}$$

Let $\chi_q^2(1-\alpha)$ be the $1-\alpha$ quantile of χ_q^2, $(0 < \alpha < 1)$. Theorem 2 implies that an approximate $1-\alpha$ confidence region for β can be defined by

$$C_\alpha(\beta) = \left\{\beta \mid \widehat{R}_{\text{ad}}(\beta) \le \chi_q^2(1-\alpha)\right\}. \tag{17}$$

3. Numerical Results

In this section, we conduct several simulation experiments to illustrate the finite sample performances of the proposed method and consider a real data set analysis for further illustration.

3.1. Simulation Studies. To evaluate the performance of the proposed adjustment-based empirical likelihood (AEL) method, we present some simulation results. In the simulation, we simulated data from the following model:

$$Y = \left(2U^2 - 1\right)X + Z + \epsilon, \tag{18}$$

where $\beta = 1$, $\theta(u) = 2u^2 - 1$. We generated $n = 50, 100$ and 150 subjects, respectively, and the covariates $X \sim N(0,1)$, $Z \sim N(1, 1.5)$, and $U \sim U(-1,1)$. Y is generated according to the model with $\epsilon \sim N(0, 0.5)$. The censoring variable $C \sim N(\mu, 1)$, where $\mu = 3.5$, 2.1 and 1.3, respectively, such that the corresponding censoring rates (CR) are about 0.1, 0.3, and 0.45. We use the Epanechnikov kernel function $K(u) = 0.75(1-u^2)_+$, and the cross-validation bandwidth h_{CV} is obtained by minimizing

$$\text{CV}(h) = \sum_{i=1}^{n}\left(Y_{iG_n}^{*} - X_i^T\widehat{\theta}_{[i]}(U_i) - Z_i^T\widehat{\beta}_{[i]}\right)^2, \tag{19}$$

where $\widehat{\beta}_{[i]}$ and $\widehat{\theta}_{[i]}(\cdot)$ are estimators of β and $\theta(\cdot)$ which are computed with all of the measurements, but the ith subject is deleted. $\widehat{\beta}_{[i]}$ can be obtained by minimizing (10), $\widehat{\theta}_{[i]}(\cdot)$ can be obtained by replacing $G(\cdot)$, and β in (4) by $G_n(\cdot)$ and $\widehat{\beta}_{[i]}$, respectively.

For comparison, we consider two approaches for constructing the confidence intervals: the AEL method proposed in this paper and normality approximation (NA) method proposed by [12]. To construct confidence intervals based on the NA method, we need to estimate the asymptotic variance. However, note that the asymptotic variance of the estimator has a complicated structure; then we estimate the asymptotic variance by the bootstrap method. The average lengths and coverage probabilities of the confidence intervals are shown in Table 1. Here the nominal level is taken as $1-\alpha = 95\%$, and the simulation is computed with 1000 simulation runs. From Table 1, we can see the following observations.

(i) For any given level of CR and the size of sample, although the coverage probabilities of the confidence intervals obtained by the AEL method and the NA method are similar, the lengths of the confidence intervals obtained by the AEL method are shorter than those obtained by the NA method. This implies that the confidence intervals obtained by the AEL method outperform those obtained by the NA method.

(ii) For the given size of sample n, the performances of the confidence intervals, obtained by the AEL method under the moderate censoring rate, are all close to the nominal level 95%. This implies that the adjustment scheme is workable for the moderate censoring rate. However, the simulation results also suggest that larger samples would be needed when the censoring rate is relatively high such as CR = 0.45.

3.2. Application to CGD Data. We illustrate the applicability of our proposed method using the chronic granulomatous disease (CGD) data set from the International CGD Cooperative Study Group. This data set was designed to have a single interim analysis when the follow-up data as of July 15, 1989 were complete. The monitoring committee for the trial terminated the trial at a meeting on September 22, 1989. The treatment given each patient was unblinded at the first scheduled visit for the patient following the decision of the monitoring committee. More details for this data description and analysis can be seen in [13].

The variables contained here are X: treatment code, 1 = rIFN, 2 = placebo; Z_1: pattern of inheritance, 1 = X-linked, 2 = autosomal recessive; Z_2: age, in years; Z_3: height, in cm; Z_4: weight, in kg; Z_5: using corticosteroids at time of study entry, 1 = yes, 2 = no; Z_6: using prophylactic antibiotics at time of study entry, 1 = yes, 2 = no; Z_7: 1 = male, 2 = female; Z_8: hospital category, 1 = US-NIH, 2 = US-other, 3 = Europe, Amsterdam, 4 = Europe, other; T: elapsed time (in days) from randomization to diagnosis of a serious infection or if a censored observation: elapsed time from randomization to censoring date; δ: censoring indicator, 1 = Noncensored observation, 2 = censored observation. S: sequence number. For each patient, the infection records are in sequence number order.

Based on the generalized likelihood ratio testing, Jiang and Qian [14] show that this data set can be fitted by the following model:

$$Y = X\theta(U) + \sum_{k=0}^{8} Z_k\beta_k + \varepsilon, \tag{20}$$

where $Y = \ln(T)$, $U = S/20$, and $Z_0 = 1$ as the intercept term. The estimators and 95% confidence intervals are shown in Table 2. From Table 2, we see that the covariates Z_1, Z_5, and Z_7 have somewhat stronger associations with the response. In addition, we also can see that the confidence intervals obtained by the AEL method are workable.

Appendix

A. Proof of Theorems

For convenience and simplicity, let C denote a positive constant which may be different value at each appearance throughout this paper. Before we prove our main theorems, we list some regularity conditions which are used in this paper.

(C1) The bandwidth $h = Cn^{-1/5}$, for some constant $C > 0$. The kernel $K(\cdot)$ is a symmetric probability density function, and $\int u^4K(u)du < \infty$.

(C2) $\theta(u)$, $\sigma^2(u)$, $\Phi(u)$, and $\Psi(u)$ are twice continuously differentiable on $(0,1)$, where $\sigma^2(u) = E(\epsilon^2 \mid U = u)$.

(C3) The density function of U, says $f(u)$, is bounded away from 0 and infinity on $[0,1]$ and is continuously differentiable on $(0,1)$.

(C4) For $s \le \tau_Q = \inf\{y : Q(y) = 1\}$, $G(s)$ and $F(s)$ do not have common jumps, where $Q(y) = P(Y^* \le y)$. Moreover, we assume that

$$\int_0^{\tau_Q} \|H(Y)\|^2 \left(1 - \Lambda^G(s)\right) dG(s) < \infty,$$
$$E\left\{\frac{\|Z - \mu(U)\| \, |Y|}{[(1 - G(Y))(1 - F(Y))]^{1/2}}\right\} < \infty. \tag{A.1}$$

(C5) For given u, $\Psi(u)$ is a positive definite matrix, and $E\{[Z - \mu(U)^T X][Z - \mu(U)^T X]^T\}$ is nonsingular.

The proofs of theorems rely on the following lemmas.

Lemma 3. *Let $(X_1, Y_1), \ldots, (X_n, Y_n)$ be i.i.d. random vectors, where Y_i is scalar random variables. Assume that $\sup_x \int |y|^s f(x,y)dy < \infty$ and $E|Y_1|^s < \infty$, where $f(\cdot,\cdot)$ denotes the joint density of (X,Y). Let $K(\cdot)$ be a bounded positive function with a bounded support, satisfying a Lipschitz condition. Then*

$$\sup_x \left|\frac{1}{n}\sum_{i=1}^{n} \{K_h(X_i - x)Y_i - E[K_h(X_i - x)Y_i]\}\right|$$
$$= O_p\left(\left\{\frac{\log(1/h)}{nh}\right\}^{1/2}\right), \tag{A.2}$$

provided that $n^{2\delta-1}h \to \infty$, for some $\delta < 1 - s^{-1}$.

Proof. This follows immediately from the result that was obtained by Mack and Silverman [15]. □

Lemma 4. *Suppose that conditions (C1)-(C5) hold. Then*

$$\sup_{0<u<1} \|\tilde{\mu}(u) - \mu(u)\| = O_p(C_n),$$
$$\sup_{0<u<1} \left\|\tilde{\theta}(u) - \theta(u)\right\| = O_p(C_n), \tag{A.3}$$

where $C_n = \{\log(1/h)/nh\}^{1/2} + h^2$.

Proof. Let $S_{nl} = \sum_{i=1}^n X_iX_i^T((U_i - u)/h)^l K_h(U_i - u)$ and let $l = 0,1,2$. Similar to the proof of Lemma 7.2 in [7], a simple calculation yields

$$E(S_{nl}) = nf(u)\Psi(u)\int_0^1 s^l K(s)\,ds + o(1), \quad l = 0,1,2. \tag{A.4}$$

Note that $D_u^T\Omega_u D_u = \begin{pmatrix} S_{n0} & S_{n1} \\ S_{nl} & S_{n2} \end{pmatrix}$, by Lemma 3, we obtain

$$D_u^T\Omega_u D_u = nf(u)\Psi(u)$$
$$\otimes \begin{pmatrix} 1 & 0 \\ 0 & \int_0^1 t^2K(t)\,dt \end{pmatrix} \{1 + O_p(C_n)\}, \tag{A.5}$$

TABLE 1: The average lengths and coverage probabilities of the 95% confidence intervals for β based on the AEL method and the NA method.

n	CR	Confidence interval		Length		Coverage	
		AEL	NA	AEL	NA	AEL	NA
	0.1	(0.924, 1.247)	(0.828, 1.344)	0.323	0.516	0.914	0.916
50	0.3	(0.915, 1.259)	(0.825, 1.347)	0.344	0.522	0.896	0.894
	0.45	(0.906, 1.264)	(0.824, 1.350)	0.358	0.526	0.863	0.864
	0.1	(0.927, 1.220)	(0.913, 1.270)	0.293	0.357	0.935	0.937
100	0.3	(0.922, 1.223)	(0.907, 1.273)	0.301	0.366	0.931	0.930
	0.45	(0.916, 1.227)	(0.906, 1.276)	0.311	0.370	0.928	0.928
	0.1	(0.960, 1.161)	(0.942, 1.239)	0.201	0.297	0.948	0.948
150	0.3	(0.952, 1.199)	(0.939, 1.237)	0.247	0.298	0.945	0.943
	0.45	(0.949, 1.202)	(0.933, 1.236)	0.253	0.303	0.943	0.942

TABLE 2: Application to CGD data. The maximum empirical likelihood estimators (MELE) and 95% confidence intervals (CI) for the parametric components based on the AEL method.

	β_0	β_1	β_2	β_3	β_4
MELE	4.470	−0.288	0.013	0.004	−0.007
CI	(4.210, 4.680)	(−0.473, −0.158)	(0.002, 0.021)	(0.002, 0.005)	(−0.012, −0.004)
	β_5	β_6	β_7	β_8	
MELE	0.398	−0.097	0.305	−0.021	
CI	(0.268, 0.501)	(−0.267, 0.033)	(0.178, 0.412)	(−0.128, 0.045)	

uniformly for $u \in (0,1)$, where $\otimes$ is the Kronecker product. Using the same argument, we have

$$D_u^T \Omega_u Z = nf(u)\Phi(u) \otimes (1,0)^T \left\{1 + O_p(C_n)\right\}, \quad \text{(A.6)}$$

uniformly for $u \in (0,1)$. Combining (A.5) and (A.6) yields

$$\begin{aligned}\tilde{\mu}(u) &= \left(I_p, 0_p\right)\left(D_u^T \Omega_u D_u\right)^{-1} D_u^T \Omega_u Z \\ &= \mu(u)\left\{1 + O_p(C_n)\right\},\end{aligned} \quad \text{(A.7)}$$

uniformly for $u \in (0,1)$.

Invoking $E(Y_{iG}^* \mid X_i, Z_i, U_i) = E(Y_i \mid X_i, Z_i, U_i)$, with the similar argument, we can prove $\sup_{0<u<1}\|\tilde{\theta}(u) - \theta(u)\| = O_p(C_n)$. This completes the proof of Lemma 4. □

Lemma 5. *Suppose that conditions (C1)–(C5) hold. If β_0 is the true parameter, then*

$$\frac{1}{\sqrt{n}}\sum_{i=1}^{n}\hat{\eta}_i(\beta_0) \xrightarrow{\mathscr{L}} N(0, \Sigma(\beta_0)), \quad \text{(A.8)}$$

where $\Sigma(\beta_0)$ is defined in Theorem 1.

Proof. By (8), we have

$$\begin{aligned}\hat{\eta}_i(\beta_0) &= \check{Z}_i\left(\check{Y}_{iG}^* - \check{Z}_i^T\beta_0\right) + \check{Z}_i\left(\check{Y}_{iG_n}^* - \check{Y}_{iG}^*\right) \\ &\equiv J_{1i}(\beta_0) + J_{2i}(\beta_0).\end{aligned} \quad \text{(A.9)}$$

A simple calculation yields

$$\begin{aligned}&\frac{1}{\sqrt{n}}\sum_{i=1}^{n} J_{1i}(\beta_0) \\ &= \frac{1}{\sqrt{n}}\sum_{i=1}^{n}\left[Z_i - \mu(U_i)^T X_i\right]\left[Y_{iG}^* - X_i^T\theta(U_i) - Z_i^T\beta_0\right] \\ &+ \frac{1}{\sqrt{n}}\sum_{i=1}^{n}\left[Z_i - \mu(U_i)^T X_i\right] X_i^T\left[\theta(U_i) - \tilde{\theta}(U_i)\right] \\ &+ \frac{1}{\sqrt{n}}\sum_{i=1}^{n}\left[\mu(U_i) - \tilde{\mu}(U_i)\right]^T X_i\left[Y_{iG}^* - X_i^T\theta(U_i) - Z_i^T\beta_0\right] \\ &+ \frac{1}{\sqrt{n}}\sum_{i=1}^{n}\left[\mu(U_i) - \tilde{\mu}(U_i)\right]^T X_i X_i^T\left[\theta(U_i) - \tilde{\theta}(U_i)\right] \\ &\equiv A_1 + A_2 + A_3 + A_4.\end{aligned} \quad \text{(A.10)}$$

By the Central Limits Theorem, it is easy to prove $A_1 \xrightarrow{\mathscr{L}} N(0, \Sigma_0(\beta_0))$. Next, we prove $A_\nu = o_p(1), \nu = 2, 3, 4$. Using Abel inequality, invoking $E\{(Z_i - \mu(U_i)^T X_i)X_i^T\} = 0$, and using Lemma 4, we can prove that

$$\begin{aligned}\|A_2\| &\leq \frac{C}{\sqrt{n}}\sup_{1\leq i\leq n}\left\|\tilde{\theta}(U_i) - \theta(U_i)\right\| \\ &\quad \times \max_{1\leq k\leq n}\left\|\sum_{i=1}^{k}\left[Z_i - \mu(U_i)^T X_i\right] X_i^T\right\| \\ &= \frac{C}{\sqrt{n}} O_p(C_n) O_p(\sqrt{n}\log n),\end{aligned} \quad \text{(A.11)}$$

that is, $A_2 = o_p(1)$. Invoking $E(Y_{iG}^* \mid X_i, Z_i, U_i) = E(Y_i \mid X_i, Z_i, U_i)$, and with a similar argument, we can prove $A_3 = o_p(1)$. In addition, by Lemma 4, we have $\|A_4\| \le O_p(\sqrt{n}C_n^2) = o_p(1)$. Hence, we get that

$$\frac{1}{\sqrt{n}}\sum_{i=1}^{n} J_{1i}(\beta_0) \xrightarrow{\mathscr{L}} N\left(0, \widehat{\Sigma}_0(\beta_0)\right). \tag{A.12}$$

By Wang and Zheng [1], we have

$$\begin{aligned}\frac{1}{\sqrt{n}}\sum_{i=1}^{n} J_{2i}(\beta_0) &= \frac{1}{\sqrt{n}}\sum_{i=1}^{n}\frac{\left[Z_i - \mu(U_i)^T X_i\right] Y_i^*\delta_i}{1-G(Y_i^*)}\frac{G_n(Y_i^*)-G(Y_i^*)}{1-G(Y_i^*)} \\ &\quad + o_p(1).\end{aligned} \tag{A.13}$$

Then, using the similar argument to Yang et al. [3], we can prove that

$$\begin{aligned}\frac{1}{\sqrt{n}}\sum_{i=1}^{n} J_{2i}(\beta_0) &\xrightarrow{\mathscr{L}} N(0,\Sigma_1), \\ \frac{1}{n}\sum_{i=1}^{n} E\left\{J_{1i}J_{2i}^T\right\} &\xrightarrow{P} \Sigma_1.\end{aligned} \tag{A.14}$$

Together with (A.12), we complete the proof of Lemma 5. □

Proof of Theorem 1. From the proof of Lemma 5, it is easy to show that

$$\max_{1\le i\le n}\|\widehat{\eta}_i(\beta_0)\| = o_p\left(n^{1/2}\right). \tag{A.15}$$

Then, using the arguments similar to Owen [9], we can obtain $\|\lambda\| = O_p(n^{-1/2})$. Together with Lemma 5, applying the Taylor expansion to (10), we get that

$$\widehat{R}(\beta_0) = 2\sum_{i=1}^{n}\left\{\lambda^T\widehat{\eta}_i(\beta_0) - \frac{\left[\lambda^T\widehat{\eta}_i(\beta_0)\right]^2}{2}\right\} + o_p(1). \tag{A.16}$$

By (11), it follows that

$$\begin{aligned}0 &= \sum_{i=1}^{n}\frac{\widehat{\eta}_i(\beta_0)}{1+\lambda^T\widehat{\eta}_i(\beta_0)} \\ &= \sum_{i=1}^{n}\widehat{\eta}_i(\beta_0) - \sum_{i=1}^{n}\widehat{\eta}_i(\beta_0)\widehat{\eta}_i^T(\beta_0)\lambda \\ &\quad + \sum_{i=1}^{n}\frac{\widehat{\eta}_i(\beta_0)\left[\lambda^T\widehat{\eta}_i(\beta_0)\right]^2}{1+\lambda^T\widehat{\eta}_i(\beta_0)}.\end{aligned} \tag{A.17}$$

Then, it is easy to show that

$$\begin{aligned}\sum_{i=1}^{n}\left[\lambda^T\widehat{\eta}_i(\beta_0)\right]^2 &= \sum_{i=1}^{n}\lambda^T\widehat{\eta}_i(\beta_0) + o_p(1), \\ \lambda &= \left\{\sum_{i=1}^{n}\widehat{\eta}_i(\beta_0)\widehat{\eta}_i^T(\beta_0)\right\}^{-1}\sum_{i=1}^{n}\widehat{\eta}_i(\beta_0) + o_p\left(n^{-1/2}\right).\end{aligned} \tag{A.18}$$

Invoking (A.16)–(A.18), by some algebra calculations, we have

$$\widehat{R}(\beta_0) = \left\{\frac{1}{\sqrt{n}}\sum_{i=1}^{n}\widehat{\eta}_i(\beta_0)\right\}^T \widehat{\Sigma}_0^{-1}(\beta_0)\left\{\frac{1}{\sqrt{n}}\sum_{i=1}^{n}\widehat{\eta}_i(\beta_0)\right\}, \tag{A.19}$$

where $\widehat{\Sigma}_0(\beta_0) = n^{-1}\sum_{i=1}^{n}\widehat{\eta}_i(\beta_0)\widehat{\eta}_i^T(\beta_0)$. Invoking the proof of Lemma 5, we can obtain that $\widehat{\Sigma}_0(\beta_0) \xrightarrow{P} \Sigma_0(\beta_0)$. Hence, we get

$$\begin{aligned}\widehat{R}(\beta_0) &= \left(\Sigma^{-1/2}(\beta_0)\frac{1}{\sqrt{n}}\sum_{i=1}^{n}\widehat{\eta}_i(\beta_0)\right)^T\Gamma(\beta_0) \\ &\quad \times\left(\Sigma^{-1/2}(\beta_0)\frac{1}{\sqrt{n}}\sum_{i=1}^{n}\widehat{\eta}_i(\beta_0)\right) + o_p(1),\end{aligned} \tag{A.20}$$

where $\Gamma(\beta_0) = \Sigma^{1/2}(\beta_0)\Sigma_0^{-1}(\beta_0)\Sigma^{1/2}(\beta_0)$. Let $w_1, \ldots, w_q$ be the eigenvalues of $\Sigma_0^{-1}(\beta_0)\Sigma(\beta_0)$, and let $D = \operatorname{diag}(w_1, \ldots, w_q)$. Note that $\Sigma_0^{-1}(\beta_0)\Sigma(\beta_0)$ has the same eigenvalues as $\Gamma(\beta_0)$. Hence, there exists orthogonal matrix Q such that $Q^T D Q = \Gamma(\beta_0)$. Then, with a simple calculation we get that

$$\begin{aligned}\widehat{R}(\beta_0) &= \left(Q\Sigma^{-1/2}(\beta_0)\frac{1}{\sqrt{n}}\sum_{i=1}^{n}\widehat{\eta}_i(\beta_0)\right)^T D \\ &\quad \times\left(Q\Sigma^{-1/2}(\beta_0)\frac{1}{\sqrt{n}}\sum_{i=1}^{n}\widehat{\eta}_i(\beta_0)\right) + o_p(1).\end{aligned} \tag{A.21}$$

Note that Q is an orthogonal matrix, and by Lemma 5, Theorem 1 can be proved. □

Proof of Theorem 2. Let

$$\Xi(\beta_0) = \left\{\frac{1}{\sqrt{n}}\sum_{i=1}^{n}\widehat{\eta}_i(\beta_0)\right\}^T \widehat{\Sigma}^{-1}(\beta_0)\left\{\frac{1}{\sqrt{n}}\sum_{i=1}^{n}\widehat{\eta}_i(\beta_0)\right\}. \tag{A.22}$$

It is easy to show that $\widehat{R}_{\text{ad}}(\beta_0) = \Xi(\beta_0) + o_p(1)$. Then, using similar arguments to Yang et al. [3], it can be proved that

$$\widehat{\Sigma}(\beta_0) \xrightarrow{P} \Sigma(\beta_0). \tag{A.23}$$

Together with Lemma 5, we can prove Theorem 2. □

Acknowledgments

This research was supported by the National Natural Science Foundation of China (11101119 and 11126332), the National Social Science Foundation of China (11CTJ004), the Natural Science Foundation of Guangxi (2010GXNSFB013051), and the Philosophy and Social Sciences Foundation of Guangxi (11FTJ002).

References

[1] Q. Wang and Z. Zheng, "Asymptotic properties for the semiparametric regression model with randomly censored data," *Science in China A*, vol. 40, no. 9, pp. 945–957, 1997.

[2] Q. H. Wang and G. Li, "Empirical likelihood semiparametric regression analysis under random censorship," *Journal of Multivariate Analysis*, vol. 83, no. 2, pp. 469–486, 2002.

[3] Y. P. Yang, L. G. Xue, and W. H. Cheng, "An empirical likelihood method in a partially linear single-index model with right censored data," *Acta Mathematica Sinica*, vol. 28, no. 5, pp. 1041–1060, 2012.

[4] X. Lu and T. L. Cheng, "Randomly censored partially linear single-index models," *Journal of Multivariate Analysis*, vol. 98, no. 10, pp. 1895–1922, 2007.

[5] Q. Wang and L. Xue, "Statistical inference in partially-varying-coefficient single-index model," *Journal of Multivariate Analysis*, vol. 102, no. 1, pp. 1–19, 2011.

[6] Z. Huang, "Empirical likelihood-based inference in varying-coefficient single-index models," *Journal of the Korean Statistical Society*, vol. 40, no. 2, pp. 205–215, 2011.

[7] J. Fan and T. Huang, "Profile likelihood inferences on semiparametric varying-coefficient partially linear models," *Bernoulli*, vol. 11, no. 6, pp. 1031–1057, 2005.

[8] W. Zhang, S. Y. Lee, and X. Song, "Local polynomial fitting in semivarying coefficient model," *Journal of Multivariate Analysis*, vol. 82, no. 1, pp. 166–188, 2002.

[9] A. B. Owen, "Empirical likelihood ratio confidence regions," *Annals of Statistics*, vol. 18, no. 1, pp. 90–120, 1990.

[10] L. Xue, "Empirical likelihood local polynomial regression analysis of clustered data," *Scandinavian Journal of Statistics*, vol. 37, no. 4, pp. 644–663, 2010.

[11] Y. P. Yang, L. G. Xue, and W. H. Cheng, "Empirical likelihood for a partially linear model with covariate data missing at random," *Journal of Statistical Planning and Inference*, vol. 139, no. 12, pp. 4143–4153, 2009.

[12] X. Luo, Y. Li, Y. Ma, and Y. Zhou, "Varying-coefficient partially linear models with censored data," *Acta Mathematica Scientia*, vol. 26, pp. 79–92, 2010.

[13] T. R. Fleming and D. P. Harrington, *Counting Processes and Survival Analysis*, John Wiley and Sons, New York, NY, USA, 2005.

[14] R. Jiang and W. M. Qian, "Generalized likelihood ratio tests for varying-coefficient models with censored data," *Open Journal of Statistics*, vol. 1, no. 1, pp. 19–23, 2011.

[15] Y. P. Mack and B. W. Silverman, "Weak and strong uniform consistency of kernel regression estimates," *Zeitschrift für Wahrscheinlichkeitstheorie und Verwandte Gebiete*, vol. 61, no. 3, pp. 405–415, 1982.

14

The Inverses of Block Toeplitz Matrices

Xiao-Guang Lv and Ting-Zhu Huang

School of Mathematical Sciences, University of Electronic Science and Technology of China, Chengdu, Sichuan 611731, China

Correspondence should be addressed to Xiao-Guang Lv; xiaoguanglv@126.com

Academic Editor: Peter Grabner

We study the inverses of block Toeplitz matrices based on the analysis of the block cyclic displacement. New formulas for the inverses of block Toeplitz matrices are proposed. We show that the inverses of block Toeplitz matrices can be decomposed as a sum of products of block circulant matrices. In the scalar case, the inverse formulas are proved to be numerically forward stable, if the Toeplitz matrix is nonsingular and well conditioned.

1. Introduction

Let

$$T = \begin{bmatrix} T_0 & T_{-1} & \cdots & T_{1-n} \\ T_1 & T_0 & \ddots & T_{2-n} \\ \vdots & \ddots & \ddots & \vdots \\ T_{n-1} & T_{n-2} & \cdots & T_0 \end{bmatrix} \quad (1)$$

be an $n \times n$ block Toeplitz matrix with blocks of size $m \times m$. We use the shorthand $T = (T_{i-j})_{i,j=1}^{n}$ for a block Toeplitz matrix. The block Toeplitz systems $Tx = b$ arise in a variety of applications in mathematics, scientific computing, and engineering, for instance, image restoration problems in image processing, numerical differential equations and integral equations, time series analysis, and control theory [1–3]. If we want to solve more than one block Toeplitz linear system with the same coefficient matrix, then we usually solve four or so special block linear systems in order to determine the block Toeplitz inverse formula that expresses T^{-1} as the sum of products of block upper and block lower Toeplitz matrices. For example, Van Barel and Bultheel [4] gave an inverse formula for a block Toeplitz matrix and then derived a weakly stable algorithm to solve a block Toeplitz system of linear equations. The special structure of block Toeplitz matrices has resulted in some closed formulas for their inverses.

In the scalar case, Gohberg and Semencul [5] have shown that if the $(1,1)$st entry of the inverse of a Toeplitz matrix T is nonzero, then the first and the last columns of the inverse of the Toeplitz matrix are sufficient to reconstruct T^{-1}. In [6], an inverse formula can be obtained by the solutions of two equations (the so-called fundamental equations), where each right-hand side of them is a shifted column of the Toeplitz matrix. Later, Ben-Artzi and Shalom [7], Labahn and Shalom [8], Huckle [9], Ng et al. [10], and Heinig [11] have studied the Toeplitz matrix inverse formulas when the $(1,1)$st entry of the inverse of a Toeplitz matrix is zero. In [12], Cabay and Meleshko presented an efficient algorithm (NPADE) for numerically computing Padé approximants in a weakly stable fashion. As an application of NPADE, it has been shown that it can be used to compute stably, in a weak sense, the inverse of a Hankel or Toeplitz matrix.

When $m > 1$, additional problems are encountered in obtaining the inverse formula of a block Toeplitz matrix. A well-known formula of Gohberg and Heinig can construct T^{-1}, provided that the first and last columns together with the first and last rows of the inverse are known [13]. In [14], a set of new formulas for the inverse of a block Hankel or block Toeplitz matrix is given by Labahn et al. The formulas are expressed in terms of certain matrix Padé forms, which approximate a matrix power series associated with the block Hankel matrix. We refer the reader to [15, 16] for the computation of Padé-Hermite and simultaneous Padé systems in detail. In [7], Ben-Artzi and Shalom have proved that each inverse of a Toeplitz matrix can be constructed via three of its columns, and thus, a parametrization of the set of inverses of Toeplitz matrices is obtained. Then they generalized these results to block Toeplitz matrices; see [17]. In [18], Gemignani has shown that the representation of T^{-1} relies upon a strong structure-preserving property of the Schur complements of the nonsingular leading principal submatrices of a certain generalized Bezoutian of matrix polynomials.

In this paper, we focus our attention to the inverses of block Toeplitz matrices with the help of the block cyclic displacement. In [19], Ammar and Gader have shown that the inverse of a Toeplitz matrix can be represented as sums of products of lower triangular Toeplitz matrices and circulant matrices. The derivation of their results is based on the idea of cyclic displacement structure. In [20], Gohberg and Olshevsky also obtained new formulas for representation of matrices and their inverses in the form of sums of products of factor circulant, which are based on the analysis of the factor cyclic displacement of matrices. The results in applications to Toeplitz matrices generalized the Gohberg-Semencul, Ben-Artzi-Shalom, and Heinig-Rost formulas. Motivated by a number of related results on Toepltiz inverse formulas, we study the representation of the inverses of block Toeplitz matrices. Since block Toeplitz matrices have similar displacement structure as Toeplitz matrices, all results about Toeplitz matrices extend quite naturally to block Toeplitz matrices.

At first, for purposes of presentation, we adopt some notations that will be used throughout the paper. We denote the $m \times m$ identity matrix by I_m. E_i $(i = 0, 1, \dots, n-1)$ means a matrix of dimension $mn \times m$ in which the $(i+1)$-th block is the identity matrix I_m, and the remaining blocks are all zeros. J is called an $n \times n$ block reverse identity matrix, if J has the identity matrices on the block antidiagonal and zero elsewhere. Let $A = (A_{ij})$ be a block matrix or a block vector with block size of $m \times m$. $A' = (A_{ji})$ is said to be the block transpose of A. By $C(r, \varphi)$ we denote the block φ-circulant with the first column $r = [r_0, \dots, r_{n-1}]'$ with a block size of $m \times m$; that is, the matrix of form is defined by

$$\begin{bmatrix} r_0 & \varphi r_{n-1} & \cdots & \cdots & \varphi r_1 \\ r_1 & r_0 & \varphi r_{n-1} & \ddots & \vdots \\ \vdots & r_1 & \ddots & \ddots & \vdots \\ \vdots & \ddots & \ddots & \ddots & \varphi r_{n-1} \\ r_{n-1} & \cdots & \cdots & r_1 & r_0 \end{bmatrix}. \tag{2}$$

In this work, new formulas for the inverses of block Toeplitz matrices are proposed. As is well known, any block Toeplitz matrix T has a property of block persymmetry, that is, $JT'J = T$, where J is the block reverse identity matrix. Unfortunately, the matrix inverse does not hold this property. This can explain why the Gohberg-Heĭnig formula [13] needs the first and last columns together with the first and last rows of the inverse for constructing the inverse. For a block Toeplitz matrix T, let T_c and T_s be its block circulant and block skew-circulant parts, respectively. The representation $T = T_c \cdot I + T_s \cdot I$ shows that there will exist block circulant matrices A_i $(i = 1, 2)$ and B_i $(i = 1, 2)$ such that $T^{-1} = A_1 B_1 + A_2 B_2$. In this paper, by solving four linear equations with the same coefficient matrix T, we decompose the inverses of block Toeplitz matrices as a sum of products of block circulant matrices. From the point of view of the computation, the formulas involving block circulants instead of block upper or lower Toeplitz matrices are more attractive. For example, in the scalar case, matrix-vector product by a circulant matrix is roughly twice as fast as matrix-vector product by a triangular Toeplitz matrix of the same size. In fact, the efficient multiplication of a Toeplitz matrix and a vector is achieved by embedding the Toeplitz matrix in a circulant matrix of twice the size. The derivation of the formulas we present is based on the factor cyclic displacement of block matrices. Later on, we give some corollaries about the representations of the inverse matrices. The results generalize the Gohberg-Olshevsky, Heinig-Rost, and Ben-Artzi-Shalom formulas. In the scalar case, we consider the stability of the Toeplitz matrix inverse formulas. We show that if the Toeplitz matrix is nonsingular and well conditioned, the inverse formulas are numerically forward stable.

We summarize briefly the content of our paper. Section 2 reviews some fundamental concepts such as the factor cyclic displacement structure of a block matrix and every square block matrix can be written as a sum of products of block circulant matrices. In Section 3, we decompose the inverses of block Toeplitz matrices as a sum of products of block circulant matrices by solving four equations with the same coefficient matrix T. Some corollaries are also obtained in this section. We discuss the stability of the inverse formulas of scalar Toeplitz matrices in Section 4. Some remarks are given in Section 5.

2. The Factor Block Cyclic Displacement

The inverse formulas derived in the next section depend on some results of the factor block cyclic displacement of a block matrix. The focus of this section lies upon the definitions and the properties of the factor block cyclic displacement of a block matrix. We start with the definition of the factor block cyclic displacement.

Let Z_φ be the block φ-cyclic lower shift matrix defined by

$$\begin{bmatrix} 0 & \cdots & \cdots & 0 & \varphi I \\ I & 0 & \ddots & \ddots & 0 \\ 0 & I & \ddots & \ddots & \vdots \\ \vdots & \ddots & \ddots & \ddots & \vdots \\ 0 & \cdots & 0 & I & 0 \end{bmatrix}. \tag{3}$$

The block φ-cyclic displacement of an $n \times n$ block matrix A with blocks size of $m \times m$ is defined as

$$\nabla_\varphi (A) = A - Z_\varphi A Z'_{1/\varphi}, \quad \varphi \neq 0. \tag{4}$$

The number $\alpha = \operatorname{rank} \nabla_1 (A)$ is referred to as block φ-cyclic displacement rank of the block matrix A. Since all block

Toeplitz matrices have a special structure, it is not difficult to obtain that

$$\nabla_{\varphi}(T) = \begin{bmatrix} O & T_{-1} - \varphi T_{n-1} & \cdots & T_{1-n} - \varphi T_1 \\ T_1 - \frac{1}{\varphi} T_{1-n} & O & \cdots & O \\ \vdots & \vdots & & \vdots \\ T_{n-1} - \frac{1}{\varphi} T_{-1} & O & \cdots & O \end{bmatrix}$$

$$= E_0 \cdot \begin{bmatrix} -T_n \\ T_{-1} - \varphi T_{n-1} \\ \vdots \\ T_{1-n} - \varphi T_1 \end{bmatrix}' + \begin{bmatrix} T_n \\ T_1 - \frac{1}{\varphi} T_{1-n} \\ \vdots \\ T_{n-1} - \frac{1}{\varphi} T_{-1} \end{bmatrix} \cdot E_0', \tag{5}$$

where T_n may be an arbitrary $m \times m$ matrix. Therefore, the φ-cyclic displacement rank of a block Toeplitz matrix is at most $2m$.

By straightforward computations, it is easy to see that the equality $\nabla_{\varphi}(A) = 0$ holds if and only if A is a block φ-circulant matrix. For any invertible block matrix A there exists a simple interrelation between the block φ-cyclic displacement of the inverse matrix A^{-1} and the block φ-cyclic displacement of A. This connection is given by

$$\nabla_{\varphi}\left(A^{-1}\right) = A^{-1} \nabla_{\varphi}(A) Z_{\varphi} A^{-1} Z'_{1/\varphi}. \tag{6}$$

From the previous equation, we know that the block φ-cyclic displacement rank of the inverse matrix A^{-1} is the same as that of A. Moreover, the previous equation also allows us to make use of the block φ-cyclic displacement technique for the calculation of the inverse matrix. We will study the inverses of block Toeplitz matrices with the help of the block cyclic displacement. We need the following result.

Theorem 1. *If the block φ-cyclic displacement of A is given as the sum*

$$A - Z_{\varphi} A Z'_{1/\varphi} = \sum_{m=1}^{\alpha} g_m h'_m, \tag{7}$$

where g_i and h_i $(i = 1, \ldots, m)$ are $n \times 1$ block vectors with blocks of size $m \times m$, then

$$A = C(lr, \varphi) + \frac{\varphi}{\varphi - \psi} \left(\sum_{m=1}^{\alpha} C(g_m, \psi) \cdot C\left(h_m, \frac{1}{\varphi}\right)' \right) \tag{8}$$

or

$$A = C(lc, \varphi) + \frac{\psi}{\psi - \varphi} \left(\sum_{m=1}^{\alpha} C(g_m, \varphi) \cdot C\left(h_m, \frac{1}{\psi}\right)' \right), \tag{9}$$

where $C(lr, \varphi)$ is a block φ-circulant matrix with the same last block row as that of A and $C(lc, \varphi)$ is the block φ-circulant matrix with the same last block column as that of A.

Proof. From computing the block φ-cyclic displacement of matrix A defined by (4), it then follows that

$$\begin{aligned} \nabla_{\varphi}(A) &= \nabla_{\varphi}(C(\mathrm{lr}, \varphi)) \\ &\quad + \frac{\varphi}{\varphi - \psi} \nabla_{\varphi} \left(\sum_{m=1}^{\alpha} C(g_m, \psi) \cdot C\left(h_m, \frac{1}{\varphi}\right)' \right) \\ &= \frac{\varphi}{\varphi - \psi} \sum_{m=1}^{\alpha} \nabla_{\varphi} \left(C(g_m, \psi) \cdot C\left(h_m, \frac{1}{\varphi}\right)' \right). \end{aligned} \tag{10}$$

Note that the φ-cyclic displacement of any block matrix A is a linear operator and two block φ-circulant matrices can commute; we obtain

$$\nabla_{\varphi}(A) = \frac{\varphi}{\varphi - \psi} \sum_{m=1}^{\alpha} \nabla_{\varphi}(C(g_m, \psi)) \cdot C\left(h_m, \frac{1}{\varphi}\right)'. \tag{11}$$

By computing the block φ-cyclic displacement of block ψ-cyclic matrix $C(g_m, \psi)$ for all m, we have

$$\begin{aligned} \nabla_{\varphi}(A) &= \frac{\varphi}{\varphi - \psi} \sum_{m=1}^{\alpha} \nabla_{\varphi}(C(g_m, \psi)) \cdot C\left(h_m, \frac{1}{\varphi}\right)' \\ &= \sum_{m=1}^{\alpha} g_m \cdot E_0' \cdot C\left(h_m, \frac{1}{\varphi}\right)' \\ &\quad - \sum_{m=1}^{\alpha} E_0 \cdot \tilde{g}'_m \cdot C\left(h_m, \frac{1}{\varphi}\right)', \end{aligned} \tag{12}$$

where $\tilde{g}'_m$ for $m = 1, \ldots, \alpha$ are the first row of the block φ-circulant matrices $C(g_m, \varphi)$, respectively. Since $E_0' \cdot C(h_m, 1/\varphi)' = h'_m$, it is easy to observe that the sum of the first α terms in the previous equality is equal to $\sum_{m=1}^{\alpha} g_m h'_m$.

In the following, we need to show that the sum of the last α terms in the previous equality is equal to the zero matrix. Using the fact that $Z_{\varphi}^n = \varphi I$ and $Z_{\varphi}^{-1} = Z'_{1/\varphi}$, we check

$$\sum_{k=0}^{n-1} Z_{\varphi}^k \left(T^{-1} - Z_{\varphi} T^{-1} Z_{\varphi}^{-1}\right) Z_{\varphi}^{-k} = 0. \tag{13}$$

Thus we obtain

$$\begin{aligned} &\sum_{k=0}^{n-1} Z_{\varphi}^k \left(T^{-1} - Z_{\varphi} T^{-1} Z_{\varphi}^{-1}\right) Z_{\varphi}^{-k} \\ &\quad = \sum_{k=0}^{n-1} \sum_{m=1}^{\alpha} Z_{\varphi}^k g_m h'_m Z_{\varphi}^{-k} \\ &\quad = \sum_{m=1}^{\alpha} C(g_m, \varphi) \cdot C\left(h_m, \frac{1}{\varphi}\right)' \\ &\quad = 0. \end{aligned} \tag{14}$$

By comparing $\sum_{m=1}^{\alpha} E_0 \cdot \tilde{g}'_m \cdot C(h_m, 1/\varphi)'$ with $\sum_{m=1}^{\alpha} C(g_m, \varphi) \cdot C(h_m, 1/\varphi)'$, we obtain $\sum_{m=1}^{\alpha} E_0 \cdot \tilde{g}'_m \cdot C(h_m, 1/\varphi)' = 0$.

Finally, we discuss why $C(\mathrm{lr},\varphi)$ is the block φ-circulant with the same last block rows as the matrix A. We know that the last block row of the matrix $C(g_m,\psi)$ is independent of the number ψ. In view of $\sum_{m=1}^{\alpha} C(g_m,\varphi)\cdot C(h_m,1/\varphi)=0$, the last block rows of the matrix A and $C(\mathrm{lr},\varphi)$ are the same.

Equation (8) is now completely proved. A similar argument shows that (9) can be obtained from (7). □

3. Inverses of Block Toeplitz Matrices

In this section, we focus our attention to new formulas on the inverses of block Toeplitz matrices. It is easy to find that any block Toeplitz matrix has a property of block persymmetry, that is, $JT'J=T$, where J is the block reverse identity matrix. Unfortunately, the matrix inverse does not have this property. This can explain why the Gohberg-Heĭnig formula [13] needs the first and last block columns together with the first and last block rows of the inverse to construct it. Unlike the Gohberg-Semencul formula [5], it only needs the first and the last columns of the inverse if the $(1,1)$st entry of the inverse is nonzero.

Let

$$T(x_0,x_1,\dots,x_{n-1})'=E_0, \tag{15}$$

$$(\tilde{x}_0,\tilde{x}_1,\dots,\tilde{x}_{n-1})T=E'_{n-1}, \tag{16}$$

$$T(u_0,u_1,\dots,u_{n-1})' = \left(T_n, T_1-\frac{1}{\varphi}T_{1-n},\dots,T_{n-1}-\frac{1}{\varphi}T_{-1}\right)', \tag{17}$$

$$(\tilde{u}_0,\tilde{u}_1,\dots,\tilde{u}_{n-1})T=\left(T_{n-1}-\frac{1}{\varphi}T_{-1},\dots,T_1-\frac{1}{\varphi}T_{1-n},T_n\right). \tag{18}$$

For convenience, we will write $x,\tilde{x},u$ and $\tilde{u}$ as

$$\begin{aligned} x&=(x_0,x_1,\dots,x_{n-1})', & \tilde{x}&=(\tilde{x}_0,\tilde{x}_1,\dots,\tilde{x}_{n-1})',\\ u&=(u_0,u_1,\dots,u_{n-1})', & \tilde{u}&=(\tilde{u}_0,\tilde{u}_1,\dots,\tilde{u}_{n-1})'. \end{aligned} \tag{19}$$

From (5), (6), (15), and (17), by simple computations we obtain the following equality:

$$\begin{aligned} \nabla_\varphi\left(T^{-1}\right) &= -T^{-1}\nabla_\varphi(T)Z_\varphi T^{-1}Z'_{1/\varphi} \\ &= -\begin{bmatrix} x_0\\ x_1\\ \vdots\\ x_{n-1}\end{bmatrix} \cdot\left(-T_n, T_{-1}-\varphi T_{n-1},\dots,T_{1-n}-\varphi T_1\right) \\ &\quad\cdot Z_\varphi T^{-1}Z'_{1/\varphi} \\ &\quad -\begin{bmatrix} u_0\\ u_1\\ \vdots\\ u_{n-1}\end{bmatrix}\cdot(I_m,0,\dots,0)\cdot Z_\varphi T^{-1}Z'_{1/\varphi}. \end{aligned} \tag{20}$$

Substituting (16) and (18) into the previous equality and computing, we obtain

$$\begin{aligned} \nabla_\varphi\left(T^{-1}\right) &= \varphi\begin{bmatrix} x_0\\ x_1\\ \vdots\\ x_{n-1}\end{bmatrix} \cdot\left(T_{n-1}-\frac{1}{\varphi}T_{-1},\dots,T_1-\frac{1}{\varphi}T_{1-n},T_n\right)T^{-1}Z'_{1/\varphi} \\ &\quad -\varphi\begin{bmatrix} u_0\\ u_1\\ \vdots\\ u_{n-1}\end{bmatrix}\cdot(0,\dots,0,I_m)\cdot T^{-1}Z'_{1/\varphi} \\ &= \begin{bmatrix} x_0\\ x_1\\ \vdots\\ x_{n-1}\end{bmatrix}\left(\tilde{u}_{n-1},\varphi\tilde{u}_0,\dots,\varphi\tilde{u}_{n-2}\right) \\ &\quad -\begin{bmatrix} u_0\\ u_1\\ \vdots\\ u_{n-1}\end{bmatrix}\left(\tilde{x}_{n-1},\varphi\tilde{x}_0,\dots,\varphi\tilde{x}_{n-2}\right). \end{aligned} \tag{21}$$

Theorem 2. *Let T be a block Toeplitz matrix given by (1). If (15)–(18) are solvable for some $\varphi\neq 0$ and an arbitrary matrix T_n of order m, then*

(a) *T is invertible;*

(b) *T^{-1} can be denoted as*

$$\begin{aligned} T^{-1} = \frac{\varphi}{\varphi-\psi}\Bigg(& C(x,\psi)\cdot C(J\tilde{u},\varphi) \\ & -C\left(u-\frac{\varphi-\psi}{\varphi}E_0,\psi\right)\cdot C(J\tilde{x},\varphi)\Bigg), \end{aligned} \tag{22}$$

where $\psi(\neq\varphi)$ is an arbitrary number.

Proof. (a) Assume that there exists nonzero vector p of dimension mn such that $Tp=0$. Then by computing (16) and (18), we have $(-T_n,T_{-1}-\varphi T_{n-1},\dots,T_{1-n}-\varphi T_1)Z_\varphi p=0$ and $(I_m,0,\dots,0)Z_\varphi p=0$. These imply $\nabla_\varphi(T)Z_\varphi p=0$. Since $\nabla_\varphi(T)=T-Z_\varphi TZ'_{1/\varphi}$, we obtain $TZ_\varphi p=0$. A similar argument shows that $TZ_\varphi^k p=0$ $(k=2,3,\dots,n-1)$. Premultiplying the equalities $TZ_\varphi^k p=0$ $(k=0,1,\dots,n-1)$ by $(\tilde{x}_0,\tilde{x}_1,\dots,\tilde{x}_{n-1})$, we can conclude that all elements of the vector are equal to zero. Therefore, the invertibility of T is now proved.

(b) According to Theorem 1, we get

$$T^{-1} = C(\mathrm{lr}, \varphi) + \frac{\varphi}{\varphi - \psi}\left(C(x, \psi)\cdot C\left(\varphi Z_{1/\varphi}\tilde{u}, \frac{1}{\varphi}\right)' + C(u, \psi)\cdot C\left(\varphi Z_{1/\varphi}\tilde{x}, \frac{1}{\varphi}\right)'\right). \tag{23}$$

$C(\mathrm{lr}, \varphi)$ is the block φ-circulant matrix with the same last block row as that of T^{-1}. It is not difficult to see that $C(\mathrm{lr}, \varphi) = C(\varphi Z_{1/\varphi}\tilde{x}, 1/\varphi)' = C(J\tilde{x}, \varphi)$. After computations and arrangement, we obtain the formula (22) on the inverses of block Toeplitz matrices. □

Remark 3. In fact, the inverse formula (22) is a generalization of the formula of Gohberg and Olshevsky [20], where they consider scalar Toeplitz matrices. Since the matrix inverse does not hold the property of persymmetry, in the block case, for constructing T^{-1} we need to solve four linear equations with the same coefficient matrix T. Labahn et al. also gave this conclusion in terms of certain matrix Padé forms [14].

Remark 4. For the inverse formula (22), we take an interest in two particular cases. When $\varphi = 1$ and $\psi = 0$, the inverse formula (22) can be written as

$$T^{-1} = \left\{\begin{bmatrix} x_0 & 0 & \cdots & 0 \\ x_1 & x_0 & \ddots & \vdots \\ \vdots & \ddots & \ddots & 0 \\ x_{n-1} & \vdots & x_1 & x_0 \end{bmatrix} \times \begin{bmatrix} \tilde{u}_{n-1} & \tilde{u}_0 & \cdots & \tilde{u}_{n-2} \\ \tilde{u}_{n-2} & \tilde{u}_{n-1} & \ddots & \vdots \\ \vdots & \ddots & \ddots & \tilde{u}_0 \\ \tilde{u}_0 & \vdots & \tilde{u}_{n-2} & \tilde{u}_{n-1} \end{bmatrix} - \begin{bmatrix} u_0 - I_m & 0 & \cdots & 0 \\ u_1 & u_0 - I_m & \ddots & \vdots \\ \vdots & \ddots & \ddots & 0 \\ u_{n-1} & \vdots & u_1 & u_0 - I_m \end{bmatrix} \times \begin{bmatrix} \tilde{x}_{n-1} & \tilde{x}_0 & \cdots & \tilde{x}_{n-2} \\ \tilde{x}_{n-2} & \tilde{x}_{n-1} & \ddots & \vdots \\ \vdots & \ddots & \ddots & \tilde{x}_0 \\ \tilde{x}_0 & \vdots & \tilde{x}_{n-2} & \tilde{x}_{n-1} \end{bmatrix}\right\}. \tag{24}$$

So we obtain that the inverse of a block Toeplitz matrix can be represented as a sum of products of block lower triangular Toeplitz matrices and block circulant matrices. This is a generalization of the formula of Ammar and Gader [19]. And when $\varphi = 1$ and $\psi = -1$, the inverse formula (22) can be written as a sum of products of block skew-circulant matrices and block circulant matrices.

$$T^{-1} = \frac{1}{2}\left\{\begin{bmatrix} x_0 & -x_{n-1} & \cdots & -x_1 \\ x_1 & x_0 & \ddots & \vdots \\ \vdots & \ddots & \ddots & -x_{n-1} \\ x_{n-1} & \vdots & x_1 & x_0 \end{bmatrix} \times \begin{bmatrix} \tilde{u}_{n-1} & \tilde{u}_0 & \cdots & \tilde{u}_{n-2} \\ \tilde{u}_{n-2} & \tilde{u}_{n-1} & \ddots & \vdots \\ \vdots & \ddots & \ddots & \tilde{u}_0 \\ \tilde{u}_0 & \vdots & \tilde{u}_{n-2} & \tilde{u}_{n-1} \end{bmatrix} - \begin{bmatrix} u_0 - 2I_m & -u_{n-1} & \cdots & -u_1 \\ u_1 & u_0 - 2I_m & \ddots & \vdots \\ \vdots & \ddots & \ddots & -u_{n-1} \\ u_{n-1} & \vdots & u_1 & u_0 - 2I_m \end{bmatrix} \times \begin{bmatrix} \tilde{x}_{n-1} & \tilde{x}_0 & \cdots & \tilde{x}_{n-2} \\ \tilde{x}_{n-2} & \tilde{x}_{n-1} & \ddots & \vdots \\ \vdots & \ddots & \ddots & \tilde{x}_0 \\ \tilde{x}_0 & \vdots & \tilde{x}_{n-2} & \tilde{x}_{n-1} \end{bmatrix}\right\}. \tag{25}$$

This is a generalization of the formula of Ammar and Gader [21], where they considered the Hermitian Toeplitz matrix.

Let $y = (y_0, y_1, \dots, y_{n-1})'$, $z = (z_0, z_1, \dots, z_{n-1})'$, $\tilde{y}' = (\tilde{y}_0, \tilde{y}_1, \dots, \tilde{y}_{n-1})$ and $\tilde{z}' = (\tilde{z}_0, \tilde{z}_1, \dots, \tilde{z}_{n-1})$ be the solutions of the equations $Ty = E_k$, $Tz = E_{k+1}$, $\tilde{y}'T = E'_{n-k-1}$, and $\tilde{z}'T = E'_{n-k-2}$ for some integer k ($0 \le k \le n-1$), respectively. In addition, y_{n-1} and $\tilde{y}_0$ are both nonsingular, $\tilde{y}_0^{-1}\sum_{i=1}^{n-1}\tilde{y}_i T_{-n+i} = \sum_{i=1}^{n-1} T_{-i} y_{i-1} y_{n-1}^{-1}$ for $0 \le k \le n-2$ and $(1/\varphi)\tilde{y}_0^{-1} + \tilde{y}_0^{-1}\sum_{i=1}^{n-1}\tilde{y}_i T_{-n+i} = (1/\varphi) y_{n-1}^{-1} + \sum_{i=1}^{n-1} T_{-i} y_{i-1} y_{n-1}^{-1}$ for $k = n-1$.

For $k < n-1$, we have $T_n = T_0 + (1/\varphi)\sum_{i=1}^{n-1} T_{-i} y_{i-1} y_{n-1}^{-1} = T_0 + (1/\varphi)\tilde{y}_0^{-1}\sum_{i=1}^{n-1}\tilde{k}_i T_{-n+i}$. Then we obtain

$$\begin{aligned} Tu &= T\frac{1}{\varphi}\left(Z_\varphi y - z\right) y_{n-1}^{-1} \\ &= \frac{1}{\varphi} T \begin{bmatrix} \varphi y_{n-1} \\ 0 \\ \vdots \\ 0 \end{bmatrix} y_{n-1}^{-1} \\ &\quad + \frac{1}{\varphi} T \begin{bmatrix} 0 \\ y_0 \\ \vdots \\ y_{n-2} \end{bmatrix} y_{n-1}^{-1} - \frac{1}{\varphi} T z y_{n-1}^{-1} \end{aligned}$$

$$= \begin{bmatrix} T_0 \\ T_1 \\ \vdots \\ T_{n-1} \end{bmatrix} + \frac{1}{\varphi} \begin{bmatrix} \varphi (T_n - T_0) y_{n-1} \\ -T_{-n+1} y_{n-1} \\ \vdots \\ -T_{-1} y_{n-1} \end{bmatrix} \times y_{n-1}^{-1} + \frac{1}{\varphi} E_{k+1} y_{n-1}^{-1} - \frac{1}{\varphi} E_{k+1} y_{n-1}^{-1}$$

$$= \begin{bmatrix} T_n \\ T_1 - \frac{1}{\varphi} T_{-n+1} \\ \vdots \\ T_{n-1} - \frac{1}{\varphi} T_{-1} \end{bmatrix},$$

$$\begin{aligned} \tilde{u}'T &= \frac{1}{\varphi} \tilde{y}_0^{-1} (\tilde{y}' Z_\varphi - \tilde{z}') T \\ &= \frac{1}{\varphi} \tilde{y}_0^{-1} (0, \ldots, 0, \varphi \tilde{y}_0) T \\ &\quad + \frac{1}{\varphi} \tilde{y}_0^{-1} (\tilde{y}_1, \ldots, \tilde{y}_{n-1}, 0) - \frac{1}{\varphi} \tilde{y}^{-1} \tilde{z}' T \\ &= (T_{n-1}, \ldots, T_1, T_0) \\ &\quad + \frac{1}{\varphi} \tilde{y}_0^{-1} (-\tilde{y}_0 T_{-1}, \ldots, -\tilde{y}_0 T_{-n+1}, \varphi \tilde{y}_0 (T_n - T_0)) \\ &\quad + \frac{1}{\varphi} \tilde{y}_0^{-1} E'_{n-k-1} - \frac{1}{\varphi} \tilde{y}_0^{-1} E'_{n-k-1} \\ &= \left(T_{n-1} - \frac{1}{\varphi} T_{-1}, \ldots, T_1 - \frac{1}{\varphi} T_{-n+1}, T_n \right). \end{aligned} \tag{26}$$

For $k = n - 1$, we have $T_n - T_0 = (1/\varphi) y_{n-1}^{-1} + (1/\varphi) \sum_{k=1}^{n-1} T_{-i} y_{i-1} y_{n-1}^{-1} = (1/\varphi) \tilde{y}_0^{-1} + (1/\varphi) \tilde{y}_0^{-1} \sum_{k=1}^{n-1} \tilde{y}_k T_{-n+k}$. A similar argument shows that

$$Tu = \begin{bmatrix} T_n \\ T_1 - \frac{1}{\varphi} T_{-n+1} \\ \vdots \\ T_{n-1} - \frac{1}{\varphi} T_{-1} \end{bmatrix}, \tag{27}$$

$$\tilde{u}'T = \left(T_{n-1} - \frac{1}{\varphi} T_{-1}, \ldots, T_1 - \frac{1}{\varphi} T_{-n+1}, T_n \right).$$

As a result, we obtain that $u = (1/\varphi)(Z_\varphi y - z) y_{n-1}^{-1}$ and $\tilde{u}' = (1/\varphi) y_0^{-1} (\tilde{y}' Z_\varphi - \tilde{z}')$ are the solutions of (17) and (18), respectively. This result together with Theorem 2 yields the following corollary.

Corollary 5. *Let T be the block Toeplitz matrix given by* (1). *If* (15) *and* (16), $Ty = E_k, Tz = E_{k+1}, \tilde{y}'T = E'_{n-k-1}$, *and* $\tilde{z}'T = E'_{n-k-2}$ *are solvable for some integer k* ($0 \le k \le n-1$), *in addition, y_{n-1} and $\tilde{y}_0$ are both nonsingular,* $\tilde{y}_0^{-1} \sum_{i=1}^{n-1} \tilde{y}_i T_{-n+i} = \sum_{i=1}^{n-1} T_{-i} y_{i-1} y_{n-1}^{-1}$ *for* $0 \le k \le n-2$ *and* $(1/\varphi)\tilde{y}_0^{-1} + \tilde{y}_0^{-1} \sum_{i=1}^{n-1} \tilde{y}_i T_{-n+i} = (1/\varphi) y_{n-1}^{-1} + \sum_{i=1}^{n-1} T_{-i} y_{i-1} y_{n-1}^{-1}$ *for $k = n-1$, then T is invertible and*

$$\begin{aligned} T^{-1} = \frac{1}{\varphi - \psi} \Big(& C(x, \psi) \cdot C\left(J \tilde{y}_0^{-1} \left(Z'_\varphi \tilde{y} - \tilde{z} \right), \varphi \right) \\ & - C\left(\left(Z_\psi y - z \right) y_{n-1}^{-1}, \psi \right) \cdot C(J\tilde{x}, \varphi) \Big). \end{aligned} \tag{28}$$

Corollary 5 shows how to construct the inverse of a block Toeplitz matrix, provided that the first block column and last block row, along with two successive block columns and block rows of the inverse, are given. More precisely, the choice of $k = n - 1$ gives that $E_{k+1} = E_0$ and $E_{n-k-2} = E_{n-1}$. This means that $z = x$ and $\tilde{z} = \tilde{x}$. As a consequence, we have the following result.

Corollary 6. *Let T be the block Toeplitz matrix given by* (1). *If* (15) *and* (16), $Ty = E_{n-1}$, *and* $\tilde{y}'T = E_0$, *are solvable, in addition, y_{n-1} and $\tilde{y}_0$ are both nonsingular,* $\tilde{y}_0^{-1} \sum_{k=1}^{n-1} \tilde{y}_k T_{-n+k} = \sum_{k=1}^{n-1} T_{-i} y_{i-1} y_{n-1}^{-1}$, *then T is invertible and*

$$\begin{aligned} T^{-1} = \frac{1}{\varphi - \psi} \Big(& C(x, \psi) \cdot C\left(J \tilde{y}_0^{-1} \left(Z'_\varphi \tilde{y} - \tilde{x} \right), \varphi \right) \\ & - C\left(\left(Z_\psi y - x \right) y_{n-1}^{-1}, \psi \right) \cdot C(J\tilde{x}, \varphi) \Big). \end{aligned} \tag{29}$$

Corollary 6 is a generalization of the well-known formula of Gohberg and Heĭnig. In [13], the authors have shown that the first and last block columns together with the first and last block rows of the inverse are sufficient to construct the inverse T^{-1}. In addition, we decompose the inverses of block Toeplitz matrices as a sum of products of block circulant matrices.

4. Stability Analysis

In the scalar case, the inverse formula (22) in Section 3 is just the well-known formula of Gohberg and Olshevsky; see Theorem 3.1 in [20]. Hence, solving two linear systems are sufficient to reconstruct the inverse of the Toeplitz matrix T^{-1}. Let $x = (x_0, x_1, \ldots, x_{n-1})'$ and $u = (u_0, u_1, \ldots, u_{n-1})'$ be the solutions of (15) and (17), respectively. In the scalar case, formula (22) can be written as

$$\begin{aligned} T^{-1} = \frac{\varphi}{\varphi - \psi} \Big(& C(x, \psi) \cdot C(u, \varphi) \\ & - C\left(u - \frac{\varphi - \psi}{\varphi} E_0, \psi \right) \cdot C(x, \varphi) \Big), \end{aligned} \tag{30}$$

where $\varphi \neq 0$ and $\psi (\neq \varphi)$ is an arbitrary number. For convenience, we rewrite T^{-1} as

$$T^{-1} \doteq \frac{\varphi}{\varphi - \psi} (C_{11} C_{12} - C_{21} C_{22}). \tag{31}$$

In this section, we want to show that the evaluation of some scalar inverse formulas in Section 3 is forward stable. An algorithm is called forward stable if for all well-conditioned problems, the computed solution $\hat{x}$ is close to the true solution x in the sense that the relative error $\|x - \hat{x}\|_2/\|x\|_2$ is small. In the matrix computation, roundoff errors occur. Let $A, B \in C^{n,n}$ and $\alpha \in C$; if we neglect the $O(\varepsilon^2)$ terms, then for any floating-point arithmetic with machine precision ε, there are

$$\begin{aligned} &\mathrm{fl}(\alpha A) = \alpha A + E, \\ &\|E\|_F \le \varepsilon|\alpha|\,\|A\|_F \le \varepsilon\sqrt{n}|\alpha|\,\|A\|_2, \\ &\mathrm{fl}(A+B) = A + B + E, \\ &\|E\|_F \le \varepsilon\|A+B\|_F \le \varepsilon\sqrt{n}\|A+B\|_2, \\ &\mathrm{fl}(AB) = AB + E, \qquad \|E\|_F \le \varepsilon n\|A\|_F\|B\|_F. \end{aligned} \tag{32}$$

See [22]. According to the floating-point arithmetic, we have the following result.

Theorem 7. *Let T be a nonsingular Toeplitz matrix given by (1) with $m = 1$ and let it be well conditioned then the inversion formula (30) is forward stable for $\varphi \neq 0$ and a given number $\psi(\neq \varphi)$.*

Proof. Assume that we have computed the solutions $\hat{x}$, $\hat{u}$ in the inverse formula (30) which are perturbed by the normwise relative errors bounded by $\hat{\varepsilon}$

$$\|\hat{x}\|_2 \le \|x\|_2(1+\hat{\varepsilon}), \qquad \|\hat{u}\|_2 \le \|u\|_2(1+\hat{\varepsilon}). \tag{33}$$

Let $N = \max\{1, \psi\}$ and $M = \max\{1, \varphi\}$. Then, we have

$$\begin{aligned} &\|C_{11}\|_F \le \sqrt{n}N\|x\|_2, \\ &\|C_{12}\|_F \le \sqrt{n}M\|u\|_2, \\ &\|C_{21}\|_F \le \sqrt{n\left(\frac{\varphi-\psi}{\varphi}\right)^2 + nN^2\|u\|_2^2} \\ &\qquad \le \sqrt{n}\left|\frac{\varphi-\psi}{\varphi}\right| + \sqrt{n}N\|u\|_2, \\ &\|C_{22}\|_F \le \sqrt{n}M\|x\|_2. \end{aligned} \tag{34}$$

By using the perturbed solutions $\hat{x}$, $\hat{u}$, the inverse formula (30) can be expressed as

$$\begin{aligned} \hat{T}^{-1} &= \mathrm{fl}\left(\frac{\varphi}{\varphi-\psi}\left(\hat{C}_{11}\hat{C}_{12} + \hat{C}_{21}\hat{C}_{22}\right)\right) \\ &= \mathrm{fl}\left(\frac{\varphi}{\varphi-\psi}\left((C_{11}+\Delta C_{11})(C_{12}+\Delta C_{12})\right.\right. \\ &\qquad \left.\left. + (C_{21}+\Delta C_{21})(C_{22}+\Delta C_{22})\right)\right) \\ &= T^{-1} + \frac{\varphi}{\varphi-\psi}\left(\Delta C_{11}C_{12} + C_{11}\Delta C_{12} + \Delta C_{21}C_{22}\right. \\ &\qquad \left. + C_{21}\Delta C_{22} + E + F\right) + G. \end{aligned} \tag{35}$$

Here, E is the matrix containing the error which results from computing the matrix products, F contains the error from subtracting the matrices, and G represents the error of the multiplication by $\varphi/(\varphi - \psi)$. For the error matrices ΔC_{11}, ΔC_{12}, ΔC_{21}, and ΔC_{22}, we have

$$\begin{aligned} &\|\Delta C_{11}\|_F \le \hat{\varepsilon}\|C_{11}\|_F \le \hat{\varepsilon}\sqrt{n}N\|x\|_2, \\ &\|\Delta C_{12}\|_F \le \hat{\varepsilon}\|C_{12}\|_F \le \hat{\varepsilon}\sqrt{n}M\|u\|_2, \\ &\|\Delta C_{21}\|_F \le \hat{\varepsilon}\|C_{21}\|_F \le \hat{\varepsilon}\sqrt{n}\left(\left|\frac{\varphi-\psi}{\varphi}\right| + N\|u\|_2\right), \\ &\|\Delta C_{22}\|_F \le \hat{\varepsilon}\|C_{22}\|_F = \hat{\varepsilon}\sqrt{n}M\|x\|_2. \end{aligned} \tag{36}$$

It follows that

$$\begin{aligned} \|E\|_2 &\le \|E\|_F \\ &\le \varepsilon n\left(\|C_{11}\|_F\|C_{12}\|_F + \|C_{21}\|_F\|C_{22}\|_F\right) \\ &\le \varepsilon n^2 M\|x\|_2\left(\left|\frac{\varphi-\psi}{\varphi}\right| + 2N\|u\|_2\right), \\ \|F\|_2 &\le \varepsilon\sqrt{n}\left\|T^{-1}\right\|_2, \end{aligned} \tag{37}$$

$$\begin{aligned} \|G\|_2 &\le \|G\|_F \\ &\le \varepsilon\frac{\varphi}{\varphi-\psi}\left(\|C_{11}\|_F\|C_{12}\|_F + \|C_{21}\|_F\|C_{22}\|_F\right) \\ &\le \frac{\varepsilon n M\varphi}{\varphi-\psi}\|x\|_2\left(\left|\frac{\varphi-\psi}{\varphi}\right| + 2N\|u\|_2\right). \end{aligned}$$

After adding all these error bounds, we have

$$\begin{aligned} \left\|\hat{T}^{-1} - T^{-1}\right\|_2 &\le \frac{\varphi}{\varphi-\psi}\left(2\hat{\varepsilon}n + \varepsilon n^2 + \varepsilon n\right) \\ &\quad \times M\|x\|_2\left(\left|\frac{\varphi-\psi}{\varphi}\right| + 2N\|u\|_2\right) \\ &\quad + \frac{\varepsilon\sqrt{n}\varphi}{\varphi-\psi}\left\|T^{-1}\right\|_2. \end{aligned} \tag{38}$$

Note that x and u are the solutions of (15) and (17), respectively; then $\|x\|_2 \le \|T^{-1}\|_2$ and $\|u\|_2 \le \|T^{-1}\|_2\|f\|_2$, where $f = (T_n, T_1 - (1/\varphi)T_{1-n}, \ldots, T_{n-1} - (1/\varphi)T_{-1})'$. Thus, the relative error is

$$\begin{aligned} \frac{\left\|\hat{T}^{-1} - T^{-1}\right\|_2}{\left\|T^{-1}\right\|_2} &\le \frac{\varphi}{\varphi-\psi}\left(2\hat{\varepsilon}n + \varepsilon n^2 + \varepsilon n\right) \\ &\quad \times M\left(\left|\frac{\varphi-\psi}{\varphi}\right| + 2N\left\|T^{-1}\right\|_2\|f\|_2\right) + \frac{\varepsilon\sqrt{n}\varphi}{\varphi-\psi}. \end{aligned} \tag{39}$$

As T is well conditioned, thus, $\|T^{-1}\|_2$ is finite. Obviously, $\|f\|_2$ is finite. Therefore, the formula (30) presented in this section is forward stable. □

For example, there are two particular cases of the inverse formula (30). When $\varphi = 1$ and $\psi = -1$, the inverse formula $T^{-1} = (1/2)(C(x,-1)\cdot C(u,1) - C(u - 2E_0, -1)\cdot C(x,1))$, which denotes T^{-1} as a sum of products of skew-circulant and circulant matrices, is numerically forward stable for a well-conditioned and nonsingular Toeplitz matrix T. In fact, this is the formula of Ammar and Gader in [21]. And when $\varphi = 1$ and $\psi = 0$, this is the formula of Ammar and Gader in [19]. The inverse formula $T^{-1} = (C(x,0) \cdot C(u,1) - C(u - E_0, 0) \cdot C(x,1))$ in which skew-circulant matrices are replaced by upper triangular factors is also forward stable, if T is nonsingular and well conditioned.

We also find that even when $\psi = \infty$ and a given number $\varphi \neq 0$, the inverse formula (30) is still numerically forward stable. A similar argument can show that the inverse formulas in Corollaries 5 and 6 for the scalar case are all numerically forward stable when T is nonsingular and well conditioned.

5. Concluding Remarks

In this paper, with the help of the block cyclic displacement, the inverses of block Toeplitz matrices are discussed. New formulas for inversion of block Toeplitz matrices are obtained, which can be decomposed in a sum of products of block circulant matrices. In the scalar case, we analyze the stability of the inverse formulas. In addition, it should be remarked that any mention of matrix Padé forms for representing inverses of Toeplitz matrices could also be given in terms of solving block linear systems. Also the proofs of the inverses formulas in [14] could also be extended to prove the inverse formulas in this work.

Acknowledgments

This research is supported by NSFC (61170311), 973 Program (2013CB329404), and Sichuan Province Sci. & Tech. Research Project (2012GZX0080). The authors are grateful to the unknown referee for pertinent comments, especially for suggesting four references, which contributed substantially to this revised work.

References

[1] R. H.-F. Chan and X.-Q. Jin, *An Introduction to Iterative Toeplitz Solvers*, vol. 5 of *Fundamentals of Algorithms*, Society for Industrial and Applied Mathematics (SIAM), Philadelphia, Pa, USA, 2007.

[2] M. K. Ng, *Iterative Methods for Toeplitz Systems*, Numerical Mathematics and Scientific Computation, Oxford University Press, New York, NY, USA, 2004.

[3] X.-Q. Jin, *Developments and Applications of Block Toeplitz Iterative Solvers*, vol. 2 of *Combinatorics and Computer Science*, Kluwer Academic, Dordrecht, The Netherlands, 2002.

[4] M. Van Barel and A. Bultheel, "A lookahead algorithm for the solution of block Toeplitz systems," *Linear Algebra and its Applications*, vol. 266, pp. 291–335, 1997.

[5] I. C. Gohberg and A. A. Semencul, "The inversion of finite Toeplitz matrices and their continual analogues," *Matematicheskie Issledovaniya*, vol. 7, no. 2, pp. 201–223, 1972.

[6] G. Heinig and K. Rost, *Algebraic Methods for Toeplitz-Like Matrices and Operators*, vol. 19 of *Mathematical Research*, Birkhäauser, Berlin, Germany, 1984.

[7] A. Ben-Artzi and T. Shalom, "On inversion of Toeplitz and close to Toeplitz matrices," *Linear Algebra and its Applications*, vol. 75, pp. 173–192, 1986.

[8] G. Labahn and T. Shalom, "Inversion of Toeplitz matrices with only two standard equations," *Linear Algebra and its Applications*, vol. 175, pp. 143–158, 1992.

[9] T. Huckle, "Computations with Gohberg-Semencul-type formulas for Toeplitz matrices," *Linear Algebra and its Applications*, vol. 273, pp. 169–198, 1998.

[10] M. K. Ng, K. Rost, and Y.-W. Wen, "On inversion of Toeplitz matrices," *Linear Algebra and its Applications*, vol. 348, pp. 145–151, 2002.

[11] G. Heinig, "On the reconstruction of Toeplitz matrix inverses from columns," *Linear Algebra and its Applications*, vol. 350, pp. 199–212, 2002.

[12] S. Cabay and R. Meleshko, "A weakly stable algorithm for Padé approximants and the inversion of Hankel matrices," *SIAM Journal on Matrix Analysis and Applications*, vol. 14, no. 3, pp. 735–765, 1993.

[13] I. C. Gohberg and G. Haĭnig, "Inversion of finite Toeplitz matrices consisting of elements of a noncommutative algebra," *Revue Roumaine de Mathématiques Pures et Appliquées*, vol. 19, pp. 623–663, 1974.

[14] G. Labahn, D. K. Choi, and S. Cabay, "The inverses of block Hankel and block Toeplitz matrices," *SIAM Journal on Computing*, vol. 19, no. 1, pp. 98–123, 1990.

[15] S. Cabay, A. R. Jones, and G. Labahn, "Computation of numerical Padé-Hermite and simultaneous Padé systems. I. Near inversion of generalized Sylvester matrices," *SIAM Journal on Matrix Analysis and Applications*, vol. 17, no. 2, pp. 248–267, 1996.

[16] S. Cabay, A. R. Jones, and G. Labahn, "Computation of numerical Padé-Hermite and simultaneous Padé systems. II. A weakly stable algorithm," *SIAM Journal on Matrix Analysis and Applications*, vol. 17, no. 2, pp. 268–297, 1996.

[17] A. Ben-Artzi and T. Shalom, "On inversion of block Toeplitz matrices," *Integral Equations and Operator Theory*, vol. 8, no. 6, pp. 751–779, 1985.

[18] L. Gemignani, "Schur complements of Bezoutians and the inversion of block Hankel and block Toeplitz matrices," *Linear Algebra and its Applications*, vol. 253, pp. 39–59, 1997.

[19] G. Ammar and P. Gader, "A variant of the Gohberg-Semencul formula involving circulant matrices," *SIAM Journal on Matrix Analysis and Applications*, vol. 12, no. 3, pp. 534–540, 1991.

[20] I. Gohberg and V. Olshevsky, "Circulants, displacements and decompositions of matrices," *Integral Equations and Operator Theory*, vol. 15, no. 5, pp. 730–743, 1992.

[21] G. Ammar and P. Gader, "New decompositions of the inverse of a toeplitz matrix," in *Proceedings of the International Symposium MTNS-89: Signal Processing, Scattering and Operator Theory, and Numerical Methods*, vol. 3, pp. 421–428, Birkhäauser, Boston, Mass, USA, 1990.

[22] G. H. Golub and C. F. Van Loan, *Matrix Computations*, vol. 3 of *Johns Hopkins Series in the Mathematical Sciences*, Johns Hopkins University Press, Baltimore, Md, USA, 2nd edition, 1989.

Convergence and Characterization of Wavelets Associated with Dilation Matrix in Variable L^p Spaces

Devendra Kumar

Department of Mathematics, Research and Post Graduate Studies, M. M. H. College, Model Town, Ghaziabad, Uttar Pradesh 20/001, India

Correspondence should be addressed to Devendra Kumar; d_kumar001@rediffmail.com

Academic Editor: Ding-Xuan Zhou

We study the convergence of wavelet expansions associated with dilation matrix in the variable $L^{p(\cdot)}(\mathbb{R}^d)$ spaces using the approximate identity. Also, we obtain conditions for the wavelet characterizations of $L^{p(\cdot)}(\mathbb{R}^d)$ with respect to norm estimates. Moreover, the results of Izuki (2008) and Kumar (2009) have been extended.

1. Introduction

Various authors such as Cruz-Uribe et al. [1, 2], Diening [3, 4], Nekvinda [5], Izuki [6], and Kopaliani [7] studied the conditions for the boundedness of the Hardy-Littlewood maximal function M on variable L^p spaces $L^{p(\cdot)}(\Omega)$. In the case $\Omega = \mathbb{R}^d$, Diening [4] has proved that the boundedness of M on $L^{p(\cdot)}(\mathbb{R}^d)$ is equivalent to that on $L^{p'(\cdot)}(\mathbb{R}^d)$, where $p'(\cdot)$ is the conjugate exponent function. Cruz-Uribe et al. [8] used the extrapolation technique to give a different proof of the result obtained by Izuki [6].

The convergence of the wavelet series has been studied by Meyer [9], Walter [10], and Kelly et al. [11, 12]. Meyer was amongst the first to study convergence results for wavelet expansions. Wavelet theory is based on analyzing signals to their components by using a set of bases functions. An important characteristic of wavelet bases functions is that they relate to each other by simple scaling and translation. Meyer [9] showed that the regular wavelet expansions converge in L^p, $1 \le p \le \infty$, for expansions of uniformly continuous functions; the expansions of continuous functions converge everywhere. His results are based on the assumption of the regularity for basic wavelets and their derivatives. Walter [10] studied the results for regular wavelet expansions of continuous functions. Kelly et al. [11, 12] extended and obtained results analogous to those obtained by Carleson [13] and Hunt [14] for the Fourier series. The results of Kelly et al. [11, 12] only assumed that the wavelets are bounded by radial decreasing L^1-functions. The behavior of wavelet expansions outside the Lebesgue set is discussed by Reyes [15], for $1 \le p < \infty$, whereas Kelly et al. [11] proved that the wavelet expansion of a function in L^p spaces converges pointwise everywhere on the Lebesgue set of a given function. Tao [16] has extended the results of Meyer [9] and Kelly et al. [11, 12] and showed that the wavelet expansion of any L^p-function converges pointwise almost everywhere under the wavelet projections, hard sampling, and soft sampling summation methods, for $1 < p < \infty$.

The pointwise convergence of wavelet expansions associated with a dilation matrix has been studied by Manchanda et al. [17]. For the applications point of view, sometimes it is reasonable to study the characterization and convergence of wavelets associated with a dilation matrix in variable $L^{p(\cdot)}(\mathbb{R}^d)$ spaces. The construction of wavelet bases in variable exponent spaces was first studied by Sharapudinov [18, 19] who gave a basis of $L^{p(\cdot)}([0,1])$ using the Haar system. In this paper, we study the convergence and characterization of wavelet expansions associated with dilation matrix in variable $L^{p(\cdot)}(\Omega)$ spaces where $\Omega \subseteq \mathbb{R}^d$.

2. Definitions and Auxiliary Results

Let A denote a dilation matrix in $\mathbb{Z}^d$; that is, A is a $d \times d$ matrix such that

(i) $A(\mathbb{Z}^d) \subset \mathbb{Z}^d$;

(ii) all eigen values λ of A satisfy $|\lambda| > 1$.

Property (i) implies that A has integer entries, and hence, $|\det A|$ is an integer, and from property (ii) we know that $|\det A| > 1$. Considering $\mathbb{Z}^d$ as an additive group, we find that $A\mathbb{Z}^d$ is a normal subgroup of $\mathbb{Z}^d$; so, we can form the cosets of $A\mathbb{Z}^d$ in $\mathbb{Z}^d$. It is a well-known fact that the number of distinct cosets of $A\mathbb{Z}^d$ in $\mathbb{Z}^d$ is equal to $|\det A|$ [20].

2.1. Variable L^p Spaces and Hardy-Littlewood Maximal Function. First, we define variable L^p spaces on Ω, $\Omega \subseteq \mathbb{R}^d$, and introduce some known results on the boundedness of the Hardy-Littlewood maximal operator. Throughout the paper, we consider a measurable function $p : \mathbb{R}^d \to [1, \infty)$, and $p'(\cdot)$ means the conjugate exponent function of $p(\cdot)$; that is, $p'(\cdot)$ satisfies $1/p(x) + 1/p'(x) = 1$ $(x \in \Omega)$. χ_E denotes the characteristic function of a measurable set $E \subset \mathbb{R}^d$.

Definition 1. The variable L^p space $L^{p(\cdot)}(\Omega)$ consists of all measurable functions f defined on Ω such that $\rho_p(f/\lambda) < \infty$ for some constant $\lambda > 0$, where $\rho_p(f) = \int_\Omega |f(x)|^{p(x)} dx$.

$L^{p(\cdot)}(\Omega)$ is a Banach space with the norm defined by

$$\|f\|_{L^{p(\cdot)}(\Omega)} = \inf\left\{\lambda > 0 : \rho_p\left(\frac{f}{\lambda}\right) \le 1\right\}. \tag{1}$$

Now, we define two classes of exponent functions. We use the following notations:

$$\begin{aligned} p_- &= \operatorname{ess\,inf}\{p(x) : x \in \Omega\}, \\ p_+ &= \operatorname{ess\,sup}\{p(x) : x \in \Omega\}. \end{aligned} \tag{2}$$

Definition 2. The set $\wp(\Omega)$ consists of all $p(\cdot)$ satisfying $p_- > 1$ and $p_+ < \infty$.

Given a function $f \in L^1_{\text{loc}}(\Omega)$, the Hardy-Littlewood maximal operator M is defined by

$$Mf(x) = \sup_{x\in Q}\frac{1}{|Q|}\int_Q |f(y)|\,dy \tag{3}$$

$(x \in \Omega)$, where the supremum is taken over all cubes Q containing x.

Definition 3. $B(\Omega)$ is the set of $p(\cdot) \in \wp(\Omega)$ satisfying the condition that M is bounded on $L^{p(\cdot)}(\Omega)$.

Cruz-Uribe et al. [1] and Nekvinda [5] proved the following sufficient conditions independently. Note that Nekvinda [5] gave a more general condition in place of (5).

Proposition 4 (see [1, 5]). *Let $p(\cdot) \in \wp(\Omega)$. Suppose that*

$$|p(x) - p(y)| \le \frac{C}{-\log(|x-y|)} \quad \text{if } |x-y| \le \frac{1}{2}, \tag{4}$$

$$|p(x) - p(y)| \le \frac{C}{\log(e+|x|)} \quad \text{if } |y| \ge |x|, \tag{5}$$

where $C > 0$ is a constant independent of x and y. Then, one has $p(\cdot) \in B(\Omega)$. If $p(\cdot)$ satisfies (4) and (5), then so does $p'(\cdot)$. Diening [4] proved the following equivalence.

Proposition 5 (see [4, Theorem 8.1]). *Let $p(\cdot) \in \wp(\mathbb{R}^d)$. Then, the following four conditions are equivalent:*

(1) $p(\cdot) \in B(\mathbb{R}^d)$;

(2) $p'(\cdot) \in B(\mathbb{R}^d)$;

(3) *there exists a constant $q \in (1, p_-)$ such that $p(\cdot)/q \in B(\mathbb{R}^d)$;*

(4) *there exists a constant $q \in (1, p_-)$ such that $(p(\cdot)/q)' \in B(\mathbb{R}^d)$.*

Definition 6 (wavelets). Let $\{\varphi^l : l = 1, 2, \ldots, |\det A| - 1\}$ be a set of functions belonging to $L^2(\mathbb{R}^d)$. Define

$$\begin{aligned} \varphi^l_{j,k}(x) &= |\det A|^{j/2}\varphi^l\left(A^j x - k\right) \\ &= |\det A|^{j/2}\varphi^l\left(A^j x_1 - k_1, \ldots, A^j x_d - k_d\right), \\ &\qquad \left(x = (x_1, \ldots, x_d) \in \mathbb{R}^d\right) \end{aligned} \tag{6}$$

for each $l = 1, 2, \ldots, |\det A| - 1$, $j \in \mathbb{Z}$, and $k = (k_1, \ldots, k_d) \in \mathbb{Z}^d$. The sequence $\{\varphi^l : l = 1, 2, \ldots, |\det A| - 1\}$ is a wavelet set if $\{\varphi^l_{j,k} : l = 1, 2, \ldots, |\det A| - 1\}$, $j \in \mathbb{Z}$, $k \in \mathbb{Z}^d$, forms an orthonormal basis in $L^2(\mathbb{R}^d)$, and each φ^l is a wavelet.

We generally need suitable smoothness or decay on wavelets in order to obtain wavelet characterizations of function spaces. Now, we define the class $S(\mathbb{R}^d)$ of rapidly decreasing C^∞-functions on $\mathbb{R}^d$ as

$$\begin{aligned} S\left(\mathbb{R}^d\right) = \Big\{ f : \mathbb{R}^d \longrightarrow \mathbb{R}^d : \sup_{x\in\mathbb{R}^d}\left(x^n \frac{d^m}{dx^m} f\right)(x) < \infty, \\ n \in N_0,\ m \in N_0^d,\ N_0 = N \cup \{0\} \Big\}. \end{aligned} \tag{7}$$

In this paper, we use a wavelet set which consists of wavelets in $S(\mathbb{R}^d)$. We can construct it by means of a multiresolution analysis (MRA) [9, 20, 21].

Definition 7. An MRA is a sequence $\{V_j\}_{j\in\mathbb{Z}}$ of closed subspaces of $L^2(\mathbb{R}^d)$ associated with dilation matrix A such that the following conditions hold:

(1) $V_j \subset V_{j+1}$ for all $j \in \mathbb{Z}$;

(2) $\cup_{j\in\mathbb{Z}} V_j$ is dense in $L^2(\mathbb{R}^d)$;

(3) $\cap_{j\in\mathbb{Z}} V_j = \{0\}$;

(4) $f \in V_j$ holds if and only if $f(Ax) \in V_{j+1}$ for all $j \in \mathbb{Z}$;

(5) $f \in V_0$ holds if and only if $f(x-k) \in V_0$ for every $k \in \mathbb{Z}^d$;

(6) there exists a function $\phi \in V_0$ such that the system $\{\phi(x-k)\}_{k\in\mathbb{Z}^d}$ is an orthonormal basis in V_0. ϕ is called a scaling function of $\{V_j\}_{j\in\mathbb{Z}}$.

Denoting the orthogonal complement of V_j in V_{j+1} by W_j, it is easy to show that the integer translations $\{\phi(x-k) : k \in \mathbb{Z}^d\}$ of the mother wavelet ϕ form an orthonormal basis for W_0. Then, we can construct a wavelet set $\{\varphi^l : l = 1, 2, \ldots, |\det A| - 1\}$ associated with $\{V_j\}_{j\in\mathbb{Z}}$ and satisfying the assumption that each wavelet φ^l is in $S(\mathbb{R}^d)$, and $\{\varphi^l_{j,k}(x) = |\det A|^{j/2}\varphi^l(A^j x - k) : k \in \mathbb{Z}^d\}$ forms an orthonormal basis in W_j for every $j \in \mathbb{Z}$. Since $\cup_{j\in\mathbb{Z}} V_j$ is dense in $L^2(\mathbb{R}^d)$, the collection $\{\varphi^l_{j,k}\}$ is an orthonormal basis of $L^2(\mathbb{R}^d)$.

Given an MRA $\{V_j\}_{j\in\mathbb{Z}}$ with scaling function ϕ, let P_j and Q_j, respectively, be the orthogonal projections onto the spaces V_j and W_j with the kernels $P_j(x,y)$ and $Q_j(x,y)$, defined as

$$P_j(x,y) = \sum_{l=1}^{|\det A|-1} \sum_{k\in\mathbb{Z}^d} \phi^l_{j,k}(x)\overline{\phi^l_{j,k}(y)}, \tag{8}$$

where $\phi^l_{j,k}(x) = |\det A|^{j/2}\phi^l(A^j x - k)$ and

$$Q_j(x,y) = \sum_{l=1}^{|\det A|-1} \sum_{k\in\mathbb{Z}^d} \varphi^l_{j,k}(x)\overline{\varphi^l_{j,k}(y)}. \tag{9}$$

In view of $V_{j+1} = V_j \oplus W_j$, $P_j(x,y)$ can be written as

$$\begin{aligned} P_j(x,y) &= \sum_{m<j} Q_m(x,y) \\ &= \sum_{l=1}^{|\det A|-1} \sum_{m<j} \sum_{k\in\mathbb{Z}^d} \varphi^l_{m,k}(x)\overline{\varphi^l_{m,k}(y)}. \end{aligned} \tag{10}$$

The sequence of projections $\{P_j f(x)\}$ is called the multiresolution expansion of f. The scaling expansion of any function $f \in L^2(\mathbb{R}^d)$ is defined as

$$\begin{aligned} f(x) = &\sum_{l=1}^{|\det A|-1} \sum_{k\in\mathbb{Z}^d} a_{l,j,k} |\det A|^j \phi^l\left(A^j x - k\right) \\ &+ \sum_{l=1}^{|\det A|-1} \sum_{j\le m} \sum_{k\in\mathbb{Z}^d} b_{l,j,k}\varphi^l_{j,k}(x), \end{aligned} \tag{11}$$

where

$$\begin{aligned} a_{l,j,k} &= \int_{\mathbb{R}^d} f(x) |\det A|^j \phi^l\left(A^j x - k\right) dx, \\ b_{l,j,k} &= \int_{\mathbb{R}^d} f(x)\varphi^l_{j,k}(x)\, dx. \end{aligned} \tag{12}$$

Therefore the wavelet expansion associated with dilation matrix A of f is given by

$$f(x) \sim \sum_{l=1}^{|\det A|-1} \sum_{j\in\mathbb{Z}} \sum_{k\in\mathbb{Z}^d} b_{l,j,k}\varphi^l_{j,k}(x), \tag{13}$$

where the wavelet coefficients $b_{l,j,k} = \langle f, \varphi^l_{j,k}\rangle$ are defined as earlier.

By considering convergence in the sense of $L^2(\mathbb{R}^d)$, we may write

$$\begin{aligned} f(x) &= \sum_{l=1}^{|\det A|-1} \sum_{j\in\mathbb{Z}} \sum_{k\in\mathbb{Z}^d} b_{l,j,k}\varphi^l_{j,k}(x), \\ f(x) &= \sum_{l=1}^{|\det A|-1} \sum_{k\in\mathbb{Z}^d} a_{l,j,k}|\det A|^j \phi^l\left(A^j x - k\right) \\ &\quad + \sum_{l=1}^{|\det A|-1} \sum_{j\le m} \sum_{k\in\mathbb{Z}^d} b_{l,m,k}\varphi^l_{j,k}(x) \\ &= f^l_j(x) + r^l_m(x). \end{aligned} \tag{14}$$

The function $f^l_m(x)$ is in the space V_m. It is the projection of f onto V_m and can be written as

$$f^l_m(x) = \int_{\mathbb{R}^d} q^l_m(x,t) f(t)\, dt, \tag{15}$$

where

$$\begin{aligned} q^l_m(x,t) &= |\det A|^m q^l_0\left(A^m x, A^m t\right) \\ &= \lambda^d q_0(\lambda x, \lambda t), \quad (\lambda = A^m) \\ &= \sum_{l=1}^{|\det A|-1} \sum_{k\in\mathbb{Z}^d} \phi^l_{m,k}(x)\phi^l_{m,k}(t), \end{aligned} \tag{16}$$

$$q^l_0(x,t) = \sum_{l=1}^{|\det A|-1} \sum_{k\in\mathbb{Z}^d} \phi^l(x-k)\phi^l(t-k).$$

$q^l_m(x,t)$ will be called the reproducing kernel of V_m. Now, (15) takes the form

$$f^l_m(x) = \int_{\mathbb{R}^d} \lambda^d q^l_0(\lambda x, \lambda t) f(t)\, dt. \tag{17}$$

We will assume that the scaling function ϕ does not generate an orthonormal basis of V_0 or of W_0. This leads to the fact that condition (6) of MRA can be replaced by the weaker condition that $\{\phi(x-k) : k \in \mathbb{Z}^d\}$ is a Riesz basis of V_0. We also assume that Fourier transform of ϕ has a compact support. Since $\{\phi(x-k) : k \in \mathbb{Z}\}$ is a Riesz basis of V_0, it has a biorthonormal basis $\{\phi^*_k(x)\}$ such that for any function $f \in V_0$, we have

$$f(x) = \sum_{k\in\mathbb{Z}^d} \langle f, \phi^*_k\rangle \phi_{0,k}(x). \tag{18}$$

Similarly, for $f \in V_m$, we have

$$f^l_m(x) = \sum_{l=1}^{|\det A|-1} \sum_{k\in\mathbb{Z}^d} \left\langle f, \phi^{*l}_{m,k}\right\rangle \phi^l_{m,k}(x), \tag{19}$$

where $\{\phi^{*l}_{m,k}(x)\}_{k\in\mathbb{Z}^d}$ is the biorthonormal basis of $\{\phi^l_{m,k}(x)\}_{k\in\mathbb{Z}^d}$ and $\{l = 1, 2, \ldots, |\det A| - 1\}$.

Definition 8. If $\phi \in L^1(\mathbb{R}^d)$ with $\widehat{\phi}(0) = 1$ and one defines $\phi_n(x) = n^d\phi(nx)$, where $n \to \infty$, and $\widehat{\phi}$ denotes the Fourier transform of ϕ, then the sequence of functions $\{\phi_n\}_{n=1}^{\infty}$ is an approximate identity if

(1) $\int_{\mathbb{R}^d} \phi_n(x)dx = 1$ for all n,

(2) $\sup_n \int_{\mathbb{R}^d} |\phi_n(x)|dx < \infty$,

(3) $\lim_{n\to\infty} \int_{|x|>\delta} |\phi_n(x)|dx = 0$ for every $\delta > 0$.

Remark 9. If $0 \le \phi(x) \in S(\mathbb{R}^d)$, then $\phi_n(x) = n^d\phi(nx)$ is an approximate identity.

Lemma 10. *If $q_0^l(x,t) \in L^1(\mathbb{R}^d)$ with $\int_{\mathbb{R}^d} q_0^l(x,t)dt = 1$, then $q_m^l(x,t)$ is an approximate identity if*

(1) $\int_{\mathbb{R}^d} q_m^l(x,t)dt = 1$ for all m,

(2) $\sup_m \int_{\mathbb{R}^d} |q_m^l(x,t)|dt < \infty$,

(3) $\lim_{m\to\infty} \int_{|x-t|>\delta} |q_m^l(x,t)|dt = 0$ for every $\delta > 0$.

Proof. From [22, pages 64-65], we can easily obtain that $q_0^l(x,t) = q_0^l(|x-t|)$ almost everywhere for $\{l = 1, 2, \ldots, |\det A| - 1\}$. Now, following the lines of proof of [22, Lemma 2.2], the proof of Lemma 10 is immediate. □

Remark 11. Let $0 \le q_0^l(x,t) \in S(\mathbb{R}^d)$. Then, $q_m^l(x,t) = \lambda^d q_0(\lambda|x-t|)$ is an approximate identity.

Lemma 12. *If $f \in L^1(\mathbb{R}^d)$ and $q_m^l(x,t) \in S(\mathbb{R}^d)$, then $q_m^l(x,t) * f \in S(\mathbb{R}^d)$, where $*$ denotes the convolution operator.*

Proof. The proof follows on proceeding on the lines of proof of [22, Lemma 1.3]. □

Lemma 13. *Suppose that $p_+ < \infty$. Then, $S(\mathbb{R}^d)$ is dense in $L^{p(\cdot)}(\mathbb{R}^d)$.*

Proof. Set $C_c^\infty(\mathbb{R}^d) = \{\phi \in C^\infty(\mathbb{R}^d)\}$ such that $\operatorname{supp}(\phi)$ is compact. □

Obviously, $S(\mathbb{R}^d)$ is a vector space and $C_c^\infty(\mathbb{R}^d) \subseteq S(\mathbb{R}^d)$ and $S(\mathbb{R}^d)$ is dense in $L^p(\mathbb{R}^d)$. So, we have

$$C_c^\infty(\mathbb{R}^d) \subseteq S(\mathbb{R}^d) \subseteq L^p(\mathbb{R}^d), \quad 1 \le p < \infty. \tag{20}$$

It has been proved by Kováčik and Rákosník [23, Theorem 2.11] that if $p_+ < \infty$, then $C_c^\infty(\mathbb{R}^d)$ is dense in $L^{p(\cdot)}(\mathbb{R}^d)$. By virtue of (20), it follows that $S(\mathbb{R}^d)$ is dense in $L^{p(\cdot)}(\mathbb{R}^d)$.

Definition 14. Let $1 < p_0 < \infty$ be a constant, and let w be a weight on $\mathbb{R}^d$ such that $w^{-1/(p_0-1)}$ is locally integrable; w is said to be an A_{p_0} weight if w satisfies

$$A_{p_0,w} = \sup_{Q:\text{cube}} \frac{1}{|Q|}\int_Q w(x)\,dx \times \left(\frac{1}{|Q|}\int_Q w(x)^{-1/(p_0-1)}dx\right)^{p_0-1} < \infty, \tag{21}$$

where $|Q|$ means the Lebesgue measure of Q. A_{p_0} denotes the set of all A_{p_0} weights.

Definition 15. (1) A weight w is said to be an A_∞ weight if for all $0 < \alpha < 1$, there exists a constant $0 < \beta < 1$ such that for every cube Q and measurable set $E \subset Q$ satisfying $|E| \ge \alpha|Q|$,

$$\beta\int_Q w(x)\,dx \le \int_E w(x)\,dx. \tag{22}$$

A_∞ denotes the set of all A_∞ weights.

(2) A family of weights $\{w_t\}_{t>0}$ is said to be a family of A_∞ weights uniformly in t if each w_t is an A_∞ weight and the constant β is independent of t. It is well known that $A_\infty = \cup_{1<p_0<\infty} A_{p_0}$ [24]. Lerner [24] gave the following characterization of families of A_∞ weights.

Lemma 16 (see [24, Lemma 2.1]). *Let $p(\cdot)$ be a nonnegative and measurable function on $\mathbb{R}^d$. The family of weights $\{t^{p(\cdot)}\}_{t>0}$ is a family of A_∞ weights uniformly in t if and only if $p(\cdot) \equiv p_0$ for some constant p_0.*

Definition 17. Let $1 \le p_0 < \infty$ be a constant, and let w be a weight on $\mathbb{R}^d$. The weighted L^{p_0} space $L_w^{p_0}(\mathbb{R}^d)$ is the space of all measurable functions f with

$$\|f\|_{L_w^{p_0}(\mathbb{R}^d)} = \left(\int_{\mathbb{R}^d} |f(x)|^p w(x)\,dx\right)^{1/p_0} < \infty. \tag{23}$$

$L_w^{p_0}(\mathbb{R}^d)$ is a Banach space with the norm $\|.\|_{L_w^{p_0}(\mathbb{R}^d)}$.

The following lemma also holds for variable L^p spaces on an open set of $\mathbb{R}^d$.

Lemma 18 (see [25, Corollary 1.11]). *Let $p(\cdot) \in \wp(\mathbb{R}^d)$, and let F be a family of ordered pairs of non-negative and measurable functions (f,g). Suppose that*

$$\left(\frac{p(\cdot)}{p_1}\right)' \in B(\mathbb{R}^d) \text{ for some constant } p_1 \in (1, p_-), \tag{24}$$

and that there exists a constant $p_0 \in (1,\infty)$ such that for all $w \in A_{p_0,w}$ and all $(f,g) \in F$ with $f \in L_w^{p_0}(R)$,

$$\|f\|_{L_w^{p_o}(R^n)} \le C_o\|g\|_{L_w^{p_o}(R^n)}, \tag{25}$$

where C_o is a positive constant depending only on n, p_o, and $A_{p_o,w}$. Then, it follows that for all $(f,g) \in F$ such that $f \in L^{p(\cdot)}(\mathbb{R}^d)$,

$$\|f\|_{L^{p(\cdot)}(\mathbb{R}^d)} \le C_1\|g\|_{L^{p(\cdot)}(\mathbb{R}^d)}, \tag{26}$$

where C_1 is a positive constant independent of (f,g).

2.2. Banach Function Spaces

Definition 19. Let $M(\mathbb{R}^d)$ be the set of all measurable functions on $\mathbb{R}^d$.

(1) A linear space $X \subset M(\mathbb{R}^d)$ is said to be a Banach function space if there exists a functional $\|\cdot\| : M(\mathbb{R}^d) \to \mathbb{R}$ which has the norm property and satisfies the following conditions:

(a) $f \in X$ if and only if $\|f\| < \infty$;
(b) $\|f\| = \||f|\|$ for all $f \in M(\mathbb{R}^d)$;
(c) if $0 \le f_1 \le f_2 \le \cdots$ and $\lim_{n\to\infty} f_n = f$, then $\|f_1\| \le \|f_2\| \le \cdots$ and $\lim_{n\to\infty}\|f_n\| = \|f\|$;
(d) for all $E \subset \mathbb{R}^d$ with $|E| < \infty$, it follows that $\|\chi_E\| < \infty$;
(e) for all $E \subset \mathbb{R}^d$ with $|E| < \infty$, there exists a constant $C_E > 0$ such that $\int_E |f(x)|dx \le C_E\|f\|$ for all $f \in X$.

(2) Let X be a Banach function space with the norm $\|\cdot\|$. The norm $\|\cdot\|$ is said to be an absolutely continuous norm if $\lim_{j\to\infty}\|f\chi_{E_j}\| = 0$ for all $f \in X$ and all sequences of measurable sets $\{E_j\}_{j=1}$ such that $\lim_{j\to\infty}\chi_{E_j} = 0$ almost everywhere.

The following two lemmas imply that $L^{p(\cdot)}(\mathbb{R}^d)$ is a Banach function space with the absolutely continuous norm $\|\cdot\|_{L^{p(\cdot)}(\mathbb{R}^d)}$ (cf. [26]).

Lemma 20 (see [27]). *Let $(X, \|\cdot\|)$ be a Banach function space. Then, X is separable if and only if the norm $\|\cdot\|$ is absolutely continuous on X.*

Lemma 21 (see [23]). *Suppose that $p_+ < \infty$. Then, $L^{p(\cdot)}(\mathbb{R}^d)$ is separable.*

3. Main Results

First, we prove the convergence of wavelet associated with dilation matrix in $L^{p(\cdot)}(\mathbb{R}^d)$ in reference of MRA.

Theorem 22. *Let $p(\cdot) \in B(\mathbb{R}^d)$ and $\phi, \phi^* \in S(\mathbb{R}^d)$. Define*

$$P_j^* f = \sum_{k\in\mathbb{Z}^d} \langle f, \phi_{j,k}\rangle \phi_{j,k}^*. \tag{27}$$

If $\{q_j(x,t)\}_{j=1}^{\infty}$ is an approximate identity, then for every $f \in L^{p(\cdot)}(\mathbb{R}^d)$ and $j \in \mathbb{Z}$,

$$\lim_{j\to\infty} \left\|P_j^*(f) - f\right\|_{L^{p(\cdot)}(\mathbb{R}^d)} = 0. \tag{28}$$

Remark 23. (1) If ϕ is a scaling function of an MRA and $\phi = \phi^*$, then we see that $P_j^*(f) = P_j(f)$, where P_j is the orthogonal projection for each $j \in \mathbb{Z}$.

(2) In view of generalized Holder inequality [11, Theorem 2.1], we have

$$\int_{\mathbb{R}^d} |f(x)\, g(x)|\, dx \le r_p \|f\|_{L^{p(\cdot)}(\mathbb{R}^d)} \|g\|_{L^{p'(\cdot)}(\mathbb{R}^d)} \tag{29}$$

for all $f \in L^{p(\cdot)}(\mathbb{R}^d)$ and $g \in L^{p'(\cdot)}(\mathbb{R}^d)$, where $r_p = 1 + 1/p_- - 1/p_+$. Hence, the previous L^2-inner product $\langle f, \phi_{j,k}\rangle$ makes sense.

Proof of Theorem 22. The nonorthogonal projection of $f \in L^2(\mathbb{R}^d)$ on the space V_j is given by

$$\begin{aligned} P_j^*(f) &= \sum_{k\in\mathbb{Z}^d} \langle f, \phi_{j,k}\rangle \phi_{j,k}^* \\ &= \sum_{k\in\mathbb{Z}^d} \int_{\mathbb{R}^d} f(t)\, \phi_{j,k}(x)\, \phi_{j,k}^*(t)\, dt \\ &= \int_{\mathbb{R}^d} f(t) \sum_{k\in\mathbb{Z}^d} \phi_{j,k}(x)\, \phi_{j,k}^*(t)\, dt \\ &= \int_{\mathbb{R}^d} f(t)\, q_j(x,t)\, dt \\ &= \int_{\mathbb{R}^d} f(t)\, q_j(x-t)\, dt \\ &= \left(f * q_j\right)(x) \longrightarrow f(x) \ \text{(by Lemma 12)}. \end{aligned} \tag{30}$$

Thus, if $f \in L^p(\mathbb{R}^d)$ and $q_j(x,t) \in S(\mathbb{R}^d)$, then $q_j * f \in S(\mathbb{R}^d)$. Now, applying Lemma 18, we obtain that if $f \in L^{p(\cdot)}(\mathbb{R}^d)$ and $q_j(x,t) \in S(\mathbb{R}^d)$, then $q_j * f \in S(\mathbb{R}^d)$. By virtue of Lemma 13 we get $q_j * f \in L^{p(\cdot)}(\mathbb{R}^d)$. Hence, for $f \in L^{p(\cdot)}(\mathbb{R}^d)$, $q_j * f \to f \in L^{p(\cdot)}(\mathbb{R}^d)$ for $j \to \infty$. Hence, the proof is completed. □

Corollary 24. *Let $p(\cdot) \in B(\mathbb{R}^d)$ and $\phi, \phi^* \in S(\mathbb{R}^d)$. If $\{q_j(x,t)\}_{j=1}^{\infty}$ is an approximate identity, then for every $f \in L^{p(\cdot)}(\mathbb{R}^d)$, the wavelet series*

$$f_j(x) = \sum_{k\in\mathbb{Z}^d} \langle f, \phi_{j,k}\rangle \phi_{j,k}^*(x) \tag{31}$$

converges to $f(x)$ as $j \to \infty$ at every point of continuity of $f(x)$.

To characterize the wavelet associated with dilation matrix A in $L^{p(\cdot)}(\mathbb{R}^d)$ with respect to the norm $\|\cdot\|_{L^{p(\cdot)}_{\mathbb{R}^d}}$, we define a cube

$$I_{j,k} = \Pi_{i=1}^{d} \left(|\det A|^{-j} k_i, \left|\det A^{-j}\right| (k_i + 1)\right) \tag{32}$$

and denote $\chi_{j,k} = |\det A|^{j/2} \chi_{I_{j,k}}$ for $j \in \mathbb{Z}$ and $k \in \mathbb{Z}^d$. Given a wavelet set $\{\varphi^l : l = 1, 2, \ldots, |\det A| - 1\}$, we consider the following two square functions in order to obtain wavelet characterizations:

$$\begin{aligned} V^* f &= \left(\sum_{l=1}^{|\det A|-1} \sum_{j=-\infty}^{\infty} \sum_{k\in\mathbb{Z}^d} \left|\langle f, \varphi_{j,k}^l\rangle \varphi_{j,k}^l\right|^2 \right)^{1/2}, \\ W^* f &= \left(\sum_{l=1}^{|\det A|-1} \sum_{j=-\infty}^{\infty} \sum_{k\in\mathbb{Z}^d} \left|\langle f, \varphi_{j,k}^l\rangle \chi_{j,k}\right|^2 \right)^{1/2}. \end{aligned} \tag{33}$$

Lemma 25. *Let $p_0 \in (1, \infty)$ be a constant, $w \in A_{p_0}$, and $\{\varphi^l\}$ a wavelet set such that each wavelet $\varphi^l \in S(\mathbb{R}^d)$. Then, there exist* *constants* $c, C, c', C' > 0$ *depending only on* d, p_0, $\{\varphi^l\}$, *and* $A_{p_0,w}$ *such that for all* $f \in L_w^{p_0}(\mathbb{R}^d)$,

$$c\|f\|_{L_w^{p_0}(\mathbb{R}^d)} \leq \|V^* f\|_{L_w^{p_0}(\mathbb{R}^d)} \leq C\|f\|_{L_w^{p_0}(\mathbb{R}^d)},$$
$$c'\|f\|_{L_w^{p_0}(\mathbb{R}^d)} \leq \|W^* f\|_{L_w^{p_0}(\mathbb{R}^d)} \leq C'\|f\|_{L_w^{p_0}(\mathbb{R}^d)}. \tag{34}$$

Proof. In view of [28, 29], the estimates (34) hold for $\varphi^l \in \mathbb{R}^0(\mathbb{R}^d)$ [3]. The class $\mathbb{R}^0(\mathbb{R}^d)$ is contained in $S(\mathbb{R}^d)$; therefore, estimates (34) also hold for $\varphi^l \in S(\mathbb{R}^d)$. □

Now, we prove the following.

Theorem 26. *Let* $p(\cdot) \in B(\mathbb{R}^d)$, *and let* $\{\varphi^l : l = 1, 2, \ldots, |\det A| - 1\}$ *be a wavelet set such that wavelet* $\varphi^l \in S(\mathbb{R}^d)$. *Then, there exist constants* $c, C, c', C' > 0$ *such that for all* $f \in L^{p(\cdot)}(\mathbb{R}^d)$,

$$c\|f\|_{L^{p(\cdot)}(\mathbb{R}^d)} \leq \|V^* f\|_{L^{p(\cdot)}(\mathbb{R}^d)} \leq C\|f\|_{L^{p(\cdot)}(\mathbb{R}^d)},$$
$$c'\|f\|_{L^{p(\cdot)}(\mathbb{R}^d)} \leq \|W^* f\|_{L^{p(\cdot)}(\mathbb{R}^d)} \leq C'\|f\|_{L^{p(\cdot)}(\mathbb{R}^d)}. \tag{35}$$

Proof. We see that $p(\cdot) \in B(\mathbb{R}^d)$ is equivalent to condition (24) in view of Proposition 5. Thus, Lemma 25 together with Lemma 18 implies that the estimates (35) hold for all $f \in S(\mathbb{R}^d)$ and $\varphi^l \in S(\mathbb{R}^d)$. By using Lemma 13, these inequalities also hold for $f \in L^{p(\cdot)}(\mathbb{R}^d)$. □

Applying the wavelet characterization (Theorem 26) and [27, Proposition 3.6, Ch. 1], we can construct unconditional basis of $L^{p(\cdot)}(\mathbb{R}^d)$ in terms of wavelets.

Theorem 27. *Let* $p(\cdot) \in B(\mathbb{R}^d)$, *and let* $\{\varphi^l : l = 1, 2, \ldots, |\det A| - 1\}$ *be a wavelet set such that each wavelet* φ^l *is in* $S(\mathbb{R}^d)$. *Then, the wavelet basis* $\{\varphi^l_{j,k} : l = 1, 2, \ldots, |\det A| - 1, j \in \mathbb{Z}, k \in \mathbb{Z}^d\}$ *forms an unconditional basis on* $L^{p(\cdot)}(\mathbb{R}^d)$.

Proof. Let us denote $\Lambda = \{1, 2, \ldots, |\det A| - 1\} \times \mathbb{Z} \times \mathbb{Z}^d$ and

$$S_D f = \sum_{(l,j,k)\in D} \langle f, \varphi^l_{j,k} \rangle \varphi^l_{j,k} \quad \text{for } D \subset \Lambda. \tag{36}$$

In order to complete the proof, we have to check the following two conditions:

(1) there exists a constant $C > 0$ such that $\|S_D f\|_{L^{p(\cdot)}(\mathbb{R}^d)} \leq C\|f\|_{L^{p(\cdot)}(\mathbb{R}^d)}$ for all $D \subset \Lambda$ and all $f \in L^{p(\cdot)}(\mathbb{R}^d)$;

(2) span $\{\phi^l_{j,k} : (l, j, k) \in \Lambda\}$ is dense in $L^{p(\cdot)}(\mathbb{R}^d)$.

To check condition (1), using Theorem 26 and the orthonormality, we obtain for all $f \in L^{p(\cdot)}(\mathbb{R}^d)$,

$$\begin{aligned}\|S_D f\|_{L^{p(\cdot)}(\mathbb{R}^d)} &\leq C_0\|W^*(S_D f)\|_{L^{p(\cdot)}(\mathbb{R}^d)} \\ &\leq C_0\|W^* f\|_{L^{p(\cdot)}(\mathbb{R}^d)} \\ &\leq C_0 C_1\|f\|_{L^{p(\cdot)}(\mathbb{R}^d)},\end{aligned} \tag{37}$$

where $C_0, C_1 > 0$ are constants independent of f. Using monotone convergence theorem as finite sets D exhaust Λ, we get that (37) holds for $(l, j, k) \in \Lambda$.

To check condition (2), it suffices to show that $\lim_{D\to\Lambda}\|f - S_D f\|_{L^{p(\cdot)}(\mathbb{R}^d)} = 0$. By Lemmas 20 and 21, $L^{p(\cdot)}(\mathbb{R}^d)$ is a Banach function space with absolutely continuous norm $\|\cdot\|_{L^{p(\cdot)}(\mathbb{R}^d)}$. For any $f \in L^{p(\cdot)}(\mathbb{R}^d) \cap L^2(\mathbb{R}^d)$, we have $f = \sum_{(l,j,k)\in D}\langle f, \varphi^l_{j,k}\rangle\varphi^l_{j,k}$, with convergence in $L^2(\mathbb{R}^d)$, and there exists a subsequence of partial sums $\sum_{(l,j,k_i)\in D}\langle f, \varphi^l_{j,k_i}\rangle\varphi^l_{j,k_i}$ for $i \in N$ that converge almost everywhere on $\mathbb{R}^d$. Using Fatou's lemma, we get

$$\begin{aligned}&\left\| f - \sum_{i=1,(l,j,k_i)\in D}^{N} \langle f, \varphi^l_{j,k_i} \rangle \varphi^l_{j,k_i} \right\|_{L^{p(\cdot)}(\mathbb{R}^d)} \\ &\leq \lim_{m\to\infty} \inf \left\| \sum_{i=N+1,(l,j,k_i)\in D}^{N_m} \langle f, \varphi^l_{j,k_i} \rangle \varphi^l_{j,k_i} \right\|_{L^{p(\cdot)}(\mathbb{R}^d)}, \quad m \in N, \\ &\leq C_p \left\| \left(\sum_{i=N+1,(l,j,k_i)\in D}^{\infty} \left| \langle f, \varphi^l_{j,k_i} \rangle \varphi^l_{j,k_i} \right|^2 \right)^{1/2} \right\|_{L^{p(\cdot)}(\mathbb{R}^d)}.\end{aligned} \tag{38}$$

The result [27, Proposition 3.6, Ch. 1] can be applied to $L^{p(\cdot)}(\mathbb{R}^d)$ since $L^{p(\cdot)}(\mathbb{R}^d)$ is a Banach function space with the absolutely continuous norm (cf. [26]). Using this result, the last term is zero, and by virtue of (37), $\|(\sum_{(l,j,k)\in\Lambda}|\langle f, \varphi^l_{j,k}\rangle\varphi^l_{j,k}|^2)^{1/2}\|_{L^{p(\cdot)}(\mathbb{R}^d)} < \infty$; we can use the dominated convergence theorem to obtain

$$f = \sum_{i=1,(l,j,k_i)\in D}^{\infty} \langle f, \varphi^l_{j,k_i} \rangle \varphi^l_{j,k_i} \tag{39}$$

with convergence in $L^{p(\cdot)}(\mathbb{R}^d)$. Therefore, $\lim_{D\to\Lambda}\|f - S_D f\|_{L^{p(\cdot)}(\mathbb{R}^d)} = 0$.

Hence, the proof is completed. □

Acknowledgments

The author is extremely thankful to the learned reviewers for taking pains and giving fruitful comments to improve the paper. This work is supported by University Grants Commission, New Delhi, India, under the Grant of Major Research Project no. 41-792/2012(SR).

References

[1] D. Cruz-Uribe, A. Fiorenza, and C. J. Neugebauer, "The maximal function on variable L^p spaces," *Annales Academiæ Scientiarum Fennicæ*, vol. 28, no. 1, pp. 223–238, 2003.

[2] D. Cruz-Uribe, A. Fiorenza, and C. J. Neugebauer, "Corrections to: "The maximal function on variable L^p spaces"," *Annales Academiæ Scientiarum Fennicæ*, vol. 29, no. 1, pp. 247–249, 2004.

[3] L. Diening, "Maximal function on generalized Lebesgue spaces $L^{p(\cdot)}$," *Mathematical Inequalities & Applications*, vol. 7, no. 2, pp. 245–253, 2004.

[4] L. Diening, "Maximal function on Musielak-Orlicz spaces and generalized Lebesgue spaces," *Bulletin des Sciences Mathématiques*, vol. 129, no. 8, pp. 657–700, 2005.

[5] A. Nekvinda, "Hardy-Littlewood maximal operator on $L^{p(x)}(\mathbb{R})$," *Mathematical Inequalities & Applications*, vol. 7, no. 2, pp. 255–265, 2004.

[6] M. Izuki, "Wavelets and modular inequalities in variable L^p spaces," *Georgian Mathematical Journal*, vol. 15, no. 2, pp. 281–293, 2008.

[7] T. S. Kopaliani, "Greediness of the wavelet system in $L^{p(t)}(\mathbb{R})$ spaces," *East Journal on Approximations*, vol. 14, no. 1, pp. 59–67, 2008.

[8] D. V. Cruz-Uribe, J. M. Martell, and C. Pérez, *Weights, Extrapolation and the Theory of Rubio de Francia: Operator Theory: Advances and Applications*, vol. 215, Birkhäuser/Springer, Basel, Switzerland, 2011.

[9] Y. Meyer, *Wavelets and Operators*, vol. 37 of *Cambridge Studies in Advanced Mathematics*, Cambridge University Press, Cambridge, UK, 1992.

[10] G. G. Walter, "Pointwise convergence of wavelet expansions," *Journal of Approximation Theory*, vol. 80, no. 1, pp. 108–118, 1995.

[11] S. E. Kelly, M. A. Kon, and L. A. Raphael, "Pointwise convergence of wavelet expansions," *American Mathematical Society*, vol. 30, no. 1, pp. 87–94, 1994.

[12] S. E. Kelly, M. A. Kon, and L. A. Raphael, "Local convergence for wavelet expansions," *Journal of Functional Analysis*, vol. 126, no. 1, pp. 102–138, 1994.

[13] L. Carleson, "On convergence and growth of partial sums of Fourier series," *Acta Mathematica*, vol. 116, pp. 135–157, 1966.

[14] R. A. Hunt, "On the convergence of Fourier series," in *Proceedings of the Conference on Orthogonal Expansions and Their Continuous Analogues*, D. T. Haimo, Ed., pp. 235–255, Southern Illinois University Press, 1968.

[15] N. N. Reyes, "Behavior of partial sums of wavelet series," *Journal of Approximation Theory*, vol. 103, no. 1, pp. 55–60, 2000.

[16] T. Tao, "On the almost everywhere convergence of wavelet summation methods," *Applied and Computational Harmonic Analysis*, vol. 3, no. 4, pp. 384–387, 1996.

[17] P. Manchanda, A. A. S. Mukheimer, and A. H. Siddiqi, "Pointwise convergence of wavelet expansions associated with dilation matrix," *Applicable Analysis*, vol. 76, no. 3-4, pp. 301–308, 2000.

[18] I. I. Sharapudinov, "The basis property of the Haar system in the space $L^{p(t)}([0,1])$ and the principle of localization in themean," *Matematicheskiĭ Sbornik*, vol. 130 (172), no. 2, pp. 275–283, 1986.

[19] I. I. Sharapudinov, "The basis property of the Haar system in the space $L^{p(t)}([0,1])$ and the principle of localization in the mean," *Sbornik: Mathematics*, vol. 58, no. 2, pp. 279–287, 1987.

[20] P. Wojtaszczyk, *A Mathematical Introduction to Wavelets*, vol. 37 of *London Mathematical Society Student Texts*, Cambridge University Press, Cambridge, UK, 1997.

[21] E. Hernández and G. Weiss, *A first Course on Wavelets. With a Foreword by Yves Meyer*, Studies in Advanced Mathematics, CRC Press, Boca Raton, Fla, USA, 1996.

[22] D. Kumar, "Convergence of a class of non-orthogonal wavelet expansions in $L^p(\mathbb{R})$, $1 \le p \le \infty$," *Panamerican Mathematical Journal*, vol. 19, no. 4, pp. 61–70, 2009.

[23] O. Kováčik and J. Rákosník, "On spaces $L^{p(x)}$ and $W^{K,p(x)}$," *Czechoslovak Mathematical Journal*, vol. 41 (116), no. 4, pp. 592–618, 1951.

[24] A. K. Lerner, "On modular inequalities in variable L^p spaces," *Archiv der Mathematik*, vol. 85, no. 6, pp. 538–543, 2005.

[25] D. Cruz-Uribe, A. Fiorenza, J. M. Martell, and C. Pérez, "The boundedness of classical operators on variable L^p spaces," *Annales Academiæ Scientiarum Fennicæ*, vol. 31, no. 1, pp. 239–264, 2006.

[26] D. E. Edmunds, J. Lang, and A. Nekvinda, "On $L^{p(x)}$ norms," *The Royal Society of London A*, vol. 455, no. 1981, pp. 219–225, 1999.

[27] C. Bennett and R. Sharpley, *Interpolation of Operators*, vol. 129 of *Pure and Applied Mathematics*, Academic Press, Boston, Mass, USA, 1988.

[28] J. García-Cuerva and J. M. Martell, "Wavelet characterization of weighted spaces," *The Journal of Geometric Analysis*, vol. 11, no. 2, pp. 241–264, 2001.

[29] P. G. Lemarié-Rieusset, "Ondelettes et poids de Muckenhoupt," *Studia Mathematica*, vol. 108, no. 2, pp. 127–147, 1994.

Strong Convergence Theorems for a Pair of Strictly Pseudononspreading Mappings

Bin-Chao Deng and Tong Chen

School of Management, Tianjin University, Tianjin 300072, China

Correspondence should be addressed to Bin-Chao Deng; dbchao1985@tju.edu.cn

Academic Editor: Krassimir T. Atanassov

Let H be a real Hilbert space. Let $T_1, T_2 : H \to H$ be k_1-, k_2-strictly pseudononspreading mappings; let $\{\alpha_n\}$ and $\{\beta_n\}$ be two real sequences in (0,1). For given $x_0 \in H$, the sequence $\{x_n\}$ is generated iteratively by $x_{n+1} = \beta_n x_n + (1-\beta_n)T_{w_1}(\alpha_n \gamma f(x_n) + (I - \mu\alpha_n B)T_{w_2}x_n)$, $\forall n \in N$, where $T_{w_i} = (1-w_i)I + w_i T_i$ with $i = 1, 2$ and $B : H \to H$ is strongly monotone and Lipschitzian. Under some mild conditions on parameters $\{\alpha_n\}$ and $\{\beta_n\}$, we prove that the sequence $\{x_n\}$ converges strongly to the set $F_{ix}(T_1) \cap F_{ix}(T_2)$ of fixed points of a pair of strictly pseudononspreading mappings T_1 and T_2.

1. Introduction

Let H be a real Hilbert space whose inner product and norm are denoted by $\langle \cdot, \cdot \rangle$ and $\|\cdot\|$, and let C be a nonempty, closed, and convex subset of H, respectively. Recall the following definitions.

Definition 1. Let $A : C \to H$ be a nonlinear mapping.

(1) A is said to be monotone if

$$\langle Ax - Ay, x - y \rangle \geq 0, \quad \forall x, y \in C. \tag{1}$$

(2) A is said to be strongly monotone if there exists a constant $\alpha > 0$ such that

$$\langle Ax - Ay, x - y \rangle \geq \alpha \|x - y\|^2, \quad \forall x, y \in C. \tag{2}$$

For such a case, A is said to be α-strongly-monotone.

(3) A is said to be inverse strongly if there exists a constant $\alpha > 0$ such that

$$\langle Ax - Ay, x - y \rangle \geq \alpha \|Ax - Ay\|^2, \quad \forall x, y \in C. \tag{3}$$

For such a case, A is said to be α-inverse-strongly monotone.

The classical variational inequality which is denoted by $\mathrm{VI}(A, C)$ is to find $x \in C$ such that

$$\langle Ax, y - x \rangle \geq 0, \quad \forall y \in C. \tag{4}$$

The variational inequality has been extensively studied in the literature; see, for example, [1, 2] and the reference therein. Recall that T is a nonexpansive mapping of C into itself; that is,

$$\|Tx - Ty\| \leq \|x - y\|, \quad \forall x, y \in C. \tag{5}$$

The set of fixed points of T is the set $F_{ix}(T) = \{x \in H, Tx = x\}$.

In 2011, Osilike and Isiogugu [3] introduced a new class of mappings, the so-called k-strictly pseudononspreading; that is, a mapping $T : D(T) \subseteq H \to H$ is k-strictly pseudononspreading if there exists $k \in [0, 1)$ such that

$$\begin{aligned}\|Tx - Ty\|^2 \leq \|x - y\|^2 + k\|x - Tx - (y - Ty)\|^2 \\ + 2\langle x - Tx, y - Ty \rangle,\end{aligned} \tag{6}$$

for all $x, y \in D(T)$. They showed that the class of nonspreading mappings is properly contained in the class of strictly pseudononspreading mappings.

The iteration procedure of Mann's type for approximating fixed points of a nonexpansive mapping T is the following: $x_1 \in C$ and

$$x_{n+1} = \alpha_n x_n + (1 - \alpha_n) T x_n, \quad \forall n \in N, \tag{7}$$

where $\{\alpha_n\}$ is a sequence in $[0, 1]$; see [4]. For two nonexpansive mappings S and T, Takahashi and Tamura [5] considered the following iteration procedure: $x_1 \in C$ and

$$x_{n+1} = \alpha_n x_n + (1 - \alpha_n) S (\beta_n x_n + (1 - \beta_n) T x_n), \quad \forall n \in N, \tag{8}$$

where $\{\alpha_n\}$ and $\{\beta_n\}$ are sequences in $[0, 1]$.

In 2010, Tian [6] introduced the following general viscosity iterative scheme for finding an element of set of solutions to the fixed point of nonexpansive mapping in Hilbert space. Define sequence $\{x_n\}$ by

$$x_{n+1} = \alpha_n \gamma f(x_n) + (I - \mu\alpha_n B) T x_n, \quad n \geq 0, \tag{9}$$

where B is k-Lipschitzian and η-strongly monotone operator and T is a nonexpansive mapping on H; then he proved that if the sequence $\{\alpha_n\}$ satisfies appropriate conditions, the sequence $\{x_n\}$ generated by (9) converges strongly to the unique solution $x^* \in C$ of the variational inequality

$$\langle (\gamma f - \mu B) x^*, x - x^* \rangle \leq 0, \quad \forall x \in C, \tag{10}$$

where $C = \mathrm{Fix}(T)$.

In this paper, motivated by Takahashi and Tamura [5] and Tian [6], we introduce the following iteration scheme for finding a common point of the set $F_{ix}(T_1) \bigcap F_{ix}(T_2)$ of fixed points of a pair of strictly pseudononspreading mappings T_1 and T_2:

$$\begin{aligned} x_{n+1} = {} & \beta_n x_n + (1 - \beta_n) T_{\omega_1} \\ & \times (\alpha_n \gamma f(x_n) \\ & + (I - \mu\alpha_n B) T_{\omega_2} x_n), \quad \forall n \in N, \end{aligned} \tag{11}$$

where $T_{\omega_i} = (1 - \omega_i) I + \omega_i T_i$ with $i = 1, 2$ and $B : H \to H$ is η-strongly monotone and k-Lipschitzian on H with $k > 0$, $\eta > 0$. Under suitable conditions, we prove a strong convergence theorem, which is different from the results of general viscosity iterative scheme in [6].

2. Preliminaries

We need some facts and tools in Hilbert space H which are listed as in the following lemmas.

Definition 2. A mapping T is said to be demiclosed, if for any sequence $\{x_n\}$ which weakly converges to y, and if the sequence $\{Tx_n\}$ strongly converges to z, then $T(y) = z$.

Definition 3. $T : H \to H$ is called demicontractive on H, if there exists a constant $\alpha < 1$ such that

$$\|Tx - q\|^2 \leq \|x - q\|^2 + \alpha \|x - Tx\|^2, \quad \forall (x, q) \in H \times F_{ix}(T). \tag{12}$$

Remark 4. Every k-strictly pseudononspreading mapping with a nonempty fixed point set $F_{ix}(T)$ is demicontractive (see [7, 8]).

Remark 5 (see [9]). Let T be a α-demicontractive mapping on H with $F_{ix}(T) \neq \emptyset$ and $T_\omega = (1 - \omega) I + \omega T$ for $\omega \in (0, \infty)$.

(A1) $T\alpha$-demicontractive is equivalent to

$$\langle x - Tx, x - q \rangle \geq \frac{1}{2} (1 - \alpha) \|x - Tx\|^2, \quad \forall (x, q) \in H \times F_{ix}(T). \tag{13}$$

(A2) $F_{ix}(T) = F_{ix}(T_\omega)$ if $\omega \neq 0$.

Remark 6. According to $I - T_\omega = \omega(I - T)$ with T being a k-strictly pseudononspreading mapping, we obtain

$$\langle x - T_\omega x, x - q \rangle \geq \frac{\omega (1 - k)}{2} \|x - Tx\|^2, \quad \forall (x, q) \in H \times F_{ix}(T). \tag{14}$$

Proposition 7 (see [3]). *Let C be a nonempty closed convex subset of a real Hilbert space H, and let $T : C \to C$ be a k-strictly pseudononspreading mapping. If $F_{ix}(T) \neq \emptyset$, then it is closed and convex.*

Proposition 8 (see [3]). *Let C be a nonempty closed convex subset of a real Hilbert space H, and let $T : C \to C$ be a k-strictly pseudononspreading mapping. Then $(I - T)$ is demiclosed at 0.*

Lemma 9. *Let H be a real Hilbert space. The following expressions hold:*

(i) $\|tx + (1 - t) y\|^2 = t\|x\|^2 + (1 - t)\|y\|^2 - t(1 - t)\|x - y\|^2$, $\forall x, y \in H$, $\forall t \in [0, 1]$.

(ii) $\|x + y\|^2 \leq \|x\|^2 + 2\langle y, x + y \rangle$, $\forall x, y \in H$.

Lemma 10 (see [10]). *Let C be a closed convex subset of a Hilbert space H, and let $T : H \to H$ be a k-strictly pseudononspreading mapping with a nonempty fixed point set. Let $k \leq \omega < 1$ be fixed, and define $T_\omega : C \to C$ by*

$$T_\omega x = (1 - \omega) x + \omega T x, \quad \forall x \in C. \tag{15}$$

Then $F_{ix}(T_\omega) = F_{ix}(T)$.

Lemma 11 (see [11]). *Assume $\{\alpha_n\}$ is a sequence of nonnegative real numbers such that*

$$\alpha_{n+1} \leq (1 - \gamma_n) \alpha_n + \delta_n, \quad n \geq 0, \tag{16}$$

where $\{\gamma_n\}$ is a sequence in $(0,1)$ and $\{\delta_n\}$ is a sequence in $\mathbb{R}$ such that

(i) $\sum_{n=1}^{\infty} \gamma_n = \infty$,

(ii) $\limsup_{n\to\infty}(\delta_n/\gamma_n) = 0$ *or* $\sum_{n=1}^{\infty} |\delta_n| < \infty$.

Then $\lim_{n\to\infty}\alpha_n = 0$.

Lemma 12 (see [12]). *Let $\{\mathcal{T}_n\}$ be a sequence of real numbers that does not decrease at infinity, in the sense that there exists a subsequence $\{\mathcal{T}_{n_j}\}_{j\geq 0}$ of $\{\mathcal{T}_n\}$ which satisfies $\mathcal{T}_{n_j} < \mathcal{T}_{n_j+1}$ for all $j \geq 0$. Also consider that the sequence of integers $\{\delta(n)\}_{n\geq n_0}$ is defined by*

$$\delta(n) = \max\{k \leq n \mid \mathcal{T}_k < \mathcal{T}_{k+1}\}. \quad (17)$$

Then $\{\delta(n)\}_{n\geq n_0}$ is a nondecreasing sequence verifying $\lim_{n\to\infty}\delta(n) = \infty$, $\forall n \geq n_0$; it holds that $\mathcal{T}_{\delta(n)} < \mathcal{T}_{\delta(n)+1}$ and one has

$$\mathcal{T}_n < \mathcal{T}_{\delta(n)+1}. \quad (18)$$

Lemma 13. *Let K be a closed convex subset of a real Hilbert space H, given $x \in H$ and $y \in K$. Then $y = P_K x$ if and only if there holds the inequality*

$$\langle x - y, y - z\rangle \geq 0, \quad \forall z \in K. \quad (19)$$

3. Main Results

Let C be a nonempty closed convex subset of a real Hilbert space H, and let $T_1, T_2 : C \to C$ be k_1-, k_2-strictly pseudononspreading mappings with nonempty fixed point set $\Omega = F_{ix}(T_1)\bigcap F_{ix}(T_2)$. Let f be a L-Lipschitz mapping on H with coefficient $L > 0$. Assume that the set Ω is nonempty. Since Ω is closed and convex, the nearest point projection from C onto Ω is well defined. Recall that $B : H \to H$ is η-strongly monotone and k-Lipschitzian on H with $k > 0$, $\eta > 0$.

Lemma 14 (see [13]). *Let H be a real Hilbert space and let B be a k-Lipschitzian and η-strongly monotone operator with $k > 0$, $\eta > 0$. Let $0 < \mu < (2\eta/k^2)$ and $\tau = \mu(\eta - (\eta k^2/2))$. Then for $t \in \min\{1, (1/\tau)\}$, $I - t\mu B$ is in contradiction with constant $1 - t\tau$.*

Lemma 15. *Let T be a k-strictly pseudononspreading mapping on H, and $T_\omega x = (1-\omega)x + \omega Tx$ with $0 < k < \omega < 1/2$. Then*

$$\begin{aligned}&\|T_\omega x - T_\omega y\|^2\\ &\leq \|x-y\|^2 + \frac{2}{\omega}\langle x - T_\omega x, y - T_\omega y\rangle, \quad \forall x, y \in C.\end{aligned} \quad (20)$$

Proof. For $x, y \in H$, we have

$$\begin{aligned}&\|T_\omega x - T_\omega y\|^2\\
&= (1-\omega)\|x-y\|^2 + \omega\|Tx - Ty\|^2\\
&\quad - \omega(1-\omega)\|x - Tx - (y - Ty)\|^2\\
&\leq (1-\omega)\|x-y\|^2\\
&\quad + \omega\left[\|x-y\|^2 + k\|x - Tx - (y-Ty)\|^2\right.\\
&\quad \left. + 2\langle x - Tx, y - Ty\rangle\right]\\
&\quad - \omega(1-\omega)\|x - Tx - (y-Ty)\|^2\\
&= \|x-y\|^2 + 2\omega\langle x - Tx, y - Ty\rangle\\
&\quad - \omega(1-\omega)(1-k)\|x - Tx - (y-Ty)\|^2\\
&\leq \|x-y\|^2 + 2\omega\langle x - Tx, y - Ty\rangle\\
&= \|x-y\|^2 + \frac{2}{\omega}\langle x - T_\omega x, y - T_\omega y\rangle.
\end{aligned} \quad (21)$$

□

Remark 16. If $y = p \in F_{ix}(T)$ and from Lemma 15, we can easily claim that $\|T_\omega x - p\| \leq \|x - p\|$, $\forall x \in C$.

Theorem 17. *Let C be a nonempty, closed, and convex subset of a real Hilbert space H. Let $T_1, T_2 : C \to C$ be k_1-, k_2-strictly pseudononspreading mappings and demiclosed. Let f be L-Lipschitz mapping on H with coefficient $L > 0$ and let $B : H \to H$ be η-strongly monotone and k-Lipschitzian on H with $k > 0$, $\eta > 0$, $0 < k_1$, and $k_2 < \omega < 1/2$. Let the sequence of $\{\alpha_n\}$ and $\{\beta_n\}$ satisfy the following conditions:*

(c1) $\lim_{n\to\infty}\alpha_n = 0$,

(c2) $\sum_{n=0}^{\infty}\alpha_n = \infty$ *and* $\lim_{n\to\infty}\alpha_{n+1}/\alpha_n = 1$,

(c3) $0 < c \leq \beta_n \leq d < 1$.

Let $0 < \mu < 2\eta/k^2$, $0 < \gamma < \mu(\eta - (\mu k^2/2))/L = \tau/L$, and let the following sequence $\{x_n\}_{n=0}^{\infty}$ be a sequence in C generated from an arbitrary $x_1 \in C$ by

$$\begin{aligned}x_{n+1} &= \beta_n x_n + (1-\beta_n)\\ &\quad \times T_{\omega_1}\left(\alpha_n\gamma f(x_n) + (I - \mu\alpha_n B)T_{\omega_2}x_n\right),\end{aligned} \quad (22)$$

where $T_{\omega_i} = (1-\omega_i)I + \omega_i T_i$ with $0 < k_i < \omega_i < 1$, $i = 1, 2$. Then $\{x_n\}$ converges strongly to the unique element x^* in $\Omega = F_{ix}(T_1)\bigcap F_{ix}(T_2)$ verifying

$$x^* = (P_\Omega \cdot f)x^*, \quad (23)$$

which equivalently solves the following variational inequality problem:

$$x^* \in \Omega, \quad \langle(\gamma f - \mu B)x^*, v - x^*\rangle \leq 0, \quad \forall v \in \Omega. \quad (24)$$

Proof. Let $x^* \in \Omega$. According to Remark 16, we have

$$\|T_\omega x_n - x^*\| \le \|x_n - x^*\|. \quad (25)$$

Let $y_n = \alpha_n \gamma f(x_n) + (I - \mu\alpha_n B) T_{\omega_2} x_n$. From (22), we obtain

$$\begin{aligned} \|x_{n+1} - x^*\| &= \|\beta_n x_n + (1-\beta_n) T_{\omega_1} y_n - x^*\| \\ &\le \beta_n \|x_n - x^*\| + (1-\beta_n)\|T_{\omega_1} y_n - x^*\| \quad (26) \\ &\le \beta_n \|x_n - x^*\| + (1-\beta_n)\|y_n - x^*\|, \end{aligned}$$

And, from Lemma 14, we also obtain that

$$\begin{aligned} \|y_n - x^*\| &= \|\alpha_n \gamma (f(x_n) - f(x^*)) \\ &\quad + (I - \mu\alpha_n B)(T_{\omega_2} x_n - x^*) \\ &\quad + \alpha_n (\gamma f(x^*) - \mu B x^*)\| \\ &\le \alpha_n \gamma \|f(x_n) - f(x^*)\| \\ &\quad + (1 - \alpha_n \tau)\|T_{\omega_2} x_n - x^*\| \quad (27) \\ &\quad + \alpha_n \|\gamma f(x^*) - \mu B x^*\| \\ &\le \alpha_n \gamma L \|x_n - x^*\| \\ &\quad + (1 - \alpha_n \tau)\|x_n - x^*\| \\ &\quad + \alpha_n \|\gamma f(x^*) - \mu B x^*\|. \end{aligned}$$

Together with (26) and (27), we obtain

$$\begin{aligned} &\|x_{n+1} - x^*\| \\ &\le [1 - \alpha_n (1-\beta_n)(\tau - \gamma L)]\|x_n - x^*\| \quad (28) \\ &\quad + \alpha_n (1-\beta_n)\|\gamma f(x^*) - \mu B x^*\|. \end{aligned}$$

Putting $M_1 = \max\{\|x_0 - x^*\|, \|\gamma f(x^*) - x^*\|\}$, we clearly obtain $\|x_n - x^*\| \le M_1$. By induction, we can deduce that $\{x_n\}$ is bounded and the sequences $\{f(x_n)\}$ and $\{T_{\omega_i} x_n\}$, $i = 1, 2$, are also bounded.

Next, we show the following estimation:

$$\begin{aligned} \mathcal{T}_{n+1} &\le (1 - \alpha_n \delta)\mathcal{T}_n \\ &\quad - \delta\left(\| x_n - T_2 x_n\|^2 + \| T_{\omega_1} y_n - x_n\|^2\right) \quad (29) \\ &\quad + 2\alpha_n (1-\beta_n)\langle \gamma f(x_n) - \mu B x^*, y_n - x^*\rangle, \end{aligned}$$

where $\mathcal{T}_n = \| x_n - x^*\|^2$ and $\delta > 0$ is chosen so that

$$\begin{aligned} \delta \le \min\{&\omega_2 (1 - \alpha_n \tau)(1-\beta_n)(1-\omega_2)(1-k_2), \\ &\beta_n (1-\beta_n)\}. \end{aligned} \quad (30)$$

From Lemma 15, we obtain

$$\begin{aligned} &\|T_{\omega_2} x_n - x^*\|^2 \\ &\le \|x_n - x^*\|^2 - \omega_2 (1-\omega_2)(1-k_2)\|x_n - T_2 x_n\|^2. \end{aligned} \quad (31)$$

By (25) and (31), we get

$$\begin{aligned} &\|y_n - x^*\|^2 \\ &= \|\alpha_n (\gamma f(x_n) - \mu B x^*) + (I - \mu\alpha_n B)(T_{\omega_2} x_n - x^*)\|^2 \\ &\le (1 - \alpha_n \tau)\|T_{\omega_2} x_n - x^*\|^2 \\ &\quad + 2\alpha_n \langle \gamma f(x_n) - \mu B x^*, y_n - x^*\rangle \\ &\le (1 - \alpha_n \tau)\|x_n - x^*\|^2 \\ &\quad - \omega_2 (1 - \alpha_n \tau)(1-\omega_2)(1-k_2)\|x_n - T_2 x_n\|^2 \\ &\quad + 2\alpha_n \langle \gamma f(x_n) - \mu B x^*, y_n - x^*\rangle. \end{aligned} \quad (32)$$

Consequently,

$$\begin{aligned} &\|x_{n+1} - x^*\|^2 \\ &= \beta_n \|x_n - x^*\|^2 + (1-\beta_n)\|T_{\omega_1} y_n - x^*\|^2 \\ &\quad - \beta_n (1-\beta_n)\|T_{\omega_1} y_n - x_n\|^2 \\ &\le \beta_n \|x_n - x^*\|^2 - \beta_n (1-\beta_n)\|T_{\omega_1} y_n - x_n\|^2 \\ &\quad + (1-\beta_n)\|y_n - x^*\|^2 \\ &\le \beta_n \|x_n - x^*\|^2 - \beta_n (1-\beta_n)\|T_{\omega_1} y_n - x_n\|^2 \\ &\quad + (1-\beta_n)(1-\alpha_n \tau)\|x_n - x^*\|^2 \\ &\quad - \omega_2 (1-\alpha_n \tau)(1-\beta_n)(1-\omega_2)(1-k_2) \\ &\quad \times \|x_n - T_2 x_n\|^2 \\ &\quad + 2\alpha_n (1-\beta_n)\langle \gamma f(x_n) - \mu B x^*, y_n - x^*\rangle \\ &\le (1 - \alpha_n \tau (1-\beta_n))\|x_n - x^*\|^2 \\ &\quad - \beta_n (1-\beta_n)\|T_{\omega_1} y_n - x_n\|^2 \\ &\quad - \omega_2 (1-\alpha_n \tau)(1-\beta_n)(1-\omega_2)(1-k_2) \\ &\quad \times \|x_n - T_2 x_n\|^2 \\ &\quad + 2\alpha_n (1-\beta_n)\langle \gamma f(x_n) - \mu B x^*, y_n - x^*\rangle, \end{aligned} \quad (33)$$

and the desired inequality (29) follows.

Finally, we show $\lim_{n\to\infty} \mathcal{T}_n = 0$ by considering two possible cases. □

Case 1. $\{\mathcal{T}_n\}$ is nonincreasing sequence with some $n \ge n_0$. In this case, $\{\mathcal{T}_n\}$ is then convergent because it is also nonnegative (hence it is bounded form blow). From (29), we have

$$\delta\left(\|x_n - T_2 x_n\|^2 + \|T_{\omega_1} y_n - x_n\|^2\right) \le \alpha_n M + (\mathcal{T}_n - \mathcal{T}_{n+1}), \quad (34)$$

where $0 < M = 2\langle \gamma f(x_n) - \mu Bx^*, y_n - x^*\rangle$. Consequently, both $\|x_n - T_2 x_n\|$ and $\|T_{\omega_1} y_n - x_n\|$ converge to zero. From (22) and (25), $\lim_{n\to\infty}\alpha_n = 0$ and the sequence $\{x_n\}$ is bounded, we obtain

$$\begin{aligned}\|y_n - x_n\| &= \left\|\alpha_n\left(\gamma f(x_n) - \mu Bx_n\right) + (I - \mu\alpha_n B)\left(T_{\omega_2}x_n - x_n\right)\right\| \\ &\le \alpha_n\|\gamma f(x_n) - \mu Bx_n\| + (1-\alpha_n\tau)\left\|T_{\omega_2}x_n - x_n\right\| \\ &\le \alpha_n\|\gamma f(x_n) - \mu Bx_n\| + (1-\alpha_n\tau)\,\omega_2\|T_2x_n - x_n\| \longrightarrow 0,\end{aligned} \tag{35}$$

which implies

$$\begin{aligned}\left\|x_n - T_{\omega_1}x_n\right\|^2 &\le \left\|x_n - T_{\omega_1}y_n\right\|^2 + \left\|T_{\omega_1}y_n - T_{\omega_1}x_n\right\|^2 \\ &\le \left\|x_n - T_{\omega_1}y_n\right\|^2 + \left\|x_n - y_n\right\|^2 \\ &\quad + \frac{2}{\omega_1}\left\langle x_n - T_{\omega_1}x_n, y_n - T_{\omega_1}y_n\right\rangle \\ &\le \left\|x_n - T_{\omega_1}y_n\right\|^2 + \left\|x_n - y_n\right\|^2 \\ &\quad + \frac{2}{\omega_1}\left\langle x_n - T_{\omega_1}x_n, y_n - x_n\right\rangle \\ &\quad + \frac{2}{\omega_1}\left\langle x_n - T_{\omega_1}x_n, x_n - T_{\omega_1}y_n\right\rangle.\end{aligned} \tag{36}$$

From $\lim_{n\to\infty}\|T_{\omega_1}y_n - x_n\| = 0$, $\lim_{n\to\infty}\|y_n - x_n\| = 0$, and $x_n - T_{\omega_1}x_n = \omega_1(x_n - T_1x_n)$, we obtain

$$\lim_{n\to\infty}\|x_n - T_1x_n\| = \lim_{n\to\infty}\left\|x_n - T_{\omega_1}x_n\right\| = 0. \tag{37}$$

Using the demiclosedness principle (Proposition 8) and (35), we know that $\omega_w(x_n) \in \Omega$; hence

$$\begin{aligned}&\limsup_{n\to\infty}\langle \gamma f(x_n) - \mu Bx^*, y_n - x^*\rangle \\ &\quad = \limsup_{n\to\infty}\langle \gamma f(x_n) - \mu Bx^*, x_n - x^*\rangle \\ &\quad = \max_{w\in\omega_w(x_n)}\langle \gamma f(w) - \mu Bx^*, w - x^*\rangle \le 0.\end{aligned} \tag{38}$$

It then follows from (29) that

$$\begin{aligned}\mathcal{T}_{n+1} &\le (1-\alpha_n\delta)\mathcal{T}_n \\ &\quad + 2\alpha_n(1-\beta_n)\langle \gamma f(x_n) - \mu Bx^*, y_n - x^*\rangle.\end{aligned} \tag{39}$$

According to Lemma 11, we obtain $\lim_{n\to\infty}\mathcal{T}_n = 0$.

Case 2. Suppose there exists a subsequence $\{\mathcal{T}_{n_k}\}_{k\ge0}$ of $\{\mathcal{T}_n\}_{n\ge0}$ such that $\mathcal{T}_{n_k} \le \mathcal{T}_{n_{k+1}}$ for all $k \ge 0$. In this situation, we consider the sequence of indices $\{\delta(n)\}$ as defined in Lemma 12. It follows that $\mathcal{T}_{\delta(n+1)} - \mathcal{T}_{\delta(n)} > 0$, which by (29) amounts to

$$\delta\left(\left\|x_{\delta(n)} - T_2x_{\delta(n)}\right\|^2 + \left\|T_{\omega_1}y_{\delta(n)} - x_{\delta(n)}\right\|^2\right) \le \alpha_{\delta(n)}M. \tag{40}$$

In a similar way to Case 1, we obtain

$$\limsup_{n\to\infty}\langle \gamma f(x_{\delta(n)}) - \mu Bx^*, y_{\delta(n)} - x^*\rangle \le 0. \tag{41}$$

From Lemma 12 and (29)

$$\mathcal{T}_{\delta(n)} \le 2\left(1-\beta_{\delta(n)}\right)\langle \gamma f(x_{\delta(n)}) - \mu Bx^*, y_{\delta(n)} - x^*\rangle, \tag{42}$$

for all $n > n_0$. Taking lim sup in this inequality, we obtain $\lim_{n\to\infty}\mathcal{T}_{\delta(n)} = 0$. Moreover, it follows from (22) that

$$\begin{aligned}\mathcal{T}^2_{\delta(n)+1} &= \left\|\left(x_{\delta(n)} - x^*\right) - \left(x_{\delta(n)} - x_{\delta(n)+1}\right)\right\|^2 \\ &\le \mathcal{T}^2_{\delta(n)} + \left\|x_{\delta(n)} - x_{\delta(n)+1}\right\|^2 \\ &\le \mathcal{T}^2_{\delta(n)} + \left\|x_{\delta(n)} - T_{\omega_1}y_{\delta(n)}\right\|^2,\end{aligned} \tag{43}$$

which together with (40) implies $\lim_{n\to\infty}\mathcal{T}_{\delta(n)+1} = 0$. Consequently, from Lemma 12, we obtain $\lim_{n\to\infty}\mathcal{T}_n = 0$.

In addition, from (38), we obtain

$$\begin{aligned}&\limsup_{n\to\infty}\langle \gamma f(x_n) - \mu Bx^*, x_n - x^*\rangle \\ &\quad = \limsup_{n\to\infty}\langle \gamma f(x_n) - \gamma f(x^*), x_n - x^*\rangle \\ &\qquad + \limsup_{n\to\infty}\langle \gamma f(x^*) - \mu Bx^*, x_n - x^*\rangle \\ &\quad = \gamma L\lim_{n\to\infty}\mathcal{T}_n + \limsup_{n\to\infty}\langle \gamma f(x^*) - \mu Bx^*, x_n - x^*\rangle \\ &\quad = \langle \gamma f(x^*) - \mu Bx^*, \overline{y} - x^*\rangle, \quad \overline{y} \in \Omega;\end{aligned} \tag{44}$$

then variational inequality (44) can be written as

$$\langle (I - \mu B + \gamma f)x^* - x^*, \overline{y} - x^*\rangle \ge 0, \quad \overline{y} \in \Omega. \tag{45}$$

So, by Lemma 13, it is equivalent to the fixed point equation

$$P_{F_{ix}(T)}(I - \mu B + \gamma f)x^* = x^*. \tag{46}$$

Remark 18. For a nonspreading mapping T, we have $\varrho = 0$ in Theorem 17 to obtain the following corollary.

Corollary 19. *Let C be a nonempty, closed, and convex subset of a real Hilbert space H. Let $T_1, T_2 : H \to H$ be nonspreading mappings and demiclosed. Let f be L-Lipschitz mapping on H with coefficient $L > 0$ and let $B : H \to H$ be η-strongly monotone and k-Lipschitzian on H with $k > 0$, $\eta > 0$. Let the sequence of $\{\alpha_n\}$ and $\{\beta_n\}$ satisfy the following conditions:*

(c1) $\lim_{n\to\infty}\alpha_n = 0$,

(c2) $\sum_{n=0}^{\infty}\alpha_n = \infty$ *and* $\lim_{n\to\infty}\alpha_{n+1}/\alpha_n = 1$,

(c3) $0 < c \le \beta_n \le d < 1$.

Let the following sequence $\{x_n\}_{n=0}^{\infty}$ be a sequence in C generated from an arbitrary $x_1 \in C$ by

$$x_{n+1} = \beta_n x_n + (1-\beta_n)T_{\omega_1}\left(\alpha_n\gamma f(x_n) + (I - \mu\alpha_n B)T_{\omega_2}x_n\right), \tag{47}$$

where $T_{\omega_i} = (1-\omega_i)I + \omega_i T_i$, $\omega_i \in (0,1)$, $i = 1,2$. *Then* $\{x_n\}$ *converges strongly to the unique element* x^* *in* $\Omega = F_{ix}(T_1)\bigcap F_{ix}(T_2)$ *verifying*

$$x^* = (P_\Omega \cdot f)\, x^*, \tag{48}$$

which equivalently solves the following variational inequality problem:

$$x^* \in \Omega \quad \langle (\gamma f - \mu B)\, x^*, v - x^* \rangle \le 0, \quad \forall v \in \Omega. \tag{49}$$

Remark 20. If $B = I$ and $\mu = 1$ in Corollary 19, we obtain the following corollary.

Corollary 21. *Let* C *be a nonempty, closed, and convex subset of a real Hilbert space* H. *Let* $T_1, T_2 : H \to H$ *be nonspreading mappings and demiclosed. Let* f *be L-Lipschitz mapping on* H *with coefficient* $L > 0$. *Let the sequence of* $\{\alpha_n\}$ *and* $\{\beta_n\}$ *satisfy the following conditions:*

(c1) $\lim_{n\to\infty}\alpha_n = 0$,

(c2) $\sum_{n=0}^{\infty}\alpha_n = \infty$ *and* $\lim_{n\to\infty}\alpha_{n+1}/\alpha_n = 1$,

(c3) $0 < c \le \beta_n \le d < 1$.

Let the following sequence $\{x_n\}_{n=0}^{\infty}$ *be a sequence in* C *generated from an arbitrary* $x_1 \in C$ *by*

$$x_{n+1} = \beta_n x_n + (1-\beta_n)\, T_{\omega_1}\left(\alpha_n \gamma f(x_n) + (1-\alpha_n)\, T_{\omega_2} x_n\right), \tag{50}$$

where $T_{\omega_i} = (1-\omega_i)I + \omega_i T_i$, $\omega_i \in (0,1)$, $i = 1,2$. *Then* $\{x_n\}$ *converges strongly to the unique element* x^* *in* $\Omega = F_{ix}(T_1)\bigcap F_{ix}(T_2)$ *verifying*

$$x^* = (P_\Omega \cdot f)\, x^*, \tag{51}$$

which equivalently solves the following variational inequality problem:

$$x^* \in \Omega, \quad \langle (\gamma f - I)\, x^*, v - x^* \rangle \le 0, \quad \forall v \in \Omega. \tag{52}$$

4. Application

In this section, we constructed a numerical example to illustrate that our main results are well defined. The following example is introduced by Deng et al. [14]. They prove that T is $(1/3)$-strictly pseudononspreading.

Example 22. Let $X = l^2$ with the norm $\|\cdot\|$ defined by

$$\|x\| = \sqrt{\sum_{i=1}^{\infty} x_i^2}, \quad \forall x = (x_1, x_2, \ldots, x_n, \ldots) \in X, \tag{53}$$

$C = \{x = (x_1, x_2, \ldots, x_n, \ldots) \mid x_i \in R^1, i = 1, 2, \ldots\}$, and let C be an orthogonal subspace of X (i.e., $\forall x, y \in C$, we have $\langle x, y\rangle = 0$). Then it is obvious that C is a nonempty closed convex subset of X. Now, for any $x = (x_1, x_2, \ldots, x_n, \ldots) \in C$, define a mapping $T : C \to C$ as follows:

$$Tx = \begin{cases} (x_1, x_2, \ldots, x_n, \ldots), & \prod_{i=1}^{\infty} x_i < 0, \\ (-2x_1, -2x_2, \ldots, -2x_n, \ldots), & \prod_{i=1}^{\infty} x_i \ge 0. \end{cases} \tag{54}$$

From the definition of this example, let $x = (x_1, x_2) \in R^2$ and $f(x) = 8(\sin(x_1), \cos(x_2))$ be a Lipschitz mapping on R^2. Assume that $T_1(x) = (0, 4x_1)$ and $T_2(x) = (4x_2, 0)$ be $(1/3)$-strictly pseudononspreading mappings. Let $\alpha_n = \beta_n = 1/5$, $\gamma = 1/8$, $\mu = \omega_1 = \omega_2 = 1$, and $B = I$. According to (22), we can obtain the following algorithm:

$$x_{n+1} = \frac{1}{5}x_n + \frac{4}{200}T_1\left(f(x_n)\right) + \frac{16}{25}T_1\left(T_2(x_n)\right). \tag{55}$$

From Theorem 17, we can easily know that algorithm (55) converges to the unique point in R^2. Let $x_0 = (5,4)$ and $|x_{n+1} - x_n| \le 10^{-5}$, and let x^* be the fixed point of the algorithm (55). Using the software of MATLAB, we obtain $x^* = (0,0)$.

Acknowledgments

This work is supported in part by National Natural Science Foundation of China (71272148), the Ph.D. Programs Foundation of Ministry of Education of China (20120032110039), and China Postdoctoral Science Foundation (Grant no. 20100470783).

References

[1] S. Plubtieng and R. Punpaeng, "A new iterative method for equilibrium problems and fixed point problems of nonexpansive mappings and monotone mappings," *Applied Mathematics and Computation*, vol. 197, no. 2, pp. 548–558, 2008.

[2] L. C. Zeng, S. Schaible, and J. C. Yao, "Iterative algorithm for generalized set-valued strongly nonlinear mixed variational-like inequalities," *Journal of Optimization Theory and Applications*, vol. 124, no. 3, pp. 725–738, 2005.

[3] M. O. Osilike and F. O. Isiogugu, "Weak and strong convergence theorems for nonspreading-type mappings in Hilbert spaces," *Nonlinear Analysis: Theory, Methods & Applications*, vol. 74, no. 5, pp. 1814–1822, 2011.

[4] W. R. Mann, "Mean value methods in iteration," *Proceedings of the American Mathematical Society*, vol. 4, pp. 506–510, 1953.

[5] W. Takahashi and T. Tamura, "Convergence theorems for a pair of nonexpansive mappings," *Journal of Convex Analysis*, vol. 5, no. 1, pp. 45–56, 1998.

[6] M. Tian, "A general iterative algorithm for nonexpansive mappings in Hilbert spaces," *Nonlinear Analysis: Theory, Methods & Applications*, vol. 73, no. 3, pp. 689–694, 2010.

[7] T. L. Hicks and J. D. Kubicek, "On the Mann iteration process in a Hilbert space," *Journal of Mathematical Analysis and Applications*, vol. 59, no. 3, pp. 498–504, 1977.

[8] S. A. Naimpally and K. L. Singh, "Extensions of some fixed point theorems of Rhoades," *Journal of Mathematical Analysis and Applications*, vol. 96, no. 2, pp. 437–446, 1983.

[9] P. E. Maing, "The viscosity approximation process for quasi-nonexpansive mappings in Hilbert spaces," *Computers & Mathematics with Applications*, vol. 59, no. 1, pp. 74–79, 2010.

[10] N. Petrot and R. Wangkeeree, "A general iterative scheme for strict pseudononspreading mapping related to optimization problem in Hilbert spaces," *Journal of Nonlinear Analysis and Optimization*, vol. 2, no. 2, pp. 329–336, 2011.

[11] H. K. Xu, "Iterative algorithms for nonlinear operators," *Journal of the London Mathematical Society*, vol. 66, no. 1, pp. 240–256, 2002.

[12] P. E. Maing, "Strong convergence of projected subgradient methods for nonsmooth and nonstrictly convex minimization," *Set-Valued Analysis*, vol. 16, no. 7-8, pp. 899–912, 2008.

[13] H. Piri, "A general iterative method for finding common solutions of system of equilibrium problems, system of variational inequalities and fixed point problems," *Mathematical and Computer Modelling*, vol. 55, no. 3-4, pp. 1622–1638, 2012.

[14] B. C. Deng, T. Chen, and Z. F. Li, "Viscosity iteration algorithm for a ϱ-strictly pseudononspreading mapping in a Hilbert space," *Journal of Inequalities and Applications*, vol. 2013, article 80, 2013.

17

A Crossing Lemma for Annular Regions and Invariant Sets with an Application to Planar Dynamical Systems

Anna Pascoletti and Fabio Zanolin

Dipartimento di Matematica e Informatica, Università di Udine, Via delle Scienze 206, 33100 Udine, Italy

Correspondence should be addressed to Fabio Zanolin; fabio.zanolin@uniud.it

Academic Editor: Alfredo Peris

We present a topological result, named *crossing lemma*, dealing with the existence of a continuum which crosses a topological space between a pair of "opposite" sides. This topological lemma allows us to obtain some fixed point results. In the works of Pascoletti et al., 2008, and Pascoletti and Zanolin, 2010, we have widely exposed the crossing lemma for planar regions homeomorphic to a square, and we have also presented some possible applications to the theory of topological horseshoes and to the study of chaotic-like dynamics for planar maps. In this work, we move from the framework of the generalized rectangles to two other settings (annular regions and invariant sets), trying to obtain similar results. An application to a model of fluid mixing is given.

1. Introduction

In 1883-1884, Poincaré [1, 2] introduced a generalization of the Bolzano theorem to continuous vector fields defined on a hypercube of $\mathbb{R}^N$. Such result, usually known as the Poincaré-Miranda theorem [3–5], in dimension two, reads as follows.

Theorem 1. *Let $\mathcal{R} := [a_1, b_1] \times [a_2, b_2]$ be a planar rectangle, and let $f := (f_1, f_2) : \mathcal{R} \to \mathbb{R}^2$ be a continuous vector field such that f_1 changes its sign from the left to the right side of $\mathcal{R}$ and f_2 changes its sign from the lower to the upper side of $\mathcal{R}$. Then, there exists at least one point $w \in \mathcal{R}$ such that $f(w) = 0$.*

A heuristic proof of this result, *"pour faire comprendre comment on peut démontrer ce théorème,"* quoting Poincaré's words [2], can be described as follows. (The more formal proof in [2] makes use of Kronecker integral, a forerunner of topological degree theory (see [5, 6]). We also recommend the recent article [7] for a modern perspective about this topic.) The "curve" $f_2(x_1, x_2) = 0$ starts at some point of the left side $x_1 = a_1$, and it ends at some point of the right side $x_1 = b_1$. Similarly, the "curve" $f_1(x_1, x_2) = 0$ starts at some point of the lower side $x_2 = a_2$, and it ends at some point of the upper side $x_2 = b_2$. Hence, these "curves" must intersect at some point of the rectangle. Clearly, in modern language, using in this context the term "curve" may be a little misleading, although it well explains the geometry of the problem. Yet the argument of the proof is safe if we note that the set $f_2(x_1, x_2) = 0$ contains a continuum joining the left to the right side and, similarly, the set $f_1(x_1, x_2) = 0$ contains a continuum joining the lower to the upper side. The existence of at least one intersection point between these continua follows from a result of plane topology (see Lemma 7).

In order to prove the existence of such continua, one can observe that the set $f_2(x_1, x_2) = 0$ crosses any path from the lower to the upper side of the rectangle (and the same happens for the set $f_1(x_1, x_2) = 0$ with respect to the paths connecting the left to the right side of the rectangle). In [8] we called such kind of results as *crossing lemmas* and proposed new applications to the study of the dynamics of some planar maps. In an abstract setting we can describe these results as follows. We start from a closed set $\mathcal{S}$ which cuts all the paths connecting two given sets, and we look for a *closed connected* subset $\mathcal{C}$ of $\mathcal{S}$ with the same cutting property. In many applications, it turns out that the set $\mathcal{C}$ crosses the original domain "from one side to the other." A typical example in this direction is presented in Lemma 10 in the framework of *topological rectangles*, which are topological spaces which are homeomorphic to a planar rectangle.

Important theorems, where some forms of these crossing properties are considered, appear in dimension theory with the results of Hurewicz and Wallman [9], in topological games [10, 11], and in proofs of the existence of solutions to nonlinear differential equations [12].

In some applications, typically the set $\mathcal{S}$ is a set of solutions of a nonlinear equation depending on a parameter (or equivalently a set of fixed points for a family of parameter-dependent operators). From this point of view, results in this direction, in spite of the fact that they may look quite elementary, present a great usefulness in different areas of nonlinear analysis, especially in connection with bifurcation theory (see [12–15]).

When the operators whose fixed points correspond to the element of $\mathcal{S}$ have some special symmetries, it is likely that some of these symmetries are inherited by the set $\mathcal{S}$ itself. In this case, it would be desirable to prove that also the continuum $\mathcal{C} \subset \mathcal{S}$ inherits the symmetries of $\mathcal{S}$.

In the present paper we move from the framework of generalized rectangles already discussed in previous articles [8, 16] to two other settings, namely, the study of annular regions and the case of invariants sets. More precisely, for the former case, we just collect in Section 2 some classical separation results already applied in [8, 16] to prove the existence of fixed points, periodic points, and chaotic dynamics for planar dynamical systems. In Section 3 and Section 4, respectively, we discuss the variants to annular regions and invariant sets, and, as a possible application of our results, we reconsider in Section 5 an example of planar map arising in fluid mixing previously studied by Kennedy and Yorke in [17].

We conclude this introduction with some preliminary results which will be used in what follows. Throughout the paper, by a *path* in a topological space X, we mean a continuous map $\gamma : I \to X$, with $I := [0,1]$. We also set $\overline{\gamma} := \gamma(I)$. Any homeomorphic image of I will be referred to as an *arc*. The sets $B(0,r)$ and $B[0,r]$ are, respectively, the open and the closed balls of center of origin and radius r in $\mathbb{R}^N$. In the plane $\mathbb{R}^2$ (usually identified with $\mathbb{C}$) with the Euclidean norm, we denote by $E^2 := B[0,1]$ the closed unit disc and by $S^1 = \partial E^2$ its boundary.

Slightly modifying an analogous definition of Berarducci et al. [18, Definition 2.1], we give the following.

Definition 2. Let X be a topological space, and let $A, B \subset X$ be two nonempty disjoint sets. Let also $S \subset X$. One says that S cuts the paths between A and B if $S \cap \overline{\gamma} \neq \emptyset$, for every path $\gamma : I \to X$ such that $\gamma(0) \in A$ and $\gamma(1) \in B$.

In order to simplify the statements of the next results, we write $S : A \nmid B$ to express the fact that S cuts the paths between A and B. If $A = \{a\}$ is a singleton, we write $S : a \nmid B$ instead of $S : \{a\} \nmid B$. Similar notation is used if B is a singleton. In order to have Definition 2 meaningful, we implicitly assume that there exists at least a path γ in X connecting A with B (otherwise, we could take $S = \emptyset$ or S any subset of X). Clearly, if a set S satisfies the cutting property of Definition 2, then also its closure $\operatorname{cl} S$ cuts the paths between A and B. Therefore, in the sequel and without loss of generality, we usually assume S is closed. Such an assumption is also well suited for proving the existence of minimal (closed) sets satisfying Definition 2. Indeed, we have the following.

Lemma 3. *Let X be a topological spac,e and let $A, B \subset X$ be two nonempty disjoint sets which are connected by at least one path in X. Let $S \subset X$ be a (nonempty) closed set which cuts the paths between A and B. Then there exists a nonempty, closed set $C \subset S$ which is minimal with respect to the property of cutting the paths between A and B.*

Proof. The proof is a standard application of Zorn's lemma. Let $\mathcal{F}$ be the set of all the nonempty closed subsets F of S such that $F : A \nmid B$, with the elements of $\mathcal{F}$ ordered by inclusion. Clearly, $\mathcal{F}$ is nonempty for at least $S \in \mathcal{F}$. Let $(F_\alpha)_{\alpha \in J}$ be a totally ordered subset of $\mathcal{F}$. We claim that $F^* := \bigcap_{\alpha \in J} F_\alpha \in \mathcal{F}$. Indeed, let $\gamma : I \to X$ be a path such that $\gamma(0) \in A$ and $\gamma(1) \in B$. The family of compact sets $(\overline{\gamma} \cap F_\alpha)_{\alpha \in J}$ has the finite intersection property, and, therefore, $\overline{\gamma} \cap \bigcap_{\alpha \in J} F_\alpha = \overline{\gamma} \cap F^* \neq \emptyset$. This proves our claim, and the conclusion follows from Zorn's lemma. □

In [19], Dolcher studied a similar minimality problem, dealing with closed sets separating two points. The definition of separation is the standard one; that is, a set $S \subset X$ separates two points (or, in general, two nonempty sets) if the two points belong to different components of the complement $X \setminus S$. In this respect, we reconsider the following example from [19]. Let $X \subset \mathbb{R}^2$, with the topology of the plane, be defined by

$$X := \left\{(x, y) : x \geq 0, y = \frac{x}{n}, n \in \mathbb{N}_0\right\} \cup (\mathbb{R}^+ \times \{0\}), \quad (1)$$

with $\mathbb{R}^+ := [0, +\infty)$ and $\mathbb{N}_0 := \mathbb{N} \setminus \{0\}$.

Let $A := \{(0,0)\}$, $B := \{(2,0)\}$, and $S := \{(x,y) \in X : x = 1\}$. Clearly, $S : A \nmid B$, and also S separates A and B in X. As shown in [19], there is no subset of S which is minimal for the property of separating A and B in X. On the other hand, $C := \{(1,0)\}$ is the minimal subset of S which cuts the paths between A and B.

In general, a set S satisfying $S : A \nmid B$, even if minimal with respect to such property, need not be connected. For an elementary example, one can take $X := S^1$ (with the topology of the plane), $A := \{(-1,0)\}$, $B := \{(1,0)\}$, and $S := \{(0,-1),(0,1)\}$. In this case, S is a closed set, minimal with respect to the property of cutting the paths between A and B, which is not connected. The connectivity of S (or of a minimal subset of it) is, however, an important property for the proof of the existence of fixed points or of zeros for maps in Euclidean spaces. Such connectivity properties have been employed recently in [20–22] in connection with the theory of *topological horseshoes*. In order to recall some main results from the above quoted papers and to propose some further developments, we introduce, in the next section, some main definitions which play a crucial role in our approach.

2. Generalized Rectangles

Definition 4. A topological space X is called a *generalized rectangle* if it is homeomorphic to the unit square $Q := I^2$.

For a generalized rectangle X and a homeomorphism $\eta : Q \to \eta(Q) = X$, the set $\eta(\partial Q)$ is independent of the choice of the homeomorphism η. We call the set $\eta(\partial Q)$ the *contour of* X and denote it by ϑX. Clearly, for a generalized rectangle X embedded in $\mathbb{R}^2$, the contour of X coincides with the boundary of X.

Definition 5. An *oriented rectangle* is a pair $\widetilde{X} := (X, X^-)$ such that X is a generalized rectangle and $X^- \subset \vartheta X$ is the union of two disjoint arcs X_l^- and X_r^-. One says that X_l^- and X_r^- are, respectively, the left- and the right-hand sides of the (generalized) rectangle X. Sometimes one will also write $X := |\widetilde{X}|$.

Remark 6. Given an oriented rectangle, $\widetilde{X} = (X, X^-)$, we can choose a homeomorphism $\eta : Q \to \eta(Q) = X$ in such a manner that

$$X_l^- = \eta(\{0\} \times I), \qquad X_r^- = \eta(\{1\} \times I) \tag{2}$$

(this can be proved using the Schoenflies theorem [23]). Then we can also introduce the sets

$$X_b^+ = \eta(I \times \{0\}), \qquad X_t^+ = \eta(I \times \{1\}), \tag{3}$$

which are called the bottom and the top sides of $\widetilde{X}$ and define

$$X^+ := X_b^+ \cup X_t^+. \tag{4}$$

Observe that, for any homeomorphism $\eta_1 : Q \to \eta_1(Q) = X$ such that $\eta_1(\{0,1\} \times I) = X^-$, it follows that $\eta_1(I \times \{0,1\}) = X^+$. Thus, the choice of X^- determines also the choice of X^+ (which is also the closure of $\vartheta X \setminus X^-$). Given a pair (X^-, X^+), the order in which we choose to call as "right" and "left" to the two components of X^- (as well as to label the two components of X^+) is completely irrelevant for what follows. As a consequence of such remark, we have that, to each oriented rectangle $\widetilde{X}$, a "dual" oriented rectangle $\widetilde{X}'$ is associated with $|\widetilde{X}| = |\widetilde{X}'| = X$ and with X^+ as the $[\cdot]^-$-set for $\widetilde{X}'$.

2.1. Crossing Properties for Generalized Rectangles. The next (classical) result guarantees the fact that continua connecting opposite sides of an oriented rectangle must cross each other. It can be proved as a consequence of the Jordan curve theorem, and the proof is omitted (see [16] for the details and also [24] for an application to ordinary differential equations).

Lemma 7. *Let $\widetilde{X} = (X, X^-)$ be an oriented rectangle, and let $C_1, C_2 \subset X$ be closed connected sets such that*

$$C_1 \cap X_l^- \neq \emptyset \neq C_1 \cap X_r^-, \qquad C_2 \cap X_b^+ \neq \emptyset \neq C_2 \cap X_t^+. \tag{5}$$

Then $C_1 \cap C_2 \neq \emptyset$.

The connectedness of C_1 and C_2 is not enough to guarantee the existence of a nonempty intersection (see [25, Example 10.4]).

We present now some results about sets separating the opposite faces of an oriented rectangle. We have shown their role in the proof of the existence of fixed points and periodic points for continuous maps in [8, 16, 26]. Some of these results can be extended to higher dimension using the topological degree, the fixed point index, or other index theories (see [22] and the references therein). Since the applications in the present paper will be all related to planar maps, we prefer to confine ourselves to the use of a more direct tool: Alexander's lemma. Such result, named after Alexander [27, 28], as shown both in Newman's book [29] and in Sanderson's article [30], is quite useful in proving a broad range of theorems of plane topology. Quoting Smith [31] *"this lemma, the proof of which requires but a few lines, is shown [⋯] to be one of the sharpest tools in the theory of separation, if skilfully handled."* Results based on applications of Alexander's lemma or to other related theorems in [29] have been fruitfully applied to differential equations by Hastings [32–34], McLeod and Serrin [35], Turner [36], and others. For more recent applications see [15, 37, 38].

The following version of Alexander's lemma will be used in our next results. The proof requires only an elementary modification of the standard one [39] and, therefore, is omitted.

Lemma 8. *Let $\widetilde{X} = (X, X^-)$ be an oriented rectangle, and let K_1, K_2 be closed disjoint subsets of X. Assume that there exist two paths $\gamma_1, \gamma_2 : I \to X$, with $\gamma_1(0), \gamma_2(0) \in X_l^-$ and $\gamma_1(1), \gamma_2(1) \in X_r^-$ such that*

$$\overline{\gamma}_1 \cap K_1 = \emptyset, \qquad \overline{\gamma}_2 \cap K_2 = \emptyset. \tag{6}$$

Then, there exists a path $\gamma : I \to X$, with $\gamma(0) \in X_l^-$ and $\gamma(1) \in X_r^-$ such that $\overline{\gamma} \cap (K_1 \cup K_2) = \emptyset$.

Remark 9. The original version of Alexander's lemma is a variant of Lemma 8 in which we assume that all the paths have the same initial point $p \in X_l^-$ and the same end point $q \in X_r^-$.

The next result is perhaps one of the most classical and useful consequences of the previous lemma (see [29]) and is usually expressed by the fact that *"if a closed set separates the plane, then some component of this set separates the plane"* [34, page 131].

Lemma 10. *Let $\widetilde{X} = (X, X^-)$ be an oriented rectangle, and let $S \subset X$ be a closed set such that $S : X_l^- \nmid X_r^-$. Then, there exists a compact, connected set $C \subset S$ such that $C : X_l^- \nmid X_r^-$.*

Proof. By Lemma 3 there exists a closed set $C \subset S$ such that $C : X_l^- \nmid X_r^-$, with C minimal with respect to the cutting property. Suppose, by contradiction, that C is not connected, and let $C_1, C_2 \subset C$ be two closed nonempty disjoint sets with $C_1 \cup C_2 = C$. Since C is minimal and C_1, C_2 are proper subsets of C, there exist two paths γ_1, γ_2 in X which connects X_l^- to X_r^-, such that γ_i avoids C_i (for $i = 1, 2$). Then, by Lemma 8, there exists a path $\gamma : I \to X$ with $\gamma(0) \in X_l^-$ and $\gamma(1) \in X_r^-$

with $\overline{\gamma} \cap C = \emptyset$, contradicting the assumption that $C : X_l^- \pitchfork X_r^-$. □

The cutting property for C obtained in Lemma 10 can be equivalently expressed as follows.

Lemma 11. *Let $\widetilde{X} = (X, X^-)$ be an oriented rectangle, and let $C \subset X$ be a closed connected set. Then, $C : X_l^- \pitchfork X_r^-$ if and only if $C \cap X_b^+ \neq \emptyset \neq C \cap X_t^+$.*

Proof. If $C : X_l^- \pitchfork X_r^-$, then, necessarily, C must cut the upper and the lower sides of $\widetilde{X}$ which are the images of particular paths connecting X_l^- to X_r^-. On the other hand, if $\gamma : I \to X$ is any path with $\gamma(0) \in X_l^-$ and $\gamma(1) \in X_r^-$, then $\overline{\gamma}$ and C are two continua connecting the opposite sides of the oriented rectangle and, therefore, $\overline{\gamma} \cap C \neq \emptyset$ by Lemma 7. This proves that $C : X_l^- \pitchfork X_r^-$. □

Lemma 11, as well as Lemma 7, can be seen as a continuous version of the so-called Hex theorem asserting that Hex cannot end in a tie [11].

The combination of Lemma 10 and Lemma 11 gives the so-called *Crossing Lemma* asserting the fact that *if a closed set intersects all the paths from the left to the right side in an oriented rectangle, then it contains a continuum connecting the two other sides.* Note also that such a continuum can be taken as *irreducible between X_b^+ and X_t^+* (by using some classical results from [13, 40]). See [16] for a recent survey on this subject and its connections with various different results of plane topology, as well as for a different proof based on Whyburn's lemma.

3. Annular Regions

In this section, we are moving our attention from generalized rectangles to planar annuli, trying to develop analogous results of Section 2 in this new setting. First of all, we present some basic definitions.

A closed planar annulus (of inner radius $a > 0$ and outer radius $b > a$) is the set

$$A[a,b] := \{(x,y) \in \mathbb{R}^2 : a \leq \|(x,y)\| \leq b\}. \quad (7)$$

Definition 12. A topological space X is called a *topological annulus* if it is homeomorphic to a planar annulus.

Let X be a topological annulus, and let $\eta : A[1,2] \to \eta(A[1,2]) = X$ be a homeomorphism. As a consequence of the Schoenflies theorem, the set $\eta(\partial A[1,2])$ is independent of the choice of the homeomorphism η. We call the set $\eta(\partial A[1,2])$ the *contour of* X and denote it by ϑX. Clearly, for a topological annulus X embedded in $\mathbb{R}^2$, the contour of X coincides with the boundary of X. The contour of X consists of two connected components which are closed arcs (the Jordan curves) since they are homeomorphic to S^1. We call such closed arcs X_i and X_o. For a planarly embedded topological annulus, they could be chosen as the inner and the outer boundaries of the annulus. In such a special case, the bounded component of $\mathbb{R}^2 \setminus X$ turns out to be an open simply connected set $D = D(X)$ with $\partial D = X_i$ and $\operatorname{cl} D = D \cup X_i$ homeomorphic to the closed unit disc. On the other hand, in the general setting, speaking of inner and outer boundaries is meaningless, yet we keep this terminology. Finally, we define the *interior of* X as

$$\operatorname{int} X := X \setminus \vartheta X. \quad (8)$$

In the sequel, we use the standard covering projection of $\mathbb{R} \times \mathbb{R}_0^+$ onto $\mathbb{R}^2 \setminus \{0\}$, defined by the polar coordinates

$$\Pi(\theta,\rho) = (\rho\cos\theta, \rho\sin\theta). \quad (9)$$

In the space $\mathbb{R} \times \mathbb{R}_0^+$, we consider also the translation

$$\eth : (\theta,\rho) \longmapsto (\theta + 2\pi, \rho). \quad (10)$$

3.1. Crossing Properties for Annular Regions. Our aim now is to reconsider the results obtained for topological rectangles in Section 2 and adapt them to a form which may be better suited to deal with the new setting of topological annuli. We start with a version of Alexander's lemma which reads as follows.

Lemma 13. *Let X be a topological annulus, and let K_1, K_2 be closed subsets of X such that $K_1 \cap K_2$ is connected (possibly empty). Assume that there exist two paths $\gamma_1, \gamma_2 : I \to X$, with $\gamma_1(0), \gamma_2(0) \in X_i$ and $\gamma_1(1), \gamma_2(1) \in X_o$ such that*

$$\overline{\gamma}_1 \cap K_1 = \emptyset, \qquad \overline{\gamma}_2 \cap K_2 = \emptyset. \quad (11)$$

Then, there exists a path $\gamma : I \to X$, with $\gamma(0) \in X_i$ and $\gamma(1) \in X_o$ such that $\overline{\gamma} \cap (K_1 \cup K_2) = \emptyset$.

This result is an elementary variant of the version of Alexander's lemma in [29]; see also [38, page 3040] and [37, Corollary 3].

We say that a set $C \subset X$ is *essentially embedded in* X if the inclusion

$$i_C : C \longrightarrow X, \qquad i_C(x) = x, \quad \forall x \in C \quad (12)$$

is not homotopic to a constant map.

The next result is a corollary of the Borsuk separation theorem [41, Theorem 6-47] adapted to our situation. We give a proof, for completeness, following [41].

Lemma 14. *Let X be a topological annulus, and let $S \subset X$ be a closed set. Then, S is essentially embedded in X if and only if $S : X_i \pitchfork X_o$.*

Proof. Up to a homeomorphism defining the annulus X, we can assume $X = A[a,b]$ with

$$0 < a < b < 1. \quad (13)$$

In this case, $X_i = \partial B(0,a) = aS^1$ and $X_o = \partial B(0,b) = bS^1$.

Suppose that $S : X_i \pitchfork X_o$, and let $\mathscr{C}(0)$ be the connected component of $\mathbb{R}^2 \setminus S$ containing the origin. By the assumption, $\mathscr{C}(0) \cup S$ is closed and

$$B[0,a] \subset \mathscr{C}(0) \cup S \subset B[0,b]. \quad (14)$$

Assume, by contradiction, that S is not essentially embedded in X, and, therefore, the inclusion $i_S : S \to X$ is homotopic in X to a constant map, say $X \ni x \mapsto p$, for all $x \in X$, for a suitable point $p \in X$. From this, it follows immediately that the mapping $f : S \to S^1$ defined by $x \mapsto x/\|x\|$ is homotopic to a constant, which is *inessential*. Then, by [41, Theorem 4-5], there exists a continuous and inessential extension $\tilde{f}$ of f with $\tilde{f}$ defined on $\mathscr{C}(0) \cup S$. We define now the map $r : E^2 \to S^1$ by

$$r(x) := \begin{cases} \tilde{f}(x), & \text{for } x \in \mathscr{C}(0) \cup S \\ \dfrac{x}{\|x\|}, & \text{for } x \notin \mathscr{C}(0) \cup S \end{cases} \tag{15}$$

which is continuous. Thus, we are led to a contradiction since $r(\cdot)$ turns out to be a retraction of E^2 onto S^1.

Suppose now that S is essentially embedded in X, and also assume, by contradiction, that there exists a path $\gamma : I \to X$ with $\|\gamma(0)\| = a$, $\|\gamma(1)\| = b$ and that $\gamma(t) \notin S$, for all $t \in I$. Without loss of generality (taking possibly an arc inside $\overline{\gamma}$), we can suppose that γ is one to one and that $a < \|\gamma(t)\| < b$ for each $t \in]0,1[$.

Passing to the covering space $\mathscr{H} := \mathbb{R} \times [a,b]$ of $X = A[a,b]$, we have that the path γ lifts to a family of paths $\tilde{\gamma}_n : I \to \mathscr{H}$ with $\tilde{\gamma}_n(t) = \tilde{\gamma}_0(t) + (2n\pi, 0)$ and that $\tilde{\gamma}_n(t) \cap \Pi^{-1}(S) = \emptyset$, for all $t \in I$ and every $n \in \mathbb{Z}$. By the properties of γ we have that $\tilde{\gamma}_0 : I \to \mathscr{H}$ is one to one and, moreover, $\tilde{\gamma}_0(t') - \tilde{\gamma}_0(t'') \notin 2\pi\mathbb{Z}$ for $t' \neq t''$ (because $\Pi \circ \tilde{\gamma}_0 = \gamma$ is one to one). We have thus obtained an arc $\Gamma_0 := \tilde{\gamma}_0(I) \subset \mathscr{H}$ connecting the lines $\rho = a$ and $\rho = b$ and avoiding $\Pi^{-1}(S)$ such that Γ_0 intersects the line $\{\rho = a\}$ exactly at one point and the same happens with respect to $\{\rho = b\}$. Furthermore,

$$\Gamma_0 \cap ((2\pi n, 0) + \Gamma_0) = \emptyset, \quad \forall n \in \mathbb{Z} \setminus \{0\}. \tag{16}$$

Let $\Gamma_0 \cap \{\rho = a\} = \{P_a\}$, $\Gamma_0 \cap \{\rho = b\} = \{P_b\}$, and $\Gamma_1 := (2\pi, 0) + \Gamma_0$. Note that $\Gamma_0 \cap \Gamma_1 = \emptyset$. Let $\mathscr{J}$ be the Jordan curve obtained by joining (in the counterclockwise sense) the point P_a to $(2\pi, 0) + P_a$ along the line $\{\rho = a\}$, the point $(2\pi, 0) + P_a$ to $(2\pi, 0) + P_b$ along Γ_1, the point $(2\pi, 0) + P_b$ to P_b along the line $\{\rho = b\}$, and, finally, the point P_b to P_a along Γ_0. The curve $\mathscr{J}$ is the boundary of an open bounded domain D with $\mathrm{cl}\, D = D \cup \mathscr{J}$ homeomorphic to the unit square Q. Roughly speaking, $\mathrm{cl}\, D$ is the set of all the points of the strip $\mathscr{H}$ between Γ_0 and Γ_1, with the boundary arcs included. Let $\eta : Q \to \eta(Q) = \mathrm{cl}\, D$ be a homeomorphism mapping the left side of Q to Γ_0, the lower side of Q to the segment $\{P_a + (\theta, 0) : \theta \in [0, 2\pi]\}$, the right side of Q to Γ_1, and the upper side of Q to the segment $\{P_b + (\theta, 0) : \theta \in [0, 2\pi]\}$. By the construction of the topological rectangle $\mathrm{cl}\, D$ and since $\Gamma_0 \cap \Pi^{-1}(S) = \emptyset$, we have that the set

$$S' := \Pi^{-1}(S) \cap \eta(]0,1[\times [0,1]) \subset \mathrm{cl}\, D \tag{17}$$

is mapped homeomorphically onto S by the covering projection Π. Now we choose $\varepsilon \in]0, 1/2[$ sufficiently small such that

$$S' \subset \Pi^{-1}(S) \cap \eta([\varepsilon, 1-\varepsilon] \times [0,1]), \tag{18}$$

and we also introduce the set

$$B := \Pi(\eta([\varepsilon, 1-\varepsilon] \times [0,1])). \tag{19}$$

By construction, the set B is a topological rectangle contained in $A[a,b]$ and containing the set S. The continuous map

$$\begin{aligned} &([\varepsilon, 1-\varepsilon] \times [0,1]) \times [0,1] \ni (z, \lambda) \\ &\longmapsto \Pi\left(\eta\left((1-\lambda) z + \lambda\left(\frac{1}{2}, \frac{1}{2}\right)\right)\right) \subset A[a,b] \end{aligned} \tag{20}$$

restricted to $S \times I$ provides a homotopy between the identity i_S and a constant map. This contradicts the assumption that S is essentially embedded in $A[a,b]$, and the proof is complete.

A variant of the above proof has been suggested by the referee, and we wish to report it here, since it may be more interesting from a geometrical point of view. The argument is as follows.

Having obtained (as above) $\gamma : I \to A[a,b] = X$ which is one to one, avoids the set S, and intersects the inner and the outer boundaries of X only at its extreme ends $\gamma(0)$ and $\gamma(1)$, we consider a homeomorphism h of the annulus onto itself such that $h(\gamma(I)) = \{(x, 0) : x \in [a,b]\}$, $h(\gamma(0)) = (a, 0)$, and $h(\gamma(1)) = (b, 0)$. Moreover, the segment $[a,b] \times \{0\}$ avoids the set $h(S)$. We lift the segment $[a,b] \times \{0\}$ to sequence of parallel vertical lines $\bigcup_{n \in \mathbb{Z}} \{2n\pi\} \times [a,b] \subset \mathscr{H}$ which avoid the set $\Pi^{-1}(h(S))$. Now the topological rectangle $\mathrm{cl}\, D$ of the preceding part of the proof turns out to be just the standard rectangle $[0, 2\pi] \times [a,b]$, and the set $T' := \Pi^{-1}(h(S)) \cap (]0, 2\pi[\times [a,b])$ (which is mapped homeomorphically by Π onto $h(S)$) is actually contained in $[\varepsilon, 2\pi - \varepsilon] \times [a,b]$ for some $\varepsilon > 0$ and sufficiently small. Now we conclude precisely as before by showing that $i_{h(S)}$ is homotopic to a constant in the rectangle $\Pi([\varepsilon, 2\pi - \varepsilon] \times [a,b]) \subset A[a,b]$. □

A version of Lemma 10 for a topological annulus reads as follows.

Lemma 15. *Let X be a topological annulus, and let $S \subset X$ be a closed set such that $S : X_i \nmid X_o$. Then, there exists a compact, connected set $C \subset S$ such that $C : X_i \nmid X_o$ (and, therefore, C is essentially embedded in X).*

Proof. The proof follows the same argument of the one of Lemma 10. We give the details for completeness. By Lemma 3, there exists a closed set $C \subset S$ such that $C : X_i \nmid X_o$, with C minimal with respect to the cutting property. Suppose, by contradiction, that C is not connected, and let $C_1, C_2 \subset C$ be two closed nonempty disjoint sets with $C_1 \cup C_2 = C$. Since C is minimal and C_1, C_2 are proper subsets of C, there exist two paths γ_1, γ_2 in X which connect X_i to X_o such that γ_i avoids C_i (for $i = 1, 2$). Then, by Lemma 13, there exists a path $\gamma : I \to X$ with $\gamma(0) \in X_i$ and $\gamma(1) \in X_o$ with $\overline{\gamma} \cap C = \emptyset$, contradicting the assumption that $C : X_i \nmid X_o$. The continuum C is also essentially embedded in X by Lemma 14. □

The result in Lemma 15 has been proved using a minimality argument. In some cases, the minimality of the set C may be useful for the proof of some topological properties of the continuum. An example in this direction is given in the next lemma.

Lemma 16. *Let X be a topological annulus, and let $C \subset X$ be a compact connected set which is minimal with respect to the property of cutting all the paths in X from X_i to X_o. Let $f : C \to \mathbb{R}$ be a nonconstant continuous function. Then, for every $k \in]\min f(C), \max f(C)[$ there exist at least two points $w, z \in C$ with $w \neq z$ such that $f(w) = f(z) = k$.*

Proof. Without loss of generality, we can assume that $k = 0$ and f changes sign on C. The existence of at least a zero for $f|_C$ follows from Bolzano's theorem. Suppose, by contradiction, that there is only one point say $z \in C$ such that $f(z) = 0$. Consider the two nonempty compact sets $K_1 := \{x \in C : f(x) \leq 0\}$ and $K_2 := \{x \in C : f(x) \geq 0\}$. By the assumption, we have that $\{z\} = K_1 \cap K_2$ and $K_1 \neq C$ as well as $K_2 \neq C$. By the assumption of minimality of C, it follows that there exists a path γ_1 connecting X_i to X_o in X and avoiding K_1 and, similarly, there exists a path γ_2 connecting X_i to X_o in X and avoiding K_2. From Lemma 13 we know that there exists a path γ in X connecting X_i to X_o and avoiding $C = K_1 \cup K_2$. This contradicts the cutting property of C. □

We observe that the existence of two solutions is not guaranteed if the minimality of the set C is not assumed (see [42, Example 2.8]).

In the case of topological rectangles, there is a complete symmetry between the fact that a set cuts the paths between a given pair of opposite sides or it cuts the paths connecting a complementary pair of opposite sides. Thus, once we have achieved a result like Lemma 11, also its dual version, involving the other pair of sides, is guaranteed. In the case of topological annuli, the situation is different. We have just proved a result which expresses the fact that a compact set which crosses all the paths from the inner to the outer boundary must contain a continuum which nontrivially winds around the annulus. A dual result should express the fact that if a compact set intersects all the nontrivial loops of the annulus, then it must contain a continuum joining the inner and the outer boundaries of the annulus. This is precisely the content of the next lemma. To this end, we first recall some basic facts from homotopy theory. Let $\omega : I \to X$ be a *loop*, which is a continuous path such that $\omega(0) = \omega(1)$. We say that ω is (homotopically) trivial in X if it is homotopic in X to the constant loop $e_{x_0} : I \to x_0$ with $x_0 = \omega(0) = \omega(1)$. Since a loop in X (up to a change in the parameter) may be also seen as a continuous map $\omega : S^1 \to X$, triviality of ω can be also expressed by the fact that there is a continuous extension $\alpha : E^2 \to X$ with $\alpha_{S^1} = \omega$. We say that a loop ω in X is *nontrivial* if it is not homotopically trivial in X. Then we have the following.

Lemma 17. *Let X be a topological annulus, and let $S \subset X$ be a closed set such that $S \cap \overline{\omega} \neq \emptyset$ for each nontrivial loop ω in X. Then, there exists a compact, connected set $C \subset S$ such that*

$$C \cap X_i \neq \emptyset \neq C \cap X_o. \tag{21}$$

Proof. Without loss of generality (up to a homeomorphism), we suppose that $X = A[a,b]$ with $0 < a < b$, so that X has inner boundary $X_i = aS^1$ and outer boundary $X_o = bS^1$. By the assumption of crossing the nontrivial loops, we know that $S_i := S \cap X_i \neq \emptyset$ and also $S_o := S \cap X_o \neq \emptyset$.

Suppose, by contradiction, that S does not contain any compact connected set C satisfying (21). Then, by the Kuratowski-Whyburn lemma [13, 40], it follows that S splits as the disjoint union of two compact sets S', S'' with $S' \supset S_i$ and $S'' \supset S_o$. We pass now to the covering space $\mathcal{H} := \mathbb{R} \times [a,b]$ of $X = A[a,b]$ and consider the closed subsets of $\mathcal{H}$,

$$W' := \Pi^{-1}(S') \cup (\mathbb{R} \times \{a\}), \quad W'' := \Pi^{-1}(S'') \cup (\mathbb{R} \times \{b\}). \tag{22}$$

By definition, $\Pi(W' \cup W'') \supset S$. Moreover, $W' \cap W'' = \emptyset$, and both such sets are invariant with respect to the translation $\mathcal{D}$ defined in (10). We define

$$\delta := \operatorname{dist}(W', W'') = \inf\{\|w' - w''\| : w' \in W', w'' \in W''\}. \tag{23}$$

It is clear that $\delta > 0$, and it is actually a minimum (this follows from a standard compactness argument, using the periodicity). Then, we define the two closed ε-tubular neighborhoods of W' and W'' as $W'[\varepsilon] := \{y = (\theta, \rho) \in \mathcal{H} : \operatorname{dist}(y, W') \leq \varepsilon\}$ and $W''[\varepsilon] := \{y = (\theta, \rho) \in \mathcal{H} : \operatorname{dist}(y, W'') \leq \varepsilon\}$, for

$$0 < \varepsilon \leq \frac{\delta}{3}. \tag{24}$$

We modify now a pigeonhole argument used in the proof of Theorem 1 in [10] as follows. Let us fix a positive integer $N > (b-a)/\varepsilon$ and consider the rectangle

$$\mathcal{R} := [0, 2N\pi] \times [a,b]. \tag{25}$$

The sets

$$K_1 := W'[\varepsilon] \cap \mathcal{R}, \qquad K_2 := W''[\varepsilon] \cap \mathcal{R} \tag{26}$$

are closed and disjoint. The lower edge $[0, 2N\pi] \times \{a\}$ is the image of a path connecting the left to the right side of $\mathcal{R}$ and avoiding the set K_2. Similarly, via the upper edge $[0, 2N\pi] \times \{b\}$ we find a path connecting the left to the right side of $\mathcal{R}$ and avoiding the set K_1. Alexander's lemma (Lemma 8) guarantees the existence of a path $\gamma = (\theta(t), \rho(t)) : I \to \mathcal{R}$ with $\theta(0) = 0$ and $\theta(1) = 2N\pi$, such that $\gamma(t) \notin K_1 \cup K_2$ for all $t \in I$. For each $i = 0, \dots, N$, let

$$t_i := \min\{t \in I : \theta(t) = 2i\pi\}, \tag{27}$$

so that

$$0 = t_0 < t_1 < \cdots < t_N \leq 1 \tag{28}$$

and the $N+1$-tuple of points $(\rho(t_0), \rho(t_1), \dots, \rho(t_N))$ in $]a,b[$ is well defined. Clearly, by the choice of N such that $N\varepsilon > (b-a)$, there exists at least a pair of points (t_j, t_k) with $j < k$ such that $|\rho(t_j) - \rho(t_k)| < \varepsilon$. Since

$$\min_{t \in [0,1]} \{\operatorname{dist}(\gamma(t), W' \cup W'')\} > \varepsilon, \tag{29}$$

we conclude that the segment joining the points $(2k\pi, \rho(t_j))$ and $(2k\pi, \rho(t_k))$ does not intersect the set $W' \cup W''$ (of course, such statement is trivial if $\rho(t_j) = \rho(t_k)$). We can now define the path

$$\tilde{\omega}(s) := \begin{cases} \gamma\left(t_j + 2s\left(t_k - t_j\right)\right), & \text{for } 0 \leq s \leq \frac{1}{2}, \\ \left(2k\pi, \rho\left(t_k\right) + (2s-1)\left(\rho\left(t_j\right) - \rho\left(t_k\right)\right)\right), & \text{for } \frac{1}{2} \leq s \leq 1, \end{cases} \tag{30}$$

which takes values in $\mathcal{R} \setminus (W' \cup W'')$ and is such that

$$\tilde{\omega}(0) = \left(2j\pi, \rho\left(t_j\right)\right), \qquad \tilde{\omega}(1) = \left(2k\pi, \rho\left(t_j\right)\right). \tag{31}$$

Hence, the projection $\omega := \Pi \circ \tilde{\omega} : I \rightarrow X$ is a nontrivial loop in X (in fact, it corresponds to $k - j \in \mathbb{Z} \setminus \{0\}$ in the fundamental group of X) and, by construction, $\omega(t) \notin S$, for all $t \in I$. This contradicts the hypothesis, and hence, the conclusion follows. □

Our last result in this section can be seen as a continuous version of the no-tie theorem for the Hex game on the annulus. See [10] for a discrete version of this result.

Lemma 18. *Let X be a topological annulus, and let $C_1, C_2 \subset X$ be closed connected sets such that*

$$C_1 \cap X_i \neq \emptyset \neq C_1 \cap X_o, \qquad C_2 \text{ is essentially embedded in } X. \tag{32}$$

Then, $C_1 \cap C_2 \neq \emptyset$.

Proof. Assume, by contradiction, that $C_1 \cap C_2 = \emptyset$, and let $\mathrm{dist}(C_1, C_2) = \delta > 0$. In a $\delta/2$-neighborhood of C_1, we can find the image of a path $\gamma : I \rightarrow X$ with $\gamma(0) \in X_i$ and $\gamma(1) \in X_o$. By construction, $\gamma(t) \notin C_2$, for all $t \in I$. This proves that it is not true that $C_2 : X_i \nmid X_o$, and therefore (by Lemma 14) C_2 is not essentially embedded in X, thus contradicting one of our assumptions. □

4. A Crossing Lemma for Invariant Sets

Let X be a topological space, and let $h : X \rightarrow X$ be a homeomorphism. Our first result is a version of Lemma 3 for h-invariant sets. Indeed, we have the following.

Lemma 19. *Let $A, B \subset X$ be two nonempty disjoint sets which are connected by at least one path in X. Let $S \subset X$ be a closed set which satisfies $S : A \nmid B$ and is invariant for h. Then, there exists a nonempty, closed set $C \subset S$ which is minimal with respect to the property of cutting the paths between A and B and is invariant for h.*

Proof. Let $\mathcal{F}$ be the set of all the nonempty closed subsets F of S such that $F : A \nmid B$ and $h(F) = F$, with the elements of $\mathcal{F}$ ordered by inclusion. Let $(F_\alpha)_{\alpha \in J}$ be a totally ordered subset of $\mathcal{F}$, and define $F^* := \bigcap_{\alpha \in J} F_\alpha$. From $h(F_\alpha) = F_\alpha$ for all $\alpha \in J$, it follows that $h(F^*) = F^*$. The proof that $F^* : A \nmid B$ is the same as that of Lemma 3. Thus, we obtain $F^* \in \mathcal{F}$, and the conclusion follows from Zorn's lemma. □

Our goal now is to develop a result analog of Lemma 10 in the frame of h-invariant sets. Note that if in Lemma 10 we have a continuum $C : X_l^- \nmid X_r^-$ which is also h-invariant, then by Lemma 11 we also have an h-invariant continuum intersecting X_b^+ and X_t^+.

In order to fix the ideas and before moving to the class of generalized and oriented rectangles, we consider for a moment a planar rectangle $\mathcal{R} = [a, b] \times [c, d]$ oriented in the standard manner and suppose that $h : \mathcal{R} \rightarrow \mathcal{R}$ is a *homeomorphism*. Let $S \subset \mathcal{R}$ also be a compact set which intersects all the paths in $\mathcal{R}$ joining $\mathcal{R}_l^-$ to $\mathcal{R}_r^-$ such that $h(S) = S$. We are looking for the existence of a continuum $C \subset S$ with $C : \mathcal{R}_l^- \nmid \mathcal{R}_r^-$ and $h(C) = C$. It is not difficult to see that, in general, the answer is negative, as shown by the following elementary example.

Example 20. Let $\mathcal{R} = [-2, 2] \times [0, 1]$ and $S = \{(\pm 1, y) : y \in [0, 1]\}$. Clearly, S has the property of intersecting all the paths in $\mathcal{R}$ joining the left edge to the right edge. Moreover, consistently with Lemma 10, S contains two continua $\{-1\} \times [0, 1]$ and $\{1\} \times [0, 1]$ connecting the lower and the upper sides of $\mathcal{R}$. However, if we take as a homeomorphism $h(x, y) = (-x, y)$, that is, the symmetry with respect to the y-axis, we have that $h(S) = S$, but there is no connected subset of S which is invariant under h.

Observe that in Example 20 the set S cannot be split as the union of two disjoint closed (nonempty) subsets which are invariant. This property is also expressed by saying that S, (although not connected) is *invariantly connected* (see [43, Definition 4.2]). If we are allowed to replace the word "connected" with "invariantly connected," we can get a full extension of Lemma 10, as follows.

Lemma 21. *Let $\tilde{X} = (X, X^-)$ be an oriented rectangle, let $h : X \rightarrow X$ be a homeomorphism, and let $S \subset X$ be a closed set such that*

$$S : X_l^- \nmid X_r^-, \qquad h(S) = S. \tag{33}$$

Then, there exists a compact, invariantly connected set $C \subset S$ such that

$$C : X_l^- \nmid X_r^-. \tag{34}$$

Proof. By Lemma 19 there exists a closed set $C \subset S$ such that $C : X_l^- \nmid X_r^-$, with C invariant for h and minimal with respect to the cutting property. Suppose, by contradiction, that C is not invariantly connected, and let $C_1, C_2 \subset C$ be two closed nonempty disjoint sets with $C_1 \cup C_2 = C$ and $h(C_i) = C_i$ for $i = 1, 2$. Now we conclude as in the proof of Lemma 10. Indeed, since C is minimal and C_1, C_2 are proper subsets of C, there exist two paths γ_1, γ_2 in X which connect X_l^- to X_r^-, such that γ_i avoids C_i (for $i = 1, 2$). Then, by Lemma 8, there exists a path $\gamma : I \rightarrow X$ with $\gamma(0) \in X_l^-$ and $\gamma(1) \in X_r^-$ with $\overline{\gamma} \cap C = \emptyset$, contradicting the assumption that $C : X_l^- \nmid X_r^-$. □

In order to have the connectedness of the set C, we propose a partial extension of Lemma 10 by adding a further assumption on h which prevents the possibility of a situation like the one described in Example 20. For simplicity in the exposition, we confine ourselves to the case of a planar rectangle $\mathscr{R} = [a,b] \times [c,d]$. Here, we do not require h to be a homeomorphism.

Lemma 22 (h-invariant crossing lemma). *Let $\mathscr{R} = [a,b] \times [c,d]$, and let $h : \mathscr{R} \to \mathscr{R}$ be a continuous map such that*

$$h(\mathscr{R}_b^+) \subset \mathscr{R}_b^+, \qquad h(\mathscr{R}_t^+) \subset \mathscr{R}_t^+. \tag{35}$$

Suppose that there exists a path $\sigma : I \to \mathscr{R}$ with $\sigma(0) \in \mathscr{R}_l^-$ and $\sigma(1) \in \mathscr{R}_r^-$ such that $\forall t \in I$, $\exists s \geq t : h(\sigma(t)) = \sigma(s)$.

Assume that there exists a compact set $S \subset \mathscr{R}$ which cuts the paths between $\mathscr{R}_l^-$ and $\mathscr{R}_r^-$ and satisfies

$$h(S) \subset S. \tag{36}$$

Then, there exists a compact connected set $C \subset S$ such that

(i) $h(C) = C$,

(ii) $C \cap \mathscr{R}_b^+ \neq \emptyset \neq C \cap \mathscr{R}_t^+$,

(iii) *C cuts the paths between $\mathscr{R}_l^-$ and $\mathscr{R}_r^-$.*

Proof. As a first step, we apply Lemmas 10 and 11 which guarantee the existence of a continuum contained in S which joins the lower and the upper sides of the rectangle and cuts the paths from the left to the right side in $\mathscr{R}$. We call such a continuum C^0. Note that C^0 is not necessarily invariant.

In order to obtain an invariant set, we define a sequence of continua by naming

$$C^{i+1} = h\left(C^i\right), \quad \forall i \geq 0. \tag{37}$$

Since $h(S) \subset S$, we also know that $C^i \subset S$, for all $i \geq 0$. By the cutting property of C^0, we have that

$$C^0 \cap \overline{\sigma} \neq \emptyset, \tag{38}$$

so that there exists $t_0 \in I$ with $\sigma(t_0) \in C^0$. Clearly, $h(\sigma(t_0)) \in C^1$. On the other hand, by the hypothesis on σ, there exists $t_1 \in I$ such that $t_1 \geq t_0$ and $h(\sigma(t_0)) = \sigma(t_1) \in C^1$. Proceeding by induction and using step by step the hypothesis on σ, we obtain a monotone sequence $t_0 \leq t_1 \leq \cdots t_i \leq t_{i+1} \leq \cdots$ in $[0,1]$ such that $\sigma(t_i) \in C^i$, for all $i \geq 0$. Let $t_i \nearrow t^* \in [0,1]$. Then, passing to the limit in $\sigma(\cdot)$ and $h \circ \sigma$, we obtain

$$h\left(\sigma\left(t^*\right)\right) = \sigma\left(t^*\right) \in \operatorname{Li} C^i, \tag{39}$$

where $\operatorname{Li} C^i$ is the set of all the points x such that for every open neighborhood U of x it holds that $U \cap C^i \neq \emptyset$ for each i large enough. Therefore, by a classical result from [40, Theorem 6, Chapter 5, Section 47,II], we have that

$$C := \operatorname{Ls} C^i \tag{40}$$

is a nonempty continuum (where $\operatorname{Ls} C^i$ is the set of all the points x such that for every open neighborhood U of x it holds that $U \cap C^i \neq \emptyset$ for infinitely many i). Recall that $z \in C$ if and only if there exists a sequence z_k with $z_k \in C^{i_k}$ for $(i_k)_k$ an increasing sequence of indexes such that $z_k \to z$. Then, $h(z_k) \to h(z)$, with $h(z_k) \in C^{1+i_k}$, and, therefore, $h(z) \in C$. Thus, we have proved that $h(C) \subset C$. Conversely, for $z \in C$ and $(z_k)_k \to z$, as above, let us fix $k \geq 2$, and, from $z_k \in C^{i_k} = h(C^{-1+i_k})$, let us take $y_k \in C^{-1+i_k}$ such that $h(y_k) = z_k$. By compactness, $(y_k)_k$ has a convergent subsequence. Let us set $(y_{k_n})_n \to w$. By definition, we have that $w \in C$. On the other hand, $h(w) = \lim h(y_{k_n}) = \lim z_{k_n} = z$. This proves that $h(C) = C$.

Since we know that $C^0 \cap \mathscr{R}_b^+ \neq \emptyset$, we have

$$\emptyset \neq h\left(C^0 \cap \mathscr{R}_b^+\right) \subset h\left(C^0\right) \cap h\left(\mathscr{R}_b^+\right) \subset C^1 \cap \mathscr{R}_b^+. \tag{41}$$

From this, by induction, we obtain that

$$C^i \cap \mathscr{R}_b^+ \neq \emptyset, \quad \forall i \geq 0, \tag{42}$$

and then, by compactness, we conclude that $C \cap \mathscr{R}_b^+ \neq \emptyset$. The fact that $C \cap \mathscr{R}_t^+ \neq \emptyset$ is proved in the same manner. Having proved that C is a continuum intersecting the horizontal edges of the rectangle, we conclude that it cuts all the paths between $\mathscr{R}_l^-$ and $\mathscr{R}_r^-$. □

Remark 23. Observe that the same result holds true (with an obvious modification in the proof) if we replace condition (35) with

$$h(\mathscr{R}_b^+) \subset \mathscr{R}_t^+, \qquad h(\mathscr{R}_t^+) \subset \mathscr{R}_b^+. \tag{43}$$

Remark 24. By the assumption $h(\mathscr{R}_b^+) \subset \mathscr{R}_b^+$ in (35), we have that

$$h(x,c) = \left(f(x), c\right), \quad \forall x \in [a,b], \tag{44}$$

where $f : [a,b] \to [a,b]$ is a suitable continuous function. It easily follows that if f is monotone nondecreasing, then the path $\sigma(t) := (a + t(b-a), c)$ satisfies the hypothesis of Lemma 22. A similar observation holds for $\mathscr{R}_t^+$. Hence, we easily obtain the following.

Corollary 25. *Let $\mathscr{R}$ be as in Lemma 22, and let $h : \mathscr{R} \to \mathscr{R}$ be a continuous map satisfying* (35), *and suppose also that at least one between $h|_{\mathscr{R}_b^+}$ and $h|_{\mathscr{R}_t^+}$ is monotone nondecreasing. Assume that there exists a compact set $S \subset \mathscr{R}$ which cuts the paths between $\mathscr{R}_l^-$ and $\mathscr{R}_r^-$ and satisfies $h(S) \subset S$. Then, the same conclusion of Lemma 22 holds.*

Notice that in Example 20 the function h is decreasing along both horizontal edges of the rectangle.

A trivial case of a continuous map which is monotone nondecreasing along the horizontal lines is the identity. In such a case, Corollary 25 reduces to the crossing lemma in Section 2.

A useful property of the continua connecting two compact disjoint sets is the minimality, often named as irreducibility [13]. Indeed, a stronger version of Lemma 10 holds, guaranteeing the existence of compact connected set $C \subset S$

which is irreducible between X_l^- and X_r^- (see [13, Proposition 3]).

Then, as a next step, we look for the existence of irreducible invariant continua in this new setting. Such a result, indeed, is of general nature and independent of the fact that in the present paper we consider only the case of planar continua.

Lemma 26. *Let X be a compact Hausdorff space, and let $h : X \to X$ be a homeomorphism. Assume that $A, B \subset X$ are closed disjoint sets which are invariant for h, and let $C \subset X$ be a continuum such that*

$$h(C) = C, \quad C \cap A \neq \emptyset \neq C \cap B. \tag{45}$$

Then, there exists $\mathscr{C} \subset C$ satisfying (45) and minimal with respect to such property.

Proof. Let $\mathscr{F}$ be the family of all the continua $\mathscr{C} \subset C$ which are invariant for the homeomorphism h and intersect both A and B, with $\mathscr{F}$ ordered by inclusion. The family $\mathscr{F}$ is nonempty since $C \in \mathscr{F}$. The existence of a minimal subcontinuum of C satisfying (45) will be ensured by Zorn's lemma.

Given any chain $(\mathscr{C}_j)_{j\in\mathscr{L}}$ in $\mathscr{F}$, we observe that $\bigcap_{j\in\mathscr{J}} \mathscr{C}_j \neq \emptyset$ for every finite subset of indices $\mathscr{J} \subset \mathscr{L}$. Hence, by compactness, $C^* := \bigcap_{j\in\mathscr{L}} C_j$ is nonempty. Moreover C^* is compact and intersects both A and B. Using the fact that h is a homeomorphism, we obtain the invariance of C^*. Thus, if we prove that C^* is connected, we have that $C^* \in \mathscr{F}$, and Zorn's lemma allows to conclude the proof.

Assume, by contradiction, that C^* is not connected. Then, there exist C', C'' nonempty compact sets such that $C^* = C' \cup C''$ and $C' \cap C'' = \emptyset$. Then, there are also two open disjoint sets A', A'' with $A' \supset C'$ and $A'' \supset C''$.

We claim that there exists an index j^* such that the set $C_{j^*} \subset A' \cup A''$. Otherwise, it would happen that $C_j \not\subset A' \cup A''$ which implies that $D_j := C_j - (A' \cup A'') \neq \emptyset$, $\forall j \in \mathscr{L}$. The family $(D_j)_{j\in\mathscr{L}}$ is a family of closed sets with the finite-intersection property, and, therefore, we obtain

$$\exists \hat{x} \in \bigcap_{j\in\mathscr{L}} D_j \subset X - \left(A' \cup A''\right). \tag{46}$$

This is a contradiction to $\bigcap_{j\in\mathscr{L}} D_j \subset \bigcap_{j\in\mathscr{L}} C_j = C^* \subset A' \cup A''$.

To conclude, we observe that $C^* \subset C_{j^*} \subset A' \cup A''$ with C_{j^*} a connected set. Since $C^* = C' \cup C''$ with $C' \subset A'$ and $C'' \subset A''$, we have that $C_{j^*} \cap A' \neq \emptyset$ and $C_{j^*} \cap A'' \neq \emptyset$, which contradicts the fact that C_{j^*} is connected (in fact, A' and A'' are disjoint). □

5. Applications

We conclude this paper by presenting a possible application of the above-described classical separation results to an example which is inspired by a model arising from the theory of fluid mixing previously considered by Kennedy and Yorke in [17]. Following [17, Section 2], we consider a planar map $\Phi : \mathbb{R}^2 \to \mathbb{R}^2$ which is the composition of a squeeze map J_λ and a rotation map R_{θ_0}. In [17], these two maps are defined as follows:

$$J_\lambda(x, y) = \left(\lambda x, \frac{y}{\lambda}\right), \quad \text{with } \lambda > 1. \tag{47}$$

In order to define the map R_{θ_0} we pass to the polar coordinates (θ, ρ) and require that R_{θ_0} is a counterclockwise rotation which leaves all the concentric circumferences $\rho > 0$ invariant and keeps all the points of the plane with $\rho \geq 1$ still, and, moreover,

$$\lim_{\rho\to 0^+} \Theta(\theta, \rho) = \theta_0 > 0, \tag{48}$$

where $\Theta(\theta, \rho)$ is the angular displacement performed by R_{θ_0} on the point $(\rho\cos\theta, \rho\sin\theta)$.

In [17], under the hypotheses that R_{θ_0} is a diffeomorphism and that $\theta_0 > \pi/2$, with θ_0 not an odd multiple of $\pi/2$, the authors prove the existence of a Smale horseshoe if $\lambda > 0$ is sufficiently large. In particular, there exists an invariant Cantor set on which the map $F := R_{\theta_0} \circ J_\lambda$ is conjugate to an m-shift.

We are going to prove a result which, although not so sharp like that in [17], makes use of weaker conditions. To be more specific, from now on, the following assumptions are made.

Let $J : \mathbb{R}^2 \to \mathbb{R}^2$ be a continuous map with

$$J(x, y) = (J_1(x, y), J_2(x, y)) \tag{49}$$

which satisfies the following properties:

(i) $J(0) = 0$, $J(z) \neq 0$ for $z \neq 0$, and $J(Q_i) \subset Q_i$ for $i = 1, \ldots, 4$, where with Q_i we denote the ith closed quadrant of the plane,

(ii) $J_2(x, 0) = 0$ and $J_1(x, 0) = a(x)x$, with $a(x) > 1$ for $x \neq 0$,

(iii) $J_1(0, y) = 0$ and $J_2(0, y) = b(y)y$, with $0 < b(y) < 1$ for $y \neq 0$.

Plainly speaking, J leaves the quadrants invariant and moves the points of the x-axis away from the origin, while it pushes the points of the y-axis toward the origin. In the sequel it will be convenient to express the map J (restricted to $\mathbb{R}^2 \setminus \{0\}$) via its lifting to the covering space $\mathbb{R} \times \mathbb{R}_0^+$ as

$$\begin{gathered}\tilde{J} : (\theta, \rho) \longmapsto (\theta', \rho'), \\ \theta' = \theta + \Theta_J(\theta, \rho), \quad \rho' = \mathscr{R}_J(\theta, \rho),\end{gathered} \tag{50}$$

with Θ_J and $\mathscr{R}_J$ continuous functions which are 2π-periodic in the θ-variable. Note that the assumption $J(Q_i) \subset Q_i$, for $i = 1, \ldots, 4$, reflects the fact that

$$\left|\Theta_J(\theta, \rho)\right| \leq \frac{\pi}{2}. \tag{51}$$

As a second map, we consider a continuous counterclockwise rotation R around the origin such that $R(0) = 0$ and $R(z) \neq 0$

for $z \neq 0$. We also express R (restricted to $\mathbb{R}^2 \setminus \{0\}$) by means of polar coordinates through its lifting

$$\widetilde{R} : (\theta, \rho) \longmapsto (\theta', \rho'),$$
$$\theta' = \theta + \Theta_R(\theta, \rho), \quad \rho' = \mathscr{R}_R(\theta, \rho), \tag{52}$$

with Θ_R and $\mathscr{R}_R$ continuous functions which are 2π-periodic in the θ-variable, and assume the following conditions.

There exist $r_0 \in]0, 1[$ and θ_0 such that

(i) $\Theta_R(\theta, r_0) \geq \theta_0$, $\Theta_R(\theta, 1) = 0$, for all $\theta \in \mathbb{R}$,

(ii) $\mathscr{R}_R(\theta, \rho) = \rho$, for all $\theta \in \mathbb{R}$ and $\rho \in [r_0, 1]$.

According to the above hypotheses, the map R leaves the circumferences of center the origin and radius $\rho \in [r_0, 1]$ invariant. Moreover, the points with $\rho = r_0$ are rotated in the counterclockwise sense by an angle larger or equal to θ_0, while the points with $\rho = 1$ are kept still.

Under the previous assumptions on J and R, the following result holds.

Theorem 27. *Let R be a homeomorphism of the annulus $\mathscr{A} := A[r_0, 1]$ onto itself, and suppose also that $\theta_0 > 2\pi + \pi/2$. Then, the map $\Psi := J \circ R$ has at least four fixed points in the interior of the annulus. Such result is stable with respect to small continuous perturbations of the map Ψ.*

Proof. Let us denote by $\mathscr{A}_i$ and $\mathscr{A}_o$ the inner and outer boundaries of $\mathscr{A}$. We also restrict the map Ψ to the annulus $\mathscr{A}$ and consider its lifting $\widetilde{\Psi}$ the covering space $\mathbb{R} \times [r_0, 1]$ as

$$\widetilde{\Psi}(\theta, \rho) \longmapsto (\theta'', \rho''),$$
$$\theta'' = \theta + \Theta_\Psi(\theta, \rho), \quad \rho'' = \mathscr{R}_\Psi(\theta, \rho), \tag{53}$$

with Θ_Ψ and $\mathscr{R}_\Psi$ continuous functions which are 2π-periodic in the θ-variable. By the above positions for $\widetilde{J}$ and $\widetilde{R}$ we have that

$$\theta'' = \theta' + \Theta_J(\theta', \rho'),$$
$$\text{with } \theta' = \theta + \Theta_R(\theta, \rho), \quad \rho' = \mathscr{R}_R(\theta, \rho). \tag{54}$$

We also introduce a set $S \subset \mathscr{A}$ defined as

$$S := \Pi(\{(\theta, \rho) : \Theta_\Psi(\theta, \rho) = 2\pi\}), \tag{55}$$

where Π is the standard covering projection associated with the polar coordinates. The set S is a compact subset of the annulus consisting of the points which are rotated by an angle of exactly 2π under the action of Ψ.

Suppose that $\gamma : I \to \mathscr{A}$ is a path with $\gamma(0) \in \mathscr{A}_i$ and $\gamma(1) \in \mathscr{A}_o$. We express the points of $\gamma(t)$ in polar coordinates as

$$\gamma(t) = (\rho(t)\cos\theta(t), \rho(t)\sin\theta(t)) \tag{56}$$

and consider the angular displacement for the map R along the points of $\gamma(t)$, using the function

$$\omega_\gamma(t) : I \ni t \longmapsto \Theta_R(\theta(t), \rho(t)). \tag{57}$$

By the assumptions, $\omega_\gamma(0) > 2\pi + \pi/2$ and $\omega_\gamma(1) = 0$. Now we apply the map J to the points of $R(\gamma(t))$. Using condition (51), we have that the angular displacement Θ_Ψ along the curve $\gamma(t)$ can be expressed as

$$\Theta_\Psi(\theta(t), \rho(t)) = \omega_\gamma(t) + \Delta(t), \tag{58}$$

where $\Delta(t)$ is a continuous function satisfying

$$|\Delta(t)| \leq \frac{\pi}{2}, \quad \forall t \in [0, 1]. \tag{59}$$

Recalling the properties of ω_γ, we find

$$\Theta_\Psi(\theta(0), \rho(0)) > 2\pi,$$
$$\Theta_\Psi(\theta(1), \rho(1)) \leq \frac{\pi}{2} < 2\pi, \tag{60}$$

and, therefore, by the continuity of γ, we can conclude that $\overline{\gamma} \cap S \neq \emptyset$. Now we can apply Lemma 15 which ensures the existence of a compact connected set $C \subset S$ which is essentially embedded into $\mathscr{A}$. It is also clear that C (as well as S) is contained in the interior of $\mathscr{A}$.

Let us consider now the intersection of $\mathscr{A}$ with the first quadrant Q_1. The boundary of such intersection consists of two segments $L_1 := [r_0, 1] \times \{0\}$, $L_2 := \{0\} \times [r_0, 1]$ and two arcs $C_1 := r_0 S^1 \cap Q_1$, $C_2 := S^1 \cap Q_1$. We also define

$$\mathscr{B} := R^{-1}(\mathscr{A} \cap Q_1). \tag{61}$$

The set $\mathscr{B}$ is a topological rectangle for which we give an orientation by setting

$$\mathscr{B}_l^- := R^{-1}(L_1),$$
$$\mathscr{B}_r^- := R^{-1}(L_2),$$
$$\mathscr{B}_t^+ := R^{-1}(C_1),$$
$$\mathscr{B}_b^+ := R^{-1}(C_2). \tag{62}$$

By the assumptions, we see that

$$\|\Psi(z)\| > \|z\|, \quad \forall z \in \mathscr{B}_l^-,$$
$$\|\Psi(z)\| < \|z\|, \quad \forall z \in \mathscr{B}_r^-. \tag{63}$$

Hence, on each path with values in $\mathscr{B}$ connecting $\mathscr{B}_l^-$ with $\mathscr{B}_r^-$ there is some point where $\|\Psi(z)\| = \|z\|$. Lemma 11 ensures the existence of a continuum $C' \subset \mathscr{B}$ with $C' \cap \mathscr{B}_b^+ \neq \emptyset$, $C' \cap \mathscr{B}_t^+ \neq \emptyset$, and $\|\Psi(z)\| = \|z\|$, for each $z \in C$.

We are now in a position to apply Lemma 18 which guarantees that $C' \cap C \neq \emptyset$. By definition of C and C' we conclude that any point $w \in C' \cap C$ is a fixed point for Ψ with $\Psi(w) = w \in \operatorname{int}(Q_1 \cap \mathscr{A})$. Repeating the same argument for the other quadrants, we find the remaining three fixed points.

We end the proof with a final comment about the "stability" of the existence result with respect to small perturbations. By this assertion, we mean that the existence of at least four fixed points is preserved if, instead Ψ, we take a continuous map which is sufficiently near to Ψ in the $\|\cdot\|_\infty$-norm on $\mathscr{A}$.

The reason for this fact comes from the strict inequalities that we have in (60) and (63). More precisely, the map $\Theta_\psi(z) - 2\pi$ is strictly positive on $\mathscr{A}_i$ and strictly negative on $\mathscr{A}_o$. Similarly, $\|\Psi(z)\| - \|z\|$, is strictly positive on one side of $\mathscr{B}$ and strictly negative on an opposite side of $\mathscr{B}$. Hence, the maps $\Theta_\psi(z) - 2\pi$ and $\|\Psi(z)\| - \|z\|$ can be seen as the two components of a vector field, and such components change their sign passing from one side to the opposite side of a topological rectangle. In such a case, we can prove that the topological degree (the Brouwer degree) associated with such a vector field is nonzero (this is exactly what happens when we give a proof of the Poincaré-Miranda theorem using the Brouwer degree and assume that the vector field does not vanish on the boundary of the rectangle in order to have the topological degree well defined). In case of nonzero degree (which, one could prove, is our situation) it is known that the results about the existence of zeros (or, resp., fixed points) are stable with respect to small perturbations of the vector field. This is our case when the above strict inequalities hold. □

Following the proof, it is clear that if $\theta_0 > 2j\pi + \pi/2$, for some positive integer j, then there are at least $4j$ fixed points.

For a different application of these results to the existence of fixed points and periodic points to planar maps arising from ordinary differential equations, we refer also to [42].

Acknowledgments

The authors are sincerely grateful and indebted to the referee of their paper who checked very carefully their work and gave them extremely useful suggestions on how to correct some bugs in the original proofs and enhance the presentation.

References

[1] H. Poincaré, "Sur certaines solutions particulières du problème des trois corps," *Comptes Rendus de l'Académie des Sciences*, vol. 97, pp. 251–252, 1883.

[2] H. Poincaré, "Sur certaines solutions particulières du problème des trois corps," *Bulletin Astronomique*, vol. 1, pp. 65–74, 1884.

[3] F. E. Browder, "Fixed point theory and nonlinear problems," *Bulletin of the American Mathematical Society*, vol. 9, no. 1, pp. 1–39, 1983.

[4] W. Kulpa, "The Poincaré-Miranda theorem," *The American Mathematical Monthly*, vol. 104, no. 6, pp. 545–550, 1997.

[5] J. Mawhin, "Poincaré's early use of Analysis situs in nonlinear differential equations: variations around the theme of Kronecker's integral," *Philosophia Scientiae*, vol. 4, no. 1, pp. 103–143, 2000.

[6] H. W. Siegberg, "Some historical remarks concerning degree theory," *The American Mathematical Monthly*, vol. 88, no. 2, pp. 125–139, 1981.

[7] J. Mawhin, "Variations on Poincaré's-Miranda's Theorem," *Advanced Nonlinear Studies*, vol. 13, pp. 209–217, 2013.

[8] A. Pascoletti, M. Pireddu, and F. Zanolin, "Multiple periodic solutions and complex dynamics for second order ODEs via linked twist maps," in *The 8th Colloquium on the Qualitative Theory of Differential Equations*, vol. 8, Szeged, 2008, Electronic Journal of Qualitative Theory of Differential Equations, Szeged, vol. 14, pp. 1–32, 2008.

[9] W. Hurewicz and H. Wallman, *Dimension Theory*, vol. 4 of *Princeton Mathematical Series*, Princeton University Press, Princeton, NJ, USA, 1941.

[10] S. Alpern and A. Beck, "Hex games and twist maps on the annulus," *The American Mathematical Monthly*, vol. 98, no. 9, pp. 803–811, 1991.

[11] D. Gale, "The game of Hex and the Brouwer fixed-point theorem," *The American Mathematical Monthly*, vol. 86, no. 10, pp. 818–827, 1979.

[12] C. Conley, "An application of Ważewski's method to a nonlinear boundary value problem which arises in population genetics," *Journal of Mathematical Biology*, vol. 2, no. 3, pp. 241–249, 1975.

[13] J. C. Alexander, "A primer on connectivity," in *Fixed Point Theory*, vol. 886 of *Lecture Notes in Math*, pp. 455–483, Springer, Berlin, Germany, 1981.

[14] G. J. Butler, "Rapid oscillation, nonextendability, and the existence of periodic solutions to second order nonlinear ordinary differential equations," *Journal of Differential Equations*, vol. 22, no. 2, pp. 467–477, 1976.

[15] C. Rebelo and F. Zanolin, "On the existence and multiplicity of branches of nodal solutions for a class of parameter-dependent Sturm-Liouville problems via the shooting map," *Differential and Integral Equations*, vol. 13, no. 10–12, pp. 1473–1502, 2000.

[16] A. Pascoletti and F. Zanolin, "A path crossing lemma and applications to nonlinear second order equations under slowly varying perturbations," *Le Matematiche*, vol. 65, no. 2, pp. 121–168, 2010.

[17] J. A. Kennedy and J. A. Yorke, "The topology of stirred fluids," *Topology and Its Applications*, vol. 80, no. 3, pp. 201–238, 1997.

[18] A. Berarducci, D. Dikranjan, and J. Pelant, "Uniform quasi components, thin spaces and compact separation," *Topology and Its Applications*, vol. 122, pp. 51–64, 2002.

[19] M. Dolcher, "Questioni di minimo per insiemi chiusi sconnettenti uno spazio topologico," *The Mathematical Journal of the University of Padova*, vol. 19, pp. 159–171, 1950.

[20] D. Papini and F. Zanolin, "Fixed points, periodic points, and coin-tossing sequences for mappings defined on two-dimensional cells," *Fixed Point Theory and Applications*, vol. 2004, no. 2, pp. 113–134, 2004.

[21] A. Pascoletti and F. Zanolin, "Example of a suspension bridge ODE model exhibiting chaotic dynamics: a topological approach," *Journal of Mathematical Analysis and Applications*, vol. 339, no. 2, pp. 1179–1198, 2008.

[22] M. Pireddu and F. Zanolin, "Cutting surfaces and applications to periodic points and chaotic-like dynamics," *Topological Methods in Nonlinear Analysis*, vol. 30, no. 2, pp. 279–319, 2007.

[23] E. E. Moise, *Geometric Topology in Dimensions 2 and 3*, vol. 47 of *Graduate Texts in Mathematics*, Springer, New York, NY, USA, 1977.

[24] J. S. Muldowney and D. Willett, "An elementary proof of the existence of solutions to second order nonlinear boundary value problems," *SIAM Journal on Mathematical Analysis*, vol. 5, pp. 701–707, 1974.

[25] B. R. Gelbaum and J. M. H. Olmsted, *Counterexamples in Analysis*, Dover, Mineola, NY, USA, 2003.

[26] A. Pascoletti and F. Zanolin, "Chaotic dynamics in periodically forced asymmetric ordinary differential equations," *Journal of Mathematical Analysis and Applications*, vol. 352, no. 2, pp. 890–906, 2009.

[27] J. W. Alexander, "A proof of Jordan's theorem about a simple closed curve," *Annals of Mathematics*, vol. 21, no. 3, pp. 180–184, 1920.

[28] J. W. Alexander, "A proof and extension of the Jordan-Brouwer separation theorem," *Transactions of the American Mathematical Society*, vol. 23, no. 4, pp. 333–349, 1922.

[29] M. H. A. Newman, *Elements of the Topology of Plane Sets of Points*, Cambridge University Press, New York, NY, USA, 1961.

[30] D. E. Sanderson, "Advanced plane topology from an elementary standpoint," *Mathematics Magazine*, vol. 53, no. 2, pp. 81–89, 1980.

[31] P. A. Smith, "Book review: elements of the topology of plane sets of points," *Bulletin of the American Mathematical Society*, vol. 45, no. 11, pp. 822–824, 1939.

[32] S. P. Hastings, "An existence theorem for a problem from boundary layer theory," *Archive for Rational Mechanics and Analysis*, vol. 33, pp. 103–109, 1969.

[33] S. Hastings, "The existence of periodic solutions to Nagumo's equation," *The Quarterly Journal of Mathematics*, vol. 25, pp. 369–378, 1974.

[34] S. P. Hastings, "On the existence of homoclinic and periodic orbits for the Fitzhugh-Nagumo equations," *The Quarterly Journal of Mathematics*, vol. 27, no. 105, pp. 123–134, 1976.

[35] J. B. McLeod and J. Serrin, "The existence of similiar solutions for some laminar boundary layer problems," *Archive for Rational Mechanics and Analysis*, vol. 31, pp. 288–303, 1969.

[36] R. E. L. Turner, "Nonlinear eigenvalue problems with applications to elliptic equations," *Archive for Rational Mechanics and Analysis*, vol. 42, pp. 184–193, 1971.

[37] P. Le Calvez, M. Martens, C. Tresser, and P. A. Worfolk, "Stably nonsynchronizable maps of the plane," *Nonlinearity*, vol. 12, no. 1, pp. 9–18, 1999.

[38] M. Kallipoliti and P. Papasoglu, "Simply connected homogeneous continua are not separated by arcs," *Topology and Its Applications*, vol. 154, no. 17, pp. 3039–3047, 2007.

[39] M. Henle, *A Combinatorial Introduction to Topology*, Dover, New York, NY, USA, 1994.

[40] K. Kuratowski, *Topology. Vol. I*, Academic Press, New York, NY, USA, 1966.

[41] J. G. Hocking and G. S. Young, *Topology*, Dover, New York, NY, USA, 2nd edition, 1988.

[42] A. Pascoletti and F. Zanolin, "A topological approach to bend-twist maps with applications," *International Journal of Differential Equations*, vol. 2011, Article ID 612041, 20 pages, 2011.

[43] J. P. LaSalle, *The Stability of Dynamical Systems*, SIAM, Philadelphia, Pa, USA, 1976.

Restoring Poissonian Images by a Combined First-Order and Second-Order Variation Approach

Le Jiang,[1,2] Jin Huang,[1] Xiao-Guang Lv,[2] and Jun Liu[1]

[1] *School of Mathematical Sciences, University of Electronic Science and Technology of China, Chengdu, Sichuan 610054, China*
[2] *School of Science, Huaihai Institute of Technology, Lianyungang, Jiangsu 222005, China*

Correspondence should be addressed to Xiao-Guang Lv; xiaoguanglv@126.com

Academic Editor: Kaleem R. Kazmi

The restoration of blurred images corrupted by Poisson noise is an important topic in imaging science. The problem has recently received considerable attention in recent years. In this paper, we propose a combined first-order and second-order variation model to restore blurred images corrupted by Poisson noise. Our model can substantially reduce the staircase effect, while preserving edges in the restored images, since it combines advantages of the first-order and second-order total variation. We study the issues of existence and uniqueness of a minimizer for this variational model. Moreover, we employ a gradient descent method to solve the associated Euler-Lagrange equation. Numerical results demonstrate the validity and efficiency of the proposed method for Poisson noise removal problem.

1. Introduction

Image restoration problem has been widely studied in the areas of image processing. The goal of image restoration is to reconstruct an approximation of an original image from a blurred and noisy one [1–4]. Over the last decades, most of the literature deal with the additive noise and multiplicative noise model [5–11]. Given a blurred and noisy image $g = Ku + v$ or $g = Kuv$, where u is the original image, v represents the noise, and the blurring operator K is a point spread function (PSF); the restoration problem involving the additive noise or multiplicative noise is to recover u from the observed image g. The additive noise and multiplicative noise models have been extensively studied. We refer the reader to [12–19] and references therein for a review of various methods. In fact, there are other types of noise such as Poisson noise. In this paper, we consider the problem of seeking the approximations of original images from blurred images corrupted by Poisson noise. The restoration of blurred images corrupted by Poisson noise is an important task in various applications such as astronomical imaging, electronic microscopy, single-particle emission-computed tomography (SPECT), and positron emission tomography (PET) [20–24].

In a Poisson noise model, the recorded image g is a realization of a random variable $\widehat{g}$. A statistical model for $\widehat{g}$ is given by

$$\widehat{g} \sim \text{Poiss}\,(Ku + b)\,, \tag{1}$$

where $b > 0$ is the expected value of the background which is assumed to be constant. The given data of the problem are the imaging operator K, the expected value of the background b, and the detected image g. It is known that the major difficulty for finding the original image u of (1) is the models and algorithms developed for deconvolution of additive noise, or multiplicative noise images cannot be directly applied to the restoration of Poissonian images. Therefore, it is important to devise efficient and reliable algorithms for restoring blurred images corrupted by Poisson noise. The problem of restoration of Poissonian images has received considerable attention in recent years.

We note that, in a discrete setting, the intensity g_i in the pixel i is a random variable that follows a Poisson law of mean $(Ku + b)_i$ where the structured matrix K related to the boundary conditions is called a blurring matrix. Considering that these random variables are independent, the probability distribution of g, for given K, u, and b, can be written as

$$\mathscr{L}(u) = P(g \mid u) = \prod_i \frac{[(Ku+b)_i]^{g_i}}{g_i!} \exp^{-(Ku+b)_i}. \quad (2)$$

The use of Stirlings formula leads to the well-known Csiszár [25] I-divergence measure between $Kx+b$ and g. Indeed, by taking the negative logarithm of the likelihood function, we obtain

$$\mathscr{T}_0(u) = -\log[L(u)] = \sum_i (Ku+b)_i - g_i + g_i \log \frac{g_i}{(Ku+b)_i}. \quad (3)$$

This measure generalizes the Kullback-Leibler divergence or cross-entropy measure to accommodate functions whose integrals are not constant, as they would be if they were probability distributions. Dropping the terms independent of u in $\mathscr{L}(u)$, we obtain

$$\mathscr{J}_0(u) = \sum_i (Ku+b)_i - g_i \log (Ku+b)_i. \quad (4)$$

Therefore, the maximum likelihood estimator of the original image is the minimizer with respect to u of the negative-log Poisson likelihood functional $\mathscr{J}_0(u)$.

As is well known, the structured matrix K has many singular values of different orders of magnitude close to the origin. This makes the minimizer of (4) very sensitive to the noise in the right-hand side g. Thus a regularization method may be employed to compute the approximate solution that is less sensitive to noise than the naive solution. In general, regularization methods formulate the image restoration problem as a minimization problem of the form:

$$\mathscr{J}(u) = \mathscr{J}_0(u) + \alpha \mathscr{J}_R(u), \quad (5)$$

where $\mathscr{J}_R(u)$ is a regularization function and $\alpha > 0$ is the regularization parameter that controls the balance between $\mathscr{J}_0$ and $\mathscr{J}_R$. We note that the regularization parameter α is an important quantity which controls the properties of the regularized solution, and α should therefore be chosen with care. Throughout the years a variety of parameter choice strategies such as the discrepancy principle, the L-curve, and generalized cross-validation (GCV) have been proposed for Poisson noise removal problem [26–28].

It is shown in [29] that the choice of the regularization function $\mathscr{J}_R(u)$ is related to the features of the images to be restored. The use of different regularization functions is an active research area. Probably one of the most popular regularization functions is the Tikhonov regularization function $\mathscr{J}_R(u) = (1/2)\|Lu\|_2^2$, where L is an auxiliary operator chosen among the identity and low-order differential operators; see [30] for more details. In [31], Landi and Piccolomini proposed a quasi-Newton projection method for deblurring Poisson-corrupted images after formulating the image restoration problem as a nonnegatively constrained minimization problem where the objective function is the sum of the Kullback-Leibler divergence, used to express fidelity to the data in the presence of Poisson noise, and of a Tikhonov regularization term. Their numerical results show the potential of the method both in terms of relative error reduction and computational efficiency. In [32, 33], the authors developed a cost functional which incorporates the statistics of the noise in the image data and Tikhonov regularization to induce stability and employed an efficient hybrid gradient projection-reduced Newton (GPRNCG) method for the problem of restoration of Poissonian images. Finally, they gave a comparison between the Richardson-Lucy [34, 35] iterative regularization method and GPRNCG algorithm for the restoration of blurred images corrupted by Poisson noise. By the measures of performance used in this comparison, GPRNCG was more efficient than Richardson-Lucy iteration.

It is known that Tikhonov regularization estimate is similar to low-pass filtering; therefore, it produces a smoothing effect on the restored images; that is, it penalizes edges, which is not a good approximation of the original image if it contains edges. To overcome this shortcoming, Rudin et al. [5] proposed a total variation- (TV-) based regularization technique, which preserves the edge information in the restored image. In the case of TV regularization, the estimated solution is obtained by minimizing the object function:

$$\mathscr{J}(u) = \mathscr{J}_0(u) + \alpha \|u\|_{\mathrm{TV}}, \quad (6)$$

where $\|u\|_{\mathrm{TV}} = \|\nabla u\|_1$ is the discrete TV regularization term. It is shown in [36] that the TV penalty function acts more like a model selection device by identifying a small number of critical points at which u is allowed to jump. The number of these selected jump points is controlled by α. Especially, if u is piecewise constant with a finite number of jump discontinuities, then $\|u\|_{\mathrm{TV}}$ gives the sum of magnitudes of the jumps.

In [21], Bardsley and Luttman considered the TV regularization for poissonian images and showed that the problem of computing the minimizer of the resulting TV-penalized Poisson likelihood functional is well posed. They then proved that as the errors in the data and in the blurring operator tend to zero, the resulting minimizer converges to the minimizer of the exact likelihood function. Finally, the practical effectiveness of the approach is demonstrated on synthetically generated data, and a nonnegatively constrained, projected quasi-Newton method was introduced. In [22], Figueiredo and Bioucas-Dias proposed an approach to deconvolving Poissonian images with the TV function, which is based upon an alternating direction optimization method. Using standard convex analysis tools, they presented sufficient conditions for existence and uniqueness of solutions of these optimization problems. In [23], Setzer et al. considered the restoration of blurred images corrupted by Poisson noise by minimizing an energy functional consisting of the Kullback-Leibler divergence as similarity term and the TV regularization term. Their minimizing algorithm uses alternating split Bregman techniques which can be reinterpreted as Douglas-Rachford splitting applied to the dual problem. In contrast to recently developed iterative algorithms, their algorithm contains no inner iterations and produces nonnegative images. The high efficiency of their algorithm in comparison to other recently developed algorithms is demonstrated by artificial and real-world numerical examples. We refer the reader to [24, 37] and references therein for other efficient methods solving the restoration of blurred images corrupted by Poisson noise with the TV function.

Although the total variation regularization is extremely popular in a variety of applications, it also gives rise to some undesired effects. Both from a theoretical and experimental point of view, it has been shown that the TV norm transforms smooth signal into piecewise constants, the so-called staircase effect. Staircase solutions fail to satisfy the evaluation of visual quality, and they can develop false edges that do not exist in the true image. To attenuate the staircase effect, there is a growing interest in the literature for replacing the TV norm by a high-order TV norm. The motivation behind such attempt is to restore potentially a wider range of images, which comprise more than merely piecewise constant regions. The majority of the high-order norms involve second-order differential operators because piecewise-vanishing second-order derivatives lead to piecewise-linear solutions that better fit smooth intensity changes; see [38] for more details.

Second-order regularization schemes have been considered so far in the literature mainly for dealing with the staircase effect while preserving the edge information in the restored image. There are two classes of second-order regularization methods for image restoration problems. The first class combines a second-order regularizer with the TV norm. For example, a technique in [39, 40] combining the TV filter with a fourth-order PDE filter was proposed to preserve edges and at the same time avoid the staircase effect in smooth regions for noise removal. In [41], Papafitsoros and Schönlieb considered a high-order model involving convex functions of the TV and the TV of the first derivatives for image restoration problems and used the split Bregman method to solve numerically the corresponding discretized problem. Chan et al. proposed a second-order model to substantially reduce the staircase effect, while preserving sharp jump discontinuities for image restoration in [42]. The second class employs a second-order regularizer in a standalone way. The second-order regularizer was first proposed in [43] for additive noise removal in medical magnetic resonance images. The LLT model presented in [43] can overcome the staircase effect that occurs with the TV norm filter and better preserve the fine details in nonblocky images. In order to test its practical potential the authors have applied their method to a wide range of real images, including structural and functional MRI data sets. The main strength of their method was the ability to process signals with a smooth change in the intensity value. Numerical results have shown that, compared with some related partial differential equation (PDE) models, the LLT model is rather robust in removing noise and handling edges. In [44, 45], the authors reconsidered the fourth-order PDE model for additive noise removal and employed the dual algorithm of Chambolle for solving the high-order problems.

In order to get restored images with edge preserved and simultaneously smooth region reconstructed for Poisson noise removal, it is natural to utilize a combined first-order and second-order total variation technique. In this paper, we consider to modify the total variation model by adding a high-order functional for restoring blurred images corrupted by Poisson noise. Our model can substantially reduce the staircase effect, while preserving edges in the restored images, since it combines advantages of the first-order and second-order total variation. We study the issues of existence and uniqueness of a minimizer for this variational model. Moreover, we employ a gradient descent method to solve the associated Euler-Lagrange equation. Some numerical results demonstrate the validity and efficiency of the proposed method for Poisson noise removal problem.

The organization of this paper is outlined as follows. In the next section, we present a combined first-order and second-order variation model to restore blurred images corrupted by Poisson noise and study the issues of existence and uniqueness of a minimizer for this variational model. A gradient descent method to solve the associated Euler-Lagrange equation is presented in Section 3. Some numerical experiments are given to illustrate the performance of the proposed algorithm in Section 4. We give concluding remarks in the last section.

2. A Combined First-Order and Second-Order Variation Model

In the Bayesian approach, a prior probability density $P_u(u)$ for u is also specified, and the posterior density:

$$P_u(u \mid g) = \frac{P_g(g \mid u) P_u(u)}{P_g(g)}, \tag{7}$$

given by Bayes's law, is maximized with respect to u. In (7), $P_g(g)$ defines the prior probability distribution of the measurements g, which are given by measurements, and the likelihood function $P_g(g \mid u)$ is the conditional probability of observing a fixed g for a variable u. The maximizer of $P_u(u \mid g)$ is called the maximum a posteriori (MAP) estimate. Since $P_g(g)$ does not depend on u, it is not needed in the computation of the MAP estimate and thus can be ignored. Taking the natural logarithm of (7) and dropping the terms independent of u, we obtain that maximizing (7) is equivalent to minimizing

$$E(u) = \mathcal{J}_0(u) - \log P_u(u) \tag{8}$$

with respect to u, where $\mathcal{J}_0(u) = \int_\Omega ((Ku+b) - g\log(Ku+b))dxdy$. It is clear that $-\log P_u(u)$ corresponds to the regularization term in the classical penalized likelihood approach to regularization. However in the Bayesian setting, $P_u(u)$ is the probability density known as the prior from which the unknown u is assumed to arise. Thus the prior knowledge regarding the characteristics of u can be formulated in the form of a probability density $P_u(u)$, and this yields to a natural, and statistically rigorous, motivation for the regularization method [27].

Standard Tikhonov regularization corresponds to the following choice of the prior:

$$P_u(u) = \exp^{-(\alpha/2)\|u\|_2^2}. \tag{9}$$

This corresponds to the assumption that the prior for u is a zero-mean Gaussian random variable with covariance matrix $\alpha^{-1}I$, which has the effect of penalizing reconstructions with large L_2 norm. For L_2 norm of the gradient regularization, the penalty has the similar form

$$P_u(u) = \exp^{-(\alpha/2)\|\nabla u\|_2^2}, \quad (10)$$

where ∇ is the gradient operator. The use of this regularization function has the effect of penalizing reconstructions that are not smooth. As an alternative to Tikhonov regularization including standard Tikhonov regularization and gradient regularization for Poisson noise removal problem, total variation regularization is another regularization technique that allows for the presence of sharp edges in the resulting reconstruction. In the case of total variation regularization, we have

$$P_u(u) = \exp^{-\alpha\|\nabla u\|_1}, \quad (11)$$

where $\|\nabla u\|_1 = \int_\Omega \sqrt{u_x^2 + u_y^2}dxdy$.

Optimization techniques such as alternation direction method of multipliers (ADMM) and alternating split Bregman algorithm were proposed for the restoration of blurred images corrupted by Poisson noise by minimizing an energy functional consisting of the Kullback-Leibler divergence as the similarity term and the TV regularization term; see [21–24] for more details. In the other hand, some Newton-related gradient methods were proposed for deblurring Poisson-corrupted images by formulating the image restoration problem as a nonnegatively constrained minimization problem where the objective function is the sum of the Kullback-Leibler divergence, used to express fidelity to the data in the presence of Poisson noise, and of a Tikhonov regularization term; see [31–33] for more details.

To attenuate the staircase effect in the TV regularization, we consider a combined first-order and second-order variation model to restore blurred images corrupted by Poisson noise in this work. In the proposed model, the prior probability density is of the form:

$$P_u(u) = \exp^{-\alpha(\theta\|\nabla u\|_1 + (1-\theta)\|\nabla^2 u\|_1)}, \quad (12)$$

where $\|\nabla^2 u\|_1 = \int_\Omega \sqrt{u_{xx}^2 + u_{xy}^2 + u_{yx}^2 + u_{yy}^2}dxdy$. It leads to present the following functional for restoring blurred images corrupted with Poisson noise:

$$\min_u \left\{ E(u) = \mathcal{J}_0(u) + \alpha \mathcal{J}_R(u) = \int_\Omega \left((Ku+b) - g\log(Ku+b) + \alpha\left(\theta|\nabla u| + (1-\theta)\left|\nabla^2 u\right|\right)\right) dxdy \right\}, \quad (13)$$

where the parameter $\theta \in [0,1]$ is used to control the balance between the edges and the smooth surface and α is the regularization parameter which measures the tradeoff between the fidelity term and the regularized term. The parameter θ can be computed by using the information of the edges and smooth regions of the resulting image obtained by smoothing the observed image g with the low-pass filters such as the median filter and the Gauss filter.

The functional $E(u)$ in (13) is defined on the set of $u \in BV(\Omega) \cap BV^2(\Omega)$ such that $\log u \in L_1(\Omega)$; in particular, u must be positive almost everywhere. Some basic notations and properties concerning the BV space and BV^2 space can be found in [39, 40]. Motivated by [46], we have the following result to show the existence and uniqueness of the minimizer for the model (13).

Theorem 1. *Let Ω be a bounded, open subset of $\mathbb{R}^2$ with Lipschitz boundary. Assume that g is a positive bounded function and K is positive definite. Then $E(u)$ for $u \in BV(\Omega) \cap BV^2(\Omega)$ such that $\log u \in L_1(\Omega)$ has a unique minimizer for the model* (13).

Proof. First, a straightforward computation can show that $\mathcal{J}_0(u)$ is bounded below by $g - g\log g$. Since the penalty function $\mathcal{J}_R(u) \geq 0$, we obtain that $E(u)$ is bounded below. Thus, we can choose a sequence $\{u^n\}$ such that $E(u^n) \geq E(u^{n+1})$. Hence, $\mathcal{J}_0(u^n)$ and $\mathcal{J}_R(u^n)$ are bounded. Using Jensen's inequality, we have

$$\mathcal{J}_0(u^n) \geq \|u^n\|_1 - \|g\| \log \|u^n\|_1, \quad (14)$$

which indicates that $\|u^n\|_1$ is bounded. This and the boundedness of $\mathcal{J}_R(u^n)$ yields that the sequence $\{u^n\}$ is bounded in $BV(\Omega) \cap BV^2(\Omega)$. Following the compactness of $L_1(\Omega)$ in $BV(\Omega) \cap BV^2(\Omega)$, we deduce that there exists $u^* \in BV(\Omega) \cap BV^2(\Omega)$ such that a subsequence $\{u^{n_k}\}$ converges to u^* pointwise almost everywhere in L_1. By the lower semicontinuity of the $BV(\Omega) \cap BV^2(\Omega)$ space, we obtain that $\liminf \mathcal{J}_R(u^{n_k}) \geq \mathcal{J}_R(u^*)$. Since $\mathcal{J}_0(u_{n_k})$ is bounded below, we get that $\lim \mathcal{J}_0(u^{n_k}) \geq \mathcal{J}_0(u^*)$ by using Fatou's lemma. Therefore, $\liminf E(u^{n_k}) \geq E(u^*)$ and u^* minimizes $E(u)$ for all $u \in BV(\Omega) \cap BV^2(\Omega)$ such that $\log u \in L_1(\Omega)$.

Since K is positive definite and g is positive, it immediately follows that $\mathcal{J}_0(u)$ is strictly convex. Obviously, $\mathcal{J}_R(u)$ is a convex function. We conclude that $E(u)$ is strictly convex, as the sum of a convex function and of a strictly convex function. Therefore, the minimizer u^* is unique. □

3. Computational Method

In this section, our aim is to propose a time-marching gradient descent method for computing the minimizer of the model (13). From [5, 39, 40] we know that minimizing (13) for a given constant θ yields the associate Euler-Lagrange equation:

$$\begin{aligned} 0 = & -\theta\nabla\cdot\left(\frac{\nabla u}{|\nabla u|}\right) \\ & +(1-\theta)\left[\left(\nabla\cdot\left(\frac{\nabla u_x}{|\nabla^2 u|}\right)\right)_x+\left(\nabla\cdot\left(\frac{\nabla u_y}{|\nabla^2 u|}\right)\right)_y\right] \\ & +K^T\frac{Ku+b-g}{\alpha(Ku+b)} \end{aligned} \quad (15)$$

with the boundary conditions

$$\begin{gathered} \nabla u\cdot N=0, \qquad \nabla u_x\cdot N=0, \qquad \nabla u_y\cdot N=0, \\ \nabla\cdot\left(\frac{\nabla u_x}{|\nabla^2 u|}\right)n_1=0, \qquad \nabla\cdot\left(\frac{\nabla u_y}{|\nabla^2 u|}\right)n_2=0, \end{gathered} \quad (16)$$

where $N=(n_1,n_2)$ denotes the unit outernormal vector of $\partial\Omega$.

Using the gradient descent method, we are able to derive the associated heat flow for the model (13):

$$\begin{aligned} \frac{\partial u}{\partial t} = & \,\theta\nabla\cdot\left(\frac{\nabla u}{|\nabla u|}\right)-(1-\theta) \\ & \times\left[\left(\frac{u_{xx}}{|\nabla^2 u|}\right)_{xx}+\left(\frac{u_{xy}}{|\nabla^2 u|}\right)_{yx}\right. \\ & \left.+\left(\frac{u_{yx}}{|\nabla^2 u|}\right)_{xy}+\left(\frac{u_{yy}}{|\nabla^2 u|}\right)_{yy}\right]-K^T\frac{Ku+b-g}{\alpha(Ku+b)}. \end{aligned} \quad (17)$$

In this work, we use the finite difference scheme to discretize (17); see [5, 39, 40] for more details. Denoting the step space by $h=1$, we employ the following discretization used in the implementations; see Table 1.

In the discretization, the notation $m[a,b]=((\operatorname{sgn} a+\operatorname{sgn} b)/2)\cdot\min(|a|,|b|)$ and $\delta>0$ is near 0. Denoting the time step by τ, we get the following explicit computational scheme:

$$\begin{aligned} u^{k+1} = & \,u^k+\tau\theta\left[D_x^-\frac{D_x^+u^k}{|D_xu^k|}+D_y^-\frac{D_y^+u^k}{|D_yu^k|}\right]+\tau(1-\theta) \\ & \times\left[D_{xx}\left(\frac{D_{xx}u^k}{|D^2u^k|}\right)+D_{yx}^-\left(\frac{D_{xy}^+u^k}{|D^2u^k|}\right)\right. \\ & \left.+D_{xy}^+\left(\frac{D_{yx}^-u^k}{|D^2u^k|}\right)+D_{yy}\left(\frac{D_{yy}u^k}{|D^2u^k|}\right)\right] \\ & -\tau K^T\frac{Ku^k+b-g}{\alpha(Ku^k+b)}. \end{aligned} \quad (18)$$

Now we discuss the choice of the weighting parameter θ in (18). Due to the strengths and weaknesses of the first-order and second-order variation approach, it is desirable

TABLE 1

$D_x^{\pm}u_{i,j}$	$\pm(u_{i\pm1,j}-u_{i,j})$
$D_y^{\pm}u_{i,j}$	$\pm(u_{i,j\pm1}-u_{i,j})$
$D_{xx}u_{i,j}$	$D_x^{\pm}u_{i,j}-D_x^{\pm}u_{i-1,j}$
$D_{xy}^{\pm}u_{i,j}$	$\pm[D_x^{\pm}u_{i,j}-D_x^{\pm}u_{i-1,j}]$
$D_{yx}^{\pm}u_{i,j}$	$\pm[D_y^{\pm}u_{i\pm1,j}-D_y^{\pm}(u_{i,j})]$
$D_{yy}u_{i,j}$	$D_x^{\pm}u_{i,j}-D_x^{\pm}u_{i,j-1}$
$\lvert D_xu_{i,j}\rvert$	$\sqrt{(D_x^+u_{i,j})^2+(m[D_y^+u_{i,j},D_y^-u_{i,j}])^2+\delta}$
$\lvert D_yu_{i,j}\rvert$	$\sqrt{(m[D_x^+u_{i,j},D_x^-u_{i,j}])^2+(D_y^+u_{i,j})^2+\delta}$
$\lvert D^2u_{i,j}\rvert$	$\sqrt{(D_{xx}u_{i,j})^2+(D_{xy}^+u_{i,j})^2+(D_{yx}^+u_{i,j})^2+(D_{yy}u_{i,j})^2+\delta}$

that the weighting parameter $\theta=1$ along edges and in flat regions, emphasizing the restoration properties for the first-order total variation. To emphasize the restoration properties of the second-order total variation in smooth regions, we want $0\le\theta<1$. Specially, the resulting algorithm of (18) is just the TV regularization method for Poisson noise removal problem when $\theta=1$, while the resulting algorithm is the high-order TV regularization method when $\theta=0$. Usually, we may compute the parameter θ by using the information of the edges and smooth regions of the resulting image obtained by smoothing the observed image g with the low-pass filters such as the median filter and the Gauss filter. We have carried out some numerical experiments. We observe that the fixed θ obtained by computing the observed image can give very good results. In this work, in order to detect edges and smoothing regions much better, we adopt the method for updating θ as proposed in [40]. The results obtained by carrying out various numerical examples show that the updating procedure behaves better for our model than the fixed θ. It is because, as the iteration proceeds, the edges and smoothing regions of recovered image are closer to the original image, then the parameter θ computed by the updating scheme can be better suitable for restoration. So we employ the method in [40] for updating θ in our numerical experiments. More precisely, suppose that u^k is the kth iterative solution, and we update the parameter θ as follows:

$$\theta_{i,j} = \begin{cases} 1, & \text{if } \dfrac{|\nabla u_{i,j}^k|}{\max_{i,j}\left(|\nabla u_{i,j}^k|\right)} \ge c, \\ \dfrac{1}{2}\cos\left(\dfrac{2\pi|\nabla u_{i,j}^k|}{c\max_{i,j}\left(|\nabla u_{i,j}^k|\right)}\right)+\dfrac{1}{2}, & \text{otherwise,} \end{cases} \quad (19)$$

where $0\ll c<1$, and it means that only the absolute largest jumps are unaffected of the high-order regularization. As reported in [40], for large and small values of $|\nabla u_{i,j}^k|$ the parameter θ is closer to 1, and for intermediate values of $|\nabla u_{i,j}^k|$ the parameter θ approaches 0, which means that

(1) Input g, K, b, MaxIter, α and τ. Initialize u^0
(2) For $k = 1$: MaxIter
compute u^k using the explicit scheme (18).
update the parameter θ according to (19).
end for

ALGORITHM 1: The time-marching gradient descent algorithm for solving (13).

(a) True image
(b) Degraded image
(c) Tikhonov
(d) TV
(e) HTV
(f) Hybrid TV

FIGURE 1: The "Cameraman" image is degraded by the Gaussian blur and Poisson noise.

the high-order filter dominates the computation and the staircase effect is suppressed. Since only small jumps should be suppressed with the high-order regularization, it is a good choice for $c = 1/8$.

We are now in a position to describe the time-marching gradient descent algorithm (Algorithm 1) for restoring blurred images corrupted by Poisson noise.

4. Numerical Experiments

In this section, we present some numerical results to illustrate the performance of the proposed model for Poisson noise removal problem. We compare our model with the one proposed in [21] (TV method), the one proposed in [31] (Tikhonov method), and the one proposed in [47] (HTV method). All computations of the present paper were carried out in Matlab 7.10. The results were obtained by running the Matlab codes on an Intel(R) Core(TM) i3 CPU (2.27 GHz, 2.27 GHz) computer with RAM of 2048 M.

The quality of the restoration results with different methods is compared quantitatively by using the signal-to-noise ratio (SNR), the relative error (ReErr), and structural SIMilarity index (SSIM). They are defined as follows:

$$\mathrm{SNR} = 20\log_{10}\left(\frac{\|u_{\mathrm{true}}\|_2}{\|u^k - u_{\mathrm{true}}\|_2}\right),$$
$$\mathrm{ReErr} = \frac{\|u^k - u_{\mathrm{true}}\|_2}{\|u_{\mathrm{true}}\|_2}, \tag{20}$$

where u_{true} and u^k are the ideal image and the restored image, respectively, and

$$\mathrm{SSIM} = \frac{(2\mu_1\mu_2 + C_1)(2\sigma_{12} + C_2)}{(\mu_1^2 + \mu_2^2 + C_1)(\sigma_1^2 + \sigma_2^2 + C_2)}, \tag{21}$$

where μ_1 and μ_2 are averages of u_{true} and u^k, respectively, σ_f and $\sigma_{\bar{f}}$ are the variance of u_{true} and u^k, respectively, and σ_{12} is the covariance of u_{true} and u^k. The positive constants C_1 and C_2 can be thought of as stabilizing constants for near-zero denominator values. The SSIM is a well-known quality metric used to measure the similarity between two images. The SSIM method was developed by Wang et al. [48] and is based on three specific statistical measures that are much closer to how the human eye perceives differences between images. In the following experiments, we will use SSIM map to reveal areas

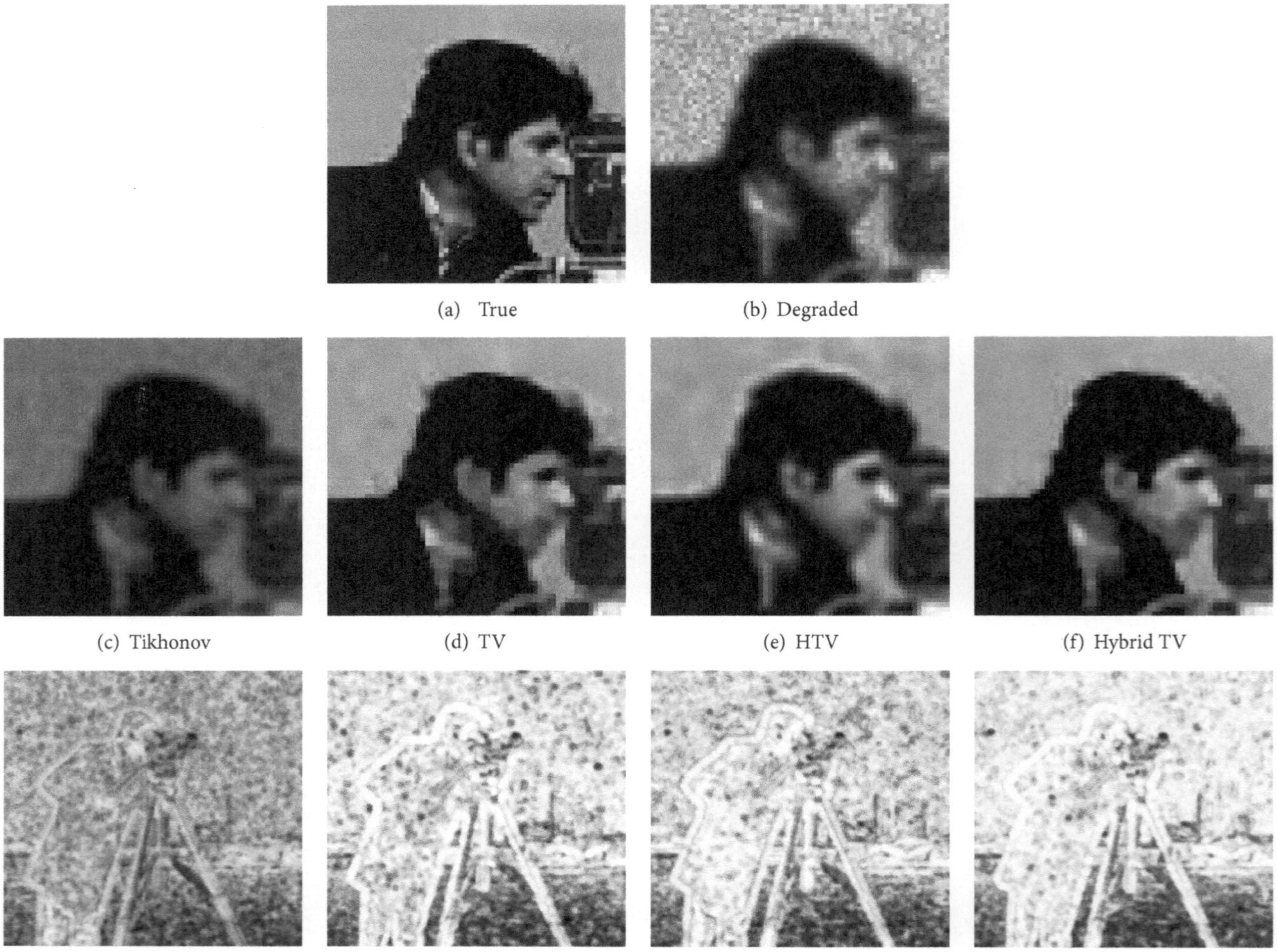

(a) True (b) Degraded

(c) Tikhonov (d) TV (e) HTV (f) Hybrid TV

(g) Tikhonov (h) TV (i) HTV (j) Hybrid TV

FIGURE 2: Comparison of a small portion of the restored images and SSIM maps for Example 1.

of high or low similarity between two images, the whiter SSIM map, and the closer between the two images.

In order to give a good comparison, we employ the time-marching gradient descent algorithm for the four models. In the time-marching gradient descent algorithm, the parameter $\epsilon = 10^{-4}$ is introduced to avoid divisions by zero. The initial guess is chosen to be the degraded image in all tests. We set the step size τ as 0.1 in order to obtain a stable iterative procedure. We terminate the iterations for the methods and accept u^k as the computed approximation of the ideal image as soon as the maximum number of allowed outer iterations has been carried out or the relative differences between consecutive iterates $u^1, u^2, u^3, \ldots$ satisfy

$$\frac{\left\| u^{k+1} - u^k \right\|_2}{\left\| u^{k+1} \right\|_2} < 10^{-4}. \tag{22}$$

In addition, we carry out many experiments with different regularization parameters in the models, andthe one with the best results is presented in this work. In this way, we have a fair comparison since we compare the restorations for four different methods.

In the first experiment, we consider the "Cameraman" image of size 256×256. The degraded image in Figure 1(b) is obtained by performing the blurring operation *psfGauss* $(5, 1)$ proposed in [49] on the original image in Figure 1(a) with the background $b = 10$ and adding the Poisson noise. Note that there is no parameter associated with the Poisson noise, but the noise magnitude depends on the absolute image intensities. The amount of noise in a region of the image increases with the intensity of the image there. The restored images by all four methods are shown in Figures 1(c)–1(f). From these figures, compared with the tikhonov regularization method, the TV regularization method, and the high TV regularization method, the proposed approach yields better results in image restoration since it avoids the staircase effect of the general TV methods, and the edges are preserved as well as the general TV methods. In Figures 2(c)–2(f), we have enlarged some details of the four restored images. As it is seen in the zoomed parts, the proposed method outperforms the other three methods. In Figures 2(c)–2(f), we also present the SSIM maps of the restored images in Figures 2(g)–2(j). Note that the SSIM maps of the restored image by the proposed method are whiter than the

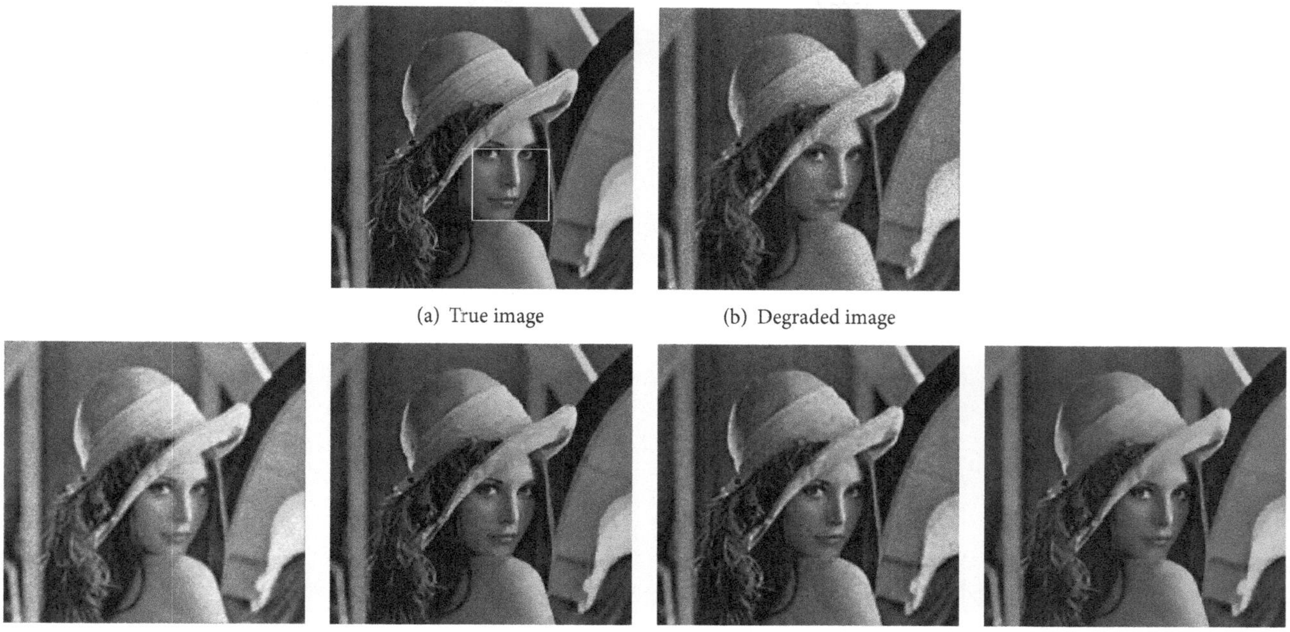

(a) True image (b) Degraded image

(c) Tikhonov (d) TV (e) HTV (f) Hybrid TV

FIGURE 3: The "Barbara" image is degraded by the Moffat blur and Poisson noise.

TABLE 2: Restoration results with different methods for Example 1, Example 2, and Example 3.

	Method	Parameter	SNR	ReErr
Example 1	Tikhonov	$\alpha = 400$	8.725	0.2657
	TV	$\alpha = 65$	14.321	0.0895
	HTV	$\alpha = 80$	14.001	0.0930
	Hybrid TV	$\alpha = 60$	14.538	0.0871
Example 2	Tikhonov	$\alpha = 500$	9.728	0.2223
	TV	$\alpha = 30$	15.811	0.0758
	HTV	$\alpha = 80$	16.022	0.0740
	Hybrid TV	$\alpha = 40$	16.323	0.0718
Example 3	Tikhonov	$\alpha = 400$	6.924	0.4369
	TV	$\alpha = 50$	13.675	0.2007
	HTV	$\alpha = 15$	11.566	0.2559
	Hybrid TV	$\alpha = 70$	14.912	0.1690

other methods; that is, our method can get better results. In Table 2, we compare their restoration results in SNRs and ReErrs. We observe from Table 2 that both the SNR and ReErr values of the restored images by the proposed method are better than those by the other three methods.

Moffat blur is considered in the second example. It is known that the PSF of an astronomical telescope is often modeled by the Moffat function. The 256×256 "Lena" image shown in Figure 3(a) is degraded by the Moffat blur *psfMoffat* $(5, 2, 5)$ proposed in [50] with $b = 10$ and the Poisson noise. The information of restored images by the four methods is displayed in Figures 3(c)–3(f). From the visual quality of restored images, the proposed regularization method is better than the other three methods. In Figures 4(c)–4(f), we display the zoom parts of the restored images (the part shown as the white rectangle in Figure 3(a)). From Figures 4(c)–4(f), we see that the restored image obtained by our method has more details than those by the other methods. The comparison of SSIM maps shown in Figures 4(g)–4(j) also proves that our method can get better results. We report the SNR and RelErr values in Table 2. From the table, we know that our method behaves much better.

In the third example, the 256×256 original astronomical object shown in Figure 5(a) is degraded by the given PSF in Figure 5(b) and the Poisson noise. The degraded image is displayed in Figure 5(c). The relative error between the noisy image and the original image is 0.3150.

In Figures 5(e)–5(h), we show the restored images by the four different methods. From these figures, we observe that the restored image by our model is more better than the other models in terms of the staircase effect and edge preservation. In addition, we plot the choice of θ in Figure 5(d). From

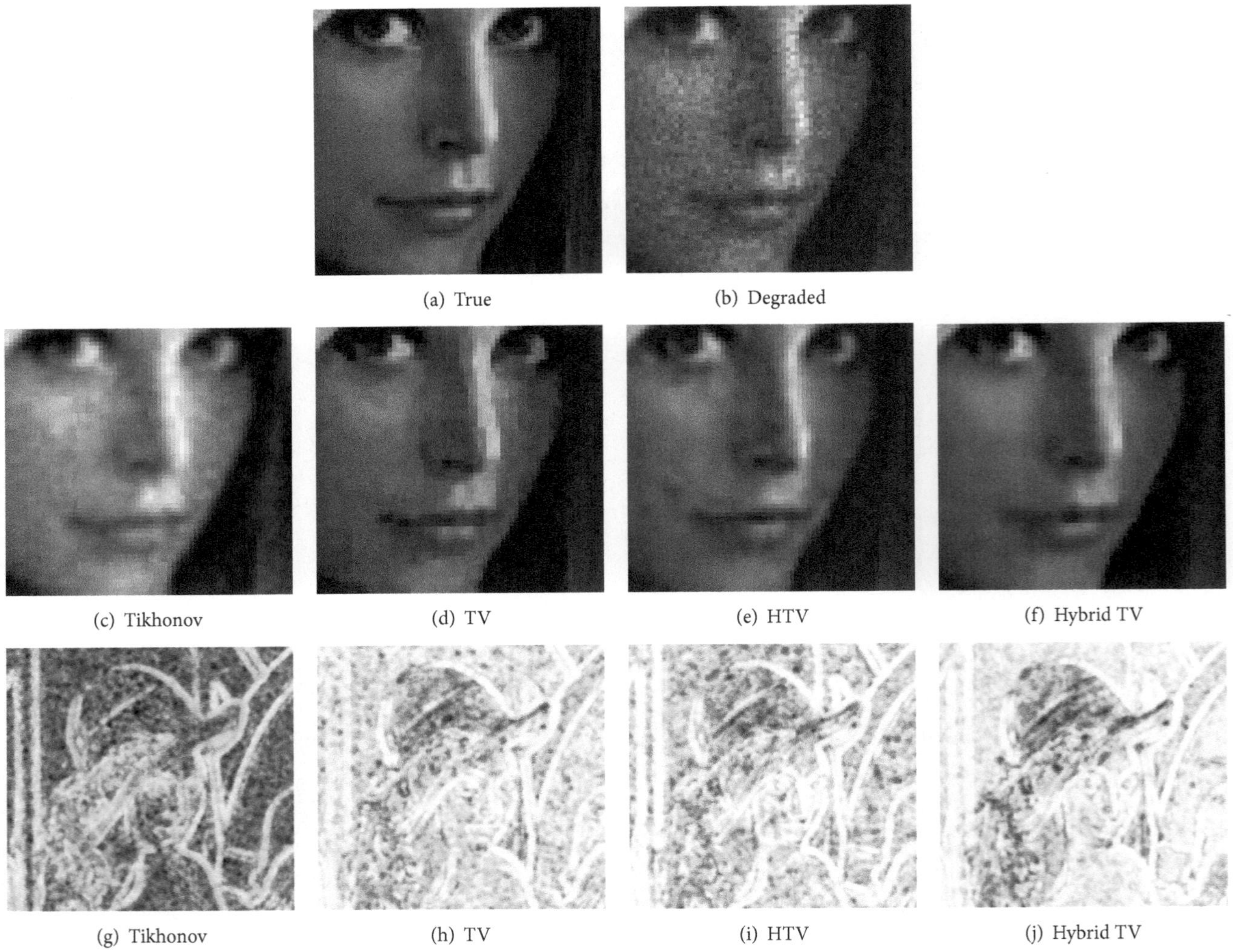

FIGURE 4: Comparison of a small portion of the restored images and SSIM maps for Example 2.

the figure, we see that only small jumps are suppressed with the second-order regularization. The SNRs and relative errors are reported in Table 2. The SNR by our method is higher than those by the other methods, and the relative error by our method is smaller than those by the other methods. From these figures and Table 2, we observe that the proposed method outperforms the tikhonov regularization method, the TV regularization method, and the high TV regularization method.

5. Conclusions

In this paper, we propose a new variational model to restore blurred images corrupted by Poisson noise. Based on the good feature of high-order functional, we propose a model by adding an extra high-order functional term in the total variation model. Our model combines advantages of the first-order and second-order total variation. It can substantially reduce the staircase effect, while preserving edges in the restored images. The issues of existence and uniqueness of a minimizer for this variational model is discussed. At last, we employ a gradient descent method to solve the associated Euler-Lagrange equation. The numerical experiments show that the proposed method outperforms some existing restoration methods in terms of the SNR, relative error, and SSIM map for Poisson noise removal problem. The comparisons between the reconstructed images obtained by the four methods show that the proposed one can alleviate the staircase effect significantly while preserve edges.

Conflict of Interests

All of the coauthors do not have a direct financial relation with the trademarks mentioned in our paper that might lead to a conflict of interests for any of the coauthors.

Acknowledgment

L. Jiang and J. Huang are supported by NSFC (10871034), X.-G. Lv and J. Liu are supported by NSFC (60973015 and 61170311), Sichuan Province Sci. & Tech. Research

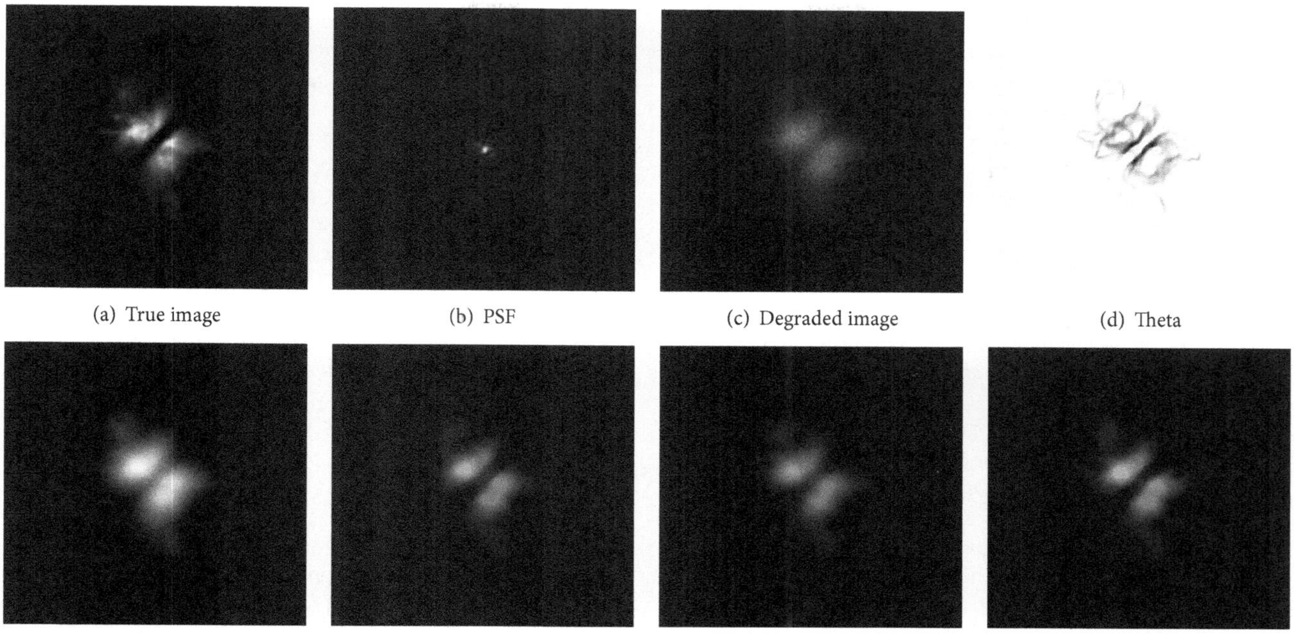

(a) True image (b) PSF (c) Degraded image (d) Theta

(e) Tikhonov (f) TV (g) HTV (h) Hybrid TV

FIGURE 5: Results of different methods for Example 3.

Project (2011JY0002 and 12ZC1802), and Chinese Universities Specialized Research Fund for the Doctoral Program (20110185110020).

References

[1] H. Andrew and B. Hunt, *Digital Image Restoration,*, Prentice Hall, Englewood Cliffs, NJ, USA, 1977.

[2] M. Bertero and P. Boccacci, *Introduction to Inverse Problems in Imaging*, Institute of Physics Publishing, London, UK, 1998.

[3] T. F. Chan and J. Shen, *Image Processing and Analysis: Variational, PDE, Wavelet, and Stochastic Methods*, SIAM, Philadelphia, Pa, USA, 2005.

[4] M. Banham and A. Katsaggelos, "Digital image restoration," *IEEE Signal Processing Magazine*, vol. 14, no. 2, pp. 24–41, 1997.

[5] L. Rudin, S. Osher, and E. Fatemi, "Nonlinear total variation based noise removal algorithms," *Physical D*, vol. 60, no. 1–4, pp. 259–268, 1992.

[6] A. Chambolle and P. L. Lions, "Image recovery via total variation minimization and related problems," *Numerische Mathematik*, vol. 76, no. 2, pp. 167–188, 1997.

[7] L. Rudin, P. L. Lions, and S. Osher, "Multiplicative denoising and deblurring: theory and algorithms," in *Geometric Level Set Methods in Imaging, Vision, and Graphics*, S. Osher and N. Paragios, Eds., pp. 103–119, Springer, New York, NY, USA, 2003.

[8] G. Aubert and J. F. Aujol, "A variational approach to removing multiplicative noise," *SIAM Journal on Applied Mathematics*, vol. 68, no. 4, pp. 925–946, 2008.

[9] Y. M. Huang, M. K. Ng, and Y. W. Wen, "A new total variation method for multiplicative noise removal," *SIAM Journal on Imaging Sciences*, vol. 2, no. 1, pp. 20–40, 2009.

[10] B. Chen, J. L. Cai, W. S. Chen, and Y. Li, "A multiplicative noise removal approach based on partial differential equation model," *Mathematical Problems in Engineering*, vol. 2012, Article ID 242043, 14 pages, 2012.

[11] S. Correia, M. Carbillet, P. Boccacci, M. Bertero, and L. Fini, "Restoration of interferometric images—I. The software package AIRY," *Astronomy and Astrophysics*, vol. 387, no. 2, pp. 733–743, 2002.

[12] A. Chambolle, "An algorithm for total variation minimization and applications," *Journal of Mathematical Imaging and Vision*, vol. 20, no. 1-2, pp. 89–97, 2004.

[13] T. Chan, S. Esedoglu, and F. Park, "Image decomposition combining staircase reduction and texture extraction," *Journal of Visual Communication and Image Representation*, vol. 18, no. 6, pp. 464–486, 2007.

[14] S. Becker, J. Bobin, and E. J. Candès, "NESTA: a fast and accurate first-order method for sparse recovery," *SIAM Journal on Imaging Sciences*, vol. 4, no. 1, pp. 1–39, 2011.

[15] A. Beck and M. Teboulle, "Fast gradient-based algorithms for constrained total variation image denoising and deblurring problems," *IEEE Transactions on Image Processing*, vol. 18, no. 11, pp. 2419–2434, 2009.

[16] Y. L. Wang, J. F. Yang, W. T. Yin, and Y. Zhang, "A new alternating minimization algorithm for total variation image reconstruction," *SIAM Journal on Imaging Sciences*, vol. 1, no. 3, pp. 248–272, 2008.

[17] J. M. Bioucas-Dias and M. A. T. Figueiredo, "Multiplicative noise removal using variable splitting and constrained optimization," *IEEE Transactions on Image Processing*, vol. 19, no. 7, pp. 1720–1730, 2010.

[18] S. Yun and H. Woo, "A new multiplicative denoising variational model based on *m*th root transformation," *IEEE Transactions on Image Processing*, vol. 21, no. 5, pp. 2523–2533, 2012.

[19] H. Woo and S. Yun, "Alternating minimization algorithm for speckle reduction with a shifting technique," *IEEE Transactions on Image Processing*, vol. 21, no. 4, pp. 1701–1714, 2012.

[20] M. Bertero, P. Boccacci, G. Desiderà, and G. Vicidomini, "Image deblurring with Poisson data: from cells to galaxies," *Inverse Problems*, vol. 25, no. 12, Article ID 123006, 26 pages, 2009.

[21] J. M. Bardsley and A. Luttman, "Total variation-penalized Poisson likelihood estimation for ill-posed problems," *Advances in Computational Mathematics*, vol. 31, no. 1-3, pp. 35–59, 2009.

[22] M. A. T. Figueiredo and J. M. Bioucas-Dias, "Restoration of Poissonian images using alternating direction optimization," *IEEE Transactions on Image Processing*, vol. 19, no. 12, pp. 3133–3145, 2010.

[23] S. Setzer, G. Steidl, and T. Teuber, "Deblurring Poissonian images by split Bregman techniques," *Journal of Visual Communication and Image Representation*, vol. 21, no. 3, pp. 193–199, 2010.

[24] D. Q. Chen and L. Z. Cheng, "Deconvolving Poissonian images by a novel hybrid variational model," *Journal of Visual Communication and Image Representation*, vol. 22, no. 7, pp. 643–652, 2011.

[25] I. Csiszár, "Why least squares and maximum entropy? An axiomatic approach to inference for linear inverse problems," *The Annals of Statistics*, vol. 19, no. 4, pp. 2032–2066, 1991.

[26] J. M. Bardsley and J. Goldes, "Techniques for regularization parameter and hyper-parameter selection in PET and SPECT imaging," *Inverse Problems in Science and Engineering*, vol. 19, no. 2, pp. 267–280, 2011.

[27] J. M. Bardsley and J. Goldes, "Regularization parameter selection methods for ill-posed Poisson maximum likelihood estimation," *Inverse Problems*, vol. 25, no. 9, Article ID 095005, 18 pages, 2009.

[28] D. Q. Chen and L. Z. Cheng, "Spatially adapted regularization parameter selection based on the local discrepancy function for Poissonian image deblurring," *Inverse Problems*, vol. 28, no. 1, Article ID 015004, 24 pages, 2012.

[29] G. Landi and E. L. Piccolomini, "NPTool: a Matlab software for nonnegative image restoration with Newton projection methods," *Numerical Algorithms*, 2012.

[30] A. Tikhonov and V. Arsenin, *Solution of Ill-Poised Problems*, Winston, Washington, DC, USA, 1977.

[31] G. Landi and E. L. Piccolomini, "An improved Newton projection method for nonnegative deblurring of Poisson-corrupted images with Tikhonov regularization," *Numerical Algorithms*, vol. 60, no. 1, pp. 169–188, 2012.

[32] J. M. Bardsley and C. R. Vogel, "A nonnegatively constrained convex programming method for image reconstruction," *SIAM Journal on Scientific Computing*, vol. 25, no. 4, pp. 1326–1343, 2003/04.

[33] J. M. Bardsley and N. Laobeul, "Tikhonov regularized Poisson likelihood estimation: theoretical justification and a computational method," *Inverse Problems in Science and Engineering*, vol. 16, no. 2, pp. 199–215, 2008.

[34] W. H. Richardson, "Bayesan-based iterative methods of image restoration," *Journal of the Optical Society of America*, vol. 62, no. 1, pp. 55–59, 1972.

[35] L. B. Lucy, "An iterative technique for the rectification of observed images," *The Astronomical Journal*, vol. 79, pp. 745–754, 1974.

[36] V. Agarwal, A. V. Gribok, and M. A. Abidi, "Image restoration using L_1 norm penalty function," *Inverse Problems in Science and Engineering*, vol. 15, no. 8, pp. 785–809, 2007.

[37] G. Landi and E. L. Piccolomini, "An efficient method for nonnegatively constrained total variation-based denoising of medical images corrupted by Poisson noise," *Computerized Medical Imaging and Graphics*, vol. 36, no. 1, pp. 38–46, 2012.

[38] S. Lefkimmiatis, A. Bourquard, and M. Unser, "Hessian-based norm regularization for image restoration with biomedical applications," *IEEE Transactions on Image Processing*, vol. 21, no. 3, pp. 983–995, 2012.

[39] F. Li, C. M. Shen, J. S. Fan, and C. L. Shen, "Image restoration combining a total variational filter and a fourth-order filter," *Journal of Visual Communication and Image Representation*, vol. 18, no. 4, pp. 322–330, 2007.

[40] M. Lysker and X. C. Tai, "Iterative image restoration combining total varition minimization and a second-order functional," *International Journal of Computer Vision*, vol. 66, no. 1, pp. 5–18, 2006.

[41] K. Papafitsoros and C. B. Schönlieb, "A combined first and second order variational approach for image reconstruction," http://arxiv.org/abs/1202.6341.

[42] T. Chan, A. Marquina, and P. Mulet, "High-order total variation-based image restoration," *SIAM Journal on Scientific Computing*, vol. 22, no. 2, pp. 503–516, 2000.

[43] M. Lysaker, A. Lundervold, and X. C. Tai, "Noise removal using fourth-order partial differential equation with applications to medical magnetic resonance images in space and time," *IEEE Transactions on Image Processing*, vol. 12, no. 12, pp. 1579–1590, 2003.

[44] G. Steidl, "A note on the dual treatment of higher-order regularization functionals," *Computing*, vol. 76, no. 1-2, pp. 135–148, 2006.

[45] H. Z. Chen, J. P. Song, and X. C. Tai, "A dual algorithm for minimization of the LLT model," *Advances in Computational Mathematics*, vol. 31, no. 1–3, pp. 115–130, 2009.

[46] T. Le, R. Chartrand, and T. J. Asaki, "A variational approach to reconstructing images corrupted by Poisson noise," *Journal of Mathematical Imaging and Vision*, vol. 27, no. 3, pp. 257–263, 2007.

[47] W. F. Zhou and Q. G. Li, "Poisson noise removal scheme based on fourth-order PDE by alternating minimization algorithm," *Abstract and Applied Analysis*, vol. 2012, Article ID 965281, 14 pages, 2012.

[48] Z. Wang, A. C. Bovik, H. R. Sheikh, and E. P. Simoncelli, "Image quality assessment: from error visibility to structural similarity," *IEEE Transactions on Image Processing*, vol. 13, no. 4, pp. 600–612, 2004.

[49] J. G. Nagy, K. Palmer, and L. Perrone, "Iterative methods for image deblurring: a Matlab object-oriented approach," *Numerical Algorithms*, vol. 36, no. 1, pp. 73–93, 2004.

[50] P. C. Hansen, J. G. Nagy, and D. P. O'Leary, *Deblurring Images: Matrices, Spectra, and Filtering*, vol. 3 of *Fundamentals of Algorithms*, SIAM, Philadelphia, Pa, USA, 2006.

19

The Monoid Consisting of Kuratowski Operations

Szymon Plewik and Marta Walczyńska

Institute of Mathematics, University of Silesia, ul. Bankowa 14, 40-007 Katowice, Poland

Correspondence should be addressed to Szymon Plewik; plewik@math.us.edu.pl

Academic Editor: Nan-Jing Huang

The paper fills gaps in knowledge about Kuratowski operations which are already in the literature. The Cayley table for these operations has been drawn up. Techniques, using only paper and pencil, to point out all semigroups and its isomorphism types are applied. Some results apply only to topology, and one cannot bring them out, using only properties of the complement and a closure-like operation. The arguments are by systematic study of possibilities.

1. Introduction

Let X be a topological space. Denote by A^{-} closure of the set $A \subseteq X$. Let A^{c} be the complement of A; that is, $X \setminus A = A^{c}$. The aim of this paper is to examine monoids generated under compositions from the closure and the complement. A widely known fact due to Kuratowski [1] states that at most 14 distinct operations can be formed from such compositions. Mark them as follows. *Kuratowski operations*:

$\sigma_0(A) = A$ (the identity),

$\sigma_1(A) = A^{c}$ (the complement),

$\sigma_2(A) = A^{-}$ (the closure),

$\sigma_3(A) = A^{c-}$,

$\sigma_4(A) = A^{-c}$,

$\sigma_5(A) = A^{c-c}$ (the interior),

$\sigma_6(A) = A^{-c-}$,

$\sigma_7(A) = A^{c-c-}$,

$\sigma_8(A) = A^{-c-c}$,

$\sigma_9(A) = A^{c-c-c}$,

$\sigma_{10}(A) = A^{-c-c-}$,

$\sigma_{11}(A) = A^{c-c-c-}$,

$\sigma_{12}(A) = A^{-c-c-c}$,

$\sigma_{13}(A) = A^{c-c-c-c}$.

The following rules was found in the original paper by Kuratowski [1, pages 183-184]. In the Engelking book [2], they are commented by a hint on page 81. *Cancellation rules*:

$$A^{-c-} = A^{-c-c-c-}, \qquad A^{c-c-} = A^{c-c-c-c-}. \tag{1}$$

Kuratowski operations have been studied by several authors, for example, [3] or [4]. A list of some other authors one can find in the paper [5] by Gardner and Jackson. For the first time these operations were systematically studied in the dissertation by Kuratowski, whose results were published in [1]. Tasks relating to these operations are usually resolved at lectures or exercises with general topology. They are normally left to students for independent resolution. For example, determine how many different ways they convert a given set.

This note is organized as follows. Kuratowski operations and their marking are described in the introduction. Their properties of a much broader context than for topologies are presented in Section 2. The Cayley table, for the monoid $\mathbb{M}$ of all Kuratowski operations, has been drawn up in Section 3. We hope that this table has not yet been published in the literature. Having this table, one can create a computer program that calculates all the semigroups contained in $\mathbb{M}$. However, in Sections 4–8, we present a framework (i.e., techniques using only paper and pencil) to point out all 118 semigroups and 56 isomorphism types of them. The list of 43 semigroups which are not monoids is presented in Section 9. In this section, also isomorphism types are discussed in order of the number of elements in semigroups. Finally, we present cancellation rules (relations) motivated by some topological spaces.

2. Cancellation Rules

A map $f : P(X) \to P(X)$ is called:

(i) *increasing*, if $A \subseteq B$ implies $f(A) \subseteq f(B)$;

(ii) *decreasing*, if $A \subseteq B$ implies $f(B) \subseteq f(A)$;

(iii) *an involution*, if the composition $f \circ f$ is the identity;

(iv) *an idempotent*, if $f \circ f = f$.

Assume that $A \mapsto \sigma_0(A)$ is the identity, $A \mapsto \sigma_1(A)$ is a decreasing involution, and $A \mapsto \sigma_2(A)$ is an increasing idempotent map. Other operations σ_i let be compositions of σ_1 and σ_2 as it was with the kuratowski operations. We get the following cancellation rules.

Lemma 1. *If $B \subseteq \sigma_2(B)$, then*

$$\sigma_2 \circ \sigma_{12} = \sigma_6, \qquad \sigma_2 \circ \sigma_{13} = \sigma_7. \tag{2}$$

Proof. For clarity of this proof, use designations $\sigma_1(A) = A^c$ and $\sigma_2(A) = A^-$. Thus, we shall prove

$$A^{-c-c-c-} = A^{-c-}, \qquad A^{c-c-c-c-} = A^{c-c-}. \tag{3}$$

We start with $A^{-c-c} \subseteq A^{-c-c-}$, substituting $B = A^{-c-c}$ in $B \subseteq B^-$. This corresponds to $A^{-c-c-c} \subseteq A^{-c-cc} = A^{-c-}$, since σ_1 is a decreasing involution. Hence $A^{-c-c-c-} \subseteq A^{-c-}$, since σ_2 an increasing idempotent.

Since σ_1 is decreasing, σ_2 is increasing, and $B \subseteq B^-$, we have

$$B^{-c} \subseteq B^c \subseteq B^{c-}. \tag{4}$$

Thus $A^{-c-c} \subseteq A^{-cc-} = A^-$, if we put $A^{-c} = B$. Again using that σ_2 is increasing and σ_1 is decreasing, we obtain $A^{-c-c-} \subseteq A^-$ and then $A^{-c} \subseteq A^{-c-c-c}$. Finally, we get $A^{-c-} \subseteq A^{-c-c-c-}$.

With A^c in the place of A in the rule $A^{-c-} = A^{-c-c-c-}$ we get the second rule. □

In academic textbooks of general topology, for example, [2, Problem 1.7.1.], one can find a hint suggested to prove the above cancellation rules. Students go like this: steps $A \subseteq A^-$ and $A^c \subseteq A^{c-}$ lead to $A^{c-c} \subseteq A$; the special case $A^{-c-c-c} \subseteq A^{-c-}$ (of $A^{c-c} \subseteq A$) leads to $A^{-c-c-c-} \subseteq A^{-c-}$; steps $A^{-c-c} \subseteq A^-$ and $A^{-c-c-} \subseteq A^-$ lead to $A^{-c-} \subseteq A^{-c-c-c-}$; at the end, use the last step of the proof of Lemma 1. Note that, the above proofs do not use the axioms of topology as follows:

(i) $\emptyset = \emptyset^-$;

(ii) $(A \cup B)^- = A^- \cup B^-$.

In the literature there are articles in which Kuratowski operations are replaced by some other mappings. For example, Koenen [6] considered linear spaces and put $\sigma_2(A)$ to be the convex hull of A. In fact, Shum [7] considered σ_2 as the closure due to the algebraic operations. Add to this, that these operations can be applied to so-called Fréchet (V) spaces, which were considered in the book [8, pages 3–37].

3. The Monoid $\mathbb{M}$

Let $\mathbb{M}$ be the monoid consisting of all Kuratowski operations; that is, there are assumed cancellation rules: $\sigma_6 = \sigma_2 \circ \sigma_{12}$ and $\sigma_7 = \sigma_2 \circ \sigma_{13}$. Fill in the Cayley table for $\mathbb{M}$, where the row and column marked by the identity are omitted. Similarly as in [9], the factor that labels the row comes first, and that the factor that labels the column is second. For example, $\sigma_i \circ \sigma_k$ is in the row marked by σ_i and the column marked by σ_k

	σ_1	σ_2	σ_3	σ_4	σ_5	σ_6	σ_7	σ_8	σ_9	σ_{10}	σ_{11}	σ_{12}	σ_{13}
σ_1	σ_0	σ_4	σ_5	σ_2	σ_3	σ_8	σ_9	σ_6	σ_7	σ_{12}	σ_{13}	σ_{10}	σ_{11}
σ_2	σ_3	σ_2	σ_3	σ_6	σ_7	σ_6	σ_7	σ_{10}	σ_{11}	σ_{10}	σ_{11}	σ_6	σ_7
σ_3	σ_2	σ_6	σ_7	σ_2	σ_3	σ_{10}	σ_{11}	σ_6	σ_7	σ_6	σ_7	σ_{10}	σ_{11}
σ_4	σ_5	σ_4	σ_5	σ_8	σ_9	σ_8	σ_9	σ_{12}	σ_{13}	σ_{12}	σ_{13}	σ_8	σ_9
σ_5	σ_4	σ_8	σ_9	σ_4	σ_5	σ_{12}	σ_{13}	σ_8	σ_9	σ_8	σ_9	σ_{12}	σ_{13}
σ_6	σ_7	σ_6	σ_7	σ_{10}	σ_{11}	σ_{10}	σ_{11}	σ_6	σ_7	σ_6	σ_7	σ_{10}	σ_{11}
σ_7	σ_6	σ_{10}	σ_{11}	σ_6	σ_7	σ_6	σ_7	σ_{10}	σ_{11}	σ_{10}	σ_{11}	σ_6	σ_7
σ_8	σ_9	σ_8	σ_9	σ_{12}	σ_{13}	σ_{12}	σ_{13}	σ_8	σ_9	σ_8	σ_9	σ_{12}	σ_{13}
σ_9	σ_8	σ_{12}	σ_{13}	σ_8	σ_9	σ_8	σ_9	σ_{12}	σ_{13}	σ_{12}	σ_{13}	σ_8	σ_9
σ_{10}	σ_{11}	σ_{10}	σ_{11}	σ_6	σ_7	σ_6	σ_7	σ_{10}	σ_{11}	σ_{10}	σ_{11}	σ_6	σ_7
σ_{11}	σ_{10}	σ_6	σ_7	σ_{10}	σ_{11}	σ_{10}	σ_{11}	σ_6	σ_7	σ_6	σ_7	σ_{10}	σ_{11}
σ_{12}	σ_{13}	σ_{12}	σ_{13}	σ_8	σ_9	σ_8	σ_9	σ_{12}	σ_{13}	σ_{12}	σ_{13}	σ_8	σ_9
σ_{13}	σ_{12}	σ_8	σ_9	σ_{12}	σ_{13}	σ_{12}	σ_{13}	σ_8	σ_9	σ_8	σ_9	σ_{12}	σ_{13}

(5)

It turns out that the above table allows us to describe all semigroups contained in $\mathbb{M}$, using pencil-and-paper techniques, only. The argument will be by a systematic study of possibilities. Preparing the list of all semigroups consisting of Kuratowski operations we used following principles:

(i) minimal collection of generators is written using $\langle A, B, \ldots, Z \rangle$, where letters denote generators;

(ii) when a semigroup has a few minimal collections of generators, then its name is the first collection in the dictionary order;

(iii) all minimal collections of generators are written with the exception of some containing σ_0;

(iv) we leave to the readers verification that our list is complete, sometimes we add hints.

4. Semigroups with σ_1

Observe that each semigroup which contains σ_0 is a monoid. Since $\sigma_1 \circ \sigma_1 = \sigma_0$, a semigroup which contains σ_1 is a monoid, too.

Theorem 2. *There are three monoids containing σ_1:*

(1) $\langle \sigma_1 \rangle = \{\sigma_0, \sigma_1\}$;

(2) $\mathbb{M} = \langle \sigma_1, \sigma_i \rangle = \{\sigma_0, \sigma_1, \ldots, \sigma_{13}\}$, *where* $i \in \{2, 3, 4, 5\}$;

(3) *let* $\mathbb{M}_1 = \langle \sigma_1, \sigma_6 \rangle$. *If* $j \in \{6, 7, \ldots, 13\}$, *then*

$$\mathbb{M}_1 = \left\langle \sigma_1, \sigma_j \right\rangle = \{\sigma_0, \sigma_1\} \cup \{\sigma_6, \sigma_7, \ldots, \sigma_{13}\}. \tag{6}$$

Proof. The equality $\langle \sigma_1 \rangle = \{\sigma_0, \sigma_1\}$ is obvious.

Since $\sigma_2 = \sigma_3 \circ \sigma_1 = \sigma_1 \circ \sigma_4 = \sigma_1 \circ \sigma_5 \circ \sigma_1$, we have $\langle \sigma_1, \sigma_2 \rangle = \langle \sigma_1, \sigma_3 \rangle = \langle \sigma_1, \sigma_4 \rangle = \langle \sigma_1, \sigma_5 \rangle = \mathbb{M}$.

If $j \in \{6, 7, \ldots, 13\}$, then any composition $\sigma_j \circ \sigma_i$ or $\sigma_k \circ \sigma_j$ belongs to $\mathbb{M}_1$, and so $\langle \sigma_1, \sigma_j \rangle \subseteq \mathbb{M}_1$. We have $\sigma_7 = \sigma_6 \circ \sigma_1$,

$\sigma_8 = \sigma_1 \circ \sigma_6$, $\sigma_9 = \sigma_1 \circ \sigma_7$, $\sigma_{10} = \sigma_6 \circ \sigma_6$, $\sigma_{11} = \sigma_6 \circ \sigma_7$, $\sigma_{12} = \sigma_8 \circ \sigma_6$, and $\sigma_{13} = \sigma_8 \circ \sigma_7$, and so $\mathbb{M}_1 = \langle \sigma_1, \sigma_6 \rangle = \{\sigma_0, \sigma_1\} \cup \{\sigma_6, \sigma_7, \ldots, \sigma_{13}\}$.

Since $\sigma_6 = \sigma_7 \circ \sigma_1 = \sigma_1 \circ \sigma_8 = \sigma_1 \circ \sigma_9 \circ \sigma_1 = \sigma_{10} \circ \sigma_1 \circ \sigma_{10} = \sigma_{11} \circ \sigma_{11} \circ \sigma_1 = \sigma_1 \circ \sigma_{12} \circ \sigma_{12}$ and $\sigma_6 = \sigma_1 \circ \sigma_{13} \circ \sigma_1 \circ \sigma_{13} \circ \sigma_1$, we have $\langle \sigma_1, \sigma_j \rangle = \mathbb{M}_1$ for $j \in \{6, 7, \ldots, 13\}$. □

Consider the permutation

$$\begin{pmatrix} \sigma_0 & \sigma_1 & \sigma_2 & \sigma_3 & \sigma_4 & \sigma_5 & \sigma_6 & \sigma_7 & \sigma_8 & \sigma_9 & \sigma_{10} & \sigma_{11} & \sigma_{12} & \sigma_{13} \\ \sigma_0 & \sigma_1 & \sigma_5 & \sigma_4 & \sigma_3 & \sigma_2 & \sigma_9 & \sigma_8 & \sigma_7 & \sigma_6 & \sigma_{13} & \sigma_{12} & \sigma_{11} & \sigma_{10} \end{pmatrix}. \quad (7)$$

It determines an automorphism $\mathbb{A} : \mathbb{M} \to \mathbb{M}$.

Theorem 3. *The identity and* $\mathbb{A}$ *are the only automorphisms of* $\mathbb{M}$.

Proof. Delete rows and columns marked by σ_1 in the Cayley table for $\mathbb{M}$. Then, check that the operation σ_3 is in the row or the column marked by σ_3 only. Also, the operation σ_4 is in the row or the column marked by σ_4 only. Therefore the semigroup

$$\langle \sigma_3, \sigma_4 \rangle = \{\sigma_2, \sigma_3, \ldots, \sigma_{13}\} \quad (8)$$

has a unique minimal set of generators $\{\sigma_3, \sigma_4\}$. The reader is left to check this with the Cayley table for $\mathbb{M}$.

Suppose $\mathbb{G}$ is an automorphism of $\mathbb{M}$. By Theorem 2, $\mathbb{G}$ transforms the set $\{\sigma_2, \sigma_3, \sigma_4, \sigma_5\}$ onto itself. However σ_2 and σ_5 are idempotents, but σ_3 and σ_4 are not idempotents. So, there are two possibilities: $\mathbb{G}(\sigma_3) = \sigma_3$ and $\mathbb{G}(\sigma_4) = \sigma_4$, which implies that $\mathbb{G}$ is the identity; $\mathbb{G}(\sigma_3) = \sigma_4$ and $\mathbb{G}(\sigma_4) = \sigma_3$, which implies $\mathbb{G} = \mathbb{A}$. The reader is left to check this with the Cayley table for $\mathbb{M}$. We offer hints: $\sigma_2 = \sigma_3 \circ \sigma_4$, $\sigma_5 = \sigma_4 \circ \sigma_3$, $\sigma_6 = \sigma_3 \circ \sigma_2$, $\sigma_7 = \sigma_3 \circ \sigma_3$, $\sigma_8 = \sigma_4 \circ \sigma_4$, $\sigma_9 = \sigma_4 \circ \sigma_5$, $\sigma_{10} = \sigma_6 \circ \sigma_6$, $\sigma_{11} = \sigma_6 \circ \sigma_7$, $\sigma_{12} = \sigma_8 \circ \sigma_6$, and $\sigma_{13} = \sigma_8 \circ \sigma_7$, to verify the details of this proof. □

5. The Monoid of All Idempotents

The set $\{\sigma_0, \sigma_2, \sigma_5, \sigma_7, \sigma_8, \sigma_{10}, \sigma_{13}\}$ consists of all squares in $\mathbb{M}$. These squares are idempotents and lie on the main diagonal in the Cayley table for $\mathbb{M}$. They constitute a monoid and

$$\{\sigma_0, \sigma_2, \sigma_5, \sigma_7, \sigma_8, \sigma_{10}, \sigma_{13}\} = \langle \sigma_0, \sigma_2, \sigma_5 \rangle . \quad (9)$$

The permutation

$$\begin{pmatrix} \sigma_0 & \sigma_2 & \sigma_5 & \sigma_7 & \sigma_8 & \sigma_{10} & \sigma_{13} \\ \sigma_0 & \sigma_2 & \sigma_5 & \sigma_8 & \sigma_7 & \sigma_{10} & \sigma_{13} \end{pmatrix} \quad (10)$$

determines the bijection $\mathbb{I} : \langle \sigma_0, \sigma_2, \sigma_5 \rangle \to \langle \sigma_0, \sigma_2, \sigma_5 \rangle$, such that

$$\mathbb{I}(\alpha \circ \beta) = \mathbb{I}(\beta) \circ \mathbb{I}(\alpha), \quad (11)$$

for any $\alpha, \beta \in \langle \sigma_0, \sigma_2, \sigma_5 \rangle$. To verify this, apply equalities $\sigma_2 \circ \sigma_5 = \sigma_7$, $\sigma_5 \circ \sigma_2 = \sigma_8$, $\sigma_2 \circ \sigma_5 \circ \sigma_2 = \sigma_{10}$, and $\sigma_5 \circ \sigma_2 \circ \sigma_5 = \sigma_{13}$. Any bijection $\mathbb{I} : G \to H$, having property $\mathbb{I}(\alpha \circ \beta) = \mathbb{I}(\beta) \circ \mathbb{I}(\alpha)$, transposes Cayley tables for semigroups G and H. Several semigroups contained in $\mathbb{M}$ have this property. We leave the reader to verify this.

We shall classify all semigroups contained in the semigroup $\langle \sigma_2, \sigma_5 \rangle$. Every such semigroup can be extended to a monoid by attaching σ_0 to it. This gives a complete classification of all semigroups in $\langle \sigma_0, \sigma_2, \sigma_5 \rangle$.

The semigroup $\langle \sigma_2, \sigma_5 \rangle$ contains six groups with exactly one element.

Semigroups $\langle \sigma_2, \sigma_{10} \rangle = \{\sigma_2, \sigma_{10}\}$ and $\langle \sigma_5, \sigma_{13} \rangle = \{\sigma_5, \sigma_{13}\}$ are monoids. Both consist of exactly two elements and are not groups, so they are isomorphic.

Semigroups $\langle \sigma_7, \sigma_{10} \rangle$ and $\langle \sigma_8, \sigma_{13} \rangle$ are isomorphic, in particular $\mathbb{A}[\langle \sigma_7, \sigma_{10} \rangle] = \langle \sigma_8, \sigma_{13} \rangle$. Also, semigroups $\langle \sigma_7, \sigma_{13} \rangle$ and $\langle \sigma_8, \sigma_{10} \rangle$ are isomorphic by $\mathbb{A}$. Every of these four semigroups has exactly two elements. None of them is a monoid. They form two types of nonisomorphic semigroups, because the bijections $\{(\sigma_7, \sigma_{10}), (\sigma_{10}, \sigma_8)\}$ and $\{(\sigma_7, \sigma_8), (\sigma_{10}, \sigma_{10})\}$ are not isomorphisms.

Semigroups $\langle \sigma_2, \sigma_7 \rangle = \{\sigma_2, \sigma_7, \sigma_{10}\}$ and $\langle \sigma_5, \sigma_8 \rangle = \{\sigma_5, \sigma_8, \sigma_{13}\}$ are isomorphic. Also, semigroups $\langle \sigma_2, \sigma_8 \rangle = \{\sigma_2, \sigma_8, \sigma_{10}\}$ and $\langle \sigma_5, \sigma_7 \rangle = \{\sigma_5, \sigma_7, \sigma_{13}\}$ are isomorphic. In fact, $\mathbb{A}[\langle \sigma_2, \sigma_7 \rangle] = \langle \sigma_5, \sigma_8 \rangle$ and $\mathbb{A}[\langle \sigma_2, \sigma_8 \rangle] = \langle \sigma_5, \sigma_7 \rangle$. None of these semigroups is a monoid. They form two types of nonisomorphic semigroups. Indeed, any isomorphism between $\langle \sigma_2, \sigma_7 \rangle$ and $\langle \sigma_2, \sigma_8 \rangle$ must be the identity on the monoid $\langle \sigma_2, \sigma_{10} \rangle$. Therefore would have to be the restriction of $\mathbb{I}$. But $\mathbb{I}$ restricted to $\langle \sigma_7, \sigma_{10} \rangle$ is not an isomorphism.

Semigroups $\langle \sigma_2, \sigma_{13} \rangle = \langle \sigma_2, \sigma_7, \sigma_8 \rangle = \{\sigma_2, \sigma_7, \sigma_8, \sigma_{10}, \sigma_{13}\}$ and $\langle \sigma_5, \sigma_{10} \rangle = \langle \sigma_5, \sigma_7, \sigma_8 \rangle = \{\sigma_5, \sigma_7, \sigma_8, \sigma_{10}, \sigma_{13}\}$ are not monoids. They are isomorphic by $\mathbb{A}$.

The semigroup $\langle \sigma_2, \sigma_5 \rangle$ contains exactly one semigroup with four elements $\langle \sigma_7, \sigma_8 \rangle = \langle \sigma_{10}, \sigma_{13} \rangle = \{\sigma_7, \sigma_8, \sigma_{10}, \sigma_{13}\}$ which is not a monoid.

Note that $\langle \sigma_2, \sigma_5 \rangle$ contains twenty different semigroups with nine non-isomorphism types. These are six isomorphic groups with exactly one element $\langle \sigma_2 \rangle \cong \langle \sigma_5 \rangle \cong \langle \sigma_7 \rangle \cong \langle \sigma_8 \rangle \cong \langle \sigma_{10} \rangle \cong \langle \sigma_{13} \rangle$, two isomorphic monoids with exactly two elements $\langle \sigma_2, \sigma_{10} \rangle \cong \langle \sigma_5, \sigma_{13} \rangle$, two pairs of isomorphic semigroups with exactly two elements $\langle \sigma_7, \sigma_{10} \rangle \cong \langle \sigma_8, \sigma_{13} \rangle$ and $\langle \sigma_7, \sigma_{13} \rangle \cong \langle \sigma_8, \sigma_{10} \rangle$, two pairs of isomorphic semigroups with exactly three elements $\langle \sigma_2, \sigma_7 \rangle \cong \langle \sigma_5, \sigma_8 \rangle$ and $\langle \sigma_2, \sigma_8 \rangle \cong \langle \sigma_5, \sigma_7 \rangle$, a semigroup with exactly four elements $\langle \sigma_7, \sigma_8 \rangle$, and two isomorphic semigroups with exactly five elements $\langle \sigma_2, \sigma_{13} \rangle \cong \langle \sigma_5, \sigma_{10} \rangle$ and also $\langle \sigma_2, \sigma_5 \rangle$. Thus, $\langle \sigma_2, \sigma_5 \rangle$ contains twenty different semigroups with nine isomorphism types. But $\langle \sigma_0, \sigma_2, \sigma_5 \rangle$ contains forty-one different semigroups with seventeen isomorphism types. Indeed, adding σ_0 to semigroups contained in $\langle \sigma_2, \sigma_5 \rangle$, which are not monoids, we get twenty monoids with eight non-isomorphism types. Adding σ_0 to a group contained in $\langle \sigma_2, \sigma_5 \rangle$ we get a monoid isomorphic to $\langle \sigma_2, \sigma_{10} \rangle$.

6. The Semigroup Consisting of $\{\sigma_6, \sigma_7, \ldots, \sigma_{13}\}$

Using the Cayley table for $\mathbb{M}$, check that

$$\mathbb{A}\left[\{\sigma_6, \sigma_7, \ldots, \sigma_{13}\}\right] = \{\sigma_6, \sigma_7, \ldots, \sigma_{13}\} . \quad (12)$$

Similarly, check that the semigroup $\langle \sigma_6, \sigma_9 \rangle = \{\sigma_6, \sigma_7, \ldots, \sigma_{13}\}$ can be represented as $\langle \sigma_6, \sigma_{13} \rangle$, $\langle \sigma_7, \sigma_{12} \rangle$, $\langle \sigma_8, \sigma_{11} \rangle$,

$\langle\sigma_9,\sigma_{10}\rangle$, or $\langle\sigma_{11},\sigma_{12}\rangle$. Also $\langle\sigma_6,\sigma_9\rangle = \langle\sigma_6,\sigma_7,\sigma_8\rangle = \langle\sigma_7,\sigma_8,\sigma_9\rangle = \langle\sigma_{10},\sigma_{11},\sigma_{13}\rangle = \langle\sigma_{10},\sigma_{12},\sigma_{13}\rangle$. These representations exhaust all minimal collections of the Kuratowski operations which generate $\langle\sigma_6,\sigma_9\rangle$. Other semigroups included in $\langle\sigma_6,\sigma_9\rangle$ have one or two minimal collection of generators. One generator has groups $\langle\sigma_6\rangle = \{\sigma_6,\sigma_{10}\}$, $\langle\sigma_9\rangle = \{\sigma_9,\sigma_{13}\}$, $\langle\sigma_{11}\rangle = \{\sigma_7,\sigma_{11}\}$, and $\langle\sigma_{12}\rangle = \{\sigma_8,\sigma_{12}\}$. Each of them has exactly two elements, so they are isomorphic. Semigroups $\langle\sigma_7,\sigma_{10}\rangle$, $\langle\sigma_7,\sigma_{13}\rangle$, $\langle\sigma_8,\sigma_{10}\rangle$, $\langle\sigma_8,\sigma_{13}\rangle$, and $\langle\sigma_7,\sigma_8\rangle$ are discussed in the previous section. Contained in $\langle\sigma_6,\sigma_9\rangle$ and not previously discussed semigroups are $\langle\sigma_6,\sigma_7\rangle$, $\langle\sigma_6,\sigma_8\rangle$,$\langle\sigma_7,\sigma_9\rangle$, and $\langle\sigma_8,\sigma_9\rangle$. We leave the reader to verify that the following are all possible pairs of Kuratowski operations which constitute a minimal collection of generators for semigroups contained in $\langle\sigma_6,\sigma_9\rangle$, but different from the whole. One has

(i) $\langle\sigma_6,\sigma_7\rangle = \langle\sigma_6,\sigma_{11}\rangle = \langle\sigma_{10},\sigma_{11}\rangle = \{\sigma_6,\sigma_7,\sigma_{10},\sigma_{11}\}$;

(ii) $\langle\sigma_6,\sigma_8\rangle = \langle\sigma_6,\sigma_{12}\rangle = \langle\sigma_{10},\sigma_{12}\rangle = \{\sigma_6,\sigma_8,\sigma_{10},\sigma_{12}\}$;

(iii) $\langle\sigma_7,\sigma_9\rangle = \langle\sigma_9,\sigma_{11}\rangle = \langle\sigma_{11},\sigma_{13}\rangle = \{\sigma_7,\sigma_9,\sigma_{11},\sigma_{13}\}$;

(iv) $\langle\sigma_8,\sigma_9\rangle = \langle\sigma_9,\sigma_{12}\rangle = \langle\sigma_{12},\sigma_{13}\rangle = \{\sigma_8,\sigma_9,\sigma_{12},\sigma_{13}\}$.

Proposition 4. *Semigroups $\langle\sigma_6,\sigma_7\rangle$ and $\langle\sigma_8,\sigma_9\rangle$ are isomorphic, and also semigroups $\langle\sigma_6,\sigma_8\rangle$ and $\langle\sigma_7,\sigma_9\rangle$ are isomorphic, but semigroups $\langle\sigma_6,\sigma_7\rangle$, $\langle\sigma_6,\sigma_8\rangle$ are not isomorphic.*

Proof. Isomorphisms are defined by $\mathbb{A}$. Suppose $J : \langle\sigma_6,\sigma_7\rangle \rightarrow \langle\sigma_6,\sigma_8\rangle$ is an isomorphism. Thus $J[\langle\sigma_7,\sigma_{10}\rangle] = \langle\sigma_8,\sigma_{10}\rangle$. Given $J(\sigma_7) = \sigma_8$, we get $\sigma_{10} = J(\sigma_{10}) = J(\sigma_7 \circ \sigma_{10}) \neq \sigma_8 \circ \sigma_{10} = \sigma_8$. But $J(\sigma_7) = \sigma_{10}$ implies $\sigma_{10} = J(\sigma_7) = J(\sigma_{10} \circ \sigma_7) \neq \sigma_8 \circ \sigma_{10} = \sigma_8$. Both possibilities lead to a contradiction. □

So, $\langle\sigma_6,\sigma_9\rangle$ contains eighteen different semigroups with eight non-isomorphism types. Indeed, these are four isomorphic groups with exactly one element $\langle\sigma_7\rangle \cong \langle\sigma_8\rangle \cong \langle\sigma_{10}\rangle \cong \langle\sigma_{13}\rangle$, four isomorphic groups with exactly two elements $\langle\sigma_6\rangle \cong \langle\sigma_9\rangle \cong \langle\sigma_{11}\rangle \cong \langle\sigma_{12}\rangle$, two pairs of isomorphic semigroups with exactly two elements $\langle\sigma_7,\sigma_{10}\rangle \cong \langle\sigma_8,\sigma_{13}\rangle$ and $\langle\sigma_7,\sigma_{13}\rangle \cong \langle\sigma_8,\sigma_{10}\rangle$, and five semigroups with exactly four elements $\langle\sigma_6,\sigma_7\rangle \cong \langle\sigma_8,\sigma_9\rangle$, $\langle\sigma_6,\sigma_8\rangle \cong \langle\sigma_7,\sigma_9\rangle$, $\langle\sigma_7,\sigma_8\rangle$, and also $\langle\sigma_6,\sigma_9\rangle$.

7. Remaining Semigroups in $\langle\sigma_3,\sigma_4\rangle$

We have yet to discuss semigroups included in $\langle\sigma_3,\sigma_4\rangle$, not included in $\langle\sigma_2,\sigma_5\rangle$ and containing at least one of Kuratowski operation σ_2, σ_3, σ_4, or σ_5. It will be discussed up to the isomorphism $\mathbb{A}$. Obviously, $\langle\sigma_2\rangle = \{\sigma_2\}$ and $\langle\sigma_5\rangle = \{\sigma_5\}$ are groups.

7.1. Extensions of $\langle\boldsymbol{\sigma_2}\rangle$ and $\langle\boldsymbol{\sigma_5}\rangle$ with Elements of $\langle\boldsymbol{\sigma_6},\boldsymbol{\sigma_9}\rangle$. Monoids $\langle\sigma_2,\sigma_6\rangle = \{\sigma_2,\sigma_6,\sigma_{10}\}$ and $\langle\sigma_2,\sigma_{10}\rangle = \{\sigma_2,\sigma_{10}\}$ have different numbers of elements. Also, $\langle\sigma_2,\sigma_6\rangle$ is isomorphic to $\langle\sigma_0,\sigma_6\rangle$. Nonisomorphic semigroups $\langle\sigma_2,\sigma_7\rangle$ and $\langle\sigma_2,\sigma_8\rangle$ are discussed above. The following three semigroups:

(i) $\langle\sigma_2,\sigma_{11}\rangle = \langle\sigma_2,\sigma_6,\sigma_7\rangle = \{\sigma_2,\sigma_6,\sigma_7,\sigma_{10},\sigma_{11}\}$,

(ii) $\langle\sigma_2,\sigma_{12}\rangle = \langle\sigma_2,\sigma_6,\sigma_8\rangle = \{\sigma_2,\sigma_6,\sigma_8,\sigma_{10},\sigma_{12}\}$,

(iii) $\langle\sigma_2,\sigma_{13}\rangle = \langle\sigma_2,\sigma_7,\sigma_8\rangle = \{\sigma_2,\sigma_7,\sigma_8,\sigma_{10},\sigma_{13}\}$

are not monoids. They are not isomorphic. Indeed, any isomorphism between these semigroups would lead an isomorphism between $\langle\sigma_6,\sigma_7\rangle$, $\langle\sigma_6,\sigma_8\rangle$, or $\langle\sigma_7,\sigma_8\rangle$. This is impossible, by Proposition 4 and because $\langle\sigma_7,\sigma_8\rangle$ consists of idempotents, but σ_6 is not an idempotent. The nine-element semigroup on the set $\{\sigma_2,\sigma_6,\sigma_7,\ldots,\sigma_{13}\}$ is represented as $\langle\sigma_2,\sigma_9\rangle$. So, we have added four new semigroups, which are not isomorphic with the semigroups previously discussed. These are $\langle\sigma_2,\sigma_{11}\rangle$, $\langle\sigma_2,\sigma_{12}\rangle$, $\langle\sigma_2,\sigma_{13}\rangle$, and $\langle\sigma_2,\sigma_9\rangle$.

Using $\mathbb{A}$, we have described eight semigroups—each one isomorphic to a semigroup previously discussed—which contains σ_5 and elements (at least one) of $\langle\sigma_6,\sigma_9\rangle$. Collections of generators: $\langle\sigma_2,\sigma_6,\sigma_{13}\rangle$, $\langle\sigma_2,\sigma_7,\sigma_{12}\rangle$, $\langle\sigma_2,\sigma_8,\sigma_{11}\rangle$, $\langle\sigma_2,\sigma_{11},\sigma_{12}\rangle$, $\langle\sigma_2,\sigma_{11},\sigma_{13}\rangle$, and $\langle\sigma_2,\sigma_{12},\sigma_{13}\rangle$ are minimal in $\langle\sigma_2,\sigma_9\rangle$. Also, collections of generators: $\langle\sigma_5,\sigma_7,\sigma_{12}\rangle$, $\langle\sigma_5,\sigma_8,\sigma_{11}\rangle$, $\langle\sigma_5,\sigma_9,\sigma_{10}\rangle$, $\langle\sigma_5,\sigma_{10},\sigma_{11}\rangle$, $\langle\sigma_5,\sigma_{10},\sigma_{12}\rangle$, and $\langle\sigma_5,\sigma_{11},\sigma_{12}\rangle$ are minimal in $\langle\sigma_5,\sigma_6\rangle$.

7.2. Extensions of $\langle\boldsymbol{\sigma_3}\rangle$ and $\langle\boldsymbol{\sigma_4}\rangle$ by Elements from the Semigroup $\langle\boldsymbol{\sigma_6},\boldsymbol{\sigma_9}\rangle$. Semigroups $\langle\sigma_3\rangle = \{\sigma_3,\sigma_7,\sigma_{11}\}$ and $\langle\sigma_4\rangle = \{\sigma_4,\sigma_8,\sigma_{12}\}$ are isomorphic by $\mathbb{A}$. They are not monoids. The semigroup $\langle\sigma_3\rangle$ can be extended using elements of $\langle\sigma_6,\sigma_9\rangle$, in three following ways:

(i) $\langle\sigma_3,\sigma_6\rangle = \langle\sigma_3,\sigma_{10}\rangle = \{\sigma_3,\sigma_6,\sigma_7,\sigma_{10},\sigma_{11}\}$;

(ii) $\langle\sigma_3,\sigma_8\rangle = \langle\sigma_3,\sigma_{12}\rangle = \{\sigma_3,\sigma_6,\sigma_7,\ldots,\sigma_{13}\}$;

(iii) $\langle\sigma_3,\sigma_9\rangle = \langle\sigma_3,\sigma_{13}\rangle = \{\sigma_3,\sigma_7,\sigma_9,\sigma_{11},\sigma_{13}\}$.

Semigroups $\langle\sigma_3,\sigma_6\rangle$ and $\langle\sigma_3,\sigma_9\rangle$ are not isomorphic. Indeed, suppose $J : \langle\sigma_3,\sigma_6\rangle \rightarrow \langle\sigma_3,\sigma_9\rangle$ is an isomorphism. Thus, J is the identity on $\langle\sigma_3\rangle$, $J(\sigma_6) = \sigma_9$, and $J(\sigma_{10}) = \sigma_{13}$. This gives a contradiction, since $\sigma_3 \circ \sigma_6 = \sigma_{10}$ and $\sigma_3 \circ \sigma_9 = \sigma_7$.

Neither $\langle\sigma_3,\sigma_6\rangle$ nor $\langle\sigma_3,\sigma_9\rangle$ has a minimal collection of generators with three elements, so they give new isomorphism types. Also, $\langle\sigma_2,\sigma_9\rangle$ is not isomorphic to $\langle\sigma_3,\sigma_8\rangle$, since $\langle\sigma_2,\sigma_9\rangle$ has a unique pair of generators and $\langle\sigma_3,\sigma_8\rangle = \langle\sigma_3,\sigma_{12}\rangle$.

Using $\mathbb{A}$, we get—isomorphic to previously discussed ones—semigroups $\langle\sigma_4,\sigma_9\rangle = \langle\sigma_4,\sigma_{13}\rangle = \{\sigma_4,\sigma_8,\sigma_9\ \sigma_{12},\sigma_{13}\}$, $\langle\sigma_4,\sigma_6\rangle = \langle\sigma_4,\sigma_{10}\rangle = \{\sigma_4,\sigma_6,\sigma_8\ \sigma_{10},\sigma_{12}\}$, and $\langle\sigma_4,\sigma_7\rangle = \langle\sigma_4,\sigma_{11}\rangle = \{\sigma_4,\sigma_6,\sigma_7,\ldots,\sigma_{13}\}$. There exist minimal collections of generators, such as follows:

$$\begin{aligned} \langle\sigma_3,\sigma_8\rangle &= \langle\sigma_3,\sigma_6,\sigma_9\rangle = \langle\sigma_3,\sigma_6,\sigma_{13}\rangle \\ &= \langle\sigma_3,\sigma_9,\sigma_{10}\rangle = \langle\sigma_3,\sigma_{10},\sigma_{13}\rangle, \\ \langle\sigma_4,\sigma_7\rangle &= \langle\sigma_4,\sigma_6,\sigma_9\rangle = \langle\sigma_4,\sigma_9,\sigma_{10}\rangle \\ &= \langle\sigma_4,\sigma_6,\sigma_{13}\rangle = \langle\sigma_4,\sigma_{10},\sigma_{13}\rangle. \end{aligned} \tag{13}$$

7.3. More Generators from the Set $\{\boldsymbol{\sigma_2},\boldsymbol{\sigma_3},\boldsymbol{\sigma_4},\boldsymbol{\sigma_5}\}$. Now we check that $\langle\sigma_2,\sigma_3\rangle = \{\sigma_2,\sigma_3,\sigma_6,\sigma_7,\sigma_{10}\ \sigma_{11}\}$ and $\langle\sigma_4,\sigma_5\rangle = \{\sigma_4,\sigma_5,\sigma_8,\sigma_9,\sigma_{12}\ \sigma_{13}\} = \mathbb{A}[\langle\sigma_2,\sigma_3\rangle]$ and also $\langle\sigma_2,\sigma_4\rangle = \{\sigma_2,\sigma_4,\sigma_6,\sigma_8,\sigma_{10}\ \sigma_{12}\}$ and $\langle\sigma_3,\sigma_5\rangle = \{\sigma_3,\sigma_5,\sigma_7,\sigma_9,\sigma_{11}\ \sigma_{13}\} = \mathbb{A}[\langle\sigma_2,\sigma_4\rangle]$ are two pairs of isomorphic semigroups which give two new isomorphism types.

Each of these semigroups has six element, so in $\mathbb{M}$ there are five six-element semigroups of three isomorphism types, since $\langle \sigma_2, \sigma_5 \rangle$ has 6 elements which are idempotents.

In $\langle \sigma_3, \sigma_4 \rangle$ there are seven semigroups which have three generators and have not two generators. These are

(1) $\langle \sigma_2, \sigma_3, \sigma_8 \rangle = \langle \sigma_2, \sigma_3, \sigma_9 \rangle = \langle \sigma_2, \sigma_3, \sigma_{12} \rangle = \langle \sigma_2, \sigma_3, \sigma_{13} \rangle$;

(2) $\langle \sigma_4, \sigma_5, \sigma_6 \rangle = \langle \sigma_4, \sigma_5, \sigma_7 \rangle = \langle \sigma_4, \sigma_5, \sigma_{10} \rangle = \langle \sigma_4, \sigma_5, \sigma_{11} \rangle$;

(3) $\langle \sigma_2, \sigma_4, \sigma_7 \rangle = \langle \sigma_2, \sigma_4, \sigma_9 \rangle = \langle \sigma_2, \sigma_4, \sigma_{11} \rangle = \langle \sigma_2, \sigma_4, \sigma_{13} \rangle$;

(4) $\langle \sigma_3, \sigma_5, \sigma_6 \rangle = \langle \sigma_3, \sigma_5, \sigma_8 \rangle = \langle \sigma_3, \sigma_5, \sigma_{10} \rangle = \langle \sigma_3, \sigma_5, \sigma_{12} \rangle$;

(5) $\langle \sigma_2, \sigma_5, \sigma_6 \rangle = \langle \sigma_2, \sigma_5, \sigma_9 \rangle = \langle \sigma_2, \sigma_5, \sigma_{11} \rangle = \langle \sigma_2, \sigma_5, \sigma_{12} \rangle$;

(6) $\langle \sigma_2, \sigma_3, \sigma_5 \rangle$;

(7) $\langle \sigma_2, \sigma_4, \sigma_5 \rangle$.

8. Semigroups which Are Contained in $\langle \sigma_3, \sigma_4 \rangle$

Groups $\langle \sigma_2 \rangle \cong \langle \sigma_5 \rangle \cong \langle \sigma_7 \rangle \cong \langle \sigma_8 \rangle \cong \langle \sigma_{10} \rangle \cong \langle \sigma_{13} \rangle$ have one element and are isomorphic.

Groups $\langle \sigma_6 \rangle \cong \langle \sigma_9 \rangle \cong \langle \sigma_{11} \rangle \cong \langle \sigma_{12} \rangle$, monoids $\langle \sigma_2, \sigma_{10} \rangle \cong \langle \sigma_5, \sigma_{13} \rangle$, and also semigroups $\langle \sigma_7, \sigma_{10} \rangle \cong \langle \sigma_8, \sigma_{13} \rangle$ and $\langle \sigma_7, \sigma_{13} \rangle \cong \langle \sigma_8, \sigma_{10} \rangle$ have two elements and Cayley tables as follows:

	A	B
A	B	A
B	A	B

	A	B
A	A	B
B	B	A

	A	B
A	A	B
B	A	B

	A	B
A	A	A
B	B	B

(14)

Monoids $\langle \sigma_2, \sigma_6 \rangle \cong \langle \sigma_5, \sigma_9 \rangle$, and also semigroups $\langle \sigma_2, \sigma_7 \rangle \cong \langle \sigma_5, \sigma_8 \rangle$, $\langle \sigma_2, \sigma_8 \rangle \cong \langle \sigma_5, \sigma_7 \rangle$, and $\langle \sigma_3 \rangle \cong \langle \sigma_4 \rangle$ have three elements and Cayley tables as follows:

	A	B	C
A	A	B	C
B	B	C	B
C	C	B	C

	A	B	C
A	A	B	C
B	C	B	C
C	C	B	C

	A	B	C
A	A	C	C
B	B	B	B
C	C	C	C

	A	B	C
A	B	C	B
B	C	B	C
C	B	C	B

(15)

Semigroups $\langle \sigma_6, \sigma_7 \rangle \cong \langle \sigma_8, \sigma_9 \rangle$, $\langle \sigma_6, \sigma_8 \rangle \cong \langle \sigma_7, \sigma_9 \rangle$, and $\langle \sigma_7, \sigma_8 \rangle$ have four elements and Cayley tables as follows:

	A	B	C	D
A	C	D	A	B
B	A	B	C	D
C	A	B	C	D
D	C	D	A	B

	A	B	C	D
A	C	A	A	C
B	D	B	B	D
C	A	C	C	A
D	B	D	D	B

	A	B	C	D
A	A	C	C	A
B	D	B	B	D
C	A	C	C	A
D	D	B	B	D

(16)

Semigroups $\langle \sigma_2, \sigma_{11} \rangle \cong \langle \sigma_5, \sigma_{12} \rangle$, $\langle \sigma_2, \sigma_{12} \rangle \cong \langle \sigma_5, \sigma_{11} \rangle$, $\langle \sigma_2, \sigma_{13} \rangle \cong \langle \sigma_5, \sigma_{10} \rangle$, $\langle \sigma_3, \sigma_6 \rangle \cong \langle \sigma_4, \sigma_9 \rangle$, and $\langle \sigma_3, \sigma_9 \rangle \cong \langle \sigma_4, \sigma_6 \rangle$ have five elements. Semigroups $\langle \sigma_2, \sigma_3 \rangle \cong \langle \sigma_4, \sigma_5 \rangle$, $\langle \sigma_2, \sigma_4 \rangle \cong \langle \sigma_3, \sigma_5 \rangle$, and $\langle \sigma_2, \sigma_5 \rangle$ have six elements. Construction of Cayley tables for these semigroups, as well as other semigroups, we leave it to the readers. One can prepare such tables removing from the Cayley table for $\mathbb{M}$ some columns and rows. Then change Kuratowski operations onto letters of the alphabet. For example, $\langle \sigma_0, \sigma_2, \sigma_5 \rangle$ has the following Cayley table:

	0	A	B	C	D	E	F
0	0	A	B	C	D	E	F
A	A	A	C	C	E	E	C
B	B	D	B	F	D	D	F
C	C	E	C	C	E	E	C
D	D	D	F	F	D	D	F
E	E	E	C	C	E	E	C
F	F	D	F	F	D	D	F

(17)

Preparing the Cayley table for $\langle \sigma_0, \sigma_2, \sigma_5 \rangle$ we put $\sigma_0 = 0$, $\sigma_2 = A$, $\sigma_5 = B$, $\sigma_7 = C$, $\sigma_8 = D$, $\sigma_{10} = E$, and $\sigma_{13} = F$. This table immediately shows that semigroups $\langle \sigma_2, \sigma_5 \rangle$ and $\langle \sigma_0, \sigma_2, \sigma_5 \rangle$ have exactly two automorphisms. These are identities and restrictions of $\mathbb{A}$.

Semigroup $\langle \sigma_6, \sigma_9 \rangle$ has eight elements. Semigroups $\langle \sigma_2, \sigma_9 \rangle \cong \langle \sigma_5, \sigma_6 \rangle$ and $\langle \sigma_3, \sigma_8 \rangle \cong \langle \sigma_4, \sigma_7 \rangle$ have nine elements. Semigroups $\langle \sigma_2, \sigma_3, \sigma_8 \rangle \cong \langle \sigma_4, \sigma_5, \sigma_7 \rangle$, $\langle \sigma_2, \sigma_4, \sigma_7 \rangle \cong \langle \sigma_3, \sigma_5, \sigma_8 \rangle$, and $\langle \sigma_2, \sigma_5, \sigma_6 \rangle$ have ten elements. Semigroups $\langle \sigma_2, \sigma_3, \sigma_5 \rangle \cong \langle \sigma_2, \sigma_4, \sigma_5 \rangle$ have eleven elements. In the end, the semigroup $\langle \sigma_3, \sigma_4 \rangle$ has twelve elements.

Thus, the semigroup $\langle \sigma_3, \sigma_4 \rangle$ includes fifty-seven semigroups, among which there are ten groups, fourteen monoids, and also forty-three semigroups which are not monoids. These semigroups consist of twenty-eight types of nonisomorphic semigroups, two non-isomorphism types of groups, two non-isomorphism types of monoids which are not groups, and twenty four non-isomorphism types of semigroups which are not monoids.

9. Viewing Semigroups Contained in $\mathbb{M}$

9.1. Descriptive Data on Semigroups which Are Contained in $\mathbb{M}$. There are one hundred eighteen, that is, $118 = 2 \cdot 57 +$

4, semigroups which are contained in $\mathbb{M}$. These are fifty-seven semigroups contained in $\langle\sigma_3,\sigma_4\rangle$, fifty-seven monoids formed by adding σ_0 to a semigroup contained in $\langle\sigma_3,\sigma_4\rangle$, groups $\langle\sigma_0\rangle$, $\langle\sigma_1\rangle$, and monoids $\langle\sigma_1,\sigma_6\rangle = \mathbb{M}_1$ and $\langle\sigma_1,\sigma_2\rangle = \mathbb{M}$.

There are fifty-six types of nonisomorphic semigroups in $\mathbb{M}$. These are twenty-eight non-isomorphism types of semigroups contained in $\langle\sigma_3,\sigma_4\rangle$, twenty-six types of non-isomorphic monoids formed by adding σ_0 to a semigroup contained in $\langle\sigma_3,\sigma_4\rangle$, and also $\mathbb{M}_1$ and $\mathbb{M}$. Indeed, adding σ_0 to a semigroup which is not a monoid we obtain a monoid. In this way, we get twenty-four types of nonisomorphic monoids. Adding σ_0 to a monoid contained in $\langle\sigma_3,\sigma_4\rangle$ we get two new non-isomorphism types of monoids. But, adding σ_0 to a group contained in $\langle\sigma_3,\sigma_4\rangle$ we get no new type of monoid, since we get a monoid isomorphic to $\langle\sigma_2,\sigma_{10}\rangle$ or $\langle\sigma_2,\sigma_6\rangle$. The other two types are $\mathbb{M}_1$ and $\mathbb{M}$.

9.2. Semigroups which Are Not Monoids. Below we have reproduced, using the smallest number of generators and the dictionary order, a list of all 43 semigroups, which are included in the $\mathbb{M}$ as follows:

(1) $\langle\sigma_2,\sigma_3\rangle = \{\sigma_2,\sigma_3,\sigma_6,\sigma_7,\sigma_{10},\sigma_{11}\}$.

(2) $\langle\sigma_2,\sigma_3,\sigma_5\rangle = \{\sigma_2,\sigma_3,\sigma_5,\sigma_6,\dots,\sigma_{13}\}$.

(3) $\langle\sigma_2,\sigma_3,\sigma_8\rangle = \langle\sigma_2,\sigma_3,\sigma_9\rangle = \langle\sigma_2,\sigma_3,\sigma_{12}\rangle = \langle\sigma_2,\sigma_3,\sigma_{13}\rangle = \{\sigma_2,\sigma_3,\sigma_6,\sigma_7,\dots,\sigma_{13}\}$.

(4) $\langle\sigma_2,\sigma_4\rangle = \{\sigma_2,\sigma_4,\sigma_6,\sigma_8,\sigma_{10},\sigma_{12}\}$.

(5) $\langle\sigma_2,\sigma_4,\sigma_5\rangle = \{\sigma_2,\sigma_4,\sigma_5,\dots,\sigma_{13}\}$.

(6) $\langle\sigma_2,\sigma_4,\sigma_7\rangle = \langle\sigma_2,\sigma_4,\sigma_9\rangle = \langle\sigma_2,\sigma_4,\sigma_{11}\rangle = \langle\sigma_2,\sigma_4,\sigma_{13}\rangle = \{\sigma_2,\sigma_4,\sigma_6,\sigma_7\dots,\sigma_{13}\}$.

(7) $\langle\sigma_2,\sigma_5\rangle = \{\sigma_2,\sigma_5,\sigma_7,\sigma_8,\sigma_{10},\sigma_{13}\}$.

(8) $\langle\sigma_2,\sigma_5,\sigma_6\rangle = \langle\sigma_2,\sigma_5,\sigma_9\rangle = \langle\sigma_2,\sigma_5,\sigma_{11}\rangle = \langle\sigma_2,\sigma_5,\sigma_{12}\rangle = \{\sigma_2,\sigma_5,\sigma_6,\dots,\sigma_{13}\}$.

(9) $\langle\sigma_2,\sigma_7\rangle = \{\sigma_2,\sigma_7,\sigma_{10}\}$.

(10) $\langle\sigma_2,\sigma_8\rangle = \{\sigma_2,\sigma_8,\sigma_{10}\}$.

(11) $\langle\sigma_2,\sigma_9\rangle = \langle\sigma_2,\sigma_6,\sigma_{13}\rangle = \langle\sigma_2,\sigma_7,\sigma_{12}\rangle = \langle\sigma_2,\sigma_8,\sigma_{11}\rangle = \langle\sigma_2,\sigma_{11},\sigma_{12}\rangle = \langle\sigma_2,\sigma_{11},\sigma_{13}\rangle = \langle\sigma_2,\sigma_{12},\sigma_{13}\rangle = \{\sigma_2,\sigma_6,\sigma_7,\dots,\sigma_{13}\}$.

(12) $\langle\sigma_2,\sigma_{11}\rangle = \langle\sigma_2,\sigma_6,\sigma_7\rangle = \{\sigma_2,\sigma_6,\sigma_7,\sigma_{10},\sigma_{11}\}$.

(13) $\langle\sigma_2,\sigma_{12}\rangle = \langle\sigma_2,\sigma_6,\sigma_8\rangle = \{\sigma_2,\sigma_6,\sigma_8,\sigma_{10},\sigma_{12}\}$.

(14) $\langle\sigma_2,\sigma_{13}\rangle = \langle\sigma_2,\sigma_7,\sigma_8\rangle = \{\sigma_2,\sigma_7,\sigma_8,\sigma_{10},\sigma_{13}\}$.

(15) $\langle\sigma_3\rangle = \{\sigma_3,\sigma_7,\sigma_{11}\}$.

(16) $\langle\sigma_3,\sigma_4\rangle = \{\sigma_2,\sigma_3,\dots,\sigma_{13}\}$.

(17) $\langle\sigma_3,\sigma_5\rangle = \{\sigma_3,\sigma_5,\sigma_7,\sigma_9,\sigma_{11},\sigma_{13}\}$.

(18) $\langle\sigma_3,\sigma_5,\sigma_6\rangle = \langle\sigma_3,\sigma_5,\sigma_8\rangle = \langle\sigma_3,\sigma_5,\sigma_{10}\rangle = \langle\sigma_3,\sigma_5,\sigma_{12}\rangle = \{\sigma_3,\sigma_5,\sigma_6,\dots,\sigma_{13}\}$.

(19) $\langle\sigma_3,\sigma_6\rangle = \langle\sigma_3,\sigma_{10}\rangle = \{\sigma_3,\sigma_6,\sigma_7,\sigma_{10},\sigma_{11}\}$.

(20) $\langle\sigma_3,\sigma_8\rangle = \langle\sigma_3,\sigma_{12}\rangle = \{\sigma_3,\sigma_6,\sigma_7,\dots,\sigma_{13}\}$.

(21) $\langle\sigma_3,\sigma_9\rangle = \langle\sigma_3,\sigma_{13}\rangle = \{\sigma_3,\sigma_7,\sigma_9,\sigma_{11},\sigma_{13}\}$.

(22) $\langle\sigma_4\rangle = \{\sigma_4,\sigma_8,\sigma_{12}\}$.

(23) $\langle\sigma_4,\sigma_5\rangle = \{\sigma_4,\sigma_5,\sigma_8,\sigma_9,\sigma_{12},\sigma_{13}\}$.

(24) $\langle\sigma_4,\sigma_5,\sigma_6\rangle = \langle\sigma_4,\sigma_5,\sigma_7\rangle = \langle\sigma_4,\sigma_5,\sigma_{10}\rangle = \langle\sigma_4,\sigma_5,\sigma_{11}\rangle = \{\sigma_4,\sigma_5,\dots,\sigma_{13}\}$.

(25) $\langle\sigma_4,\sigma_6\rangle = \langle\sigma_4,\sigma_{10}\rangle = \{\sigma_4,\sigma_6,\sigma_8,\sigma_{10},\sigma_{12}\}$.

(26) $\langle\sigma_4,\sigma_7\rangle = \langle\sigma_4,\sigma_{11}\rangle = \{\sigma_4,\sigma_6,\sigma_7,\dots,\sigma_{13}\}$.

(27) $\langle\sigma_4,\sigma_9\rangle = \langle\sigma_4,\sigma_{13}\rangle = \{\sigma_4,\sigma_8,\sigma_9,\sigma_{12},\sigma_{13}\}$.

(28) $\langle\sigma_5,\sigma_6\rangle = \langle\sigma_5,\sigma_7,\sigma_{12}\rangle = \langle\sigma_5,\sigma_8,\sigma_{11}\rangle = \langle\sigma_5,\sigma_9,\sigma_{10}\rangle = \langle\sigma_5,\sigma_{10},\sigma_{11}\rangle = \langle\sigma_5,\sigma_{10},\sigma_{12}\rangle = \langle\sigma_5,\sigma_{11},\sigma_{12}\rangle = \{\sigma_5,\sigma_6,\dots,\sigma_{13}\}$.

(29) $\langle\sigma_5,\sigma_7\rangle = \{\sigma_5,\sigma_7,\sigma_{13}\}$.

(30) $\langle\sigma_5,\sigma_8\rangle = \{\sigma_5,\sigma_8,\sigma_{13}\}$.

(31) $\langle\sigma_5,\sigma_{10}\rangle = \langle\sigma_5,\sigma_7,\sigma_8\rangle = \{\sigma_5,\sigma_7,\sigma_8,\sigma_{10},\sigma_{13}\}$.

(32) $\langle\sigma_5,\sigma_{11}\rangle = \langle\sigma_5,\sigma_7,\sigma_9\rangle = \{\sigma_5,\sigma_7,\sigma_9,\sigma_{11},\sigma_{13}\}$.

(33) $\langle\sigma_5,\sigma_{12}\rangle = \langle\sigma_5,\sigma_8,\sigma_9\rangle = \{\sigma_5,\sigma_8,\sigma_9,\sigma_{12},\sigma_{13}\}$.

(34) $\langle\sigma_6,\sigma_7\rangle = \langle\sigma_6,\sigma_{11}\rangle = \langle\sigma_{10},\sigma_{11}\rangle = \{\sigma_6,\sigma_7,\sigma_{10},\sigma_{11}\}$.

(35) $\langle\sigma_6,\sigma_8\rangle = \langle\sigma_6,\sigma_{12}\rangle = \langle\sigma_{10},\sigma_{12}\rangle = \{\sigma_6,\sigma_8,\sigma_{10},\sigma_{12}\}$.

(36) $\langle\sigma_6,\sigma_9\rangle = \langle\sigma_6,\sigma_{13}\rangle = \langle\sigma_7,\sigma_{12}\rangle = \langle\sigma_8,\sigma_{11}\rangle = \langle\sigma_9,\sigma_{10}\rangle = \langle\sigma_{11},\sigma_{12}\rangle = \langle\sigma_6,\sigma_7,\sigma_8\rangle = \langle\sigma_7,\sigma_8,\sigma_9\rangle = \langle\sigma_{10},\sigma_{11},\sigma_{13}\rangle = \langle\sigma_{10},\sigma_{12},\sigma_{13}\rangle = \{\sigma_6,\sigma_7,\dots,\sigma_{13}\}$.

(37) $\langle\sigma_7,\sigma_8\rangle = \{\sigma_7,\sigma_8,\sigma_{10},\sigma_{13}\}$.

(38) $\langle\sigma_7,\sigma_9\rangle = \{\sigma_7,\sigma_9,\sigma_{11},\sigma_{13}\}$.

(39) $\langle\sigma_7,\sigma_{10}\rangle = \{\sigma_7,\sigma_{10}\}$.

(40) $\langle\sigma_7,\sigma_{13}\rangle = \{\sigma_7,\sigma_{13}\}$.

(41) $\langle\sigma_8,\sigma_9\rangle = \{\sigma_8,\sigma_9,\sigma_{12},\sigma_{13}\}$.

(42) $\langle\sigma_8,\sigma_{10}\rangle = \{\sigma_8,\sigma_{10}\}$.

(43) $\langle\sigma_8,\sigma_{13}\rangle = \{\sigma_8,\sigma_{13}\}$.

9.3. Isomorphism Types of Semigroups Contained in $\mathbb{M}$. Systematize the list of all isomorphism types of semigroups contained in the monoid $\mathbb{M}$. Isomorphisms, which are restrictions of the isomorphism $\mathbb{A}$, will be regarded as self-evident, and therefore they will not be commented on.

(i) The monoid $\mathbb{M}$ contains 12 groups with 2 isomorphism types. These are 7 one-element groups and 5 two-element groups.

(ii) The monoid $\mathbb{M}$ contains 8 two-element monoids with 2 isomorphism types. These are σ_0 added to 6 one-element groups and $\langle\sigma_2,\sigma_{10}\rangle \cong \langle\sigma_5,\sigma_{13}\rangle$. Also, it contains 4 two-element semigroups, not monoids, with 2 isomorphism types. These are $\langle\sigma_7,\sigma_{10}\rangle \cong \langle\sigma_8,\sigma_{13}\rangle$ and $\langle\sigma_8,\sigma_{10}\rangle \cong \langle\sigma_7,\sigma_{13}\rangle$.

(iii) The monoid $\mathbb{M}$ contains 12 three-element monoids with 5 isomorphism types. These are σ_0 added to 4 two-element groups and also $\langle\sigma_0,\sigma_2,\sigma_{10}\rangle \cong \langle\sigma_0,\sigma_5,\sigma_{13}\rangle$, $\langle\sigma_0,\sigma_7,\sigma_{10}\rangle \cong \langle\sigma_0,\sigma_8,\sigma_{13}\rangle$, $\langle\sigma_0,\sigma_8,\sigma_{10}\rangle \cong \langle\sigma_0,\sigma_7,\sigma_{13}\rangle$, and $\langle\sigma_2,\sigma_6\rangle \cong \langle\sigma_5,\sigma_9\rangle$.

(iv) The monoid $\mathbb{M}$ contains 6 three-element semigroups, not monoids, with 3 isomorphism types. These are $\langle\sigma_2,\sigma_7\rangle \cong \langle\sigma_5,\sigma_8\rangle$, $\langle\sigma_2,\sigma_8\rangle \cong \langle\sigma_5,\sigma_7\rangle$, and $\langle\sigma_3\rangle \cong \langle\sigma_4\rangle$.

(v) The monoid $\mathbb{M}$ contains 8 four-element monoids, each contains σ_0, with 4 isomorphism types. These are semigroups from two preceding items that can be substantially extended by σ_0.

(vi) The monoid $\mathbb{M}$ contains 5 four-element semigroups, not monoids, with 3 isomorphism types. These are $\langle\sigma_6,\sigma_7\rangle \cong \langle\sigma_8,\sigma_9\rangle$, $\langle\sigma_6,\sigma_8\rangle \cong \langle\sigma_7,\sigma_9\rangle$ and $\langle\sigma_7,\sigma_8\rangle$. These semigroups extended by σ_0 yield 5 monoids, all which consist of five elements, with 3 isomorphism types.

(vii) The monoid $\mathbb{M}$ contains 10 five-element semigroups—not monoids, with 5 isomorphism types. These are $\langle\sigma_2,\sigma_{11}\rangle \cong \langle\sigma_5,\sigma_{12}\rangle$, $\langle\sigma_2,\sigma_{12}\rangle \cong \langle\sigma_5,\sigma_{11}\rangle$, $\langle\sigma_2,\sigma_{13}\rangle \cong \langle\sigma_5,\sigma_{10}\rangle$, $\langle\sigma_3,\sigma_6\rangle \cong \langle\sigma_4,\sigma_9\rangle$, and $\langle\sigma_3,\sigma_9\rangle \cong \langle\sigma_4,\sigma_6\rangle$. These semigroups extended by σ_0 yield 10 monoids, all of which consist of six elements, with 5 isomorphism types. We get 10 new isomorphism types, since the semigroups are distinguished by semigroups $\langle\sigma_3\rangle$ and $\langle\sigma_4\rangle$, and by nonisomorphic semigroups $\langle\sigma_6,\sigma_7\rangle$, $\langle\sigma_6,\sigma_8\rangle$, and $\langle\sigma_7,\sigma_8\rangle$.

(viii) The monoid $\mathbb{M}$ contains 5 six-element semigroups, not monoids, with 3 isomorphism types. These are $\langle\sigma_2,\sigma_3\rangle \cong \langle\sigma_4,\sigma_5\rangle$, $\langle\sigma_2,\sigma_4\rangle \cong \langle\sigma_3,\sigma_5\rangle$, and $\langle\sigma_2,\sigma_5\rangle$. These semigroups extended by σ_0 yield 5 monoids, all of which consist of seven elements, with 3 isomorphism types. We get 6 new isomorphism types, since the semigroups are distinguished by non isomorphic semigroups $\langle\sigma_6,\sigma_7\rangle$, $\langle\sigma_6,\sigma_8\rangle$, and $\langle\sigma_7,\sigma_8\rangle$.

(ix) The monoid $\mathbb{M}$ contains no seven-element semigroup, not a monoid, no eight-element monoid and the only semigroup $\langle\sigma_6,\sigma_9\rangle$ with exactly eight elements and the only monoid $\langle\sigma_0,\sigma_6,\sigma_9\rangle$ with exactly nine elements.

(x) The monoid $\mathbb{M}$ contains 4 nine-element semigroups, not monoids, with 2 isomorphism types. These are $\langle\sigma_2,\sigma_9\rangle \cong \langle\sigma_5,\sigma_6\rangle$ and $\langle\sigma_3,\sigma_8\rangle \cong \langle\sigma_4,\sigma_7\rangle$. The semigroup $\langle\sigma_2,\sigma_9\rangle$ does not contain a semigroup isomorphic to $\langle\sigma_3\rangle$, hence it is not isomorphic to $\langle\sigma_3,\sigma_8\rangle$. These semigroups extended by σ_0 yield 4 monoids, all of which consist of ten elements, with 2 isomorphism types.

(xi) The monoid $\mathbb{M}$ contains 6 ten-element semigroups, not monoids, with 4 isomorphism types. These are $\langle\sigma_1,\sigma_6\rangle$, $\langle\sigma_2,\sigma_3,\sigma_8\rangle \cong \langle\sigma_4,\sigma_5,\sigma_6\rangle$, $\langle\sigma_2,\sigma_4,\sigma_7\rangle \cong \langle\sigma_3,\sigma_5,\sigma_6\rangle$, and $\langle\sigma_2,\sigma_5,\sigma_6\rangle$. These semigroups (except $\langle\sigma_1,\sigma_6\rangle$) extended by σ_0 yield 5 monoids, all of which consist of ten elements, with 3 isomorphism types. We get 6 new isomorphism types, since the semigroups are distinguished by not isomorphic semigroups $\langle\sigma_2,\sigma_3\rangle$, $\langle\sigma_2,\sigma_4\rangle$, and $\langle\sigma_2,\sigma_5\rangle$.

(xii) The monoid $\mathbb{M}$ contains 2 isomorphic semigroups, not monoids, which consist of 11 elements, that is, $\langle\sigma_2,\sigma_3,\sigma_5\rangle \cong \langle\sigma_2,\sigma_4,\sigma_5\rangle$. These semigroups extended by σ_0 yield 2 isomorphic monoids, which consist of 12 elements. The monoid $\mathbb{M}$ contains no larger semigroup with the exception of itself.

10. Cancellation Rules Motivated by Some Topological Properties

10.1. Some Consequences of the Axiom $\mathbf{\emptyset} = \mathbf{\emptyset}^-$. So far, we used only the following relations (above named cancellation rules): $\sigma_2\circ\sigma_2 = \sigma_2$, $\sigma_1\circ\sigma_1 = \sigma_0$, $\sigma_2\circ\sigma_{12} = \sigma_6$, and $\sigma_2\circ\sigma_{13} = \sigma_7$. When one assumes $X \neq \emptyset = \emptyset^-$, then

$$\begin{aligned} X &= \sigma_0(X) = \sigma_2(X) = \sigma_5(X) = \sigma_7(X) \\ &= \sigma_8(X) = \sigma_{10}(X) = \sigma_{13}(X), \\ \emptyset &= \sigma_1(X) = \sigma_3(X) = \sigma_4(X) = \sigma_6(X) \\ &= \sigma_9(X) = \sigma_{11}(X) = \sigma_{12}(X). \end{aligned} \tag{18}$$

Using the substitution $A \mapsto A^c$, one obtains equivalent relations between operations from the set $\{\sigma_1,\sigma_3,\sigma_4,\sigma_6,\sigma_9,\sigma_{11},\sigma_{12}\}$, and conversely. Therefore, cancellation rules are topologically reasonable only between the operations from the monoid as follows:

$$\langle\sigma_0,\sigma_2,\sigma_5\rangle = \{\sigma_0,\sigma_2,\sigma_5,\sigma_7,\sigma_8,\sigma_{10},\sigma_{13}\}. \tag{19}$$

Chapman, see [3], considered properties of subsets with respect to such relations. Below, we are going to identify relations that are determined by some topological spaces; compare [10, 11].

10.2. The Relation $\boldsymbol{\sigma_0} = \boldsymbol{\sigma_2}$. If a topological space X is discrete, then there exist two Kuratowski operation, only. These are σ_0 and σ_1. So, the monoid of Kuratowski operations reduced to the group $\langle\sigma_1\rangle$.

The relation $\sigma_0 = \sigma_2$ is equivalent to any relation $\sigma_0 = \sigma_i$, where $i \in \{5,7,8,10,13\}$. Any such relation implies that every subset of X has to be closed and open; that is, X has to be discrete. However, one can check these using (only) the facts that σ_1 is an involution and σ_2 is an idempotent and the cancellation rules, that is, the Cayley table for $\mathbb{M}$. So, $\sigma_0 = \sigma_2$ follows

$$\sigma_0 = \sigma_2 = \sigma_5 = \sigma_7 = \sigma_8 = \sigma_{10} = \sigma_{13}. \tag{20}$$

10.3. The Relation $\boldsymbol{\sigma_2} = \boldsymbol{\sigma_5}$. Topologically, $\sigma_2 = \sigma_5$ means that X must be discrete. This is so because $A^{c-c} \subseteq A \subseteq A^-$ for any $A \subseteq X$.

10.4. The Relation $\boldsymbol{\sigma_2} = \boldsymbol{\sigma_7}$. Topologically, the relation $\sigma_2 = \sigma_7$ implies $\sigma_0 = \sigma_2$. But it requires the use of topology axioms $\emptyset = \emptyset^-$ and $C^- \cup B^- = (C \cup B)^-$ for each C and B.

Lemma 5. *For any topological space* $\sigma_2 = \sigma_7$ *implies* $\sigma_2 = \sigma_8$.

Proof. If $\sigma_2 = \sigma_7$, then $A \neq \emptyset \Rightarrow A^{c-c} \neq \emptyset$, for any $A \subseteq X$. Indeed, if $A^{c-c} = \emptyset$, then $\sigma_7(A) = \emptyset^- = \emptyset$. Since $A \neq \emptyset$, then $\sigma_2(A) \neq \emptyset$. Hence $\sigma_2(A) \neq \sigma_7(A)$, a contradiction.

The axiom $C^- \cup B^- = (C \cup B)^-$ implies that always

$$\left(A^- \cap A^{-c-}\right)^{c-c} = \emptyset. \tag{21}$$

Thus, the additional assumption $\sigma_2 = \sigma_7$ follows that always $A^- \cap A^{-c-} = \emptyset$. Therefore, always $A^{-c-} = A^{-c}$, but this means that any closed set has to be open; in other words, $\sigma_2 = \sigma_8$. □

Proposition 6. *For any topological space $\sigma_2 = \sigma_7$ implies $\sigma_0 = \sigma_2$.*

Proof. But the relation $\sigma_2 = \sigma_7$ is equivalent with $\sigma_5 = \sigma_8$. By Lemma 5, we get $\sigma_2 = \sigma_5$. Finally $\sigma_0 = \sigma_2$. □

The relation $\sigma_2 = \sigma_7$ has interpretation without the axiom $\emptyset = \emptyset^-$. Indeed, suppose $X = \{a, b\}$. Put

$$\sigma_2(\emptyset) = \{a\} = \sigma_2(\{a\}), \qquad X = \sigma_2(X) = \sigma_2(\{b\}). \quad (22)$$

Then, check that $\sigma_2 = \sigma_7$ and

$$\sigma_8(\emptyset) = \sigma_5(\{a\}) = \sigma_4(\{b\}) = \sigma_1(X) = \emptyset; \quad (23)$$

in other words, $\sigma_2 = \sigma_7$ and $\sigma_2 \neq \sigma_8$. However, $\sigma_2 = \sigma_7$ is equivalent to $\sigma_5 = \sigma_8$.

This relation implies $\sigma_2 = \sigma_7 = \sigma_{10}$, $\sigma_5 = \sigma_8 = \sigma_{13}$, $\sigma_3 = \sigma_6 = \sigma_{11}$, and $\sigma_4 = \sigma_9 = \sigma_{12}$. For this interpretation, the monoid $\mathbb{M}/(\sigma_2 = \sigma_7)$, consisting of Kuratowski operation over a such X, has six elements, only. In $\mathbb{M}/(\sigma_2 = \sigma_7)$, there are relations covered by the following proposition, only.

Proposition 7. *For any monoid with the Cayley table as for $\mathbb{M}$, the relation $\sigma_2 = \sigma_7$ implies*

(i) $\sigma_2 = \sigma_7 = \sigma_{10} = \sigma_7 \circ \sigma_2$;

(ii) $\sigma_5 = \sigma_1 \circ \sigma_2 \circ \sigma_1 = \sigma_1 \circ \sigma_7 \circ \sigma_1 = \sigma_8 = \sigma_5 \circ \sigma_2 = \sigma_5 \circ \sigma_7 = \sigma_{13}$;

(iii) $\sigma_3 = \sigma_2 \circ \sigma_1 = \sigma_7 \circ \sigma_1 = \sigma_6 = \sigma_{10} \circ \sigma_1 = \sigma_{11}$;

(iv) $\sigma_4 = \sigma_1 \circ \sigma_2 = \sigma_1 \circ \sigma_7 = \sigma_9 = \sigma_1 \circ \sigma_{10} = \sigma_{12}$.

Thus, the Cayley table does not contain the complete information resulting from the axioms of topology.

10.5. The Relation $\boldsymbol{\sigma_2 = \sigma_8}$. Topologically, the relation $\sigma_2 = \sigma_8$ means that any closed set is open, too. Thus, if $X = \{a, b, c\}$ is a topological space with the open sets X, $\emptyset$, $\{a, b\}$, and $\{c\}$, then $\mathbb{M}/(\sigma_2 = \sigma_8)$ is the monoid of all Kuratowski operations over X. Relations $\sigma_2 = \sigma_8$ and $\sigma_5 = \sigma_7$ are equivalent. They imply relations: $\sigma_2 = \sigma_8 = \sigma_{10}$, $\sigma_5 = \sigma_7 = \sigma_{13}$, $\sigma_3 = \sigma_9 = \sigma_{11}$, and $\sigma_4 = \sigma_6 = \sigma_{12}$. The permutation

$$\begin{pmatrix} \sigma_0 & \sigma_1 & \sigma_2 & \sigma_3 & \sigma_4 & \sigma_5 \\ \sigma_0 & \sigma_1 & \sigma_5 & \sigma_4 & \sigma_3 & \sigma_2 \end{pmatrix} \quad (24)$$

determines the isomorphism between monoids $\mathbb{M}/(\sigma_2 = \sigma_7)$ and $\mathbb{M}/(\sigma_2 = \sigma_8)$.

10.6. The Relations $\boldsymbol{\sigma_2 = \sigma_{10}}$ *and* $\boldsymbol{\sigma_2 = \sigma_{13}}$. Topologically, the relation $\sigma_2 = \sigma_8$ means that any nonempty closed set has nonempty interior. For each A, the closed set $A^- \cap A^{-c-}$ has empty interior, so the relation $\sigma_2 = \sigma_{10}$ implies that A^{-c} is closed. Hence, any open set has to be closed, so it implies $\sigma_2 = \sigma_8$. The relation $\sigma_2 = \sigma_{13}$ follows that each closed set has to be open, so it implies $\sigma_2 = \sigma_8$, too.

10.7. The Relation $\boldsymbol{\sigma_7 = \sigma_8}$. Using the Cayley table for $\mathbb{M}$, one can check that the relations $\sigma_7 = \sigma_8$ and $\sigma_{10} = \sigma_{13}$ are equivalent. Each of them gives $\sigma_7 = \sigma_8 = \sigma_{10} = \sigma_{13}$ and $\sigma_6 = \sigma_9 = \sigma_{11} = \sigma_{12}$. If $X = \{a, b\}$ is a topological space with the open sets X, $\emptyset$, and $\{a\}$, then $\mathbb{M}/(\sigma_7 = \sigma_8)$ is the monoid of all Kuratowski operations over X and consists of 8 elements.

10.8. The Relation $\boldsymbol{\sigma_7 = \sigma_{10}}$. Using the Cayley table for $\mathbb{M}$, one can check that the relations $\sigma_7 = \sigma_{10}$, $\sigma_8 = \sigma_{13}$, $\sigma_6 = \sigma_{11}$, and $\sigma_9 = \sigma_{12}$ are equivalent. If X is a sequence converging to the point g and $g \in X$, then $\mathbb{M}/(\sigma_7 = \sigma_{10})$ is the monoid of all Kuratowski operations over X and consists of 10 elements.

10.9. The Relation $\boldsymbol{\sigma_7 = \sigma_{13}}$. Using the Cayley table for $\mathbb{M}$, one can check that the relations $\sigma_7 = \sigma_{13}$ and $\sigma_8 = \sigma_{10}$ are equivalent. Also $\sigma_6 = \sigma_{12}$ and $\sigma_9 = \sigma_{11}$. These relations give the monoid with the following Cayley table, where the row and column marked by the identity are omitted:

	σ_1	σ_2	σ_3	σ_4	σ_5	σ_6	σ_7	σ_8	σ_9
σ_1	σ_0	σ_4	σ_5	σ_2	σ_3	σ_8	σ_9	σ_6	σ_7
σ_2	σ_3	σ_2	σ_3	σ_6	σ_7	σ_6	σ_7	σ_8	σ_9
σ_3	σ_2	σ_6	σ_7	σ_2	σ_3	σ_8	σ_9	σ_6	σ_7
σ_4	σ_5	σ_4	σ_5	σ_8	σ_9	σ_8	σ_9	σ_6	σ_7
σ_5	σ_4	σ_8	σ_9	σ_4	σ_5	σ_6	σ_7	σ_8	σ_9
σ_6	σ_7	σ_6	σ_7	σ_8	σ_9	σ_8	σ_9	σ_6	σ_7
σ_7	σ_6	σ_8	σ_9	σ_6	σ_7	σ_6	σ_7	σ_8	σ_9
σ_8	σ_9	σ_8	σ_9	σ_6	σ_7	σ_6	σ_7	σ_8	σ_9
σ_9	σ_8	σ_6	σ_7	σ_8	σ_9	σ_8	σ_9	σ_6	σ_7

(25)

If a space X is extremally disconnected, then the closures of open sets are open; compare [2, page 452]. It follows that $\sigma_6 = \sigma_{12}$. The space $X = \{a, b\}$ with the open sets X, $\emptyset$, and $\{a\}$ is extremally disconnected. But it contains a one-element open and dense set $\{a\}$, and it follows that $\sigma_7 = \sigma_8$. Similar is for the space βN; see [2, pages 228 and 453] to find the definition and properties of βN. There are Hausdorff extremally disconnected spaces which are dense in itself. For example, the Stone space of the complete Boolean algebra of all regular closed subsets of the unit interval, compare [12]. For such spaces $\sigma_7 \neq \sigma_8$ and $\sigma_7 = \sigma_{13}$. To see this, suppose a Hausdorff X is extremally disconnected and dense in itself. Let $X = U \cup V \cup W$, where sets U, V, and W are closed and open. Consider a set $A = A^{c-c} \cup B \cup C$, such that

(i) $C^{c-c} = \emptyset$ and $C^- = W$;

(ii) $\emptyset \neq B \subseteq V$ and $B^{-c-c} = \emptyset$;

(iii) $U = A^{c-c-} \neq A^{c-c}$.

Then check that

(i) $\sigma_0(A) = A$ and $\sigma_1(A) = X \setminus (A^{c-c} \cup B \cup C)$;

(ii) $\sigma_2(A) = U \cup B^- \cup W$ and $\sigma_3(A) = X \setminus A^{c-c}$;

(iii) $\sigma_4(A) = V \setminus B^-$ and $\sigma_5(A) = A^{c-c}$;

(iv) $\sigma_6(A) = \sigma_{12}(A) = V$ and $\sigma_7(A) = \sigma_{13}(A) = U$;

(v) $\sigma_8(A) = \sigma_{10}(A) = U \cup W$ and $\sigma_9(A) = \sigma_{11}(A) = V \cup W$.

Hence we have that $\sigma_7 \neq \sigma_8$. Note that, if $W = \emptyset$, then $\sigma_7(A) = \sigma_8(A)$. This is the case of subsets of βN.

References

[1] K. Kuratowski, "Sur l'opération $\overline{A}$ de l'analysis situs," *Fundamenta Mathematicae*, vol. 3, pp. 182–199, 1922.

[2] R. Engelking, *General Topology*, PWN-Polish Scientific Publishers, Warsaw, Poland, 1977.

[3] T. A. Chapman, "A further note on closure and interior operators," *The American Mathematical Monthly*, vol. 69, no. 6, pp. 524–529, 1962.

[4] D. Sherman, "Variations on Kuratowski's 14-set theorem," *American Mathematical Monthly*, vol. 117, no. 2, pp. 113–123, 2010.

[5] B. J. Gardner and M. Jackson, "The Kuratowski closure-complement theorem," *New Zealand Journal of Mathematics*, vol. 38, pp. 9–44, 2008.

[6] W. Koenen, "The Kuratowski closure problem in the topology of convexity," *The American Mathematical Monthly*, vol. 73, pp. 704–708, 1966.

[7] K.-P. Shum, "Closure functions on the set of positive integers," *Science in China. Series A*, vol. 39, no. 4, pp. 337–346, 1996.

[8] W. Sierpiński, *General Topology*, Mathematical Expositions, University of Toronto Press, Toronto, Canada, 1956.

[9] A. Cayley, "On the theory of groups," *American Journal of Mathematics*, vol. 11, no. 2, pp. 139–157, 1889.

[10] C. E. Aull, "Classification of topological spaces," *Bulletin de l'Académie Polonaise des Sciences. Série des Sciences Mathématiques, Astronomiques et Physiques*, vol. 15, pp. 773–778, 1967.

[11] C. E. Aull, "Corrigendumml: "Classification of topological spaces"," *Bulletin de l'Académie Polonaise des Sciences. Série des Sciences Mathématiques, Astronomiques et Physiques*, vol. 16, no. 6, 1968.

[12] A. M. Gleason, "Projective topological spaces," *Illinois Journal of Mathematics*, vol. 2, pp. 482–489, 1958.

Some Fixed Point Theorems for Prešić-Hardy-Rogers Type Contractions in Metric Spaces

Satish Shukla,[1] Stojan Radenović,[2] and Slaviša Pantelić[2]

[1] *Department of Applied Mathematics, Shri Vaishnav Institute of Technology and Science, Gram Baroli, Sanwer Road, Indore 453331, India*
[2] *Faculty of Mechanical Engineering, University of Belgrade, Kraljice Marije 16, 11120 Beograd, Serbia*

Correspondence should be addressed to Satish Shukla; satishmathematics@yahoo.co.in

Academic Editor: NanJing Huang

We introduce some generalizations of Prešić type contractions and establish some fixed point theorems for mappings satisfying Prešić-Hardy-Rogers type contractive conditions in metric spaces. Our results generalize and extend several known results in metric spaces. Some examples are included which illustrate the cases when new results can be applied while old ones cannot.

1. Introduction

The well-known Banach contraction mapping principle states that if (X, d) is a complete metric space and $T : X \to X$ is a self-mapping such that

$$d(Tx, Ty) \leq \lambda d(x, y) \quad (1)$$

for all $x, y \in X$, where $0 \leq \lambda < 1$, then there exists a unique $x \in X$ such that $Tx = x$. This point x is called the fixed point of mapping T.

On the other hand, for mappings $T : X \to X$, Kannan [1] introduced the contractive condition:

$$d(Tx, Ty) \leq \lambda [d(x, Tx) + d(y, Ty)], \quad (2)$$

for all $x, y \in X$, where $\lambda \in [0, 1/2)$ is a constant and proved a fixed point theorem using (2) instead of (1). The conditions (1) and (2) are independent, as it was shown by two examples in [2].

Reich [3], for mappings $T : X \to X$, generalized Banach and Kannan fixed point theorems, using contractive condition:

$$d(Tx, Ty) \leq \alpha d(x, y) + \beta d(x, Tx) + \gamma d(y, Ty), \quad (3)$$

for all $x, y \in X$, where α, β, γ are nonnegative constants with $\alpha + \beta + \gamma < 1$. An example in [3] shows that the condition (3) is a proper generalization of (1) and (2).

For mapping $T : X \to X$ Chatterjea [4] introduced the contractive condition:

$$d(Tx, Ty) \leq \lambda [d(x, Ty) + d(y, Tx)], \quad (4)$$

for all $x, y \in X$, where $\lambda \in [0, 1/2)$ is a constant and proved a fixed point result using (4).

Ćirić [5], for mappings $T : X \to X$, generalized all above mappings, using contractive condition:

$$d(Tx, Ty) \leq \alpha d(x, y) + \beta d(x, Tx) + \gamma d(y, Ty) + \delta [d(x, Ty) + d(y, Tx)], \quad (5)$$

for all $x, y \in X$, where $\alpha, \beta, \gamma, \delta$ are nonnegative constants with $\alpha + \beta + \gamma + 2\delta < 1$. A mapping satisfying (5) is called Generalized contraction.

Hardy and Rogers [6], for mappings $T : X \to X$, used the contractive condition:

$$d(Tx, Ty) \leq \alpha d(x, y) + \beta d(x, Tx) + \gamma d(y, Ty) + \delta d(x, Ty) + \mu d(y, Tx), \quad (6)$$

for all $x, y \in X$, where $\alpha, \beta, \gamma, \delta, \mu$ are nonnegative constants with $\alpha + \beta + \gamma + \delta + \mu < 1$ and proved fixed point result. Note that condition (6) generalizes all the previous conditions.

In 1965, Prešić [7, 8] extended Banach contraction mapping principle to mappings defined on product spaces and proved the following theorem.

Theorem 1. *Let (X,d) be a complete metric space, k a positive integer, and $f : X^k \to X$ a mapping satisfying the following contractive type condition:*

$$d\left(f\left(x_1,x_2,\ldots,x_k\right),f\left(x_2,x_3,\ldots,x_{k+1}\right)\right) \leq \sum_{i=1}^{k} q_i d\left(x_i,x_{i+1}\right), \tag{7}$$

for every $x_1,x_2,\ldots,x_{k+1} \in X$, where $q_1,q_2,\ldots,q_k$ are non-negative constants such that $q_1+q_2+\cdots+q_k<1$. Then there exists a unique point $x \in X$ such that $f(x,x,\ldots,x) = x$. Moreover if $x_1,x_2,\ldots,x_k$ are arbitrary points in X and for $n \in \mathbb{N}$,

$$x_{n+k} = f\left(x_n,x_{n+1},\ldots,x_{n+k-1}\right), \tag{8}$$

then the sequence $\{x_n\}$ is convergent and $\lim x_n = f(\lim x_n, \lim x_n,\ldots,\lim x_n)$.

Note that condition (7) in the case $k = 1$ reduces to the well-known Banach contraction mapping principle. So, Theorem 1 is a generalization of the Banach fixed point theorem. Some generalizations and applications of Prešić theorem can be seen in [9–18].

The k-step iterative sequence given by (8) represents a nonlinear difference equation and the solution of this equation can be assumed to be a fixed point of f; that is, solution of (8) is a point $x^* \in X$ such that $x^* = f(x^*,x^*,\ldots,x^*)$. The Prešić theorem insures the convergence of the sequence $\{x_n\}$ defined by (8) and provides a sufficient condition for the existence of solution of (8) in the case when mapping f satisfies the condition (7). A condition, independent from (7); namely, the Prešić-Kannan condition, is considered in [11] (for the proof of independency of these conditions in case $k = 1$, we refer [1, 2]). In this paper, we introduce some generalizations of Prešić type contractions in metric spaces and use a more general condition; namely, the Prešić-Hardy-Rogers type condition, to prove the existence of fixed point of f in metric spaces. We note that this condition generalizes the result of Prešić [7, 8], Păcurar [11], Hardy and Rogers [6], and several known results in metric spaces. Some examples are included which illustrate the cases when new results can be applied while old ones cannot.

2. Some Generalizations of Presić Type Contractions

In this section, we introduced some Prešić type contractions in metric spaces.

Let (X,d) be a metric space, k a positive integer, and $f : X^k \to X$ be a mapping.

(i) f is said to be a Prešić contraction if f satisfies the condition (7).

(ii) f is said to be a Prešić-Kannan contraction (see [11] for detail) if f satisfies following condition:

$$d\left(f\left(x_1,x_2,\ldots,x_k\right),f\left(x_2,x_3,\ldots,x_{k+1}\right)\right) \leq \beta\sum_{i=1}^{k+1} d\left(x_i,f\left(x_i,x_i,\ldots,x_i\right)\right) \tag{9}$$

for all $x_1,x_2,\ldots,x_k,x_{k+1} \in X$, where

$$0 \leq \beta k\left(k+1\right) < 1. \tag{10}$$

(iii) f is said to be a Prešić-Reich contraction if f satisfies following condition:

$$d\left(f\left(x_1,x_2,\ldots,x_k\right),f\left(x_2,x_3,\ldots,x_{k+1}\right)\right) \leq \sum_{i=1}^{k} \alpha_i d\left(x_i,x_{i+1}\right) + \sum_{i=1}^{k+1} \beta_i d\left(x_i,f\left(x_i,x_i,\ldots,x_i\right)\right) \tag{11}$$

for all $x_1,x_2,\ldots,x_k,x_{k+1} \in X$, where α_i,β_i are non-negative constants such that

$$\sum_{i=1}^{k}\alpha_i + k\sum_{i=1}^{k+1}\beta_i < 1. \tag{12}$$

(iv) f is said to be a Prešić-Chatterjea contraction if f satisfies following condition:

$$d\left(f\left(x_1,x_2,\ldots,x_k\right),f\left(x_2,x_3,\ldots,x_{k+1}\right)\right) \leq \gamma \sum_{i=1,i\neq j}^{k+1}\sum_{j=1}^{k+1} d\left(x_i,f\left(x_j,x_j,\ldots,x_j\right)\right) \tag{13}$$

for all $x_1,x_2,\ldots,x_k,x_{k+1} \in X$, where

$$0 \leq \gamma k^2\left(k+1\right) < 1. \tag{14}$$

(v) f is said to be a Generalized-Prešić contraction if f satisfies following condition:

$$d\left(f\left(x_1,x_2,\ldots,x_k\right),f\left(x_2,x_3,\ldots,x_{k+1}\right)\right) \leq \sum_{i=1}^{k} \alpha_i d\left(x_i,x_{i+1}\right) + \sum_{i=1}^{k+1} \beta_i d\left(x_i,f\left(x_i,x_i,\ldots,x_i\right)\right) + \beta \sum_{i=1,i\neq j}^{k+1}\sum_{j=1}^{k+1} d\left(x_i,f\left(x_j,x_j,\ldots,x_j\right)\right) \tag{15}$$

for all $x_1, x_2, \ldots, x_k, x_{k+1} \in X$, where α_i, β_i, β are nonnegative constants such that

$$\sum_{i=1}^{k} \alpha_i + k \sum_{i=1}^{k+1} \beta_i + \beta k^2 (k+1) < 1. \tag{16}$$

(vi) f is said to be a Prešić-Hardy-Rogers contraction if f satisfies following condition:

$$\begin{aligned} d(f(x_1, x_2, \ldots, x_k), f(x_2, x_3, \ldots, x_{k+1})) &\leq \sum_{i=1}^{k} \alpha_i d(x_i, x_{i+1}) \\ &+ \sum_{i=1}^{k+1} \sum_{j=1}^{k+1} \beta_{i,j} d(x_i, f(x_j, x_j, \ldots, x_j)) \end{aligned} \tag{17}$$

for all $x_1, x_2, \ldots, x_k, x_{k+1} \in X$, where $\alpha_i, \beta_{i,j}$ are nonnegative constants such that

$$\sum_{i=1}^{k} \alpha_i + k \sum_{i=1}^{k+1} \sum_{j=1}^{k+1} \beta_{i,j} < 1. \tag{18}$$

Remark 2. Note that for $\beta_{i,j} = \beta$ for all $i, j \in \{1, 2, \ldots, k, k+1\}$ with $i \neq j$ and $\beta_{i,i} = \beta_i$ for all $i \in \{1, 2, \ldots, k, k+1\}$, the Prešić-Hardy-Rogers contraction reduces into the Generalized-Prešić contraction. With $\beta = 0$, the Generalized-Prešić contraction reduces into the Prešić-Reich contraction and with $\alpha_i = 0$ for all $i \in \{1, 2, \ldots, k\}$, $\beta_i = 0$ for all $i \in \{1, 2, \ldots, k, k+1\}$, and $\beta = \gamma$, the Generalized-Prešić contraction reduces into the Prešić-Chatterjea contraction. With $\alpha_i = 0$ for all $i \in \{1, 2, \ldots, k\}$, the Prešić-Reich contraction reduces into the Prešić-Kannan contraction and with $\beta_i = 0$ for all $i \in \{1, 2, \ldots, k, k+1\}$, the Prešić-Reich contraction reduces into the Prešić contraction. Therefore among all above definitions, the Prešić-Hardy-Rogers contraction is the most general contraction.

Remark 3. It is easy to see that for $k = 1$, Prešić-Hardy-Rogers contraction reduces into Hardy-Rogers contraction and for $k = 1$, Generalized-Prešić contraction reduces into Generalized contraction and so forth; therefore, the comparison as considered in [19] shows that the above generalization is proper.

Now, we shall prove some fixed point results for Prešić-Hardy-Rogers type contractions in metric spaces.

3. Main Results

The following theorem is the fixed point result for Prešić-Hardy-Rogers type contractions and the main result of this paper.

Theorem 4. *Let (X, d) be any complete metric space, k a positive integer. Let $f : X^k \to X$ be a Prešić-Hardy-Rogers contraction, then f has a unique fixed point in X.*

Proof. Let $x_0 \in X$ be arbitrary. Define a sequence $\{x_n\}$ in X by

$$x_{n+1} = f(x_n, \ldots, x_n), \quad n \geq 0. \tag{19}$$

If $x_n = x_{n+1}$ for any n then x_n is a fixed point of f. Therefore we assume $x_n \neq x_{n+1}$ for all n.

We shall show that this sequence is a Cauchy sequence in X.

For simplicity, set

$$d_i = d(x_i, x_{i+1}), \qquad D_{i,j} = d(x_i, f(x_j, \ldots, x_j)) \quad \forall i, j \geq 1. \tag{20}$$

For any $n \geq 0$, we obtain

$$\begin{aligned} d_{n+1} &= d(x_{n+1}, x_{n+2}) \\ &= d(f(x_n, \ldots, x_n), f(x_{n+1}, \ldots, x_{n+1})) \\ &\leq d(f(x_n, \ldots, x_n), f(x_n, \ldots, x_n, x_{n+1})) \\ &\quad + d(f(x_n, \ldots, x_n, x_{n+1}), f(x_n, \ldots, x_n, x_{n+1}, x_{n+1})) \\ &\quad + \cdots + d(f(x_n, x_{n+1}, \ldots, x_{n+1}), f(x_{n+1}, \ldots, x_{n+1})), \end{aligned} \tag{21}$$

using (17), it follows from above inequality that

$$\begin{aligned} d_{n+1} \leq & \left\{ \alpha_k d_n + \left[\sum_{j=1}^{k} \beta_{1,j} + \sum_{j=1}^{k} \beta_{2,j} + \cdots + \sum_{j=1}^{k} \beta_{k,j} \right] D_{n,n} \right. \\ & + \left[\sum_{i=1}^{k} \beta_{i,k+1} \right] D_{n,n+1} + \left[\sum_{j=1}^{k} \beta_{k+1,j} \right] D_{n+1,n} \\ & \left. + \beta_{k+1,k+1} D_{n+1,n+1} \right\} \\ & + \left\{ \alpha_{k-1} d_n \right. \\ & + \left[\sum_{j=1}^{k-1} \beta_{1,j} + \sum_{j=1}^{k-1} \beta_{2,j} + \cdots + \sum_{j=1}^{k-1} \beta_{k-1,j} \right] D_{n,n} \\ & + \left[\sum_{i=1}^{k-1} \beta_{i,k} + \sum_{i=1}^{k-1} \beta_{i,k+1} \right] D_{n,n+1} \\ & + \left[\sum_{j=1}^{k-1} \beta_{k,j} + \sum_{j=1}^{k-1} \beta_{k+1,j} \right] D_{n+1,n} \\ & \left. + \left[\sum_{j=k}^{k+1} \beta_{k,j} + \sum_{j=k}^{k+1} \beta_{k+1,j} \right] D_{n+1,n+1} \right\} \\ & + \cdots \end{aligned}$$

$$
\begin{aligned}
&+\left\{\alpha_1 d_n+\beta_{1,1}D_{n,n}\right.\\
&\quad+\left[\sum_{j=2}^{k+1}\beta_{1,j}\right]D_{n,n+1}+\left[\sum_{i=2}^{k+1}\beta_{i,1}\right]D_{n+1,n}\\
&\quad\left.+\left[\sum_{j=2}^{k+1}\beta_{2,j}+\sum_{j=2}^{k+1}\beta_{3,j}+\cdots+\sum_{j=2}^{k+1}\beta_{k+1,j}\right]D_{n+1,n+1}\right\},
\end{aligned}\tag{22}
$$

that is,

$$
\begin{aligned}
d_{n+1}\le{}&\left[\sum_{i=1}^{k}\alpha_i\right]d_n\\
&+\left\{\left[\sum_{i=1}^{k}\sum_{j=1}^{k}\beta_{i,j}\right]D_{n,n}+\left[\sum_{i=1}^{k}\beta_{i,k+1}\right]D_{r,n+1}\right.\\
&\quad\left.+\left[\sum_{j=1}^{k}\beta_{k+1,j}\right]D_{n+1,n}+\beta_{k+1,k+1}D_{n+1,n+1}\right\}\\
&+\left\{\left[\sum_{i=1}^{k-1}\sum_{j=1}^{k-1}\beta_{i,j}\right]D_{n,n}+\left[\sum_{i=1}^{k-1}\sum_{j=k}^{k+1}\beta_{i,j}\right]D_{n,n+1}\right.\\
&\quad\left.+\left[\sum_{i=k}^{k+1}\sum_{j=1}^{k-1}\beta_{i,j}\right]D_{n+1,n}+\left[\sum_{i=k}^{k+1}\sum_{j=k}^{k+1}\beta_{i,j}\right]D_{n+1,n+1}\right\}\\
&+\cdots\\
&+\left\{\beta_{1,1}D_{n,n}+\left[\sum_{j=2}^{k+1}\beta_{1,j}\right]D_{n,n+1}\right.\\
&\quad\left.+\left[\sum_{i=2}^{k+1}\beta_{i,1}\right]D_{n+1,n}+\left[\sum_{i=2}^{k+1}\sum_{j=2}^{k+1}\beta_{i,j}\right]D_{n+1,n+1}\right\},
\end{aligned}\tag{23}
$$

that is,

$$
\begin{aligned}
&d_{n+1}\\
&\le\left[\sum_{i=1}^{k}\alpha_i\right]d_n\\
&+\left[\sum_{i=1}^{k}\sum_{j=1}^{k}\beta_{i,j}+\sum_{i=1}^{k-1}\sum_{j=1}^{k-1}\beta_{i,j}+\cdots+\sum_{i=1}^{2}\sum_{j=1}^{2}\beta_{i,j}+\beta_{1,1}\right]D_{n,n}\\
&+\left[\sum_{i=1}^{k}\beta_{i,k+1}+\sum_{i=1}^{k-1}\sum_{j=k}^{k+1}\beta_{i,j}+\cdots+\sum_{i=1}^{2}\sum_{j=3}^{k+1}\beta_{i,j}+\sum_{j=2}^{k+1}\beta_{1,j}\right]D_{n,n+1}\\
&+\left[\sum_{j=1}^{k}\beta_{k+1,j}+\sum_{i=k}^{k+1}\sum_{j=1}^{k-1}\beta_{i,j}+\cdots+\sum_{i=3}^{k+1}\sum_{j=1}^{2}\beta_{i,j}+\sum_{i=2}^{k+1}\beta_{i,1}\right]D_{n+1,n}\\
&+\left[\sum_{i=2}^{k+1}\sum_{j=2}^{k+1}\beta_{i,j}+\sum_{i=3}^{k+1}\sum_{j=3}^{k+1}\beta_{i,j}+\cdots+\sum_{i=k}^{k+1}\sum_{j=k}^{k+1}\beta_{i,j}+\beta_{k+1,k+1}\right]D_{n+1,n+1}\\
&=Ad_n+BD_{n,n}+CD_{n,n+1}+ED_{n+1,n}+FD_{n+1,n+1},
\end{aligned}\tag{24}
$$

where A, B, C, E, and F are the coefficients of $d_n, D_{n,n}, D_{n,n+1}, D_{n+1,n}$, and $D_{n+1,n+1}$, respectively, in the above inequality.

By definition, $D_{n,n}=d(x_n,f(x_n,\ldots,x_n))=d(x_n,x_{n+1})=d_n$, $D_{n,n+1}=d(x_n,f(x_{n+1},\ldots,x_{n+1}))=d(x_n,x_{n+2})$, $D_{n+1,n}=d(x_{n+1},f(x_n,\ldots,x_n))=d(x_{n+1},x_{n+1})=0$, $D_{n+1,n+1}=d(x_{n+1},f(x_{n+1},\ldots,x_{n+1}))=d(x_{n+1},x_{n+2})=d_{n+1}$, therefore

$$
\begin{aligned}
d_{n+1}&\le Ad_n+Bd_n+Cd\left(x_n,x_{n+2}\right)+Fd_{n+1}\\
&\le Ad_n+Bd_n+Cd\left(x_n,x_{n+1}\right)+Cd\left(x_{n+1},x_{n+2}\right)+Fd_{n+1}\\
&=\left(A+B+C\right)d_n+\left(C+F\right)d_{n+1},
\end{aligned}\tag{25}
$$

that is,

$$
\left(1-C-F\right)d_{n+1}\le\left(A+B+C\right)d_n.\tag{26}
$$

Again, as $d_{n+1}=d(x_n,x_{n+1})=d(x_{n+1},x_n)$, interchanging the role of x_n and x_{n+1}, and repeating above process, we obtain

$$
\left(1-E-B\right)d_{n+1}\le\left(A+F+E\right)d_n.\tag{27}
$$

It follows from (26) and (27) that

$$
\begin{aligned}
\left(2-B-C-E-F\right)d_{n+1}&\le\left(2A+B+C+E+F\right)d_n,\\
d_{n+1}&\le\frac{2A+B+C+E+F}{2-B-C-E-F}d_n,\\
d_{n+1}&\le\lambda d_n,
\end{aligned}\tag{28}
$$

where $\lambda=(2A+B+C+E+F)/(2-B-C-E-F)$.

Using (18), we obtain

$$
\begin{aligned}
&A+B+C+E+F\\
&\quad=\sum_{i=1}^{k}\alpha_i+\sum_{i=1}^{k}\sum_{j=1}^{k}\beta_{i,j}+\sum_{i=1}^{k-1}\sum_{j=1}^{k-1}\beta_{i,j}+\cdots+\sum_{i=1}^{2}\sum_{j=1}^{2}\beta_{i,j}+\beta_{1,1}\\
&\qquad+\sum_{i=1}^{k}\beta_{i,k+1}+\sum_{i=1}^{k-1}\sum_{j=k}^{k+1}\beta_{i,j}+\cdots+\sum_{i=1}^{2}\sum_{j=3}^{k+1}\beta_{i,j}+\sum_{j=2}^{k+1}\beta_{1,j}\\
&\qquad+\sum_{j=1}^{k}\beta_{k+1,j}+\sum_{i=k}^{k+1}\sum_{j=1}^{k-1}\beta_{i,j}+\cdots+\sum_{i=3}^{k+1}\sum_{j=1}^{2}\beta_{i,j}+\sum_{i=2}^{k+1}\beta_{i,1}\\
&\qquad+\sum_{i=2}^{k+1}\sum_{j=2}^{k+1}\beta_{i,j}+\sum_{i=3}^{k+1}\sum_{j=3}^{k+1}\beta_{i,j}+\cdots+\sum_{i=k}^{k+1}\sum_{j=k}^{k+1}\beta_{i,j}+\beta_{k+1,k+1}\\
&\quad=\sum_{i=1}^{k}\alpha_i+k\sum_{i=1}^{k+1}\sum_{j=1}^{k+1}\beta_{i,j}\\
&\quad<1.
\end{aligned}\tag{29}
$$

So $0 \le \lambda < 1$. By (28), we obtain

$$d_{n+1} \le \lambda^{n+1} d_0 \quad \forall n \ge 0. \tag{30}$$

Suppose $n, m \in \mathbb{N}$ with $m > n$. Then

$$\begin{aligned} d(x_n, x_m) & \\ &\le d(x_n, x_{n+1}) + d(x_{n+1}, x_{n+2}) + \cdots + d(x_{m-1}, x_m) \\ &= d_n + d_{n+1} + \cdots + d_{m-1} \\ &\le \lambda^n d_0 + \lambda^{n+1} d_0 + \cdots + \lambda^{m-1} d_0 \\ &\le \frac{\lambda^n}{1-\lambda} d_0, \end{aligned} \tag{31}$$

as $0 \le \lambda < 1$, it follows from the above inequality that $\lim_{n\to\infty} d(x_n, x_m) = 0$. Therefore $\{x_n\}$ is a Cauchy sequence. By completeness of X, there exists $u \in X$ such that $\lim_{n\to\infty} x_n = u$.

We shall show that u is the fixed point of f. Note that

$$\begin{aligned} d(u, f(u, \ldots, u)) & \\ &\le d(u, x_{n+1}) + d(x_{n+1}, f(u, \ldots, u)) \\ &= d(u, x_{n+1}) + d(f(x_n, \ldots, x_n), f(u, \ldots, u)), \end{aligned} \tag{32}$$

using a similar process as used in the calculation of d_{n+1}, we obtain

$$\begin{aligned} d(u, f(u, \ldots, u)) & \\ &\le d(u, x_{n+1}) + Ad(x_n, u) + Bd(x_n, f(x_n, \ldots, x_n)) \\ &\quad + Cd(x_n, f(u, \ldots, u)) + Ed(u, f(x_n, \ldots, x_n)) \\ &\quad + Fd(u, f(u, \ldots, u)) \\ &\le d(u, x_{n+1}) + Ad(x_n, u) + Bd(x_n, x_{n+1}) + Cd(x_n, u) \\ &\quad + Cd(u, f(u, \ldots, u)) + Ed(u, x_{n+1}) \\ &\quad + Fd(u, f(u, \ldots, u)), \end{aligned} \tag{33}$$

that is,

$$\begin{aligned} d(u, f(u, \ldots, u)) & \\ &\le \frac{A+B+C}{1-C-F} d(x_n, u) + \frac{1+B+E}{1-C-F} d(x_{n+1}, u). \end{aligned} \tag{34}$$

Using the fact that $\lim_{n\to\infty} x_n = u$, it follows from the above inequality that

$$d(u, f(u, \ldots, u)) = 0 \quad \text{that is,} \quad f(u, \ldots, u) = u. \tag{35}$$

Thus u is a fixed point of f. For uniqueness, let v be another fixed point of f, that is, $f(v, \ldots, v) = v$. Again using a similar process as used in the calculation of d_{n+1}, we obtain

$$\begin{aligned} d(u, v) &\le Ad(u, v) + Bd(u, f(u, \ldots, u)) \\ &\quad + Cd(u, f(v, \ldots, v)) + Ed(v, f(u, \ldots, u)) \\ &\quad + Fd(v, f(v, \ldots, v)) \\ &= (A + C + E) d(u, v), \end{aligned} \tag{36}$$

as $A + B + C + E + F < 1$, we obtain $d(u, v) = 0$, that is, $u = v$. Thus fixed point is unique. □

Remark 5. For $k = 1$ in the above theorem, we obtain the result of Hardy and Rogers [6]. For $\beta_{i,j} = 0$ for all $i, j \in \{1, 2, \ldots, k, k+1\}$, we obtain the fixed point result of Prešić. Therefore, above theorem is a generalization of the results of Hardy and Rogers and Prešić.

With Remark 2, the following corollaries are obtained.

Corollary 6. *Let (X, d) be any complete metric space, k a positive integer, and $f : X^k \to X$ a Generalized Prešić contraction. Then f has a unique fixed point in X.*

For $k = 1$ in above corollary, we obtain the fixed point result of Ćirić [5].

Corollary 7. *Let (X, d) be any complete metric space, k a positive integer, and $f : X^k \to X$ a Prešić-Reich contraction. Then f has a unique fixed point in X.*

For $k = 1$ in the above corollary, we obtain the fixed point result of Reich [3].

Corollary 8. *Let (X, d) be any complete metric space, k a positive integer, and $f : X^k \to X$ a Prešić-Kannan contraction. Then f has a unique fixed point in X.*

For $k = 1$ in above the corollary, we obtain the fixed point result of Kannan [2].

Corollary 9. *Let (X, d) be any complete metric space, k a positive integer, and $f : X^k \to X$ a Prešić-Chatterjea contraction. Then f has a unique fixed point in X.*

For $k = 1$ in above corollary, we obtain the fixed point result of Chatterjea [4].

The following are some examples which illustrate the cases when known results are not applicable, while our new results can be used to conclude the existence of fixed point of mapping.

Example 10. Let $X = [0,1]$ with usual metric. For $k = 2$ define $f : X^2 \to X$ by

$$f(x,y) = \begin{cases} \frac{1}{5}, & \text{if } x = y = 1; \\ \frac{x+y}{5}, & \text{otherwise.} \end{cases} \tag{37}$$

Then

(i) f is a Prešić-Reich contraction with $\alpha_1 = \alpha_2 = 1/5$, $\beta_1 = \beta_2 = \beta_3 = 1/11$;

(ii) f is not a Prešić contraction;

(iii) f is not a Prešić-Kannan contraction.

Proof. (i) Note that for $x_1, x_2, x_3 \in [0,1)$ with $x_1 \leq x_2 \leq x_3$,

$$\begin{aligned} d(f(x_1,x_2), f(x_2,x_3)) &= d\left(\frac{x_1+x_2}{5}, \frac{x_2+x_3}{5}\right) = \frac{x_3-x_1}{5} \\ &= \frac{1}{5}[(x_2-x_1)+(x_3-x_2)] \\ &= \frac{1}{5}[d(x_1,x_2)+d(x_2,x_3)] \\ &= \frac{1}{5}\sum_{i=1}^{2} d(x_i,x_{i+1}). \end{aligned} \tag{38}$$

Therefore conditions (11) and (12) are satisfied for $\alpha_1 = \alpha_2 = 1/5$ and $\beta_1, \beta_2, \beta_3$ with $\beta_1 + \beta_2 + \beta_3 \in [0, 3/10)$.

If any one of x_1, x_2, x_3 is 1 then proof is similar. If any two of x_1, x_2, x_3 are 1, for example, if $x_1 \in [0,1)$ and $x_2 = x_3 = 1$, then

$$\begin{aligned} d(f(x_1,x_2), f(x_2,x_3)) &= d(f(x_1,1), f(1,1)) = d\left(\frac{x_1+1}{5}, \frac{1}{5}\right) \\ &= \frac{x_1+1}{5} - \frac{1}{5} = \frac{x_1}{5} \\ \frac{1}{11}\sum_{i=1}^{3} d(x_i, f(x_i,x_i)) &= \frac{1}{11}[d(x_1, f(x_1,x_1)) + d(x_2, f(x_2,x_2)) \\ &\quad + d(x_3, f(x_3,x_3))] \\ &= \frac{1}{11}\left[\frac{3x_1}{5} + \frac{4}{5} + \frac{4}{5}\right] = \frac{1}{55}[3x_1+8]. \end{aligned} \tag{39}$$

As $x_1 \in [0,1)$, so conditions (11) and (12) are satisfied for $\beta_1 = \beta_2 = \beta_3 = 1/11$ and α_1, α_2 with $\alpha_1 + \alpha_2 \in [0, 5/11)$.

Similarly in all possible cases conditions (11) and (12) are satisfied with $\alpha_1 = \alpha_2 = 1/5$, $\beta_1 = \beta_2 = \beta_3 = 1/11$. Therefore f is a Prešić-Reich contraction. All other conditions of Corollary 7 are satisfied and 0 is the unique fixed point of f.

(ii) Note that for $x_1 = 9/10$ and $x_2 = x_3 = 1$

$$\begin{aligned} d(f(x_1,x_2), f(x_2,x_3)) &= d\left(\frac{19}{50}, \frac{1}{5}\right) = \frac{9}{50} \\ \sum_{i=1}^{2} \alpha_i d(x_i,x_{i+1}) &= \alpha_1 d(x_1,x_2) + \alpha_2 d(x_2,x_3) \\ &= \alpha_1 d\left(\frac{9}{10}, 1\right) + \alpha_2 d(1,1) = \frac{1}{10}\alpha_1. \end{aligned} \tag{40}$$

Therefore, we cannot find nonnegative constants α_1, α_2 such that condition (7) is satisfied with $\alpha_1 + \alpha_2 < 1$. So f is not a Prešić contraction.

(iii) Again for $x_1 = x_2 = 0$, $x_3 = 1$

$$\begin{aligned} d(f(x_1,x_2), f(x_2,x_3)) &= d\left(0, \frac{1}{5}\right) = \frac{1}{5} \\ \beta\sum_{i=1}^{3} d(x_i, f(x_i,x_i)) &= \beta[d(x_1, f(x_1,x_1)) + d(x_2, f(x_2,x_2)) \\ &\quad + d(x_3, f(x_3,x_3))] \\ &= \beta\left[d(0,0) + d(0,0) + d\left(1, \frac{1}{5}\right)\right] = \frac{4}{5}\beta. \end{aligned} \tag{41}$$

Therefore, we cannot find nonnegative constant β such that conditions (9) and (10) are satisfied. So f is not a Prešić-Kannan contraction. □

Remark 11. In the above example, we cannot apply the result of Prešić [7, 8] and Păcurar [11] to conclude the existence of fixed point of f. But Corollary 7 is applicable which insures the existence of unique fixed point of f.

Example 12. Let $X = [0,1]$ with usual metric. For $k = 2$, define $f : X^2 \to X$ by

$$f(x,y) = \begin{cases} \frac{4}{15}, & \text{if } x = y = 1; \\ 0, & \text{otherwise.} \end{cases} \tag{42}$$

Then

(i) f is a Prešić-Chatterjea contraction with $\gamma \in [1/13, 1/12)$;

(ii) f is not a Prešić contraction;

(iii) f is not a Prešić-Kannan contraction.

Proof. (i) Note that if $x_1, x_2, x_3 \in [0,1)$ or any one of x_1, x_2, x_3 is 1, then conditions (17) and (18) are satisfied trivially.

If any two of x_1, x_2, x_3 are 1, for example, if $x_1 \in [0,1)$, $x_2 = x_3 = 1$, then

$$\begin{aligned} &d(f(x_1,x_2), f(x_2,x_3)) \\ &\quad = d(f(x_1,1), f(1,1)) \\ &\quad = d\left(0,\frac{4}{15}\right) = \frac{4}{15} \\ &\gamma \sum_{i=1,i\neq j}^{3} \sum_{j=1}^{3} d(x_i, f(x_j,x_j)) \\ &\quad = \gamma\left[d\left(x_1,\frac{4}{15}\right) + d\left(x_1,\frac{4}{15}\right)\right. \\ &\qquad + d(1,0) + d\left(1,\frac{4}{15}\right) \\ &\qquad \left. + d(1,0) + d\left(1,\frac{4}{15}\right)\right] \\ &\quad = \gamma\left[2\left|x_1 - \frac{4}{15}\right| + 2 + \frac{22}{15}\right] \\ &\quad = \gamma\left[2\left|x_1 - \frac{4}{15}\right| + \frac{52}{15}\right] \\ &\quad \leq \gamma\frac{52}{15}. \end{aligned} \tag{43}$$

Therefore conditions (13) and (14) are satisfied with $\gamma \in [1/13, 1/12)$. Also all other conditions of Corollary 9 are satisfied and f has a unique fixed point 0.

(ii) For $x_1 = 9/10$, $x_2 = 1$, $x_2 = 1$, we have

$$\begin{aligned} &d(f(x_1,x_2), f(x_2,x_3)) \\ &\quad = d\left(f\left(\frac{9}{10},1\right), f(1,1)\right) = d\left(0,\frac{4}{15}\right) = \frac{4}{15} \\ &\sum_{i=1}^{2}\alpha_i d(x_i,x_{i+1}) \\ &\quad = \alpha_1 d(x_1,x_2) + \alpha_2 d(x_2,x_3) = \alpha_1\frac{1}{10}. \end{aligned} \tag{44}$$

Therefore we cannot find nonnegative constants α_1, α_2 such that condition (7) is satisfied with $\alpha_1 + \alpha_2 < 1$. So f is not a Prešić contraction.

(iii) For $x_1 = 0$, $x_2 = x_3 = 1$, we have

$$\begin{aligned} &d(f(x_1,x_2), f(x_2,x_3)) \\ &\quad = d(f(0,1), f(1,1)) = d\left(0,\frac{4}{15}\right) = \frac{4}{15} \\ &\beta\sum_{i=1}^{3} d(x_i, f(x_i,x_i)) \\ &\quad = \beta\left[d(x_1, f(x_1,x_1)) + d(x_2, f(x_2,x_2))\right. \\ &\qquad \left. + d(x_3, f(x_3,x_3))\right] = \beta\frac{22}{15}. \end{aligned} \tag{45}$$

Therefore we cannot find nonnegative constant β such that conditions (9) and (10) are satisfied. So f is not a Prešić-Kannan contraction. □

Remark 13. In the above example, we cannot apply the result of Prešić [7, 8] and Păcurar [11] to conclude the existence of fixed point of f. But Corollary 9 is applicable which insures the existence of unique fixed point of f.

The following theorem is a consequence of Theorem 4 and the recent result of Aydi et al. [20].

Theorem 14. *Let (X, d) be any complete metric space and k a positive integer. Let $f : X^k \to X$ and $T : X \to X$ be two mappings such that the following condition holds:*

$$\begin{aligned} &d(Tf(x_1,x_2,\ldots,x_k), Tf(x_2,x_3,\ldots,x_{k+1})) \\ &\quad \leq \sum_{i=1}^{k}\alpha_i d(Tx_i, Tx_{i+1}) \\ &\qquad + \sum_{i=1}^{k+1}\sum_{j=1}^{k+1}\beta_{i,j} d(Tx_i, Tf(x_j,x_j,\ldots,x_j)) \end{aligned} \tag{46}$$

for all $x_1, x_2, \ldots, x_k, x_{k+1} \in X$, where $\alpha_i, \beta_{i,j}$ are nonnegative constants such that

$$\sum_{i=1}^{k}\alpha_i + k\sum_{i=1}^{k+1}\sum_{j=1}^{k+1}\beta_{i,j} < 1 \tag{47}$$

and T is continuous, injective, and sequentially convergent. Then f has a unique fixed point in X.

Proof. Define a mapping $\rho : X \times X \to [0,\infty)$ by

$$\rho(x,y) = d(Tx,Ty) \quad \forall x,y \in X. \tag{48}$$

Then (X, ρ) is a complete metric space (see [20]). Note that condition (46) reduces to the condition (17); that is, mapping f reduces to Prešić-Hardy-Rogers contraction with respect to metric ρ. So the rest of the proof followed Theorem 4. □

Acknowledgment

The first author is thankful to Mr. Rajpal Tomar for his help in typing the manuscript.

References

[1] R. Kannan, "Some results on fixed points," *Bulletin of the Calcutta Mathematical Society*, vol. 60, pp. 71–76, 1968.

[2] R. Kannan, "Some results on fixed points. II," *The American Mathematical Monthly*, vol. 76, pp. 405–408, 1969.

[3] S. Reich, "Some remarks concerning contraction mappings," *Canadian Mathematical Bulletin*, vol. 14, pp. 121–124, 1971.

[4] S. K. Chatterjea, "Fixed-point theorems," *Comptes Rendus de l'Académie Bulgare des Sciences*, vol. 25, pp. 727–730, 1972.

[5] L. B. Ćirić, "Generalized contractions and fixed-point theorems," *Publications de l'Institut Mathématique*, vol. 12, no. 26, pp. 19–26, 1971.

[6] G. E. Hardy and T. D. Rogers, "A generalization of a fixed point theorem of Reich," *Canadian Mathematical Bulletin*, vol. 16, pp. 201–206, 1973.

[7] S. B. Prešić, "Sur une classe d'inéquations aux différences finies et sur la convergence de certaines suites," *Publications de l'Institut Mathématique*, vol. 5, no. 19, pp. 75–78, 1965.

[8] S. B. Prešić, "Sur la convergence des suites," *Comptes Rendus de l'Académie des Sciences*, vol. 260, pp. 3828–3830, 1965.

[9] L. B. Ćirić and S. B. Prešić, "On Prešić type generalization of the Banach contraction mapping principle," *Acta Mathematica Universitatis Comenianae*, vol. 76, no. 2, pp. 143–147, 2007.

[10] M. Pãcurar, "A multi-step iterative method for approximating common fixed points of Prešić-Rus type operators on metric spaces," *Studia Universitatis Babes-Bolyai, Mathematica*, vol. 55, no. 1, p. 149, 2010.

[11] M. Păcurar, "Approximating common fixed points of Presić-Kannan type operators by a multi-step iterative method," *Analele Stiintifice ale Universitatii Ovidius Constanta, Seria Matematica*, vol. 17, no. 1, pp. 153–168, 2009.

[12] M. Păcurar, "Common fixed points for almost Presić type operators," *Carpathian Journal of Mathematics*, vol. 28, no. 1, pp. 117–126, 2012.

[13] M. S. Khan, M. Berzig, and B. Samet, "Some convergence results for iterative sequences of Prešić type and applications," *Advances in Difference Equations*, vol. 2012, article 38, 2012.

[14] R. George, K. P. Reshma, and R. Rajagopalan, "A generalised fixed point theorem of Presic type in cone metric spaces and application to Markov process," *Fixed Point Theory and Applications*, vol. 2011, article 85, 2011.

[15] S. Shukla, R. Sen, and S. Radenović, "Set-valued Prešić type contraction in metric spaces," *Analele Stiintifice ale Universitatii*. In press.

[16] S. Shukla and R. Sen, "Set-valued Prešić-Reich type mappings in metric spaces," *Revista de la Real Academia de Ciencias Exactas, Fisicas y Naturales A*, 2012.

[17] S. K. Malhotra, S. Shukla, and R. Sen, "A generalization of Banach contraction principle in ordered cone metric spaces," *Journal of Advanced Mathematical Studies*, vol. 5, no. 2, pp. 59–67, 2012.

[18] Y.-Z. Chen, "A Prešić type contractive condition and its applications," *Nonlinear Analysis*, vol. 71, no. 12, pp. e2012–e2017, 2009.

[19] B. E. Rhoades, "A comparison of various definitions of contractive mappings," *Transactions of the American Mathematical Society*, vol. 226, pp. 257–290, 1977.

[20] H. Aydi, E. Karapınar, and B. Samet, "Remarks on some recent fixed point theorems," *Fixed Point Theory and Applications*, vol. 2012, article 76, 2012.

Eigenvalue for Densely Defined $(S_+)_L$ Perturbations of Multivalued Maximal Monotone Operators in Reflexive Banach Spaces

Boubakari Ibrahimou

Department of Public Health, Western Kentucky University, 1906 College Heights Boulevard, Bowling Green, KY 42104, USA

Correspondence should be addressed to Boubakari Ibrahimou; boubakari.ibrahimou@wku.edu

Academic Editor: Jen-Chih Yao

Let X be a real reflexive Banach space and let X^* be its dual. Let $\Omega \subset X$ be open and bounded such that $0 \in \Omega$. Let $T : X \supset D(T) \to 2^{X^*}$ be maximal monotone with $0 \in D(T)$ and $0 \in T(0)$. Using the topological degree theory developed by Kartsatos and Quarcoo we study the eigenvalue problem $Tx + \lambda Px \ni 0$, where the operator $P : X \supset D(P) \to X^*$ is a single-valued of class $(S_+)_L$. The existence of continuous branches of eigenvectors of infinite length then could be easily extended to the case where the operator $P : X \to 2^{X^*}$ is multivalued and is investigated.

1. Preliminaries

In what follows we assume that X is a real or complex Banach space and has been renormed such that it and its dual X^* are locally uniformly convex. The normalized duality mapping is defined by

$$J(x) = \left\{x^* \in X : \|x^*\| = \|x\|, \langle x, x^* \rangle = \|x\|^2\right\}. \quad (1)$$

The mapping $T : X \supset D(T) \to 2^{X^*}$ is said to be "monotone" if for every $x, y \in D(T)$, $u \in Tx$, and $v \in Ty$ we have

$$\langle u - v, x - y \rangle \geq 0. \quad (2)$$

A monotone operator T is "maximal monotone" if $G(T)$ is maximal in $X \times X^*$, when $X \times X^*$ is partially ordered by inclusion. In our setting, a monotone operator T is maximal if and only if $R(T + \lambda J) = X^*$ for every $\lambda > 0$. If $T : X \supset D(T) \to 2^{X^*}$ is maximal monotone operator, the operator $T_t \equiv (T^{-1} + tJ^{-1})^{-1} : X \to X^*$ is called the Yosida approximant of T and the following, which can be found in [1, page 102] is true.

Lemma 1. *Let $T : X \supset D(T) \to 2^{X^*}$ be a maximal monotone operator with $0 \in D(T)$ and $0 \in T(0)$. Then*

(i) *T_t is a bounded maximal monotone mapping with $0 = T_t(0)$ for each $t > 0$;*

(ii) *$T_t x \rightharpoonup T^{\{0\}}x$ in X as $t \to 0$ for all $x \in D(T)$, where $T^{\{0\}}x$ denotes the element $y^* \in Tx$ of minimum norm;*

(iii) *$\|T_t x\| \to \infty$ as $t \to 0$ for all $x \notin \overline{D(T)}$.*

Also $T_t x \in TJ_t x$, where $J_t = I - tJ^{-1}T_t : X \to X$ and satisfies: if $x \in \overline{\text{conv}\, D(T)}$, then $J_t x \to x$ in X as $t \to 0$, where $\text{conv}\, M$ denote the convex hull of the set M.

The proof of the next lemma can be found in Kartsatos and Skrypnik [2],

Lemma 2. *Let $T : X \supset D(T) \to 2^{X^*}$ be maximal monotone such that $0 \in D(T)$ and $0 \in T(0)$. Then the mapping $(t, x) \to T_t x$ is continuous on the set $(0, \infty) \times X$.*

An operator $T : X \supset D(T) \to Y$, with Y another Banach space, is bounded if it maps bounded subsets of $D(T)$ onto bounded sets. It is compact if it is continuous and maps bounded subsets of $D(T)$ onto relatively compact subsets of Y. It is demicontinuous if for every sequence $\{x_n\}$ such that $x_n \to x_0$ we have $Tx_n \rightharpoonup Tx_0$.

We say that the operator $T : X \supset D(T) \to 2^{X^*}$ satisfies condition (S_q) on a set $A \subset D(T)$ if for every sequence $\{x_n\} \subset A$ such that $x_n \rightharpoonup x_0 \in X$ and any $y_n^* \in Tx_n$, with $y_n^* \to$ (some)$y^* \in X^*$, we have $x_n \to x_0$. If $A = D(T)$, then we say that T satisfies (S_q).

Definition 3. Let X be a separable reflexive Banach space and let L be a dense subspace of X. A mapping $P : D(P) \subset X \to X^*$ is said to be of class $(S_+)_L$ if, for any sequence of finite dimensional subspaces F_n of L with $\overline{U_{n=1}^{\infty} F_n} = X$, $h \in X^*$, $\{x_n\}_{n=1}^{\infty} \subset D(P)$ with $x_n \rightharpoonup x_0$ and

$$\limsup_{n\to\infty} \langle Px_n - h, x_n\rangle \le 0, \qquad \lim_{n\to\infty} \langle Px_n - h, v\rangle = 0 \tag{3}$$

for all $v \in U_{n=1}^{\infty} F_n$, we have $x_n \to x_0$, $x \in D(P)$ and $Px_0 = h$.

If $h = 0$, then we call P a mapping of class $(S_+)_{0,L}$.

Definition 4. Let X be a separable reflexive Banach space and let L be a dense subspace of X. A multivalued mapping $P : D(P) \subset X \to 2^{X^*}$ is said to be of class $(S_+)_L$ if it satisfies the following conditions:

(i) Px is bounded closed and convex for each $x \in D(P)$,

(ii) P is weakly upper semicontinuous in each finite dimensional space, that is, for each finite dimensional space F of L, $F \cap D(P) \ne \emptyset$, $P : F \cap D(P) \to 2^{X^*}$ is upper semicontinuous in the weak topology,

(iii) if for any sequence of finite dimensional subspaces F_n of L with $L \subset \overline{U_{n=1}^{\infty} F_n}$, $h \in X^*$, $\{x_n\}_{n=1}^{\infty} \subset D(P) \cap L$ and $x_n \rightharpoonup x_0$ such that

$$\limsup_{n\to\infty} \langle p_n - h, x_n\rangle \le 0, \qquad \lim_{n\to\infty} \langle p_n - h, v\rangle = 0 \tag{4}$$

for all $v \in U_{n=1}^{\infty} F_n$ and some $p_n \in Px_n$, then we have $x_n \to x_0$, $x \in D(P)$ and $h \in Px_0$.

If $h = 0$, then we call P a mapping of class $(S_+)_{0,L}$.

We will need the following two conditions:

(P1) there exist a subspace L of X such that $L \subset D(P)$, $\overline{L} = X$ and that the operator P satisfies condition $(S_+)_L$;

(P2) there exist a function $\phi : \mathbb{R}_+ \to \mathbb{R}_+$ which is nondecreasing such that

$$\langle p, x\rangle \ge -\phi(\|x\|), \quad p \in Px, \ x \in D(P). \tag{5}$$

The following lemma can be found in Zeidler [3, page 915].

Lemma 5. *Let $T : X \supset D(T) \to 2^{X^*}$ be maximal monotone. Thus the following are true:*

(i) *$\{x_n\} \subset D(T) x_n \to x_0$ and $Tx_n \ni y_n \rightharpoonup y_0$ imply $x_0 \in D(T)$, and $y_0 \in Tx_0$;*

(ii) *$\{x_n\} \subset D(T) x_n \rightharpoonup x_0$ and $Tx_n \ni y_n \to y_0$ imply $x_0 \in D(T)$, and $y_0 \in Tx_0$.*

We will need the following lemma from Adhikari and Kartsatos [4].

Lemma 6. *Assume that the operators $T : X \supset D(T) \to 2^{X^*}$ and $T_0 : X \supset D(T_0) \to X^*$ are maximal monotone, with $0 \in D(T) \cap D(T_0)$ and $0 \in T(0) \cap T_0(0)$. Assume further that $T + T_0$ is maximal monotone. Assume that there is a positive sequence $\{t_n\}$ such that $t_n \downarrow 0$, a sequence $\{x_n\} \subset D(T_0)$ and a sequence $w_n \in T_0 x_n$ such that $x_n \rightharpoonup x_0$ and $T_{t_n} x_n + w_n \rightharpoonup y_0^* \in X^*$. Then the following are true:*

(i) *the inequality*

$$\lim_{n\to\infty} \left\langle T_{t_n} x_n + w_n, x_n - x_0\right\rangle < 0 \tag{6}$$

is impossible;

(ii) *if*

$$\lim_{n\to\infty} \left\langle T_{t_n} x_n + w_n, x_n - x_0\right\rangle = 0, \tag{7}$$

then $x_0 \in D(T + T_0)$ and $y^* \in (T + T_0)x_0$.

The proof of the following lemma is given in [5] proof of Theorem 7 but we will repeat it here for completeness and future reference.

Lemma 7. *Let $T : X \supset D(T) \to 2^{X^*}$ be a quasibounded maximal monotone operator such that $0 \in T(0)$. Let $\{t_n\} \subset (0, \infty)$ and $\{t_n\} \subset X$ be such that*

$$\|u_n\| \le S, \quad \left\langle T_{t_n} u_n, u_n\right\rangle \le S_1, \tag{8}$$

where S, S_1 are positive constants. Then there exists a number $K > 0$ such that $\|T_{t_n} u_n\| \le K$ for all $n = 1, 2, \ldots$.

Proof. Let

$$w_n = T_{t_n} u_n = \left(T^{-1} + t_n J^{-1}\right)^{-1} u_n. \tag{9}$$

We have that

$$w_n \in T J_{t_n} u_n = T x_n, \quad t_n w_n = J\left(u_n - x_n\right), \tag{10}$$

where $x_n = J_{t_n} u_n$. Thus,

$$\begin{aligned} \langle w_n, x_n\rangle &= \left\langle w_n, u_n - t_n J^{-1} w_n\right\rangle \\ &= \langle w_n, u_n\rangle - t_n \left\langle w_n, J^{-1} w_n\right\rangle \\ &= \langle w_n, u_n\rangle - t_n \|w_n\|^2 \\ &\le \left\langle T_{t_n} u_n, u_n\right\rangle \\ &\le S_1. \end{aligned} \tag{11}$$

From (11) we obtain

$$t_n \|w_n\|^2 = \langle w_n, u_n\rangle - \langle w_n, x_n\rangle. \tag{12}$$

Now since $0 \in T(0)$ and $w_n \in Tx_n$, we have that $\langle w_n, x_n\rangle \ge 0$, which implies $t_n \|w_n\|^2 \le S_1$. We claim that $\{w_n\}$ is bounded, if not we may assume that $||w_n|| \to \infty$ and $||w_n|| \le ||w_n||^2$ for all n. Thus $t_n||w_n|| \le S_1$ and

$$t_n \|w_n\| = \|J(u_n - x_n)\| = \|u_n - x_n\| \quad (13)$$

which implies that $\{x_n\}$ is bounded. Now since T is strongly quasibounded, the boundedness of $\{x_n\}$ and $\{\langle w_n, x_n\rangle\}$ imply the boundedness of $\{w_n\}$, that is, a contradiction. It follows that $\{T_{t_n} u_n\}$ is bounded and the proof is complete. □

We denote by J_ψ the duality mapping with gauge function ψ. The function $\psi : \mathbb{R}_+ \to \mathbb{R}_+$ is continuous, strictly increasing such that $\psi(0) = 0$ and $\psi(r) \to \infty$ as $r \to \infty$. This mapping J_ψ is continuous, bounded, surjective, strictly and maximal monotone, and satisfies condition (S_+). Also,

$$\langle J_\psi x, x\rangle = \psi(\|x\|)\|x\|, \quad \|J_\psi x\| = \psi(\|x\|), \quad x \in X \quad (14)$$

for these facts we refer to Petryshyn [6, pages 32-33 and 132].

2. The Eigenvalue Result

In this section we are using the topological degree developed by Kartsatos and Quarcoo in [7]. The following result will also improve Theorem 4 of Kartsatos and Skrypnik in [8] since we are no longer assuming that the perturbation is quasibounded.

Theorem 8. *Let Ω be open and bounded with $0 \in \Omega$. Assume that the operator $T : X \supset D(T) \to 2^{X^*}$ is maximal monotone with $0 \in D(T)$ and $0 \in T(0)$. Assume that the operator $P : X \supset D(P) \to X^*$, with $L \subset D(P)$, satisfies condition $(S_+)_L$ and (P2). Let ϵ, ϵ_0, and Λ be positive numbers. Assume that*

($\mathcal{P}$) there exists $\lambda \in (0, \Lambda]$ such that

$$Tx + \lambda Px + \epsilon J_\psi x \ni 0 \quad (15)$$

has no solution in $D(T + P) \cap \Omega$.

Then

(i) there exists $(\lambda_0, x_0) \in (0, \Lambda] \times (D(T + P) \cap \partial\Omega)$ such that

$$Tx_0 + \lambda_0 Px_0 + \epsilon J_\psi x_0 \ni 0; \quad (16)$$

(ii) if $0 \notin T(D(T) \cap \partial\Omega)$, T satisfies (S_q) on $\partial\Omega$, and property ($\mathcal{P}$) is satisfied for every $\epsilon \in (0, \epsilon_0]$, then there exists $(\lambda_0, x_0) \in (0, \Lambda] \times (D(T + P) \cap \partial\Omega)$ such that $Tx_0 + \lambda_0 Px_0 \ni 0$.

Proof. (i) Assume that ($\mathcal{P}$) is true and that (16) is false. We consider the homotopy inclusion

$$H(t, x) \equiv Tx + t\Lambda Px + \epsilon J_\psi x \ni 0, \quad t \in [0, 1]. \quad (17)$$

It is clear that this inclusion has no solution $x \in D(H(t, \cdot)) \cap \partial\Omega$ for $t \in (0, 1]$, because $t\Lambda \in (0, \Lambda]$ and our assumption about (16). It is also true for $t = 0$ because $0 \in (T+\epsilon J_\psi)(0)$ and the operator $T + \epsilon J_\psi$ is strictly monotone, hence one-to-one.

We are going to show that $H(t, x)$ is an admissible homotopy for the degree in [4]. To this end, we set $T_t = T$, $P^t \equiv t\Lambda P + J_\psi$ and we recall the operators $T_{t,s} = T_s \equiv (T^{-1} + sJ^{-1})^{-1} : X \to X^*$, $s > 0$, and we set $J_{t,s} = J_s = I - sJ^{-1}T_{t,s} = I - sJ^{-1}T_s : X \to X$. We have $D(H(0, \cdot)) = D(T)$ and $D(H(t, \cdot)) = D(T + P)$, $t \in (0, 1]$. We also set $D^t = D(t\Lambda P) = D(tP)$. We have $D^0 = X$ and $D^t = D(P)$ for $t \in (0, 1]$. Let Ω be an open and bounded subset of X. We know that the equation

$$T^t x + P^t x \ni 0 \quad (18)$$

has no solution $x \in D(H(t, \cdot)) \cap \partial\Omega$ for any $t \in [0, 1]$. Now we consider the equation

$$T_s x + t\Lambda Px + \epsilon J_\psi = 0, \quad (19)$$

and we show that there exists $s_1 > 0$ such that

$$0 \notin (T_s + t\Lambda P + \epsilon J_\psi)(D^t \cap \partial\Omega), \quad (s, t) \in (0, s_1] \times [0, 1]. \quad (20)$$

Assume that this is not true, then there exist $\{s_n\} \subset (0, \infty)$ with $s_n \downarrow 0$, $\{t_n\} \subset [0, 1]$, with $t_n \to t_0$, $\{x_n\} \subset \partial\Omega$ with $x_n \rightharpoonup x_0$, and

$$T_{s_n} x_n + t_n\Lambda Px_n + \epsilon J_\psi x_n = 0. \quad (21)$$

Clearly we cannot have $t_n = 0$ for any n, since $(T_{s_n} + J_\psi)(0) = 0$ and the operator $T_{s_n} + J_\psi$ is strictly monotone, hence one-to-one. Thus $t_n > 0$, for all n. From (21) we have that

$$\begin{aligned} \langle T_{s_n} x_n, x_n\rangle &= -t_n\Lambda \langle Px_n, x_n\rangle - \epsilon \langle J_\psi x_n, x_n\rangle \\ &\le t_n\Lambda\phi(\|x_n\|) - \epsilon\psi(\|x_n\|) \\ &\le \Lambda\phi(S_1) + \epsilon\psi(S_1), \end{aligned} \quad (22)$$

where S_1 is the bound of $\{x_n\}$. Thus we have the boundedness of $\langle T_{s_n} x_n, x_n\rangle$. Using Lemma 7, we have the boundedness of $T_{s_n} x_n$. We may thus assume that $T_{s_n} x_n \rightharpoonup h_1^*$. From (21) we also have that the sequence $\{Px_n\}$ is bounded and we may assume that $Px_n \rightharpoonup h_0^*$.

If $t_0 = 0$, then from

$$\epsilon\psi(\|x_n\|)\|x_n\| = \langle \epsilon J_\psi x_n, x_n\rangle \le -t_n\Lambda \langle Px_n, x_n\rangle \longrightarrow 0 \quad (23)$$

we obtain $x_n \to 0 \in \partial\Omega$, which is a contradiction to $0 \in \Omega$. Hence it follows that $t_0 > 0$. Since $\{J_\psi x_n\}$ is bounded, we may assume that $J_\psi x_n \rightharpoonup h_2^*$ and we have $t_0\Lambda h_0^* = -h_1^* - \epsilon h_2^*$.

From (21) it follows that

$$\langle T_{s_n} x_n + t_n\Lambda Px_n + \epsilon J_\psi x_n, x\rangle = 0, \quad \forall x \in U_{n=1}^{\infty} F_n. \quad (24)$$

Thus

$$\langle T_{s_n} x_n + t_n\Lambda Px_n + \epsilon J_\psi x_n, x_n\rangle = 0,$$

$$\lim_{n\to\infty} \langle T_{s_n} x_n + t_n\Lambda Px_n + \epsilon J_\psi x_n, v\rangle = 0, \quad \forall v \in U_{n=1}^{\infty} F_n. \quad (25)$$

Now

$$\begin{aligned}\langle t_n\Lambda Px_n + h_1^* + \epsilon h_2^*, x_n\rangle &= \left\langle T_{s_n}x_n + t_n\Lambda Px_n + \epsilon J_\psi x_n, x_n\right\rangle \\ &\quad - \left\langle T_{s_n}x_n + \epsilon J_\psi x_n, x_n - x_0\right\rangle \\ &\quad - \left\langle T_{s_n}x_n + \epsilon J_\psi x_n, x_0\right\rangle \\ &\quad + \langle h_1^* + \epsilon h_2^*, x_n\rangle,\end{aligned} \tag{26}$$

hence

$$\begin{aligned}&\limsup_{n\to\infty} \langle t_n\Lambda Px_n + h_1^* + \epsilon h_2^*, x_n\rangle \\ &\quad\le \limsup_{n\to\infty} \left\langle T_{s_n}x_n + \epsilon J_\psi x_n + t_n\Lambda Px_n, x_n\right\rangle \\ &\quad - \liminf_{n\to\infty} \left\langle T_{s_n}x_n + \epsilon J_\psi x_n, x_n - x_0\right\rangle \\ &\quad - \langle h_1^* + \epsilon h_2^*, x_0\rangle + \langle h_1^* + \epsilon h_2^*, x_0\rangle \\ &\quad\le -\liminf_{n\to\infty} \left\langle T_{s_n}x_n + \epsilon J_\psi x_n, x_n - x_0\right\rangle \\ &\quad\le 0,\end{aligned} \tag{27}$$

where the last inequality follows from Lemma 6. Now since P is of class $(S_+)_L$, it follows that $x_n \to x_0$, $x_0 \in D(P)$ and $t_0\Lambda Px_0 = -h_1^* - \epsilon h_2^*$. Also we have that

$$\left\langle T_{s_n}x_n, x_n - x_0\right\rangle = 0 \tag{28}$$

and using Lemma 6 we get that $x_0 \in D(T)$ and $Tx_0 \ni h_1^* = -t_0\Lambda Px_0 - \epsilon J_\psi x_0$. This is a contradiction since $x_n \to x_0$, we have $x_0 \in \partial\Omega$. We have shown that $H(t,x)$ is an admissible homotopy for our degree. We can now work as in Theorem 3 in [8] in order to show that $d(H(t,\cdot),\Omega,0) = \text{const}$.

Thus

$$d(H(t,\cdot),\Omega,0) = d\left(T_s + \epsilon J_\psi, \Omega, 0\right) = 1, \tag{29}$$

where the last equality follows from Theorem 3, (i) of [9].

Consequently, the inclusion $H(t,x) \ni 0$ has a solution in Ω for each $t \in [0,1]$. In particular, this says that $Tx + \lambda Px + \epsilon J_\psi x \ni 0$ has a solution in Ω for every $\lambda > 0$. This of course contradicts condition ($\mathcal{P}$) and finishes the proof of (i).

(ii) Let $\lambda_n \in (0,\Lambda]$, $x_n \in D(P)\cap\partial\Omega$ be such that, for some $u_n^* \in Tx_n$,

$$u_n^* + \lambda_n Px_n + \left(\frac{1}{n}\right) J_\psi x_n = 0 \tag{30}$$

again, $\lambda_n = 0$ for any n is not possible. Since $\lambda_n > 0$, we have by property $(P2)$ that

$$\begin{aligned}\langle u_n^*, x_n\rangle &= -\lambda_n \langle Px_n, x_n\rangle - \frac{1}{n}\langle J_\psi x_n, x_n\rangle \\ &\le \lambda_n\phi(\|x_n\|) - \frac{1}{n}\psi(\|x_n\|)\|x_n\| \\ &\le \Lambda\phi(S_1) + \psi(S_1)S_1,\end{aligned} \tag{31}$$

where S_1 is the bound of $\{x_n\}$. This show that $\langle u_n^*, x_n\rangle$ is bounded and further we obtain the boundedness of $\{u_n^*\}$ by Lemma 2. We may thus assume that $\lambda_n \to \lambda_0$, $x_n \rightharpoonup x_0$, $u_n^* \rightharpoonup h_0^*$, then $\lambda_n Px_n \rightharpoonup -h_0^*$.

If $\lambda_0 = 0$, then (30) implies that

$$\lim_{n\to\infty} u_n^* = \lim_{n\to\infty}\left[-\lambda_n Px_n - \left(\frac{1}{n}\right)J_\psi x_n\right] = 0. \tag{32}$$

Since T satisfies (S_q), it follows that $x_n \to x_0 \in \partial\Omega$. Now, by the demiclosedness of T (see Lemma 6) we obtain that $x_0 \in D(T)$, $0 \in Tx_0$ and this contradicts $0 \notin T(D(T)\cap\partial\Omega)$. Hence, $\lambda_0 > 0$. Repeating the proof of (i) starting from (21), we get again that

$$\limsup_{n\to\infty} \langle \lambda_n Px_n + h_0^*, x_n\rangle \le 0 \tag{33}$$

and since P is of class $(S_+)_L$, we have $x_n \to x_0$, $-h_0^* = \lambda_0 Px_0$. Using the demiclosedness of T, we obtain $x_0 \in D(T)\cap\partial\Omega$, $h_0 \in Tx_0$ and $Tx_0 + \lambda_0 Px_0 \ni 0$. The proof is now complete. □

3. Continuous Branches of Eigenvectors

In this section we are interested in showing that the results obtained in previous sections could be used in order to obtain the existence of continuous branches of eigenvectors. We need the following definition.

Definition 9. Let $T : X \supset D(T) \to 2^{X^*}$, $P : X \supset D(P) \to X^*$ be given and consider the problem

$$Tx + \lambda Px \ni 0. \tag{34}$$

An eigenvector x is solution of (34) for some eigenvalue λ with $x \in (D(T)\cap D(P))$. We say that the nonzero eigenvectors of the problem (34) form a continuous branch of infinite length if there exists $r_0 > 0$ such that, for every $r \ge r_0$, the sphere $\partial B_r(0)$ contains at least one nonzero eigenvector of (34).

Theorem 10. *Assume that the operator $T : X \supset D(T) \to 2^{X^*}$ is maximal monotone with $0 \in D(T)$ and $0 \in T(0)$. Assume that the operator $P : X \supset D(P) \to X^*$ with $L \subset D(P)$ is of class $(S_+)_L$ and satisfies (P2). Let Λ be a positive number. Assume that $Tx \ni 0$ implies $x = 0$, T satisfies (S_q) and*

($\mathcal{P}1$) *there exists $\alpha > 0$ and $\lambda \in (0,\Lambda]$ such that*

$$|Tx + \lambda Px| \ge \alpha, \quad x \in D(T+P). \tag{35}$$

Then the nonzero eigenvectors of problem (34) form a continuous branch of infinite length with corresponding eigenvalues $\lambda \in (0,\Lambda]$.

Proof. Let $r_0 > 0$ be given. Let $\epsilon_0 > 0$ be so small that $\epsilon_0 r_0 < \alpha$. Then

$$\begin{aligned}|Tx + \lambda Px + \epsilon Jx| &\ge |Tx + \lambda Px| - |\epsilon Jx| \\ &\ge \alpha - \epsilon\|x\| \\ &\ge \alpha - \epsilon_0 r_0,\end{aligned} \tag{36}$$

$x \in D(T+P)$, which implies that the inclusion

$$Tx + \lambda Px + \epsilon Jx \ni 0 \quad (37)$$

has no solution $x \in D(T+P) \cap B_r(0)$ for any $\epsilon \in (0, \epsilon_0]$. Since $0 \notin T(\partial B_{r_0}(0))$ and T is of type (S_q), Theorem 8 implies the existence of a solution $x_{\lambda_0} \in D(T+P) \cap \partial B_{r_0}(0)$, for some $\lambda_0 \in (0, \Lambda]$. The same argument can be repeated for any $r > r_0$. This complete the proof. □

Remark 11. This result is also true when $P : X \supset D(P) \to 2^{X^*}$ is assumed to be multivalued. We have the following theorem without proof.

Theorem 12. *Assume that the operator $T : X \supset D(T) \to 2^{X^*}$ is maximal monotone with $0 \in D(T)$ and $0 \in T(0)$. Assume that the operator $P : X \supset D(P) \to 2^{X^*}$ with $L \subset D(P)$ is of class $(S_+)_L$ and satisfies (P2). Let Λ be a positive number. Assume that $Tx \ni 0$ implies $x = 0$, T satisfies (S_q), and*

($\mathscr{P}$1) there exists $\alpha > 0$ and $\lambda \in (0, \Lambda]$ such that

$$|Tx + \lambda Px| \geq \alpha, \quad x \in D(T+P). \quad (38)$$

Then the nonzero eigenvectors of problem (34) form a continuous branch of infinite length with corresponding eigenvalues $\lambda \in (0, \Lambda]$.

In the following result, we assume that the operator T is defined and bounded on all of X. In this case we demonstrate the fact that the assumption $|Tx + Px| \geq 0$ may be replaced by the assumption $\|Px\| \geq 0$ on $D(P)$.

Theorem 13. *Assume that the operator $T : X \supset D(T) \to 2^{X^*}$ is maximal monotone and bounded with $0 \in D(T)$ and $0 \in T(0)$. Assume that the operator $P : X \supset D(P) \to X^*$, with $L \subset D(P)$ satisfies condition $(S_+)_L$ and (P2). Assume that $Tx \ni 0$ implies $x = 0$, T satisfies (S_q) and there exist $\alpha > 0$ such that*

$$\|Px\| \geq \alpha, \quad x \in D(P). \quad (39)$$

Then the nonzero eigenvectors of the problem (34) form a continuous branch of infinite length.

Proof. We show that the problem (34) has eigenvectors on the set $\partial B_r(0)$ for every $r > 0$. To this end, let fix $r > 0$, $\epsilon > 0$ and show the existence of $\tilde{\lambda} > 0$ such that

$$d\left(T + \tilde{\lambda}P + \epsilon J, B_r(0), 0\right) = 0. \quad (40)$$

If this is not true, then there exist a sequence of $\{\lambda_n\} \subset (0, \infty)$ such that $\lambda_n \to \infty$ and one of the following is true:

(i) the degree $d(T + \lambda_n P + \epsilon J, B_r(0), 0)$ is not well-defined;

(ii) $d(T + \lambda_n P + \epsilon J, B_r(0), 0) \neq 0$.

In the case (i) there exist eigenvectors $x_n \in \partial B_r(0)$ such that

$$Tx_n + \lambda_n Px_n + \epsilon Jx_n \ni 0. \quad (41)$$

In the case (ii) there exist eigenvectors $x_n \in B_r(0)$ such that (41) holds. Thus in either case, there exist a sequence $\{x_n\} \subset \overline{B_r(0)}$ such that (41) is true. But this leads to a contradiction because $\|\lambda_n Px_n + \epsilon Jx_n\| \geq \alpha\lambda_n - \epsilon r \to \infty$, while the set Tx_n lie in a bounded set. Thus, (40) is true for some $\tilde{\lambda} > 0$. We consider the homotopy

$$H(t,x) \equiv Tx + t\tilde{\lambda}Px + \epsilon Jx, \quad t \in [0,1], \ x \in D(H(t,\cdot)), \quad (42)$$

either there exist $t_0 \in [0,1]$ and $x_{t_0} \in \partial B_r(0)$ such that

$$Tx_{t_0} + t_0\tilde{\lambda}Px_{t_0} + \epsilon Jx_{t_0} \ni 0, \quad (43)$$

or

$$d(H(t,\cdot)) = d(H(1,\cdot)) = 0 = d(H(0,\cdot)) = 1, \quad t \in [0,1] \quad (44)$$

that is a contradiction. The last equality in (44) follows from Theorem 3, (i) in [2]. It follows that (43) is true. Naturally, we must have $t_0 \neq 0$ in (43) because otherwise $0 \in (T + \epsilon J)(\partial B_r(0))$. This cannot happen because we already have $0 \in (T + \epsilon J)(0)$ and $T + \epsilon J$ is one-to-one. From (43) we obtain sequences $\lambda_n \in (0, \infty)$, $\{x_n\} \subset \partial B_r(0)$ such that

$$Tx_n + \lambda_n Px_n + \left(\frac{1}{n}\right) Jx_n \ni 0. \quad (45)$$

We may assume that $x_n \rightharpoonup x_0$. Again, $\{\lambda_n\}$ cannot contain a subsequence $\{\lambda_{n_k}\}$ such that $\lambda_{n_k} \to \infty$ as $k \to \infty$ because the sequence $Tx_{n_k} + (1/n_k)Jx_{n_k}$ lies in a bounded set and $\lambda_{n_k}\|Px_{n_k}\| \to \infty$ as $k \to \infty$. Also we have that the sequence $\{Px_n\}$ is bounded. Thus we may assume that $Px_n \rightharpoonup p^* \in X^*$. Since the sequence $\{\lambda_n\}$ is bounded, we may assume that $\lambda_n \to \lambda_0$. Again, $\lambda_0 \neq 0$ otherwise $\lambda_n Px_n + (1/n)Jx_n \to 0$ and the (S_q) property of T would imply that $x_n \to x_0 \in \partial\Omega$. Since T is demiclosed, Lemma 6 would imply $Tx_0 \ni 0$. This is a contradiction to our assumption that $Tx \ni 0$ implies $x = 0$. It follows that $\lambda_0 > 0$ and we may also assume that $\lambda_n > 0$ for all n.

Now proceeding as in the proof of Theorem 8, it is easy to see that

$$\limsup_{n\to\infty} \langle \lambda_n Px_n + \lambda_0 p^*, x_n \rangle \leq 0,$$
$$\lim_{n\to\infty} \langle \lambda_n Px_n + \lambda_0 p^*, v \rangle = 0, \quad \forall v \in U_{n=1}^{\infty} F_n. \quad (46)$$

Now since P is of class $(S_+)_L$, we conclude that $x_n \to x_0$, $x_0 \in D(P)$ and $Px_0 = p^*$. Thus since also $y_n^* \rightharpoonup -\lambda_0 Px_0$, $x_n \to x_0$ and the demiclosedness of T, we obtain $x_0 \in D(P) \cap \partial B_r(0)$ and $Tx_0 + \lambda_0 Px_0 \ni 0$. Since $r > 0$ is arbitrary, the nonzero eigenvectors of problem (34) form a continuous branch of infinite length. □

References

[1] D. Pascali and S. Sburlan, *Nonlinear Mappings of Monotone Type*, Martinus Nijhoff Publishers, The Hague, The Netherlands, 1978.

[2] A. G. Kartsatos and I. V. Skrypnik, "A new topological degree theory for densely defined quasibounded $(\tilde{S}_+)$-perturbations of multivalued maximal monotone operators in reflexive Banach spaces," *Abstract and Applied Analysis*, vol. 2005, no. 2, pp. 121–158, 2005.

[3] E. Zeidler, *Functional Analysis and Its Applications*, vol. II/B, Springer, New York, NY, USA, 1990.

[4] D. R. Adhikari and A. G. Kartsatos, "Topological degree theories and nonlinear operator equations in Banach spaces," *Nonlinear Analysis: Theory, Methods & Applications*, vol. 69, no. 4, pp. 1235–1255, 2008.

[5] F. E. Browder and P. Hess, "Nonlinear mappings of monotone type in Banach spaces," *Journal of Functional Analysis*, vol. 11, pp. 251–294, 1972.

[6] W. V. Petryshyn, *Approximation-Solvability of Nonlinear Functional and Differential Equations*, vol. 171 of *Monographs and Textbooks in Pure and Applied Mathematics*, Marcel Dekker, New York, NY, USA, 1993.

[7] A. G. Kartsatos and J. Quarcoo, "A new topological degree theory for densely defined $(S_+)_L$-perturbations of multivalued maximal monotone operators in reflexive separable Banach spaces," *Nonlinear Analysis: Theory, Methods & Applications*, vol. 69, no. 8, pp. 2339–2354, 2008.

[8] A. G. Kartsatos and I. V. Skrypnik, "On the eigenvalue problem for perturbed nonlinear maximal monotone operators in reflexive Banach spaces," *Transactions of the American Mathematical Society*, vol. 358, no. 9, pp. 3851–3881, 2006.

[9] J. Berkovits, "On the degree theory for densely defined mappings of class $(S_+)_L$," *Abstract and Applied Analysis*, vol. 4, no. 3, pp. 141–152, 1999.

22

Hilbert Space Representations of Generalized Canonical Commutation Relations

Asao Arai

Department of Mathematics, Hokkaido University, Sapporo 060-0810, Japan

Correspondence should be addressed to Asao Arai; arai@math.sci.hokudai.ac.jp

Academic Editor: Stefan Siegmund

We consider Hilbert space representations of a generalization of canonical commutation relations (CCRs) : $[X_j, X_k] := X_j X_k - X_k X_j = i\Theta_{jk} I$ $(j, k = 1, 2, \ldots, 2n)$, where X_j's are the elements of an algebra with identity I, i is the imaginary unit, and Θ_{jk} is a real number with antisymmetry $\Theta_{jk} = -\Theta_{kj}$ $(k, j = 1, 2, \ldots, 2n)$. Some basic aspects on Hilbert space representations of the generalized CCR (GCCR) are discussed. We define a Schrödinger-type representation of the GCCR by an analogy with the usual Schrödinger representation of the CCR with n degrees of freedom. Also, we introduce a Weyl-type representation of the GCCR. The main result of the present paper is a uniqueness theorem on Weyl representations of the GCCR.

1. Introduction

In this paper, we consider Hilbert space representations of a *generalized canonical commutation relations* (GCCRs) with n degrees of freedom ($n \in \mathbb{N} := \{1, 2, 3, \ldots\}$) of the following type:

$$[X_j, X_k] = i\Theta_{jk} I \quad (j, k = 1, \ldots, 2n), \tag{1}$$

where X_j's are elements of an algebra with identity I, $[X_j, X_k] := X_j X_k - X_k X_j$, i is the imaginary unit, and $\Theta_{jk} \in \mathbb{R}$ (the set of real numbers) with antisymmetry $\Theta_{jk} = -\Theta_{kj}$ $(j, k = 1, \ldots, 2n)$ such that, for some pair (j, k), $\Theta_{jk} \neq 0$. For convenience, we call (1) the Θ-GCCR with n degrees of freedom and the $2n \times 2n$ matrix

$$\Theta := (\Theta_{jk})_{j,k=1,\ldots,2n} \tag{2}$$

the *noncommutative factor* for $\{X_j\}_{j=1}^{2n}$.

Note that, in the case where Θ is equal to

$$J := \begin{pmatrix} 0 & I_n \\ -I_n & 0 \end{pmatrix}, \tag{3}$$

with I_n being the $n \times n$ unit matrix, (1) becomes the CCR with n degrees of freedom. Namely, if we put $Q_j := X_j$, $P_j := X_{n+j}$ $(j = 1, \ldots, n)$ in the present case, then we have

$$\begin{gathered}[Q_j, Q_k] = 0, \qquad [P_j, P_k] = 0, \\ [Q_j, P_k] = i\delta_{jk} I \quad (j, k = 1, \ldots, n),\end{gathered} \tag{4}$$

where δ_{jk} is the Kronecker delta. Thus, (1) is a natural generalization of the CCR with n degrees of freedom.

The GCCR also includes some of non-commutative space times (e.g., [1–3]), non-commutative spaces (e.g., [4]), and non-commutative phase spaces (e.g., [5–11]). In fact, one of the motivations for the present work is to investigate general structures underlying those non-commutative objects. In this paper, however, we present only some fundamental aspects of Hilbert space representations of the GCCR. The main result is to establish a uniqueness theorem on Weyl type representations of the GCCR (for the definition, see Section 4).

In Section 2, we define Hilbert space representations of the GCCR and discuss some basic facts on them. It is shown that there exists a one-to-one correspondence between representations of the GCCR and the CCR with the same degrees of freedom. In Section 3, we introduce a Schrödinger-type representation of the GCCR, whose representation space is $L^2(\mathbb{R}^n)$ as in the case of the Schrödinger representation of the CCR with n degrees of freedom. In Section 4, Weyl-type representations of the GCCR are defined by analogy with Weyl representations of CCR. In the last section, we prove the uniqueness theorem mentioned above. In Appendix, we present some basic properties of self-adjoint operators obeying generalized Weyl relations, which are used in the text.

2. Basic Facts on Hilbert Space Representations of the Θ-GCCR

Let $\mathscr{H}$ be a complex Hilbert space with inner product $\langle\cdot,\cdot\rangle$ (antilinear in the first variable and linear in the second one) and norm $\|\cdot\|$. For a linear operator A on $\mathscr{H}$, we denote its domain by $D(A)$. For linear operators $A_1,\ldots,A_p$ on $\mathscr{H}$,

$$D\left(\sum_{i=1}^{p} A_i\right) := \cap_{i=1}^{p} D(A_i),$$
$$D(A_1A_2) := \{\Psi \in D(A_2) \mid A_2\Psi \in D(A_1)\}, \tag{5}$$
$$D(A_1\cdots A_p) := D((A_1\cdots A_{p-1})A_p) \quad (p\geq 3).$$

Definition 1. Let $\mathscr{D}$ be a dense subspace of $\mathscr{H}$ and X_j, $j=1,\ldots,2n$, be symmetric (not necessarily essentially self-adjoint) operators on $\mathscr{H}$. Set $\mathbf{X} := (X_1,\ldots,X_{2n})$. We say that the triple $(\mathscr{H},\mathscr{D},\mathbf{X})$ is a *symmetric representation* of the Θ-GCCR with n degrees of freedom if $\mathscr{D}\subset \cap_{j,k=1}^{2n} D(X_jX_k)$ and (1) holds on $\mathscr{D}$.

If all the X_j's ($j=1,\ldots,2n$) are self-adjoint, we say that $(\mathscr{H},\mathscr{D},\mathbf{X})$ is a *self-adjoint representation* of the GCCR.

Remark 2. The concept of self-adjoint representation defined above is different from the one used in representation theory of $*$-algebra (e.g., [12, page 205]).

Remark 3. In each symmetric representation $(\mathscr{H},\mathscr{D},\mathbf{X})$ of the Θ-GCCR, $\mathscr{H}$ is *infinite dimensional* (if $\mathscr{H}$ were finite dimensional, then, for (j,k) such that $\Theta_{jk}\neq 0$, 0 = trace of $[X_j,X_k] = i\Theta_{jk}\dim\mathscr{H}\neq 0$, and hence one is led to a contradiction).

Remark 4. It follows from a well-known fact on commutation properties of linear operators (e.g., [13, Theorem 1.2.3]) that, for (j,k) with $\Theta_{jk}\neq 0$, at least one of X_j and X_k is *unbounded.* Hence, one has to be careful about domains of X_j's.

Remark 5. In the case of Hilbert space representations of CCR, symmetric representations, but nonself-adjoint ones, also play important roles. For example, such representations appear in mathematical theories of time operators [14] (see also [15, 16] for investigations from purely operator-theoretic points of view). Thus, it is expected that, in addition to self-adjoint representations of the Θ-GCCR, non-self-adjoint symmetric representations of it may have any importance in applications to quantum physics.

Remark 6. In the context of quantum mechanics, for a symmetric operator A and a unit vector $\psi\in D(A)$, $(\Delta A)_\psi := \|(A-\langle\psi,A\psi\rangle)\psi\|$ is called the uncertainty of A in the vector state ψ. Let $(\mathscr{H},\mathscr{D},\mathbf{X})$ be a symmetric representation of the Θ-GCCR with n degrees of freedom. Then, one has uncertainty relations of Robertson type [17]: for all unit vectors $\psi\in\mathscr{D}$ and $j,k=1,\ldots,2n$,

$$(\Delta X_j)_\psi (\Delta X_k)_\psi \geq \frac{1}{2}\left|\langle\psi,\Theta_{jk}\psi\rangle\right|. \tag{6}$$

Let $(\mathscr{H},\mathscr{D},\mathbf{X})$ be a symmetric representation of the Θ-GCCR as in Definition 1. We assume for simplicity the following:

Assumption 1. The noncommutative factor Θ is regular (invertible).

Under this assumption, Θ is a regular antisymmetric real matrix. Hence, by a well-known fact in the theory of linear algebra (e.g., [18, page 173, Problem 9]), the following fact holds.

Lemma 7. *There exists a regular $2n\times 2n$ real matrix T_0 such that ${}^tT_0\Theta T_0 = J$, where ${}^tT_0$ is the transposed matrix of T_0 and J is defined by* (3).

The matrix T_0 in Lemma 7 belongs to the set

$$M_\Theta := \{T \mid T \text{ is a } 2n\times 2n \text{ real matrix such that } {}^tT\Theta T = J\}. \tag{7}$$

It is easy to see that for each $T\in M_\Theta$, there exists a unique $2n\times 2n$ symplectic matrix W (i.e., ${}^tWJW = J$) such that $T=T_0W$. Hence,

$$M_\Theta = \{T_0W \mid {}^tWJW = J\}. \tag{8}$$

For a $2n\times 2n$ real matrix $L=(L_{jk})_{j,k=1,\ldots,2n}$, we define

$$X_j^L := \sum_{k=1}^{2n} L_{kj}X_k, \quad j=1,\ldots,2n. \tag{9}$$

We call the correspondence $\mathbf{X}\mapsto\mathbf{X}^L := (X_1^L,\ldots,X_{2n}^L)$ the *L-transform of* $\mathbf{X}$.

Let

$$\Theta_L := {}^tL\Theta L. \tag{10}$$

Proposition 8. *(i) For all $j=1,\ldots,2n$, X_j^L is a symmetric operator on $\mathscr{H}$.*

(ii) For all $j,k=1,\ldots,2n$,

$$[X_j^L, X_k^L] = i(\Theta_L)_{jk} \tag{11}$$

on $\mathscr{D}$.

(iii) For each $T\in M_\Theta$ and $j,k=1,\ldots,2n$,

$$[X_j^T, X_k^T] = iJ_{jk} \tag{12}$$

on $\mathscr{D}$.

Proof. An easy exercise. □

Proposition 8-(i) and (ii) show that $(\mathscr{H}, \mathscr{D}, \mathbf{X}^L)$ is a symmetric representation of the Θ_L-GCCR with n degrees of freedom.

Proposition 8 (iii) implies the following.

Corollary 9. *Let $T \in M_\Theta$ and*

$$Q_j := X_j^T, \quad P_j := X_{n+j}^T \quad (j = 1, \ldots, n). \tag{13}$$

Then, $(\mathscr{H}, \mathscr{D}, \{Q_j, P_j\}_{j=1}^n)$ is a symmetric representation of the CCR with n degrees of freedom.

Corollary 9 means that for each $T \in M_\Theta$, the T-transform of $\mathbf{X}$ gives a correspondence from a symmetric representation of the Θ-GCCR with n degrees of freedom to a symmetric representation of the CCR with the same degrees of freedom.

One can easily see that (9) with $L = T \in M_\Theta$ implies that

$$X_j = \sum_{k=1}^{2n} \left(T^{-1}\right)_{kj} X_k^T \tag{14}$$

on $\cap_{j=1}^{2n} D(X_j)$. Thus, every symmetric representation of the Θ-GCCR with n degrees of freedom is constructed from a symmetric representation of the CCR with the same degrees of freedom via (14).

Conversely, if a symmetric representation $(\mathscr{H}, \mathscr{D}, \{Q_j, P_j\}_{j=1}^n)$ of the CCR with n degrees of freedom is given and let

$$X_j(\mathbf{Q}, \mathbf{P}; T) := \sum_{k=1}^{n} \left(T^{-1}\right)_{kj} Q_k + \sum_{k=1}^{n} \left(T^{-1}\right)_{(n+k)j} P_k \tag{15}$$

with $\mathbf{Q} := (Q_1, \ldots, Q_n)$ and $\mathbf{P} := (P_1, \ldots, P_n)$, then $(\mathscr{H}, \mathscr{D}, \mathbf{X}(\mathbf{Q}, \mathbf{P}; T))$ is a symmetric representation of the Θ-GCCR and (9) holds with $L = T$, $X_j^T = Q_j$, $X_{n+j}^T = P_j$ $(j = 1, \ldots, n)$ and $X_j = X_j(\mathbf{Q}, \mathbf{P}; T)$ $(j = 1, \ldots, 2n)$. Hence, every symmetric representation of the CCR with n degrees of freedom is constructed from a symmetric representation of the Θ-GCCR with the same degrees of freedom. Thus, *for each $T \in M_\Theta$, there exists a one-to-one correspondence between a symmetric representation of the Θ-GCCR and a symmetric representation of the CCR with n degrees of freedom.*

3. Representations of Schrödinger Type

Let $T \in M_\Theta$. By the fact on $\mathbf{X}(\mathbf{Q}, \mathbf{P}; T)$ stated in the preceding section, we can define a class of representations of the Θ-GCCR. Let $(L^2(\mathbb{R}^n), C_0^\infty(\mathbb{R}^n), \{q_j, p_j\}_{j=1}^n)$ be the Schrödinger representation of the CCR with n degrees of freedom, that is, q_j is the multiplication operator by the jth component x_j of $\mathbf{x} = (x_1, \ldots, x_n) \in \mathbb{R}^n$ and $p_j := -iD_j$ with D_j being the generalized partial differential operator in x_j, acting in $L^2(\mathbb{R}^n)$. Let

$$X_j(\mathbf{q}, \mathbf{p}; T) = \sum_{k=1}^{n} \left(T^{-1}\right)_{kj} q_k + \sum_{k=1}^{n} \left(T^{-1}\right)_{(n+k)j} p_k, \tag{16}$$

which is (15) with $\mathbf{Q} = \mathbf{q}$ and $\mathbf{P} = \mathbf{p}$. We denote the closure of $X_j(\mathbf{q}, \mathbf{p}; T)$ by $\overline{X}_j(\mathbf{q}, \mathbf{p}; T)$ and set

$$\overline{\mathbf{X}}(\mathbf{q}, \mathbf{p}; T) := \left(\overline{X}_1(\mathbf{q}, \mathbf{p}; T), \ldots, \overline{X}_{2n}(\mathbf{q}, \mathbf{p}; T)\right). \tag{17}$$

We call the triple $\pi_S^T := (L^2(\mathbb{R}^n), C_0^\infty(\mathbb{R}^n), \overline{\mathbf{X}}(\mathbf{q}, \mathbf{p}; T))$ the *T-Schrödinger representation of the Θ-GCCR*.

It is easy to see that for all $j = 1, \ldots, 2n$, $X_j(\mathbf{q}, \mathbf{p}; T)$ is essentially self-adjoint on $C_0^\infty(\mathbb{R}^n)$ (apply, e.g., the Nelson commutator theorem [19, Theorem X.37] with dominating operator $N = \sum_{j=1}^n (q_j^2 + p_j^2) + I$) (This can be proved also by applying Proposition 16). Hence $\overline{X}_j(\mathbf{q}, \mathbf{p}; T)$ is self-adjoint. Thus, we obtain the following.

Proposition 10. *For each $T \in M_\Theta$, the T-Schrödinger representation π_S^T is a self-adjoint representation of the Θ-GCCR.*

4. Representations of Weyl Type

Based on an analogy with Weyl representations of CCR, we introduce a concept of Weyl representation for Θ-GCCR.

Definition 11. Let $\{X_j\}_{j=1}^{2n}$ be a set of self-adjoint operators on a Hilbert space $\mathscr{H}$. We say that $\{X_j\}_{j=1}^{2n}$ is a *Weyl representation of the Θ-GCCR* with n degrees of freedom if for all $s, t \in \mathbb{R}$ and $j, k = 1, \ldots, 2n$,

$$e^{itX_j} e^{isX_k} = e^{-ist\Theta_{jk}} e^{isX_k} e^{itX_j}. \tag{18}$$

We call these relations the Θ-*Weyl relations*.

For a linear operator A on a Hilbert space, we denote its spectrum by $\sigma(A)$.

Proposition 12. *Let $\{X_j\}_{j=1}^{2n}$ be a Weyl representation of the Θ-GCCR on $\mathscr{H}$. Then, there is a dense subspace $\mathscr{D}_0 \subset \mathscr{H}$ left invariant by each X_j $(j = 1, \ldots, 2n)$ such that $(\mathscr{H}, \mathscr{D}_0, \mathbf{X})$ is a self-adjoint representation of the Θ-GCCR. Moreover, for every pair (X_j, X_k) such that $\Theta_{jk} \neq 0$, X_j and X_k are purely absolutely continuous with*

$$\sigma\left(X_j\right) = \sigma\left(X_k\right) = \mathbb{R}, \quad j = 1, \ldots, 2n. \tag{19}$$

Proof. By (18), we can apply the results described in the Appendix of the present paper. In the present context, we need only to take, in the notation in the Appendix, $N = 2n$, $a_{jk} = \Theta_{jk}$ and $A_j = X_j$. By Proposition A.4-(iii) and Corollary A.5, there exists a dense subspace $\mathscr{D}_0$ left invariant by X_j $(j = 1, \ldots, 2n)$ and $[X_j, X_k] = i\Theta_{jk}$ on $\mathscr{D}_0$. Thus, the first half of the proposition is derived. The second half follows from Proposition A.1. □

Remark 13. As in the case of self-adjoint representations of CCR (e.g., [16, 20, 21]), the converse of Proposition 12 does not hold (i.e., a self-adjoint representation of the Θ-GCCR is not necessarily a Weyl one).

We recall that a set $\{Q_j, P_j\}_{j=1}^n$ of self-adjoint operators on $\mathscr{H}$ is a Weyl representation of the CCR with n degrees of freedom if for all $s, t \in \mathbb{R}$ and $j, k = 1, \ldots, n$, the following Wey relations hold:

$$\begin{aligned} e^{itQ_j}e^{isP_k} &= e^{-ist\delta_{jk}}e^{isP_k}e^{itQ_j}, \\ e^{itQ_j}e^{isQ_k} &= e^{isQ_k}e^{itQ_j}, \\ e^{itP_j}e^{isP_k} &= e^{isP_k}e^{itP_j}. \end{aligned} \tag{20}$$

Remark 14. A set $\{Q_j, P_j\}_{j=1}^n$ of self-adjoint operators on $\mathscr{H}$ is a Weyl representation of the CCR with n degrees of freedom if and only if $\{X_j\}_{j=1}^{2n}$ with $X_j := Q_j, X_{n+j} = P_j$ $(j = 1, \ldots, n)$ is a Weyl representation of the J-GCCR, where J is given by (3).

Let $T \in M_\Theta$ be arbitrarily fixed. The next proposition shows that the T-transform of each Weyl representation of the Θ-GCCR is a Weyl representation of the CCR with n degrees of freedom.

Proposition 15. *Let $\{X_j\}_{j=1}^{2n}$ be a Weyl representation of the Θ-GCCR on $\mathscr{H}$, and let $\mathbf{X}^T$ be the T-transform of $\mathbf{X}$. Then, each X_j^T is essentially self-adjoint, and $\{\overline{X_j^T}\}_{j=1}^{2n}$ is a Weyl representation of the J-GCCR.*

Proof. The essential self-adjointness of X_j^T follows from a simple application of Theorem A.6 in Appendix. Corollary A.7 in Appendix and the relation ${}^tT\Theta T = J$ imply that $\{\overline{X_j^T}\}_{j=1}^{2n}$ satisfies the J-Weyl relations. □

In the same way as in the proof of Proposition 15, we can prove the following proposition:

Proposition 16. *Let $\{Q_j, P_j\}_{j=1}^n$ be a Weyl representation of the CCR with n degrees of freedom on a Hilbert space $\mathscr{H}$. Let $X_j(\mathbf{Q}, \mathbf{P}; T)$ $(j = 1, \ldots, 2n)$ be defined by (15). Then, each $X_j(\mathbf{Q}, \mathbf{P}; T)$ is essentially self-adjoint and $\{\overline{X}_j(\mathbf{Q}, \mathbf{P}; T)\}_{j=1}^{2n}$ is a Weyl representation of Θ-GCCR with n degrees of freedom.*

This proposition shows that the converse of Proposition 15 holds too. Thus, for each $T \in M_\Theta$, there exists a one-to-one correspondence between a Weyl representation of the CCR with n degrees of freedom and that of the Θ-GCCR with the same degrees of freedom.

It is well known [22] that the Schrödinger representation $\{q_j, p_j\}_{j=1}^n$ is a Weyl representation o the CCR with n degrees of freedom. Hence, we obtain the following result.

Corollary 17. *For each $T \in M_\Theta$, the T-Schrödinger representation $\{\overline{X}_j(\mathbf{q}, \mathbf{p}; T)\}_{j=1}^{2n}$ is a Weyl representation of the Θ-GCCR.*

We say that a Weyl representation $\{X_j\}_{j=1}^{2n}$ of the Θ-GCCR on $\mathscr{H}$ is *irreducible* if every closed subspace $\mathscr{M}$ of $\mathscr{H}$ which is invariant under the action of e^{itX_j} $(t \in \mathbb{R}, j = 1, \ldots, 2n)$ is $\{0\}$ or $\mathscr{H}$.

Proposition 18. *Let $T \in M_\Theta$. Then, the T-Schrödinger representation $\{\overline{X}_j(\mathbf{q}, \mathbf{p}; T)\}_{j=1}^{2n}$ as a Weyl representation of the Θ-GCCR is irreducible.*

Proof. Let $\mathscr{M}$ be an invariant closed subspace of $e^{it\overline{X}_j(q,p;T)}$ $(t \in \mathbb{R}, j = 1, \ldots, 2n)$. We have

$$\begin{aligned} q_j &= \sum_{k=1}^{2n} T_{kj} X_k(\mathbf{q}, \mathbf{p}; T), \\ p_j &= \sum_{k=1}^{2n} T_{k(n+j)} X_k(\mathbf{q}, \mathbf{p}; T) \end{aligned} \tag{21}$$

on $\cap_{j=1}^{2n} D(X_j(\mathbf{q}, \mathbf{p}; T)) = \cap_{j=1}^n D(q_j) \cap D(p_j)$. Hence, by an application of Theorem A.6 in Appendix, e^{itq_j} and e^{itp_j} $(t \in \mathbb{R})$ can be written, respectively, as a scalar multiple of $e^{it\overline{X}_1(\mathbf{q},\mathbf{p};T)} \cdots e^{it\overline{X}_{2n}(\mathbf{q},\mathbf{p};T)}$. Hence, $\mathscr{M}$ is invariant under the action of e^{itq_j} and e^{itp_j} $(t \in \mathbb{R}, j = 1, \ldots, n)$. It is well known that $\{e^{itq_j}, e^{itp_j} \mid t \in \mathbb{R}, j = 1, \ldots, n\}$ is irreducible. Thus, $\mathscr{M} = \{0\}$ or $\mathscr{H}$. □

5. Uniqueness Theorem on Weyl Representations of the Θ-GCCR

In this section, we prove the main result of the present paper, that is, a uniqueness theorem on Weyl representations of the Θ-GCCR, which may be regarded as a GCCR version of the celebrated von Neumann uniqueness theorem of Weyl representations of CCR ([13, Theorem 4.11.1], [22], [23, Theorem VIII.14]).

Theorem 19. *Let $\{X_j\}_{j=1}^{2n}$ be a Weyl representation of the Θ-GCCR on a separable Hilbert space $\mathscr{H}$. Then, for each $T \in M_\Theta$, there exist mutually orthogonal closed subspaces $\mathscr{H}_\ell$ ($\ell = 1, \ldots, N$; $N \in \mathbb{N}$ or ∞) such that the following (i)–(iii) hold.*

(i) $\mathscr{H} = \oplus_{\ell=1}^N \mathscr{H}_\ell$.

(ii) *For each $j = 1, \ldots, 2n$, X_j is reduced by each $\mathscr{H}_\ell$, $\ell = 1, \ldots, N$. We denote by $X_j^{(\ell)}$ the reduced part of X_j to $\mathscr{H}_\ell$.*

(iii) *For each ℓ, there exists a unitary operator $U_\ell : \mathscr{H}_\ell \to L^2(\mathbb{R}^n)$ such that*

$$U_\ell X_j^{(\ell)} U_\ell^{-1} = \overline{X}_j(\mathbf{q}, \mathbf{p}; T), \quad j = 1, \ldots, 2n, \tag{22}$$

where $\{\overline{X}_j(\mathbf{q}, \mathbf{p}; T)\}_{j=1}^{2n}$ is the T-Schrödinger representation of the Θ-GCCR.

Proof. Let $T \in M_\Theta$, $\mathbf{X}^T$ be the T-transform of $\mathbf{X}$ and $Q_j := \overline{X_j^T}$, $P_j := \overline{X_{n+j}^T}$ $(j = 1, \ldots, n)$. Then, by Proposition 15 and Remark 14, $\{Q_j, P_j\}_{j=1}^n$ is a Weyl representation of the CCR with n degrees of freedom. Hence, by the von Neumann uniqueness theorem mentioned above, there exist mutually orthogonal closed subspaces $\mathscr{H}_\ell$ such that (i) given above and the following (a) and (b) hold.

(a) For each $j = 1, \dots, n$ and all $t \in \mathbb{R}$, e^{itQ_j} and e^{itP_j} leave each $\mathscr{H}_\ell$ invariant ($\ell = 1, \dots, N$).

(b) For each ℓ, there exists a unitary operator $U_\ell : \mathscr{H}_\ell \to L^2(\mathbb{R}^n)$ such that

$$U_\ell e^{itQ_j} U_\ell^{-1} = e^{itq_j},$$
$$U_\ell e^{itP_j} U_\ell^{-1} = e^{itp_j}, \quad t \in \mathbb{R}, j = 1, \dots, n. \tag{23}$$

By (14), we have $X_j = X_j(\mathbf{Q}, \mathbf{P}; T)$ on $\cap_{j=1}^{2n} D(X_j)$. Hence, $X_j \subset \overline{X}_j(\mathbf{Q}, \mathbf{P}; T)$. By Proposition 16, $\overline{X}_j(\mathbf{Q}, \mathbf{P}; T)$ is self-adjoint. Hence, $X_j = \overline{X}_j(\mathbf{Q}, \mathbf{P}; T)$. Therefore, by Theorem A.6 in Appendix, we obtain

$$e^{itX_j} = e^{it^2 \sum_{k<m}^{2n} J_{km}(T^{-1})_{kj}(T^{-1})_{mj}/2} e^{itQ_1} \cdots e^{itQ_n} e^{itP_1} \cdots e^{itP_n},$$
$$j = 1, \dots, 2n. \tag{24}$$

Hence, each e^{itX_j} leaves $\mathscr{H}_\ell$ invariant ($\ell = 1, \dots, N$). Therefore, X_j is reduced by each $\mathscr{H}_\ell$. We denote the reduced part of X_j to $\mathscr{H}_\ell$ by $X_j^{(\ell)}$. Then, we have by (23)

$$U_\ell e^{itX_j^{(\ell)}} U_\ell^{-1} = e^{it^2 \sum_{k<m}^{2n} J_{km}(T^{-1})_{kj}(T^{-1})_{mj}/2} e^{itq_1} \cdots e^{itq_n} e^{itp_1} \cdots e^{itp_n}$$
$$= e^{it\overline{X}_j(\mathbf{q}, \mathbf{p}; T)}. \tag{25}$$

Thus, (22) follows. □

Theorem 19 tells us that every Weyl representation of the Θ-GCCR on a *separable* Hilbert space is unitarily equivalent to a direct sum of the T-Schrödinger representation of the Θ-GCCR, where $T \in M_\Theta$ is arbitrary.

The next corollary immediately follows from Theorem 19.

Corollary 20. *Let $\{X_j\}_{j=1}^{2n}$ be an irreducible Weyl representation of the Θ-GCCR on a separable Hilbert space $\mathscr{H}$. Then, for each $T \in M_\Theta$, there exists a unitary operator $U : \mathscr{H} \to L^2(\mathbb{R}^n)$ such that*

$$U X_j U^{-1} = \overline{X}_j(\mathbf{q}, \mathbf{p}; T), \quad j = 1, \dots, 2n. \tag{26}$$

The following result shows that the arbitrariness of the choice of T in the T-Schrödinger representation of the Θ-GCCR is implemented by unitary operators.

Corollary 21. *Let $S, T \in M_\Theta$. Then, there exists a unitary operator V on $L^2(\mathbb{R}^n)$ such that*

$$V \overline{X}_j(\mathbf{q}, \mathbf{p}; S) V^{-1} = \overline{X}_j(\mathbf{q}, \mathbf{p}; T), \quad j = 1, \dots, 2n. \tag{27}$$

Proof. We need only to apply Corollary 20 to the case where $X_j = \overline{X}_j(\mathbf{q}, \mathbf{p}; S)$. □

Remark 22. As in the case of non-Weyl representations of CCR, for non-Weyl representations of the Θ-GCCR, the conclusion of Theorem 19 does not hold in general. Examples of such representations of the Θ-GCCR can be constructed from non-Weyl representations of CCR (e.g., [15, 16, 20, 21]). A detailed description of some examples is given in [5].

Appendix

Some Properties of Self-Adjoint Operators Satisfying Relations of Weyl Type

Let $N \geq 2$ be an integer, and let A_j ($j = 1, \dots, N$) be self-adjoint operators on a Hilbert space $\mathscr{H}$ satisfying relations of Weyl type:

$$e^{itA_j} e^{isA_k} = e^{-itsa_{jk}} e^{isA_k} e^{itA_j}, t, s \in \mathbb{R}, j, k = 1, \dots, N, \tag{A.1}$$

where a_{jk}'s are real constants. It follows that a_{jk} is antisymmetric in (j, k):

$$a_{jk} = -a_{kj}, \quad j, k = 1, \dots, N. \tag{A.2}$$

The unitarity of e^{itA_j} and functional calculus imply that

$$\exp\left(ise^{itA_j} A_k e^{-itA_j}\right) = \exp\left(is\left(A_k - ta_{jk}\right)\right), \quad s, t \in \mathbb{R}. \tag{A.3}$$

Hence, we have the operator equality

$$e^{itA_j} A_k e^{-itA_j} = A_k - ta_{jk}, \quad t \in \mathbb{R}, j, k = 1, \dots, N. \tag{A.4}$$

For a linear operator A on a Hilbert space, we denote the spectrum of A by $\sigma(A)$.

Proposition A.1. *Suppose that there exists a pair (j, k) such that $a_{jk} \neq 0$ (hence, $j \neq k$). Then,*

$$\sigma\left(A_j\right) = \mathbb{R}, \qquad \sigma\left(A_k\right) = \mathbb{R}. \tag{A.5}$$

Moreover, A_j and A_k are purely absolutely continuous.

Proof. By (A.4) and the unitary invariance of spectrum, we have $\sigma(A_k) = \sigma(A_k - ta_{jk})$ for all $t \in \mathbb{R}$. Since $a_{jk} \neq 0$, this implies the second equation of (A.5). By (A.2), we have $a_{kj} \neq 0$. Hence, by considering the case of (j, k) replaced by (k, j), we obtain the first equation of (A.5).

Relation (A.4) means that (A_k, A_j) is a weak Weyl representation of the CCR with one degree of freedom [14, 15, 24]. Hence A_j is purely absolutely continuous [14, 15]. Similarly, we can show that A_k is purely absolutely continuous. □

Proposition A.2. *Let j and k be fixed. Then, for all $\psi \in D(A_j) \cap D(A_j A_k)$, ψ is in $D(A_k A_j)$ and*

$$\left[A_j, A_k\right] \psi = ia_{jk}\psi. \tag{A.6}$$

Proof. An easy exercise (use (A.4)). □

For each function $f \in C_0^\infty(\mathbb{R}^N)$ and each vector $\psi \in \mathscr{H}$, we define a vector ψ_f by

$$\psi_f := \int_{\mathbb{R}^N} f(\mathbf{t}) e^{it_1 A_1} \cdots e^{it_N A_N} \psi d\mathbf{t}, \tag{A.7}$$

where $\mathbf{t} = (t_1, \dots, t_N) \in \mathbb{R}^N$ and the integral on the right-hand side is taken in the strong sense. We introduce

$$\mathscr{D}_0 := \operatorname{Span}\{\psi_f \mid \psi \in \mathscr{H}, f \in C_0^\infty(\mathbb{R}^N)\}, \qquad (A.8)$$

where Span{⋯} denotes the subspace algebraically spanned by the vectors in the set {⋯}. It is easy to see that $\mathscr{D}_0$ is dense in $\mathscr{H}$.

For $f : \mathbb{R}^N \to \mathbb{C}$ (the set of complex numbers), we set $\|f\|_1 := \int_{\mathbb{R}^N} |f(\mathbf{t})| d\mathbf{t}$.

Lemma A.3. *Let $f_n, f \in C_0^\infty(\mathbb{R}^N)$ such that $\|f_n - f\|_1 \to 0$ ($n \to \infty$). Then, $\|\psi_{f_n} - \psi_f\| \to 0$ ($n \to \infty$).*

Proof. Since $e^{it_jA_j}$ is unitary, we have $\|\psi_{f_n} - \psi_f\| \le \|f_n - f\|_1 \|\psi\|$. Thus, the desired result follows. □

For each $j = 1, \dots, N$, we define a function g_j on $\mathbb{R}^N$ by

$$g_j(\mathbf{t}) := \begin{cases} 0 & \text{for } j = 1 \\ \sum_{k=1}^{j-1} a_{jk} t_k & \text{for } 2 \le j \le N, \end{cases} \quad \mathbf{t} \in \mathbb{R}^N. \qquad (A.9)$$

Proposition A.4. *(i) For all $t \in \mathbb{R}$ and $j = 1, \dots, N$, e^{itA_j} leaves $\mathscr{D}_0$ invariant.*

(ii) For each $j = 1, \dots, N$, A_j leaves $\mathscr{D}_0$ invariant (i.e., $A_j\mathscr{D}_0 \subset \mathscr{D}_0$) and for all $\ell \in \mathbb{N}$,

$$A_j^\ell \psi_f = (-i)^\ell \psi_{F_j^\ell(f)}, \quad f \in C_0^\infty(\mathbb{R}^N), \qquad (A.10)$$

where $F_j : C_0^\infty(\mathbb{R}^N) \to C_0^\infty(\mathbb{R}^N)$ is defined by

$$F_j(f) := -\partial_j f - i g_j f, \quad f \in C_0^\infty(\mathbb{R}^N), \qquad (A.11)$$

and F_j^ℓ is the ℓ times composition of F_j with $F_j^0 := I$ (identity).

(iii) For all $\ell_1, \dots, \ell_N \in \mathbb{N} \cup \{0\}$,

$$A_1^{\ell_1} A_2^{\ell_2} \cdots A_N^{\ell_N} \psi_f = (-i)^{\ell_1 + \cdots + \ell_N} \psi_{F_1^{\ell_1} \cdots F_N^{\ell_N}(f)}, \quad f \in C_0^\infty(\mathbb{R}^N). \qquad (A.12)$$

Proof. (i) Let ψ_f be as above. Then, we have $e^{itA_j}\psi_f = \int_{\mathbb{R}^N} f(\mathbf{t}) e^{itA_j} e^{it_1A_1} \cdots e^{it_NA_N} \psi d\mathbf{t}$. By (A.1), we have

$$e^{itA_j} e^{it_1A_1} \cdots e^{it_NA_N} = e^{-itg_j(\mathbf{t})} e^{it_1A_1} \cdots e^{it_{j-1}A_{j-1}} e^{i(t_j+t)A_j} e^{it_{j+1}A_{j+1}} \cdots e^{it_NA_N}. \qquad (A.13)$$

Hence,

$$e^{itA_j}\psi_f = \psi_{f_j^{(t)}}. \qquad (A.14)$$

with

$$f_j^{(t)}(\mathbf{t}) := f(t_1, \dots, t_{j-1}, t_j - t, t_{j+1}, \dots, t_N) e^{-itg_j(\mathbf{t})}. \qquad (A.15)$$

It is easy to see that $f_j^{(t)}$ is in $C_0^\infty(\mathbb{R}^N)$. Hence, $\psi_{f_j^{(t)}} \in \mathscr{D}_0$. Thus, e^{itA_j} leaves $\mathscr{D}_0$ invariant.

(ii) By (A.14), we have for all $t \in \mathbb{R} \setminus \{0\}$, $(e^{itA_j} - 1)\psi_f/t = \psi_{(f_j^{(t)} - f)/t}$. It is easy to see that $\|(f_j^{(t)} - f)/t - F_j(f)\|_1 \to 0$ ($t \to 0$). Hence, by Lemma A.3,

$$\lim_{t \to 0} \frac{(e^{itA_j} - 1)\psi_f}{t} = \psi_{F_j(f)}. \qquad (A.16)$$

Therefore, ψ_f is in $D(A_j)$ and $iA_j\psi_f = \psi_{F_j(f)}$. Hence, (A.10) with $\ell = 1$ holds. Then, one can prove (A.10) by induction.

(iii) This easily follows from (ii). □

Propositions A.2 and A.4 immediately yield the following result.

Corollary A.5. *For all $j, k = 1, \dots, N$, $[A_j, A_k] = ia_{jk}$ on $\mathscr{D}_0$.*

Theorem A.6. *For all $c_j \in \mathbb{R}$, $j = 1, \dots, N$, $\sum_{j=1}^N c_jA_j$ is essentially self-adjoint on $\mathscr{D}_0$ and*

$$e^{it\overline{\sum_{j=1}^N c_jA_j}} = e^{it^2 \sum_{j<k}^N a_{jk}c_jc_k/2} e^{itc_1A_1} e^{itc_2A_2} \cdots e^{itc_NA_N}, \qquad (A.17)$$

where for a closable operator C, $\overline{C}$ denotes the closure of C.

Proof. For each $t \in \mathbb{R}$, we define an operator $U(t)$ by

$$U(t) := e^{it^2 \sum_{j<k}^N a_{jk}c_jc_k/2} e^{itc_1A_1} e^{itc_2A_2} \cdots e^{itc_NA_N}. \qquad (A.18)$$

By using (A.1), one can show that $\{U(t)\}_{t\in\mathbb{R}}$ is a strongly continuous one-parameter unitary group. Hence, by the Stone theorem, there exists a unique self-adjoint operator A on $\mathscr{H}$ such that $U(t) = e^{itA}$, $t \in \mathbb{R}$. By Proposition A.4, $U(t)$ leaves $\mathscr{D}_0$ invariant and strongly differentiable on $\mathscr{D}_0$ with

$$\left.\frac{dU(t)\psi}{dt}\right|_{t=0} = i\sum_{j=1}^N c_jA_j\psi, \quad \psi \in \mathscr{D}_0. \qquad (A.19)$$

Hence, $\mathscr{D}_0$ is a core of A (e.g., [23, Theorem VIII.10]). Hence $A\psi = \sum_{j=1}^N c_jA_j\psi$, $\psi \in \mathscr{D}_0$. Thus, the desired result follows. □

For all $c_j \in \mathbb{R}$, $j = 1, \dots, N$, we set

$$A(\mathbf{c}) := \overline{\sum_{j=1}^N c_jA_j}, \quad \mathbf{c} = (c_1, \dots, c_N) \in \mathbb{R}^N. \qquad (A.20)$$

Corollary A.7. *For all $\mathbf{c}, \mathbf{d} \in \mathbb{R}^N$ and $t, s \in \mathbb{R}$,*

$$e^{itA(\mathbf{c})} e^{isA(\mathbf{d})} = e^{-its\sum_{j,k=1}^N a_{jk}c_jd_k} e^{isA(\mathbf{d})} e^{itA(\mathbf{c})}. \qquad (A.21)$$

Proof. By direct computations using (A.17) and (A.1). □

Acknowledgment

This work is supported by the Grant-In-Aid no. 24540154 for Scientific Research from Japan Society for the Promotion of Science (JSPS).

References

[1] S. Doplicher, K. Fredenhargen, and J. Roberts, "The quantum structure of spacetime at the Planck scale and quantum fields," *Communications in Mathematical Physics* , vol. 172, no. 1, pp. 187–220, 1995.

[2] V. Gatral, J. M. Gracia-Bondia, and F. Ruiz Ruiz, "Position-dependent noncommutative products: classical construction and field theory," *Nuclear Physics B*, vol. 727, no. 3, pp. 513–536, 2005.

[3] Y.-G. Miao, H. J. W. Muller-Kirsten, and D. K. Park, "Chiral bosons in noncommutative spacetime," *Journal of High Energy Physics*, vol. 08, p. 038, 2003.

[4] Y. Habara, "A new approach to scalar field theory on noncommutative space," *Progress of Theoretical Physics*, vol. 107, no. 1, pp. 211–230, 2002.

[5] A. Arai, "Representations of a quantum phase space with general degrees of freedom," *Mathematical Physics Preprint Archive 09-122*. In press.

[6] A. Arai, "Hilbert space representations of quantum phase spaces with general degrees of freedom," *RIMS Kôkyûroku*, vol. 1705, pp. 51–62, 2010.

[7] L. Gouba and F. G. Scholtz, "On the uniqueness of unitary representations of the non-commutative Heisenberg-Weyl algebra," *Canadian Journal of Physics*, vol. 87, no. 9, pp. 995–997, 2009.

[8] L. Jonke and S. Meljanac, "Representations of non-commutative quantum mechanics and symmetries," *European Physical Journal C*, vol. 29, pp. 433–439, 2003.

[9] K. Li and J. Wang, "The topological AC effect on noncommutative phase space," *European Physical Journal C*, vol. 50, no. 4, pp. 1007–1011, 2007.

[10] L. R. Riberio, E. Passos, C. Furtado, and J. R. Nascimento, "Landau analog levels for dipoles in non-commutative space and phase space," *European Physical Journal C*, vol. 56, no. 4, pp. 597–606, 2008.

[11] J.-Z. Zhang, "Consistent deformed bosonic algebra in noncommutative quantum mechanics," *International Journal of Modern Physics A*, vol. 23, no. 09, pp. 1393–1403, 2008.

[12] K. Schmüdgen, *Unbounded Operator Algebras and Representation Theory*, Birkhäuser, Basel, Switzerland, 1990.

[13] C. R. Putnam, *Commutation Properties of Hilbert Space Operators and Related Topics*, Springer, Berlin, Germany, 1967.

[14] A. Arai, "Generalized weak Weyl relation and decay of quantum dynamics," *Reviews in Mathematical Physics*, vol. 17, no. 9, pp. 1071–1109, 2005.

[15] K. Schmüdgen, "On the Heisenberg commutation relation. I," *Journal of Functional Analysis*, vol. 50, no. 1, pp. 8–49, 1983.

[16] K. Schmüdgen, "On the Heisenberg commutation relation. II," *Journal of Functional Analysis*, vol. 19, no. 1, pp. 601–671, 1983.

[17] H. P. Robertson, "The uncertainty principle," *Physical Review*, vol. 34, no. 1, pp. 163–164, 1929.

[18] I. Satake, *Linear Algebra*, Shokabo, Tokyo, Japan, 1974.

[19] M. Reed and B. Simon, *Methods of Modern Mathematical Physics II: Fourier Analysis, Self-Adjointness*, Academic Press, New York, NY, USA, 1975.

[20] A. Arai, "Representation-theoretic aspects of two-dimensional quantum systems in singular vector potentials: canonical commutation relations, quantum algebras, and reduction to lattice quantum systems," *Journal of Mathematical Physics*, vol. 39, no. 5, pp. 2476–2498, 1998.

[21] B. Fuglede, "On the relation $PQ - QP = -iI$," *Mathematica Scandinavica*, vol. 20, pp. 79–88, 1967.

[22] J. von Neumann, "Die Eindeutigkeit der Schrödingerschen operatoren," *Mathematische Annalen*, vol. 104, no. 1, pp. 570–578, 1931.

[23] M. Reed and B. Simon, *Methods of Modern Mathematical Physics I: Functional Analysis*, Academic Press, New York, NY, USA, 1972.

[24] A. Arai and Y. Matsuzawa, "Construction of a Weyl representation from a weak Weyl representation of the canonical commutation relation," *Letters in Mathematical Physics*, vol. 83, no. 2, pp. 201–211, 2008.

23

On Asymptotically Lacunary Statistical Equivalent Set Sequences

Uğur Ulusu and Fatih Nuray

Department of Mathematics, Faculty of Science and Literature, Afyon Kocatepe University, 03200 Afyonkarahisar, Turkey

Correspondence should be addressed to Fatih Nuray; fnuray@aku.edu.tr

Academic Editor: Peijun Guo

This paper presents three definitions which are natural combination of the definitions of asymptotic equivalence, statistical convergence, lacunary statistical convergence, and Wijsman convergence. In addition, we also present asymptotically equivalent (Wijsman sense) analogs of theorems in Patterson and Savaş (2006).

1. Introduction

In 1993, Marouf presented definitions for asymptotically equivalent and asymptotic regular matrices. In 2003, Patterson extended these concepts by presenting an asymptotically statistical equivalent analog of these definitions and natural regularity conditions for nonnegative summability matrices. In 2006, Patterson and Savaş extended the definitions presented in [1] to lacunary sequences. In addition to these definitions, natural inclusion theorems were presented. The concept of Wijsman statistical convergence is implementation of the concept of statistical convergence to sequences of sets presented by Nuray and Rhoades in 2012. Similar to this concept, the concept of Wijsman lacunary statistical convergence was presented by Ulusu and Nuray in 2012. This paper extends the definitions presented in [2] to Wijsman statistical convergent sequences and Wijsman lacunary statistical convergent sequences. In addition to these definitions, natural inclusion theorems will also be presented.

2. Definitions and Notations

Definition 1 (see Marouf [3]). Two nonnegative sequences $x = (x_k)$ and $y = (y_k)$ are said to be asymptotically equivalent if

$$\lim_k \frac{x_k}{y_k} = 1 \tag{1}$$

(denoted by $x \sim y$).

Definition 2 (see Fridy [4]). The sequence $x = (x_k)$ is said to be statistically convergent to the number L if for every $\varepsilon > 0$,

$$\lim_n \frac{1}{n} \left|\{k \le n : |x_k - L| \ge \varepsilon\}\right| = 0. \tag{2}$$

In this case we write $st - \lim x_k = L$.

The next definition is natural combination of Definitions 1 and 2.

Definition 3 (see Patterson [1]). Two nonnegative sequences $x = (x_k)$ and $y = (y_k)$ are said to be asymptotically statistical equivalent of multiple L provided that for every $\varepsilon > 0$

$$\lim_n \frac{1}{n} \left|\left\{k \le n : \left|\frac{x_k}{y_k} - L\right| \ge \varepsilon\right\}\right| = 0 \tag{3}$$

(denoted by $x \overset{S_L}{\sim} y$) and simply asymptotically statistically equivalent if $L = 1$.

By a lacunary sequence we mean an increasing integer sequence $\theta = \{k_r\}$ such that $k_0 = 0$ and $h_r = k_r - k_{r-1} \to \infty$ as $r \to \infty$. Throughout this paper the intervals determined by θ will be denoted by $I_r = (k_{r-1}, k_r]$, and ratio k_r/k_{r-1} will be abbreviated by q_r.

Definition 4 (see Patterson and Savaş [2]). Let θ be a lacunary sequence; the two nonnegative sequences $x = (x_k)$ and

$y = (y_k)$ are said to be asymptotically lacunary statistical equivalent of multiple L provided that for every $\varepsilon > 0$

$$\lim_r \frac{1}{h_r} \left| \left\{ k \in I_r : \left| \frac{x_k}{y_k} - L \right| \geq \varepsilon \right\} \right| = 0 \qquad (4)$$

(denoted by $x \overset{S_\theta^L}{\sim} y$) and simply asymptotically lacunary statistically equivalent if $L = 1$.

Definition 5 (see Patterson & Savaş [2]). Let θ be a lacunary sequence; two nonnegative number sequences $x = (x_k)$ and $y = (y_k)$ are strongly asymptotically lacunary equivalent of multiple L provided that

$$\lim_r \frac{1}{h_r} \sum_{k \in I_r} \left| \frac{x_k}{y_k} - L \right| = 0 \qquad (5)$$

(denoted by $x \overset{N_\theta^L}{\sim} y$) and strongly simply asymptotically lacunary equivalent if $L = 1$.

Let (X, ρ) be a metric space. For any point $x \in X$ and any nonempty subset A of X, we define the distance from x to A by

$$d(x, A) = \inf_{a \in A} \rho(x, A). \qquad (6)$$

Definition 6 (see Baronti & Papini [5]). Let (X, ρ) be a metric space. For any nonempty closed subset $A, A_k \subseteq X$, we say that the sequence $\{A_k\}$ is Wijsman convergent to A if

$$\lim_{k \to \infty} d(x, A_k) = d(x, A) \qquad (7)$$

for each $x \in X$. In this case we write $W - \lim A_k = A$.

Definition 7 (see Nuray & Rhoades [6]). Let (X, ρ) a metric space. For any nonempty closed subset $A, A_k \subseteq X$, we say that the sequence $\{A_k\}$ is Wijsman statistically convergent to A if $\{d(x, A_k)\}$ is statistically convergent to $d(x, A)$; that is, for $\varepsilon > 0$ and for each $x \in X$,

$$\lim_{n \to \infty} \frac{1}{n} \left| \left\{ k \leq n : \left| d(x, A_k) - d(x, A) \right| \geq \varepsilon \right\} \right| = 0. \qquad (8)$$

In this case we write $st - \lim_W A_k = A$ or $A_k \to A(WS)$.

Also the concept of bounded sequence for sequences of sets was given by Nuray and Rhoades.

Definition 8 (see Nuray & Rhoades [6]). Let (X, ρ) be a metric space. For any nonempty closed subset A_k of X, we say that the sequence $\{A_k\}$ is bounded if

$$\sup_k d(x, A_k) < \infty \qquad (9)$$

for each $x \in X$. In this case we write $\{A_k\} \in L_\infty$.

Definition 9 (see Ulusu & Nuray [7]). Let (X, ρ) be a metric space and let $\theta = \{k_r\}$ be a lacunary sequence. For any nonempty closed subset $A, A_k \subseteq X$, we say that the sequence $\{A_k\}$ is Wijsman lacunarily statistically convergent to A if $\{d(x, A_k)\}$ is lacunarily statistically convergent to $d(x, A)$; that is, for $\varepsilon > 0$ and for each $x \in X$,

$$\lim_r \frac{1}{h_r} \left| \left\{ k \in I_r : \left| d(x, A_k) - d(x, A) \right| \geq \varepsilon \right\} \right| = 0. \qquad (10)$$

In this case we write $S_\theta - \lim_W = A$ or $A_k \to A(WS_\theta)$.

Following these results we introduce three new notions that are asymptotically statistical equivalent (Wijsman sense) of multiple L, asymptotically lacunary statistical equivalent (Wijsman sense) of multiple L, and strongly asymptotically lacunary equivalent (Wijsman sense) of multiple L.

Definition 10. Let (X, ρ) be a metric space. For any non-empty closed subset $A_k, B_k \subseteq X$ such that $d(x, A_k) > 0$ and $d(x, B_k) > 0$ for each $x \in X$. We say that the sequences $\{A_k\}$ and $\{B_k\}$ are asymptotically equivalent (Wijsman sense) if for each $x \in X$,

$$\lim_k \frac{d(x, A_k)}{d(x, B_k)} = 1 \qquad (11)$$

(denoted by $A_k \sim B_k$).

As an example, consider the following sequences of circles in the (x, y)-plane:

$$A_k = \left\{ (x, y) : x^2 + y^2 + 2kx = 0 \right\}, \qquad B_k = \left\{ (x, y) : x^2 + y^2 - 2kx = 0 \right\}. \qquad (12)$$

Since

$$\lim_k \frac{d(x, A_k)}{d(x, B_k)} = 1, \qquad (13)$$

the sequences $\{A_k\}$ and $\{B_k\}$ are asymptotically equivalent (Wijsman sense); that is, $A_k \sim B_k$.

Definition 11. Let (X, ρ) be a metric space. For any non-empty closed subset $A_k, B_k \subseteq X$ such that $d(x, A_k) > 0$ and $d(x, B_k) > 0$ for each $x \in X$. We say that the sequences $\{A_k\}$ and $\{B_k\}$ are asymptotically statistically equivalent (Wijsman sense) of multiple L if for every $\varepsilon > 0$ and for each $x \in X$,

$$\lim_n \frac{1}{n} \left| \left\{ k \leq n : \left| \frac{d(x, A_k)}{d(x, B_k)} - L \right| \geq \varepsilon \right\} \right| = 0 \qquad (14)$$

(denoted by $\{A_k\} \overset{WS_L}{\sim} \{B_k\}$) and simply asymptotically statistical equivalent (Wijsman sense) if $L = 1$.

As an example, consider the following sequences of circles in the (x, y)-plane:

$$A_k = \begin{cases} \left\{(x, y) : x^2 + y^2 + 2ky = 0\right\}, & \text{if } k \text{ is a square integer,} \\ \{(1, 1)\}, & \text{otherwise,} \end{cases}$$
$$B_k = \begin{cases} \left\{(x, y) : x^2 + y^2 - 2ky = 0\right\}, & \text{if } k \text{ is a square integer,} \\ \{(1, 1)\}, & \text{otherwise.} \end{cases} \quad (15)$$

Since

$$\lim_n \frac{1}{n} \left| \left\{ k \leq n : \left| \frac{d(x, A_k)}{d(x, B_k)} - 1 \right| \geq \varepsilon \right\} \right| = 0, \quad (16)$$

the sequences $\{A_k\}$ and $\{B_k\}$ are asymptotically statistically equivalent (Wijsman sense); that is, $\{A_k\} \overset{WS_1}{\sim} \{B_k\}$.

Definition 12. Let (X, ρ) be a metric space and let θ be a lacunary sequence. For any non-empty closed subset A_k, $B_k \subseteq X$ such that $d(x, A_k) > 0$ and $d(x, B_k) > 0$ for each $x \in X$. We say that the sequences $\{A_k\}$ and $\{B_k\}$ are asymptotically lacunary equivalent (Wijsman sense) of multiple L if for each $x \in X$,

$$\lim_r \frac{1}{h_r} \sum_{k \in I_r} \frac{d(x, A_k)}{d(x, B_k)} = L \quad (17)$$

(denoted by $\{A_k\} \overset{WN_\theta^L}{\sim} \{B_k\}$) and simply asymptotically lacunarily equivalent (Wijsman sense) if $L = 1$.

Definition 13. Let (X, ρ) be a metric space and let θ be a lacunary sequence. For any non-empty closed subset A_k, $B_k \subseteq X$ such that $d(x, A_k) > 0$ and $d(x, B_k) > 0$ for each $x \in X$. We say that the sequences $\{A_k\}$ and $\{B_k\}$ are strongly asymptotically lacunary equivalent (Wijsman sense) of multiple L if for each $x \in X$,

$$\lim_r \frac{1}{h_r} \sum_{k \in I_r} \left| \frac{d(x, A_k)}{d(x, B_k)} - L \right| = 0 \quad (18)$$

(denoted by $\{A_k\} \overset{[WN]_\theta^L}{\sim} \{B_k\}$) and simply strongly asymptotically lacunarily equivalent (Wijsman sense) if $L = 1$.

As an example, consider the following sequences:

$$A_k := \begin{cases} \left\{(x, y) \in \mathbb{R}^2 : \dfrac{(x - \sqrt{k})^2}{k} + \dfrac{y^2}{2k} = 1\right\}, & \text{if } k_{r-1} < k < k_{r-1} + \left[\sqrt{h_r}\right] \\ \{(1, 1)\}, & \text{otherwise,} \end{cases}$$
$$B_k := \begin{cases} \left\{(x, y) \in \mathbb{R}^2 : \dfrac{(x + \sqrt{k})^2}{k} + \dfrac{y^2}{2k} = 1\right\}, & \text{if } k_{r-1} < k < k_{r-1} + \left[\sqrt{h_r}\right] \\ \{(1, 1)\}, & \text{otherwise.} \end{cases} \quad (19)$$

Since

$$\lim_r \frac{1}{h_r} \sum_{k \in I_r} \left| \frac{d(x, A_k)}{d(x, B_k)} - 1 \right| = 0, \quad (20)$$

the sequences $\{A_k\}$ and $\{B_k\}$ are strongly asymptotically lacunarily equivalent (Wijsman sense); that is, $\{A_k\} \overset{WN_\theta^1}{\sim} \{B_k\}$.

Definition 14. Let (X, ρ) be a metric space and let θ be a lacunary sequence. For any non-empty closed subset $A_k, B_k \subseteq X$ such that $d(x, A_k) > 0$ and $d(x, B_k) > 0$ for each $x \in X$. We say that the sequences $\{A_k\}$ and $\{B_k\}$ are asymptotically lacunarily statistical equivalent (Wijsman sense) of multiple L if for every $\varepsilon > 0$ and each $x \in X$,

$$\lim_r \frac{1}{h_r} \left| \left\{ k \in I_r : \left| \frac{d(x, A_k)}{d(x, B_k)} - L \right| \geq \varepsilon \right\} \right| = 0 \quad (21)$$

(denoted by $\{A_k\} \overset{WS_\theta^L}{\sim} \{B_k\}$) and simply asymptotically lacunarily statistically equivalent (Wijsman sense) if $L = 1$.

As an example, consider the following sequences:

$$A_k := \begin{cases} \left\{(x, y) \in \mathbb{R}^2 : x^2 + (y - 1)^2 = \dfrac{1}{k}\right\}, & \text{if } k_{r-1} < k < k_{r-1} + \left[\sqrt{h_r}\right], \ k \text{ is a square integer,} \\ \{(0, 0)\}, & \text{otherwise,} \end{cases}$$
$$B_k := \begin{cases} \left\{(x, y) \in \mathbb{R}^2 : x^2 + (y + 1)^2 = \dfrac{1}{k}\right\}, & \text{if } k_{r-1} < k < k_{r-1} + \left[\sqrt{h_r}\right], \ k \text{ is a square integer,} \\ \{(0, 0)\}, & \text{otherwise.} \end{cases} \quad (22)$$

Since

$$\lim_r \frac{1}{h_r} \left| \left\{ k \in I_r : \left| \frac{d(x, A_k)}{d(x, B_k)} - 1 \right| \geq \varepsilon \right\} \right| = 0, \quad (23)$$

the sequences $\{A_k\}$ and $\{B_k\}$ are asymptotically lacunarily statistically equivalent (Wijsman sense); that is, $\{A_k\} \overset{WS_\theta^1}{\sim} \{B_k\}$.

3. Main Results

Theorem 15. *Let (X, ρ) be a metric space, let $\theta = \{k_r\}$ be a lacunary sequence, and let A_k, B_k be non-empty closed subsets of X:*

(i) (a) $\{A_k\} \overset{[WN]_\theta^L}{\sim} \{B_k\} \Rightarrow \{A_k\} \overset{WS_\theta^L}{\sim} \{B_k\}$

(b) $[WN]_\theta^L$ *is a proper subset of* WS_θ^L,

(ii) $\{A_k\} \in L_\infty$ *and* $\{A_k\} \overset{WS_\theta^L}{\sim} \{B_k\} \Rightarrow \{A_k\} \overset{[WN]_\theta^L}{\sim} \{B_k\}$,

(iii) $WS_\theta^L \cap L_\infty = [WN]_\theta^L \cap L_\infty$,

where L_∞ denotes the set of bounded sequences of sets.

Proof. (i)-(a). Let $\varepsilon > 0$ and $\{A_k\} \overset{[WN]_\theta^L}{\sim} \{B_k\}$. Then we can write

$$\sum_{k\in I_r}\left|\frac{d(x,A_k)}{d(x,B_k)} - L\right| \geq \sum_{\substack{k\in I_r \\ |d(x,A_k)/d(x,B_k)-L|\geq\varepsilon}}\left|\frac{d(x,A_k)}{d(x,B_k)} - L\right| \geq \varepsilon\cdot\left|\left\{k\in I_r : \left|\frac{d(x,A_k)}{d(x,B_k)} - L\right|\geq\varepsilon\right\}\right| \tag{24}$$

which yields the result.

(ii)-(b). Suppose that $[WN]_\theta^L \subset WS_\theta^L$. Let $\{A_k\}$ and $\{B_k\}$ be the following sequences:

$$A_k = \begin{cases} \{k\}, & \text{if } k_{r-1} < k \leq k_{r-1} + \left[\sqrt{h_r}\right] \\ & r = 1,2,\ldots \\ \{0\}, & \text{otherwise,} \end{cases} \tag{25}$$

$$B_k = \{0\} \quad \forall k.$$

Note that $\{A_k\}$ is not bounded. We have, for every $\varepsilon > 0$ and for each $x \in X$,

$$\frac{1}{h_r}\left|\left\{k\in I_r : \left|\frac{d(x,A_k)}{d(x,B_k)} - 1\right|\geq\varepsilon\right\}\right| = \frac{\left[\sqrt{h_r}\right]}{h_r} \longrightarrow 0 \quad \text{as } r\longrightarrow\infty. \tag{26}$$

That is, $\{A_k\} \overset{WS_\theta^1}{\sim} \{B_k\}$. On the other hand,

$$\frac{1}{h_r}\sum_{k\in I_r}\left|\frac{d(x,A_k)}{d(x,B_k)} - L\right| \nrightarrow 0 \quad \text{as } r\longrightarrow\infty. \tag{27}$$

Hence $\{A_k\} \overset{[WN]_\theta^L}{\nsim} \{B_k\}$.

(ii) Suppose that $\{A_k\} \in L_\infty$ and $\{A_k\} \overset{WS_\theta^L}{\sim} \{B_k\}$. Then we can assume that

$$\left|\frac{d(x,A_k)}{d(x,B_k)} - L\right| \leq M \tag{28}$$

for each $x \in X$ and all k.

Given $\varepsilon > 0$, we get

$$\begin{aligned}\frac{1}{h_r}\sum_{k\in I_r}\left|\frac{d(x,A_k)}{d(x,B_k)} - L\right| &= \frac{1}{h_r}\sum_{\substack{k\in I_r \\ |d(x,A_k)/d(x,B_k)-L|\geq\varepsilon}}\left|\frac{d(x,A_k)}{d(x,B_k)} - L\right| + \frac{1}{h_r}\sum_{\substack{k\in I_r \\ |d(x,A_k)/d(x,B_k)-L|<\varepsilon}}\left|\frac{d(x,A_k)}{d(x,B_k)} - L\right| \\ &\leq \frac{M}{h_r}\left|\left\{k\in I_r : \left|\frac{d(x,A_k)}{d(x,B_k)} - L\right|\geq\varepsilon\right\}\right| + \varepsilon.\end{aligned} \tag{29}$$

Therefore $\{A_k\} \overset{[WN]_\theta^L}{\sim} \{B_k\}$.

(iii) This is an immediate consequences of (i)and (ii). □

Theorem 16. *Let (X,ρ) be a metric space and let A_k, B_k be non-empty closed subsets of X. If $\theta = \{k_r\}$ is a lacunary sequence with* $\liminf_r q_r > 1$, *then*

$$\{A_k\} \overset{WS_L}{\sim} \{B_k\} \Longrightarrow \{A_k\} \overset{WS_\theta^L}{\sim} \{B_k\}. \tag{30}$$

Proof. Suppose first that $\liminf_r q_r > 1$, then there exists a $\lambda > 0$ such that $q_r \geq 1+\lambda$ for sufficiently large r, which implies that

$$\frac{h_r}{k_r} \geq \frac{\lambda}{1+\lambda}. \tag{31}$$

If $\{A_k\} \overset{WS_L}{\sim} \{B_k\}$, then for every $\varepsilon > 0$, for each $x \in X$, and for sufficiently large r, we have

$$\begin{aligned}\frac{1}{k_r}\left|\left\{k\leq k_r : \left|\frac{d(x,A_k)}{d(x,B_k)} - L\right|\geq\varepsilon\right\}\right| &\geq \frac{1}{k_r}\left|\left\{k\in I_r : \left|\frac{d(x,A_k)}{d(x,B_k)} - L\right|\geq\varepsilon\right\}\right| \\ &\geq \frac{\lambda}{1+\lambda}\cdot\left(\frac{1}{h_r}\left|\left\{k\in I_r : \left|\frac{d(x,A_k)}{d(x,B_k)} - L\right|\geq\varepsilon\right\}\right|\right).\end{aligned} \tag{32}$$

This completes the proof. □

Theorem 17. *Let (X,ρ) be a metric space and let A_k, B_k be non-empty closed subsets of X. If $\theta = \{k_r\}$ is a lacunary sequence with* $\limsup_r q_r < \infty$, *then*

$$\{A_k\} \overset{WS_\theta^L}{\sim} \{B_k\} \Longrightarrow \{A_k\} \overset{WS_L}{\sim} \{B_k\}. \tag{33}$$

Proof. Let $\limsup_r q_r < \infty$. Then there is an $M > 0$ such that $q_r < M$ for all $r \geq 1$. Let $\{A_k\} \overset{WS_\theta^L}{\sim} \{B_k\}$ and $\varepsilon > 0$. There exists $R > 0$ such that for every $j \geq R$

$$A_j = \frac{1}{h_j}\left|\left\{k\in I_j : \left|\frac{d(x,A_k)}{d(x,B_k)} - L\right|\geq\varepsilon\right\}\right| < \varepsilon. \tag{34}$$

We can also find $H > 0$ such that $A_j < H$ for all $j = 1, 2, \ldots$. Now let t be any integer satisfying $k_{r-1} < t \le k_r$, where $r > R$. Then we can write

$$\begin{aligned}
&\frac{1}{t}\left|\left\{k \le t : \left|\frac{d(x, A_k)}{d(x, B_k)} - L\right| \ge \varepsilon\right\}\right| \\
&\le \frac{1}{k_{r-1}}\left|\left\{k \le k_r : \left|\frac{d(x, A_k)}{d(x, B_k)} - L\right| \ge \varepsilon\right\}\right| \\
&= \frac{1}{k_{r-1}}\left\{\left|\left\{k \in I_1 : \left|\frac{d(x, A_k)}{d(x, B_k)} - L\right| \ge \varepsilon\right\}\right|\right\} \\
&+ \frac{1}{k_{r-1}}\left\{\left|\left\{k \in I_2 : \left|\frac{d(x, A_k)}{d(x, B_k)} - L\right| \ge \varepsilon\right\}\right|\right\} \\
&+ \cdots + \frac{1}{k_{r-1}} \\
&\times \left\{\left|\left\{k \in I_r : \left|\frac{d(x, A_k)}{d(x, B_k)} - L\right| \ge \varepsilon\right\}\right|\right\} \\
&= \frac{k_1}{k_{r-1}k_1}\left|\left\{k \in I_1 : \left|\frac{d(x, A_k)}{d(x, B_k)} - L\right| \ge \varepsilon\right\}\right| \\
&+ \frac{k_2 - k_1}{k_{r-1}(k_2 - k_1)} \\
&\times \left|\left\{k \in I_2 : \left|\frac{d(x, A_k)}{d(x, B_k)} - L\right| \ge \varepsilon\right\}\right| \\
&+ \cdots + \frac{k_R - k_{R-1}}{k_{r-1}(k_R - k_{R-1})} \\
&\times \left|\left\{k \in I_R : \left|\frac{d(x, A_k)}{d(x, B_k)} - L\right| \ge \varepsilon\right\}\right| \\
&+ \cdots + \frac{k_r - k_{r-1}}{k_{r-1}(k_r - k_{r-1})} \\
&\times \left|\left\{k \in I_r : \left|\frac{d(x, A_k)}{d(x, B_k)} - L\right| \ge \varepsilon\right\}\right| \\
&= \frac{k_1}{k_{r-1}}A_1 + \frac{k_2 - k_1}{k_{r-1}}A_2 + \cdots + \frac{k_R - k_{R-1}}{k_{r-1}}A_R \\
&+ \frac{k_{R+1} - k_R}{k_{r-1}}A_{R+1} + \cdots + \frac{k_r - k_{r-1}}{k_{r-1}}A_r \\
&\le \left\{\sup_{j\ge1} A_j\right\}\frac{k_R}{k_{r-1}} + \left\{\sup_{j\ge B} A_j\right\}\frac{k_r - k_R}{k_{r-1}} \\
&\le H\frac{k_B}{k_{r-1}} + \varepsilon M.
\end{aligned} \tag{35}$$

This completes the proof. □

Combining Theorems 16 and 17 we have the following.

Theorem 18. *Let (X, ρ) be a metric space and let A_k, B_k be non-empty closed subsets of X. If $\theta = \{k_r\}$ is a lacunary sequence with $1 < \liminf_r q_r \le \limsup_r q_r < \infty$, then*

$$\{A_k\} \overset{WS_\theta^L}{\sim} \{B_k\} = \{A_k\} \overset{WS_L}{\sim} \{B_k\}. \tag{36}$$

Proof. This is an immediate consequence of Theorems 16 and 17. □

References

[1] R. F. Patterson, "On asymptotically statistical equivalent sequences," *Demonstratio Mathematica*, vol. 36, no. 1, pp. 149–153, 2003.

[2] R. F. Patterson and E. Savaş, "On asymptotically lacunary statistically equivalent sequences," *Thai Journal of Mathematics*, vol. 4, pp. 267–272, 2006.

[3] M. S. Marouf, "Asymptotic equivalence and summability," *International Journal of Mathematics and Mathematical Sciences*, vol. 16, no. 4, pp. 755–762, 1993.

[4] J. A. Fridy, "On statistical convergence," *Analysis*, vol. 5, no. 4, pp. 301–313, 1985.

[5] M. Baronti and P. L. Papini, "Convergence of sequences of sets," in *Methods of Functional Analysis in Approximation Theory*, vol. 76, pp. 135–155, Birkhäuser, Basel, Switzerland, 1986.

[6] F. Nuray and B. E. Rhoades, "Statistical convergence of sequences of sets," *Fasciculi Mathematici*, vol. 49, pp. 87–99, 2012.

[7] U. Ulusu and F. Nuray, "Lacunary statistical convergence of sequence of sets," *Progress in Applied Mathematics*, vol. 4, no. 2, pp. 99–109, 2012.

Positive Solutions for a Fourth-Order Boundary Value Problem

Kun Wang and Zhilin Yang

Department of Mathematics, Qingdao Technological University, Fushun Road No. 11, Quingdao, Shandong 266033, China

Correspondence should be addressed to Kun Wang; wangkun880304@163.com

Academic Editor: Andrea Pascucci

This paper deals with the existence and multiplicity of positive solutions for the fourth-order boundary value problem $u^{(4)} = f(t, u, u', -u'', u'''), u(0) = u'(1) = u'''(0) = u''(1) = 0$. Here $f \in C([0,1]\times\mathbb{R}_+^4, \mathbb{R}_+)(\mathbb{R}_+ := [0, +\infty))$. We use fixed point index theory to establish our main results based on a priori estimates achieved by utilizing some integral identities and integral inequalities.

1. Introduction

The purpose of this paper is to study the existence and multiplicity of positive solutions for the fourth-order boundary value problem

$$
\begin{gathered}
u^{(4)} = f\left(t, u, u', -u'', u'''\right), \\
u(0) = u'(1) = u'''(0) = u''(1) = 0,
\end{gathered}
\tag{1}
$$

where $f \in C([0,1] \times \mathbb{R}_+^4, \mathbb{R}_+)$. By a positive solution of (1), we mean a function $u \in C^4([0,1], \mathbb{R}_+)$ that solves (1) and satisfies $u(t) > 0$ for all $t \in (0,1]$.

In recent years, fourth-order boundary value problems for nonlinear ordinary differential equations have been extensively studied by using diverse methods, including fixed point theorems on cones [1–6], the method of lower and upper solutions [7–10], the iterative method [11–14], critical point theory [15], and the shooting method [16]. It should be remarked that none of the results cited above involves derivatives of all orders in their nonlinearities. In [9], Ma and Yang studied the fourth-order four-point boundary value problem

$$
\begin{gathered}
u^{(4)} = f\left(t, u, -u''\right), \\
u(0) = 0, \quad u(1) = au(\eta), \\
u''(0) = 0, \quad u''(1) = bu(\xi),
\end{gathered}
\tag{2}
$$

where $\eta, \xi \in (0,1)$, $a, b \geqslant 0$. The authors used the method of upper and lower solutions and a new maximum principle to establish their existence results. In [14], Pei and Chang studied a class of fourth-order boundary value problem

$$
\begin{gathered}
y^{(4)} = f(t, y), \\
y(0) = y(1) = y'(0) = y'(1) = 0.
\end{gathered}
\tag{3}
$$

By using the monotone iterative technique, the authors proved that problem (3) has at least one symmetric positive solution under certain conditions. In [17], Pang et al. studied the existence and multiplicity of nontrivial solutions for the fourth-order boundary value problem

$$
\begin{gathered}
x^{(4)} = f\left(x, -x''\right), \\
x(0) = x(1) = x''(0) = x''(1) = 0,
\end{gathered}
\tag{4}
$$

where $f \in C(\mathbb{R}^2, \mathbb{R})$. Making use of the theory of Leray-Schauder degree, under appropriate conditions on the nonlinearity f, the authors proved that problem (4) has at least six different nontrivial solutions, including two positive ones and two negative ones.

In [18], Yang and Sun studied the fourth-order boundary value problem

$$
\begin{gathered}
u^{(4)} = f\left(t, u, u', -u'', -u'''\right), \\
u(0) = u'(1) = u''(0) = u'''(1) = 0,
\end{gathered}
\tag{5}
$$

where $f \in C([0,1]\times\mathbb{R}_+^4,\mathbb{R}_+)$. By using fixed point index theory based on a priori estimates, the authors established the main results on the existence, multiplicity and uniqueness of positive solutions for (5).

Motivated by [18], in this paper, we will discuss the existence and multiplicity of positive solutions for problem (1). As in [18], we first use the method of order reduction to transform (1) into a boundary value problem for a second-order integro-differential equation and then seek the existence and multiplicity of positive solutions for the resultant problem. To overcome the difficulties stemming from the presence of derivatives of all orders and the difference between (1) and (5) in their boundary value conditions, we have to prove that the maximum of every nonnegative concave function v can be dominated by the integrals

$$\int_0^1 v(t)(1-t)e^{1-t}dt,$$
$$\int_0^1 \left((B_1v)(t)+2(B_1v)'(t)\right)(1-t)e^{1-t}dt \quad (6)$$

(see Lemmas 5 and 6 below for more details). Based on a priori estimates achieved by utilizing some integral identities and integral inequalities, we use fixed point index theory to prove the existence and multiplicity of positive solutions for (1).

This paper is organized as follows. In Section 2, we transform (1) into a boundary value problem for a second-order integro-differential equation and then establish some basic integral identities and integral inequalities that are useful in deriving the priori estimates in the next section. Our main results, namely Theorems 10–12, are stated and proved in Section 3.

2. Preliminaries

Let $E := C^1([0,1],\mathbb{R})$ and $\|u\| := \max\{\|u\|_0, \|u'\|_0\}$, where $\|u\|_0 := \max\{|u(t)| : t\in[0,1]\}$. Furthermore, let

$$P := \left\{u\in E : u(t)\geqslant 0,\ u'(t)0,\ \forall t\in[0,1]\right\}. \quad (7)$$

Clearly $(E,\|\cdot\|)$ is a real Banach space and P is a cone in E.

Let

$$k_1(t,s) := \min\{t,s\},$$
$$k_2(t,s) := \min\{1-s,1-t\}. \quad (8)$$

Define the linear integral operators B_1 and B_2 by

$$(B_iv)(t) := \int_0^1 k_i(t,s)v(s)\,ds,\quad v\in P\ (i=1,2). \quad (9)$$

Substituting $v(t) := -u''(t)$ into (1), we transform (1) into the following second-order boundary value problem for the integro-differential equation

$$-v''(t) = f\left(t,(B_1v)(t),(B_1v)'(t),v(t),-v'(t)\right),$$
$$v'(0) = v(1) = 0, \quad (10)$$

which is equivalent to the nonlinear integral equation

$$v(t) = \int_0^1 k_2(t,s)f\left(s,(B_1v)(s),(B_1v)'(s),v(s),-v'(s)\right)ds. \quad (11)$$

Define the operator $A : P\to P$ by

$$(Av)(t) := \int_0^1 k_2(t,s)f\left(s,(B_1v)(s),(B_1v)'(s),v(s),-v'(s)\right)ds,\quad v\in P. \quad (12)$$

Now $f\in C([0,1]\times\mathbb{R}_+^4,\mathbb{R}_+)$ implies that A is a completely continuous operator. In our setting, the existence of positive solutions for (1) is equivalent to that of positive fixed points of $A : P\to P$.

To establish the priori estimates of positive solutions for some problems associated with (11), we need several integral identities and integral inequalities below.

Lemma 1. *If $v\in P\cap C^2[0,1]$, $v'(0)=v(1)=0$, then*

$$\int_0^1 -v''(t)(1-t)e^{1-t}dt = \int_0^1\left(v(t)-2v'(t)\right)(1-t)e^{1-t}dt. \quad (13)$$

Proof. Integrating by parts and using $v'(0)=v(1)=0$, we have

$$\int_0^1\left(-v''(t)\right)(1-t)\ e^{1-t}dt$$
$$=\int_0^1 v'(t)(t-2)e^{1-t}dt \quad (14)$$
$$=\int_0^1\left(-v'(t)\right)te^{1-t}dt - 2\int_0^1 v'(t)(1-t)e^{1-t}dt,$$
$$\int_0^1 -v'(t)te^{1-t}dt = \int_0^1 v(t)(1-t)e^{1-t}dt, \quad (15)$$

from which (13) follows. This completes the proof. □

Lemma 2. *If $v\in C([0,1],\mathbb{R}_+)$, then*

$$\int_0^1\left((B_1v)(t)+2(B_1v)'(t)te^tdt\right) = \int_0^1 v(t)te^tdt. \quad (16)$$

Proof. Notice $(B_1v)(0)=0$. Integrating by parts, we obtain

$$\int_0^1 (B_1v)(t)te^tdt = \int_0^1 (1-t)e^t(B_1v)'(t)\,dt, \quad (17)$$

so that

$$\int_0^1\left((B_1v)(t)+2(B_1v)'(t)\right)te^tdt = \int_0^1(1+t)e^t(B_1v)'(t)\,dt. \quad (18)$$

Notice $(B_1v)'(1) = 0$ and $(B_1v)''(t) = -v(t)$. Integrating the right hand of the above by parts again gives (16). This completes the proof. □

If both h_1 and h_2 are continuous and decreasing on $[0,1]$, then Chebyshev's inequality for h_1 and h_2 is given by

$$\int_0^1 h_1(t) h_2(t)\,dt \geqslant \int_0^1 h_1(t)\,dt \int_0^1 h_2(t)\,dt. \quad (19)$$

In addition, if both h_1 and h_2 are continuous, with one of them decreasing on $[0,1]$ and the other increasing on $[0,1]$, then Chebyshev's inequality for h_1 and h_2 becomes

$$\int_0^1 h_1(t) h_2(t)\,dt \leqslant \int_0^1 h_1(t)\,dt \int_0^1 h_2(t)\,dt. \quad (20)$$

Lemma 3. *If $v \in P, v(1) = 0$, and v' is decreasing on $[0,1]$, then*

$$\int_0^1 v(t)(1-t)e^{1-t}dt \geqslant (e-2)\int_0^1 -v'(t)(1-t)e^{1-t}dt. \quad (21)$$

Proof. Notice $v(1) = 0$ and $v' \leqslant 0$. Since $-v'$ is increasing on $[0,1]$ and $(1-t)e^{1-t}$ is decreasing on $[0,1]$, it follows from Chebyshev's inequality that

$$\int_0^1 -v'(t)(1-t)e^{1-t}dt \leqslant \int_0^1 -v'(t)\,dt \int_0^1 (1-t)e^{1-t}dt = v(0). \quad (22)$$

Notice that te^{1-t} is increasing on $[0,1]$. Chebyshev's inequality again implies

$$\int_0^1 -v'(t)te^{1-t}dt \geqslant \int_0^1 -v'(t)\,dt \int_0^1 te^{1-t}dt = (e-2)v(0). \quad (23)$$

Note that (15) is guaranteed by $v \in P, v(1) = 0$. By virtue of (15), we have

$$\int_0^1 v(t)(1-t)e^{1-t}dt \geqslant (e-2)v(0). \quad (24)$$

Combining this with (22) yields (21). This complete the proof. □

Lemma 4. *If $v \in P, v(1) = 0$, then*

$$v(0) \leqslant \int_0^1 \left(v(t) - 2v'(t)\right)(1-t)e^{1-t}dt. \quad (25)$$

Proof. By (15), we have

$$\begin{aligned}\int_0^1 \left(v(t) - 2v'(t)\right)(1-t)e^{1-t}ds &= \int_0^1 -v'(t)(2-t)e^{1-t}dt \\ &\geqslant \int_0^1 -v'(t)\,dt = v(0).\end{aligned} \quad (26)$$

This completes the proof. □

Lemma 5. *If $v \in P \cap C^2[0,1]$ and $v'' \leqslant 0$, then*

$$(e-2)\|v\|_0 \leqslant \int_0^1 v(t)(1-t)e^{1-t}dt \leqslant e\|v\|_0. \quad (27)$$

Proof. It is easy to see that $\int_0^1 v(t)te^t dt \leqslant e\|v\|_0$. The condition $v' \leqslant 0$ implies $\|v\|_0 = v \leqslant (0)$. By the concavity of v, we have

$$\begin{aligned}\int_0^1 v(t)(1-t)e^{1-t}dt &\geqslant \int_0^1 (1-t)e^{1-t}v(t\cdot 1 + (1-t)\cdot 0)\,dt \\ &\geqslant v(0)\int_0^1 (1-t)^2 e^{1-t}dt = (e-2)\|v\|_0.\end{aligned} \quad (28)$$

This completes the proof. □

Lemma 6. *If $v \in P \cap C^2[0,1]$ and $v'' \leqslant 0$, then we have*

$$\begin{aligned}\frac{3-e}{6}\|v\|_0 &\leqslant \int_0^1 \left((B_1v)(t) + 2(B_1v)'(t)\right)(1-t)e^{1-t}dt \\ &\leqslant 3e\|v\|_0.\end{aligned} \quad (29)$$

Proof. By the concavity of B_1v and v, we have

$$\begin{aligned}&\int_0^1 \left((B_1v)(t) + 2(B_1v)'(t)\right)(1-t)e^{1-t}dt \\ &\geqslant \int_0^1 (B_1v)(t)(1-t)e^{1-t}dt \\ &= \int_0^1 (B_1v)(t\cdot 1 + (1-t)\cdot 0)(1-t)e^{1-t}dt \\ &\geqslant (B_1v)(1)\int_0^1 t(1-t)e^{1-t}dt \\ &= (3-e)\int_0^1 sv(s)\,ds \\ &\geqslant (3-e)\int_0^1 s(1-s)v(0)\,ds \\ &= \frac{3-e}{6}\|v\|_0, \\ &\int_0^1 \left((B_1v)(t) + 2(B_1v)'(t)\right)(1-t)e^{1-t}dt \\ &\leqslant e\int_0^1 (B_1v)(t) + 2(B_1v)'(t)\,dt \\ &= e\int_0^1 \left(\int_0^1 k_1(t,s)v(s)\,ds + 2\int_0^1 v(s)\,ds\right)dt \\ &\leqslant 3e\|v\|_0.\end{aligned} \quad (30)$$

This completes the proof. □

Lemma 7 (see [19]). *Let E be a real Banach space and P a cone in E. Suppose that $\Omega \subset E$ is a bounded open set and that*

$T : \overline{\Omega} \cap P \to P$ is a completely continuous operator. If there exists $\omega_0 \in P \setminus \{0\}$ such that

$$\omega - T\omega \neq \lambda\omega_0, \quad \forall\lambda \geqslant 0,\ \omega \in \partial\Omega \cap P, \tag{31}$$

then $i(T, \Omega \cap P, P) = 0$, where i indicates the fixed point index on P.

Lemma 8 (see [19]). *Let E be a real Banach space and P a cone in E. Suppose that $\Omega \subset E$ is a bounded open set with $0 \in \Omega$ and that $T : \overline{\Omega} \cap P \to P$ is a completely continuous operator. If*

$$\omega - \lambda T\omega \neq 0, \quad \forall\lambda \in [0, 1],\ \omega \in \partial\Omega \cap P, \tag{32}$$

then $i(T, \Omega \cap P, P) = 1$.

3. Existence and Multiplicity of Positive Solutions for (1)

For simplicity, we denote by $x := (x_1, x_2, x_3, x_4) \in \mathbb{R}_+^4$ and $I_\rho := [0, \rho]$ for $\rho > 0$ in this section. Now we list our hypotheses.

(H1) $f \in C([0, 1] \times \mathbb{R}_+^4, \mathbb{R}_+)$.

(H2) There exist $a_1 \geqslant 0$, $b_1 \geqslant 0$, $c > 0$, such that $d_1 := a_1 (3 - e(e-2))/6e^2 + b_1 > 1$, and

$$f(t, x) \geqslant a_1 (x_1 + 2x_2) + b_1 (x_3 + 2x_4) - c \tag{33}$$

holds for all $(t, x) \in [0, 1] \times \mathbb{R}_+^4$.

(H3) There exists a function $\Phi \in C(\mathbb{R}_+, \mathbb{R}_+)$ such that

$$f(t, x) \leqslant \Phi(x_4) \tag{34}$$

for all $x \in [0, 1] \times I_m^3 \times \mathbb{R}_+$ and

$$\int_0^\infty \frac{\xi d\xi}{\Phi(\xi) + 2me} > m. \tag{35}$$

where $m := c/(d_1 - 1) > 0$ and c, d_1 are given as in (H2).

(H4) There exist $a_2 \geqslant 0$, $b_2 \geqslant 0$, $r > 0$ such that $d_2 := 3a_2 e/(e-2) + b_2 < 1$, and

$$f(t, x) \leqslant a_2 (x_1 + 2x_2) + b_2 (x_3 + 2x_4) \tag{36}$$

holds for all $(t, x) \in [0, 1] \times I_r^3$.

(H5) There exist $a_3 \geqslant 0$, $b_3 \geqslant 0$, $c > 0$ such that $d_3 := 3a_3 e/(e-2) + b_2 < 1$, and

$$f(t, x) \leqslant a_3 (x_1 + 2x_2) + b_3 (x_3 + 2x_4) + c \tag{37}$$

holds for all $(t, x) \in [0, 1] \times \mathbb{R}_+^4$.

(H6) There are nonnegative constants $a_4 \geqslant 0$, $b_4 \geqslant 0$, $r > 0$, such that $d_4 := a_4(3-e)(e-2)/6e^2 + b_4 > 1$, and

$$f(t, x) \geqslant a_4 (x_1 + 2x_2) + b_4 (x_3 + 2x_4) \tag{38}$$

holds for all $(t, x) \in [0, 1] \times I_r^4$.

(H7) There is a constant $\omega > 0$ such that $f(t, x)$ is increasing on I_ω^4 in x, and

$$\int_0^1 f(s, \omega, \omega, \omega, \omega)\, ds < \omega. \tag{39}$$

Remark 9. $f(t, x)$ is said to be increasing in x if $f(t, x) \leqslant f(t, y)$ holds for every pair $x, y \in \mathbb{R}_+^4$ with $x \leqslant y$ and for all $t \in [0, 1]$, where the partial ordering in $\leqslant \mathbb{R}^4$ is understood componentwise.

We denote $\Omega_\rho := \{v \in E : \|v\| < \rho\}$ for $\rho > 0$ in the sequel.

Theorem 10. *If (H1)–(H4) hold, then* (1) *has at least one positive solution.*

Proof. Let

$$\mathcal{M}_1 := \{v \in P : v = Av + \mu\varphi,\ \mu \geqslant 0\}, \tag{40}$$

where $\varphi(t) := (1-t)e^{t-1} \in P$. We are now going to prove that $\mathcal{M}_1$ is bounded. Indeed, if $v \in \mathcal{M}_1$, then there exist $\mu \geqslant 0$ such that $v = Av + \mu\varphi$, which can be written in the form

$$\begin{aligned} v(t) &= \int_0^1 k_2(t, s)\, f\left(s, (B_1 v)(s), (B_1 v)'(s), v(s), -v'(s)\right) ds \\ &\quad + \mu\varphi(t), \end{aligned} \tag{41}$$

which is equivalent to

$$\begin{aligned} -v''(t) &= f\left((t), (B_1 v)(t), (B_1 v)'(t), v(t), -v'(t)\right) \\ &\quad + \mu(1+t)e^{t-1}. \end{aligned} \tag{42}$$

By (H2), we have

$$\begin{aligned} -v''(t) &\geqslant a_1 \left((B_1 v)(t) + 2(B_1 v)'(t)\right) \\ &\quad + b_1 \left(v(t) - 2v'(t)\right) - c. \end{aligned} \tag{43}$$

Note (27) and (29). Multiply the above by $(1-t)e^{1-t}$, integrate over $[0,1]$, and use Lemmas 3, 5, and 6 to obtain

$$\begin{aligned}
&\int_0^1 \left(v(t)-2v'(t)\right)(1-t)e^{1-t}dt \\
&\geqslant a_1\int_0^1 \left((B_1v)(t)+2(B_1v)'(t)\right)(1-t)e^{1-t}dt \\
&\quad + b_1\int_0^1 \left(v(t)2v'(t)\right)(1-t)e^{1-t}dt - c \\
&\geqslant \frac{a_1(3-e)}{6}\|v\|_0 \\
&\quad + b_1\int_0^1 \left(v(t)-2v'(t)\right)(1-t)e^{1-t}dt - c \\
&\geqslant \frac{a_1(3-e)}{6e}\int_0^1 v(t)(1-t)e^{1-t}dt \\
&\quad + b_1\int_0^1 \left(v(t)-2v'(t)\right)(1-t)e^{1-t}dt - c \\
&\geqslant d_1\int_0^1 \left(v(t)-2v'(t)\right)(1-t)e^{1-t}dt - c,
\end{aligned} \tag{44}$$

so that

$$\int_0^1 \left(v(t)-2v'(t)\right)(1-t)e^{1-t}dt \leqslant m, \quad \forall v\in \mathscr{M}_1. \tag{45}$$

Now Lemma 4 implies

$$\|v\|_0 = v(0)\leqslant m, \quad \forall v\in\mathscr{M}_1, \tag{46}$$

where m is given in (H3). Furthermore, this estimate leads to

$$\begin{aligned}
&\|(B_1v)(t)\|_0 = (B_1v)(1) = \int_0^1 sv(s)\,ds \leqslant m, \\
&\left\|(B_1v)'(t)\right\|_0 = (B_1v)'(0) = \int_0^1 v(s)\,ds \leqslant m
\end{aligned} \tag{47}$$

for all $v\in\mathscr{M}_1$. Let

$$\begin{aligned}
\Lambda := \{\mu\geqslant 0 : &\text{ there exist some } v\in P \\
&\text{such that } v = Av+\mu\varphi\}.
\end{aligned} \tag{48}$$

Now (46) implies that $\mu\leqslant me$ for all $\mu\in\Lambda$. By (H3), there is a function $\Phi\in C(\mathbb{R}_+,\mathbb{R}_+)$ such that

$$\begin{aligned}
&f\left(t,(B_1v)(t),(B_1v)'(t),v(t),-v'(t)\right) \\
&\quad\leqslant \Phi\left(-v'(t)\right), \quad \forall v\in\mathscr{M}_1.
\end{aligned} \tag{49}$$

Thus

$$\begin{aligned}
-v''(t) &= f\left(t,(B_1v)(t),(B_1v)'(t),v(t),-v'(t)\right) \\
&\quad + \mu(1+t)e^{t-1} \\
&\leqslant \Phi\left(-v'(t)\right)+\mu(1+t)e^{t-1} \\
&\leqslant \Phi\left(-v'(t)\right)+2me, \quad \forall v\in\mathscr{M}_1,\ \mu\in\Lambda,
\end{aligned} \tag{50}$$

$$\int_0^{-v'(1)} \frac{\xi d\xi}{\Phi(\xi)+2me} \leqslant \int_0^1 -v'(t)\,dt = v(0)\leqslant m, \quad \forall v\in\mathscr{M}_1.$$

Combining the last inequality with (35), we see that there exists $M>0$ such that $-v(1)\leqslant M$ for all $v\in\mathscr{M}_1$. This means that $\mathscr{M}_1$ is bounded. Taking $R>\sup\{\|v\| : v\in\mathscr{M}_1\}$, we have

$$v\neq Av+\mu\varphi, \quad \forall v\in\Omega_R\cap P,\ \mu\geq 0. \tag{51}$$

Now Lemma 7 yields

$$i(A,\Omega_R\cap P,P)=0. \tag{52}$$

Let

$$\mathscr{M}_2 := \left\{v\in\overline{\Omega}_r\cap P : v=\lambda Av,\ \lambda\in[0,1]\right\}. \tag{53}$$

Now we want to prove that $\mathscr{M}_2=\{0\}$. In fact, if $v\in\mathscr{M}_2$, then $v\in\overline{\Omega}_r\cap P$ and there is $\lambda\in[0,1]$ such that $v=\lambda Av$, that is,

$$\begin{aligned}
v(t) = \lambda&\int_0^1 k_2(t,s) \\
&\times f\left(s,(B_1v)(s),(B_1v)'(s),v(s),-v'(s)\right)ds,
\end{aligned} \tag{54}$$

which can be written in the form

$$-v''(t)=\lambda f\left(t,(B_1v)(t),(B_1v)'(t),v(t),-v'(t)\right). \tag{55}$$

By (H4), we have

$$-v''(t)\leqslant a_2\left((B_1v)(t)+2(B_1v)'(t)\right)+b_2\left(v(t)-2v'(t)\right). \tag{56}$$

Note (27) and (29). Multiply (56) by $(1-t)e^{1-t}$, integrate over $[0,1]$, and use Lemmas 5 and 6 to obtain

$$\begin{aligned}
&\int_0^1 \left(v(t)-2v'(t)\right)(1-t)e^{1-t}dt \\
&\leqslant a_2\int_0^1\left((B_1v)(t)+2(B_1v)'(t)\right)(1-t)e^{1-t}dt \\
&\quad + b_2\int_0^1\left(v(t)-2v'(t)\right)(1-t)e^{1-t}dt \\
&\leqslant 3ea_2\|v\|_0 + b_2\int_0^1\left(v(t)-2v'(t)\right)(1-t)e^{1-t}dt \qquad (57)\\
&\leqslant \frac{3ea_2}{e-2}\int_0^1 v(t)(1-t)e^{1-t}dt \\
&\quad + b_2\int_0^1\left(v(t)-2v'(t)\right)(1-t)e^{1-t}dt \\
&\leqslant d_2\int_0^1\left(v(t)-2v'(t)\right)(1-t)e^{1-t}dt,
\end{aligned}$$

so that $\int_0^1(v(t)-2v'(t))(1-t)e^{1-t}=0$, whence $v(t)\equiv 0$ and $\mathscr{M}_2=\{0\}$, as required. A consequence of this is

$$v\neq\lambda Av,\quad \forall v\in\partial\Omega_r\cap P,\ \lambda\in[0,1]. \qquad (58)$$

Now Lemma 8 yields

$$i(A,\Omega_r\cap P,P)=1. \qquad (59)$$

This together with (52) implies

$$i\left(A,\left(\Omega_R\setminus\overline{\Omega}_r\right)\cap P,P\right)=0-1=-1. \qquad (60)$$

Therefore A has at least one fixed point on $(\Omega_R\setminus\overline{\Omega}_r)\cap P$ and thus (1) has at least one positive solution. This completes the proof. □

Theorem 11. *If (H1), (H5), and (H6) hold, then* (1) *has at least one positive solution.*

Proof. Let

$$\mathscr{M}_3:=\{v\in P: v=\lambda Av,\ \lambda\in[0,1]\}. \qquad (61)$$

We now assert that $\mathscr{M}_3$ is bounded. Indeed, if $v\in\mathscr{M}_3$, then there is $\lambda\in[0,1]$ such that $v=\lambda Av$, which is equivalent to

$$-v''(t)=\lambda f\left(t,(B_1v)(t),(B_1)'(t),v(t),-v'(t)\right). \qquad (62)$$

Now (H5) implies

$$\begin{aligned}
-v''(t)&\leqslant a_3\left((B_1v)(t)+2(B_1v)'(t)\right)\\
&\quad + b_3\left(v(t)-2v'(t)\right)+c.
\end{aligned} \qquad (63)$$

Note (27) and (29). Multiply the above by $(1-t)e^{1-t}$, integrate over $[0,1]$, and use Lemmas 5 and 6 to obtain

$$\begin{aligned}
&\int_0^1 \left(v(t)-2v'(t)\right)(1-t)e^{1-t}dt \\
&\leqslant a_3\int_0^1\left((B_1v)(t)+2(B_1v)'(t)\right)(1-t)e^{1-t}dt \\
&\quad + b_3\int_0^1\left(v(t)-2v'(t)\right)(1-t)e^{1-t}dt+c \\
&\leqslant 3ea_3\|v\|_0 + b_3\int_0^1\left(v(t)-2v'(t)\right)(1-t)e^{1-t}dt+c \\
&\leqslant d_3\int_0^1\left(v(t)-2v'(t)\right)(1-t)e^{1-t}dt+c,
\end{aligned} \qquad (64)$$

so that $\int_0^1(v(t)-2v'(t))(1-t)e^{1-t}dt\leqslant c_1$ for all $v\in\mathscr{M}_3$, where $c_1:=c/(1-d_3)>0$. By Lemma 4, we obtain

$$\begin{aligned}
\|v\|_0=v(0)&\leqslant\int_0^1\left(v(t)-2v'(t)\right)(1-t)e^{1-t}dt\\
&\leqslant c_1,\quad \forall v\in\mathscr{M}_3.
\end{aligned} \qquad (65)$$

Furthermore, this estimate leads to

$$\begin{aligned}
\|B_1v\|_0&=(B_1v)(1)=\int_0^1 sv(s)\,ds\leqslant c_1,\\
\left\|(B_1v)'\right\|_0&=(B_1v)'(0)=\int_0^1 v(s)\,ds\leqslant c_1
\end{aligned} \qquad (66)$$

for all $v\in\mathscr{M}_3$. By (63), we have

$$-v''(t)\leqslant(3a_3+b_3)c_1-2b_3v'(t)+c,\quad \forall v\in\mathscr{M}_3. \qquad (67)$$

Note $v(1)=v'(0)=0$ and $v''\leqslant 0$. Integrating the above over $[0,1]$, we obtain

$$\begin{aligned}
\left\|v'\right\|_0&=-v'(1)\leqslant(3a_3+b_3)c_1+2b_3v(0)+c\\
&\leqslant 3(a_3+b_3)c_1+c.
\end{aligned} \qquad (68)$$

This proves the boundedness of $\mathscr{M}_3$. Taking $R>\sup\{\|v\|: v\in\mathscr{M}_3\}$, we have

$$v\neq\lambda Av,\quad \forall v\in\partial\Omega_R\cap P,\ \lambda\in[0,1]. \qquad (69)$$

Now Lemma 8 yields

$$i(A,\Omega_R\cap P,P)=1. \qquad (70)$$

On the other hand, let

$$\mathscr{M}_4:=\left\{v\in\overline{\Omega}_r\cap P: v=Av+\lambda\varphi,\ \lambda\geqslant 0\right\}, \qquad (71)$$

where $\varphi\in P\setminus\{0\}$ is given by $\varphi(t):=(1-t)e^{t-1}$. Now we want to prove that $\mathscr{M}_4\subset\{0\}$. Indeed, if $v\in\mathscr{M}_4$, then there exists $\lambda\geqslant 0$ such that $v=Av+\lambda\varphi$, which can be written as

$$\begin{aligned}
-v''(t)&=f\left(t,(B_1v)(t),(B_1v)'(t),v(t),-v'(t)\right)\\
&\quad+\lambda(t+1)e^{t-1}.
\end{aligned} \qquad (72)$$

By (H6), we have

$$-v''(t)\,a_4 \geqslant \left((B_1v)(t) + 2(B_1v)'(t)\right) + b_4\left(v(t) - 2v'(t)\right). \quad (73)$$

Note (27) and (29). Multiply the above by $(1-t)e^{1-t}$, integrate over $[0,1]$, and use Lemmas 3, 5, and 6 to obtain

$$\begin{aligned}
&\int_0^1 \left(v(t) - 2v'(t)\right)(1-t)e^{1-t}dt \\
&\geqslant a_4 \int_0^1 \left((B_1v)(t) + 2(B_1v)'(t)\right)(1-t)e^{1-t}dt \\
&\quad + b_4 \int_0^1 \left(v(t) - 2v'(t)\right)(1-t)e^{1-t}dt, \\
&\geqslant \frac{a_4(3-e)}{6e}\|v\|_0 + b_4 \int_0^1 \left(v(t) - 2v'(t)\right)(1-t)e^{1-t}dt, \\
&\geqslant \frac{a_4(3-e)}{6e}\int_0^1 v(t)(1-t)e^{1-t}dt \\
&\quad + b_4 \int_0^1 \left(v(t) - 2v'(t)\right)(1-t)e^{1-t}dt, \\
&\geqslant d_4 \int_0^1 \left(v(t) - 2v'(t)\right)(1-t)e^{1-t}dt,
\end{aligned} \quad (74)$$

so that $\int_0^1 (v(t)-2v'(t))(1-t)e^{1-t}dt = 0$, whence $v(t) \equiv 0$. This proves $\mathscr{M}_4 \subset \{0\}$, as required. As a result of this, we obtain

$$v \neq Av + \lambda\varphi, \quad \forall v \in \partial\Omega_r \cap P,\ \lambda \geqslant 0. \quad (75)$$

Lemma 7 implies

$$i\left(A, \Omega_r \cap P, P\right) = 0. \quad (76)$$

Now (70) and (76) combined lead to

$$i\left(A, \left(\Omega_R \setminus \overline{\Omega}_r\right) \cap P, P\right) = 1 - 0 = 1. \quad (77)$$

Consequently, A has at least one fixed point on $(\Omega_R \setminus \overline{\Omega}_r) \cap P$. Thus (1) has at least one positive solution. This completes the proof. □

Theorem 12. *If (H1), (H2), (H3), (H6), and (H7) hold, then (1) has at least two positive solutions.*

Proof. If $v \in P$ and $\|v\| \leqslant \omega$, then we have

$$\begin{aligned}
\|B_1v\|_0 &= (B_1v)(1) = \int_0^1 sv(s)\,ds \leqslant \omega, \left\|(B_1v)'\right\|_0 = (B_1v)(0) \\
&= \int_0^1 v(s)\,ds \leqslant \omega.
\end{aligned} \quad (78)$$

By (H7), we have

$$f(t,x) \leqslant f(t,\omega,\omega,\omega,\omega), \quad \forall t \in [0,1],\ x \in I_\omega^4. \quad (79)$$

Consequently, for all $v \in \partial\Omega_\omega \cap P$, we have

$$\begin{aligned}
\|Av\|_0 &= (Av)(0) \\
&= \int_0^1 (1-s)\,f\left(s, (B_1v)(s), (B_1v)'(s)\,v(s), -v'(s)\right)ds \\
&\leqslant \int_0^1 f(s,\omega,\omega,\omega,\omega)\,ds < \omega = \|v\|,
\end{aligned} \quad (80)$$

$$\begin{aligned}
&\left\|(Av)'\right\|_0 \\
&= -(Av)'(1) \\
&= \int_0^1 f\left(s, (B_1v)(s), (B_1v)'(s), v(s), -v'(s)\right)ds \\
&\leqslant \int_0^1 f(s,\omega,\omega,\omega,\omega)\,ds < \omega = \|v\|.
\end{aligned} \quad (81)$$

The preceding inequalities imply

$$\|Av\| < \omega = \|v\|, \quad \forall v \in \partial\Omega_\omega \cap P \quad (82)$$

and thus

$$v \neq \lambda Av, \quad \forall v \in \partial\Omega_\omega \cap P,\ 0 \leqslant \lambda \leqslant 1. \quad (83)$$

Now Lemma 5 yields

$$i\left(A, \Omega_\omega \cap P, P\right) = 1. \quad (84)$$

By (H2), (H3), and (H6), we know that (52) and (76) hold (see the proofs of Theorems 10 and 11). Note that we can choose $R > \omega > r$ in (52) and (76). Combining (52), (76), and (84), we obtain

$$\begin{aligned}
i\left(A, \left(\Omega_R \setminus \overline{\Omega}_\omega\right) \cap P, P\right) &= 0 - 1 = -1, \\
i\left(A, \left(\Omega_\omega \setminus \overline{\Omega}_r\right) \cap P, P\right) &= 1 - 0 = 1.
\end{aligned} \quad (85)$$

Therefore, A has at least two fixed points, with one on $(\Omega_R \setminus \overline{\Omega}_\omega) \cap P$ and the other on $(\Omega_\omega \setminus \overline{\Omega}_r) \cap P$. Hence (1) has at least two positive solutions. This completes the proof. □

References

[1] T. Jankowski and R. Jankowski, "Multiple solutions of boundary-value problems for fourth-order differential equations with deviating arguments," *Journal of Optimization Theory and Applications*, vol. 146, no. 1, pp. 105–115, 2010.

[2] C. Zhai, R. Song, and Q. Han, "The existence and the uniqueness of symmetric positive solutions for a fourth-order boundary value problem," *Computers & Mathematics with Applications*, vol. 62, pp. 2639–2647, 2011.

[3] B. Yang, "Positive solutions for the beam equation under certain boundary conditions," *Electronic Journal of Differential Equations*, vol. 78, pp. 1–8, 2005.

[4] Q. Yao, "Positive solutions for eigenvalue problems of fourth-order elastic beam equations," *Applied Mathematics Letters*, vol. 17, no. 2, pp. 237–243, 2004.

[5] Z. Wei and C. Pang, "Multiple sign-changing solutions for fourth order m-point boundary value problems," *Nonlinear Analysis, Theory, Methods and Applications*, vol. 66, no. 4, pp. 839–855, 2007.

[6] L. Liu, P. Kang, Y. Wu, and B. Wiwatanapataphee, "Positive solutions of singular boundary value problems for systems of nonlinear fourth order differential equations," *Nonlinear Analysis, Theory, Methods and Applications*, vol. 68, no. 3, pp. 485–498, 2008.

[7] Z. Bai, "The upper and lower solution method for some fourth-order boundary value problems," *Nonlinear Analysis, Theory, Methods and Applications*, vol. 67, no. 6, pp. 1704–1709, 2007.

[8] Q. Zhang, S. Chen, and J. Lü, "Upper and lower solution method for fourth-order four-point boundary value problems," *Journal of Computational and Applied Mathematics*, vol. 196, no. 2, pp. 387–393, 2006.

[9] D. X. Ma and X. Z. Yang, "Upper and lower solution method for fourth-order four-point boundary value problems," *Journal of Computational and Applied Mathematics*, vol. 223, no. 2, pp. 543–551, 2009.

[10] J. R. Graef and L. Kong, "A necessary and sufficient condition for existence of positive solutions of nonlinear boundary value problems," *Nonlinear Analysis, Theory, Methods and Applications*, vol. 66, no. 11, pp. 2389–2412, 2007.

[11] R. P. Agarwal and Y. M. Chow, "Iterative methods for a fourth order boundary value problem," *Journal of Computational and Applied Mathematics*, vol. 10, no. 2, pp. 203–217, 1984.

[12] Q. Yao, "Monotone iterative technique and positive solutions of Lidstone boundary value problems," *Applied Mathematics and Computation*, vol. 138, no. 1, pp. 1–9, 2003.

[13] E. Alves, T. F. Ma, and M. L. Pelicer, "Monotone positive solutions for a fourth order equation with nonlinear boundary conditions," *Nonlinear Analysis, Theory, Methods and Applications*, vol. 71, no. 9, pp. 3834–3841, 2009.

[14] M. Pei and S. K. Chang, "Monotone iterative technique and symmetric positive solutions for a fourth-order boundary value problem," *Mathematical and Computer Modelling*, vol. 51, no. 9-10, pp. 1260–1267, 2010.

[15] G. Han and F. Li, "Multiple solutions of some fourth-order boundary value problems," *Nonlinear Analysis, Theory, Methods and Applications*, vol. 66, no. 11, pp. 2591–2603, 2007.

[16] P. Amster and P. P. Cárdenas Alzate, "A shooting method for a nonlinear beam equation," *Nonlinear Analysis, Theory, Methods and Applications*, vol. 68, no. 7, pp. 2072–2078, 2008.

[17] C. Pang, W. Dong, and Z. Wei, "Multiple solutions for fourth-order boundary value problem," *Journal of Mathematical Analysis and Applications*, vol. 314, no. 2, pp. 464–476, 2006.

[18] Z. Yang and J. Sun, "Positive solutions of a fourth-order boundary value problem involving derivatives of all orders," *Communications on Pure and Applied Analysis*, vol. 11, no. 5, pp. 1615–1628, 2012.

[19] D. Guo and V. Lakshmikantham, *Problems in Abstract Cones*, Academic Press, Boston, Mass, USA, 1988.

Energy- and Regularity-Dependent Stability Estimates for Near-Field Inverse Scattering in Multidimensions

M. I. Isaev

Centre de Mathématiques Appliquées, Ecole Polytechnique, 91128 Palaiseau, France

Correspondence should be addressed to M. I. Isaev; isaev.m.i@gmail.com

Academic Editor: Zindoga Mukandavire

We prove new global Hölder-logarithmic stability estimates for the near-field inverse scattering problem in dimension $d \geq 3$. Our estimates are given in uniform norm for coefficient difference and related stability efficiently increases with increasing energy and/or coefficient regularity. In addition, a global logarithmic stability estimate for this inverse problem in dimension $d = 2$ is also given.

1. Introduction

We consider the Schrödinger equation:

$$L\psi = E\psi, \qquad L = -\Delta + v(x), \quad x \in \mathbb{R}^d, \ d \geq 2, \tag{1}$$

where

$$v \text{ is real-valued}, \quad v \in \mathbb{L}^{\infty}\left(\mathbb{R}^d\right),$$
$$v(x) = O\left(|x|^{-d-\varepsilon}\right), \quad |x| \longrightarrow \infty, \text{ for some } \varepsilon > 0. \tag{2}$$

We consider the resolvent $R(E)$ of the Schrödinger operator L in $\mathbb{L}^2(\mathbb{R}^d)$:

$$R(E) = (L - E)^{-1}, \quad E \in \mathbb{C} \setminus \sigma(L), \tag{3}$$

where $\sigma(L)$ is the spectrum of L in $\mathbb{L}^2(\mathbb{R}^d)$. We assume that $R(x, y, E)$ denotes the Schwartz kernel of $R(E)$ as of an integral operator. We consider also

$$R^+(x, y, E) = R(x, y, E + i0), \quad x, y \in \mathbb{R}^d, \ E \in \mathbb{R}_+. \tag{4}$$

We recall that in the framework of (1) the function $R^+(x, y, E)$ describes scattering of the spherical waves,

$$R_0^+(x, y, E) = -\frac{i}{4}\left(\frac{\sqrt{E}}{2\pi|x - y|}\right)^{(d-2)/2} H^{(1)}_{(d-2)/2}\left(\sqrt{E}|x - y|\right), \tag{5}$$

generated by a source at y (where $H^{(1)}_{\mu}$ is the Hankel function of the first kind of order μ). We recall also that $R^+(x, y, E)$ is the Green function for $L - E$, $E \in \mathbb{R}_+$, with the Sommerfeld radiation condition at infinity.

In addition, the function

$$S^+(x, y, E) = R^+(x, y, E) - R_0^+(x, y, E),$$
$$x, y \in \partial B_r, \quad E \in \mathbb{R}_+, \quad r \in \mathbb{R}_+ \tag{6}$$

is considered as near-field scattering data for (1), where B_r is the open ball of radius r centered at 0.

We consider, in particular, the following near-field inverse scattering problem for (1).

Problem 1. Given S^+ on $\partial B_r \times \partial B_r$ for some fixed r, $E \in \mathbb{R}_+$, find v on B_r.

This problem can be considered under the assumption that v is a priori known on $\mathbb{R}^d \setminus B_r$. Actually, in the present paper we consider Problem 1 under the assumption that $v \equiv 0$ on $\mathbb{R}^d \setminus B_r$ for some fixed $r \in \mathbb{R}_+$. Below in this paper we always assume that this additional condition is fulfilled.

It is well known that the near-field scattering data of Problem 1 uniquely and efficiently determine the scattering amplitude f for (1) at fixed energy E, see [1]. Therefore, approaches of [2–12] can be applied to Problem 1 via this reduction.

In addition, it is also known that the near-field data of Problem 1 uniquely determine the Dirichlet-to-Neumann map in the case when E is not a Dirichlet eigenvalue for operator L in B_r, see [8, 13]. Therefore, approaches of [3, 8, 14–24] can be also applied to Problem 1 via this reduction.

However, in some case it is much more optimal to deal with Problem 1 directly, see, for example, logarithmic stability results of [25] for Problem 1 in dimension $d = 3$. A principal improvement of estimates of [25] was given recently in [26]: stability of [26] efficiently increases with increasing regularity of v.

Problem 1 can be also considered as an example of ill-posed problem: see [27, 28] for an introduction to this theory.

In the present paper we continue studies of [25, 26]. We give new global Hölder-logarithmic stability estimates for Problem 1 in dimension $d \geq 3$, see Theorem 1. Our estimates are given in uniform norm for coefficient difference and related stability efficiently increases with increasing energy and/or coefficient regularity. Results of such a type for the Gel'fand inverse problem were obtained recently in [15] for $d \geq 3$ and in [29] for $d = 2$.

The main feature of our new estimates is the explicit dependence on the energy E. These estimates consist of two parts, the first is Hölder and the second is logarithmic; when E increases, the logarithmic part decreases and the Hölder part becomes dominant.

In addition, we give also global logarithmic stability estimates for Problem 1 in dimension $d = 2$, see Theorem 2.

2. Stability Estimates

We recall that if v satisfies (2) and $\operatorname{supp} v \subset B_{r_1}$ for some $r_1 > 0$, then

$$S^+(E) \text{ is bounded in } \mathbb{L}^2(\partial B_r \times \partial B_r) \quad \text{for any } r > r_1, \tag{7}$$

where $S^+(E)$ is the near-field scattering data of v for (1) with $E > 0$, for more details see, for example, Section 2 of [25].

2.1. Estimates for $d \geq 3$. In this subsection we assume for simplicity that

$$v \in W^{m,1}(\mathbb{R}^d) \quad \text{for some } m > d,\ v \text{ is real-valued}, \qquad \operatorname{supp} v \subset B_{r_1} \quad \text{for some } r_1 > 0, \tag{8}$$

where

$$W^{m,1}(\mathbb{R}^d) = \left\{v : \partial^J v \in \mathbb{L}^1(\mathbb{R}^d),\ |J| \leq m\right\}, \quad m \in \mathbb{N} \cup 0, \tag{9}$$

where

$$J \in (\mathbb{N} \cup 0)^d, \quad |J| = \sum_{i=1}^{d} J_i, \qquad \partial^J v(x) = \frac{\partial^{|J|} v(x)}{\partial x_1^{J_1} \cdots \partial x_d^{J_d}}. \tag{10}$$

Let

$$\|v\|_{m,1} = \max_{|J| \leq m} \left\|\partial^J v\right\|_{\mathbb{L}^1(\mathbb{R}^d)}. \tag{11}$$

Note that (8)⇒(2).

Theorem 1. *Let $E > 0$ and $r > r_1$ be given constants. Let dimension $d \geq 3$ and potentials v_1, v_2 satisfy (8). Let $\|v_j\|_{m,1} \leq N$, $j = 1, 2$, for some $N > 0$. Let $S_1^+(E)$ and $S_2^+(E)$ denote the near-field scattering data for v_1 and v_2, respectively. Then for $\tau \in (0,1)$ and any $s \in [0, s^*]$ the following estimate holds:*

$$\|v_2 - v_1\|_{L^\infty(B_r)} \leq C_1 (1+E)^{5/2} \delta^\tau + C_2 (1+E)^{(s-s^*)/2} \left(\ln\left(3 + \delta^{-1}\right)\right)^{-s}, \tag{12}$$

where $s^* = (m-d)/d$, $\delta = \|S_1^+(E) - S_2^+(E)\|_{\mathbb{L}^2(\partial B_r \times \partial B_r)}$, and constants $C_1, C_2 > 0$ depend only on N, m, d, r, τ.

Proof of Theorem 1 is given in Section 5. This proof is based on results presented in Sections 3 and 4.

2.2. Estimates for d=2. In this subsection we assume for simplicity that

$$v \text{ is real-valued}, \quad v \in C^2\left(\overline{B}_{r_1}\right), \qquad \operatorname{supp} v \subset B_{r_1} \quad \text{for some } r_1 > 0. \tag{13}$$

Note also that (13)⇒(2).

Theorem 2. *Let $E > 0$ and $r > r_1$ be given constants. Let dimension $d = 2$ and potentials v_1, v_2 satisfy (13). Let $\|v_j\|_{C^2(B_r)} \leq N$, $j = 1, 2$, for some $N > 0$. Let $S_1^+(E)$ and $S_2^+(E)$ denote the near-field scattering data for v_1 and v_2, respectively. Then*

$$\|v_1 - v_2\|_{L^\infty(B_r)} \leq C_3 \left(\ln\left(3 + \delta^{-1}\right)\right)^{-3/4} \left(\ln\left(3 \ln\left(3 + \delta^{-1}\right)\right)\right)^2, \tag{14}$$

where $\delta = \|S_1^+(E) - S_2^+(E)\|_{\mathbb{L}^2(\partial B_r \times \partial B_r)}$ and constant $C_3 > 0$ depends only on N, m, r.

Proof of Theorem 2 is given in Section 7. This proof is based on results presented in Sections 3 and 6.

2.3. Concluding Remarks

Remark 3. The logarithmic stability estimates for Problem 1.1 of [25, 26] follow from estimate (12) for $d = 3$ and $s = s^*$. Apparently, using the methods of [19, 20] it is possible to improve estimate (12) for $s^* = m - d$.

Remark 4. In the same way as in [25, 26] for dimesnsion $d = 3$, using estimates (12) and (14), one can obtain logarithmic stability estimates for the reconstruction of a potential v from the inverse scattering amplitude f for any $d \geq 2$.

Remark 5. Actually, in the proof of Theorem 1 we obtain the following estimate (see formula (57)):

$$\|v_1 - v_2\|_{\mathbb{L}^\infty(B_r)} \le C_4(1+E)^2\sqrt{E+\rho^2}\, e^{2\rho(r+1)}\delta + C_5\left(E+\rho^2\right)^{-(m-d)/2d}, \quad (15)$$

where constants $C_4, C_5 > 0$ depend only on N, m, d, r and the parameter $\rho > 0$ is such that $E + \rho^2$ is sufficiently large: $E + \rho^2 \ge C_6(N, r, m)$. Estimate of Theorem 1 follows from estimate (15).

3. Alessandrini-Type Identity for Near-Field Scattering

In this section we always assume that assumptions of Theorems 1 and 2 are fulfilled (in the cases of dimension $d \ge 3$ and $d = 2$, resp.).

Consider the operators $\widehat{R}_j$, $j = 1, 2$, defined as follows

$$\left(\widehat{R}_j\phi\right)(x) = \int_{\partial B_r} R_j^+(x, y, E)\,\phi(y)\,dy, \quad x \in \partial B_r,\ j = 1, 2. \quad (16)$$

Note that

$$\|\widehat{R}_1 - \widehat{R}_2\|_{\mathbb{L}^2(\partial B_r)} \le \|S_1^+(E) - S_2^+(E)\|_{\mathbb{L}^2(\partial B_r)\times\mathbb{L}^2(\partial B_r)}. \quad (17)$$

We recall that (see [25]) for any functions $\phi_1, \phi_2 \in C(\mathbb{R}^d)$, sufficiently regular in $\mathbb{R}^d \setminus \partial B_r$ and satisfying

$$-\Delta\phi + v(x)\phi = E\phi, \quad \text{in } \mathbb{R}^d \setminus \partial B_r,$$
$$\lim_{|x|\to+\infty} |x|^{(d-1)/2}\left(\frac{\partial}{\partial|x|}\phi - i\sqrt{E}\phi\right) = 0, \quad (18)$$

with $v = v_1$ and $v = v_2$, respectively, the following identity holds:

$$\int_{B_r}(v_2 - v_1)\,\phi_1\phi_2\,dx = \int_{\partial B_r}\left(\frac{\partial\phi_1}{\partial\nu_+} - \frac{\partial\phi_1}{\partial\nu_-}\right)\left[\left(\widehat{R}_1 - \widehat{R}_2\right)\left(\frac{\partial\phi_2}{\partial\nu_+} - \frac{\partial\phi_2}{\partial\nu_-}\right)\right]dx, \quad (19)$$

where where ν_+ and ν_- are the outward and inward normals to ∂B_r, respectively.

Remark 6. The identity (19) is similar to the Alessandrini identity (see Lemma 1 of [14]), where the Dirichlet-to-Neumann maps are considered instead of operators $\widehat{R}_j$.

To apply identity (19) to our considerations, we use also the following lemma.

Lemma 7. *Let $E, r > 0$ and $d \ge 2$. Then, there is a positive constant C_7 (depending only on r and d) such that for any $\phi \in C(\mathbb{R}^d \setminus B_r)$ satisfying*

$$-\Delta\phi = E\phi, \quad \text{in } \mathbb{R}^d \setminus \overline{B}_r,$$
$$\lim_{|x|\to+\infty} |x|^{(d-1)/2}\left(\frac{\partial}{\partial|x|}\phi - i\sqrt{E}\phi\right) = 0, \quad (20)$$
$$\phi|_{\partial B_r} \in \mathbb{H}^1(\partial B_r),$$

the following inequality holds:

$$\left\|\left.\frac{\partial\phi}{\partial\nu_+}\right|_{\partial B_r}\right\|_{\mathbb{L}^2(\partial B_r)} \le C_7(1+E)\left\|\phi|_{\partial B_r}\right\|_{\mathbb{H}^1(\partial B_r)}, \quad (21)$$

where $\mathbb{H}^1(\partial B_r)$ denotes the standard Sobolev space on ∂B_r.

The proof of Lemma 7 is given in Section 8.

4. Faddeev Functions

In dimension $d \ge 3$, we consider the Faddeev functions h, ψ, G (see [6, 8, 30, 31]):

$$h(k, l) = (2\pi)^{-d}\int_{\mathbb{R}^d} e^{-ilx}v(x)\,\psi(x, k)\,dx, \quad (22)$$

where $k, l \in \mathbb{C}^d$, $k^2 = l^2$, $\operatorname{Im} k = \operatorname{Im} l \ne 0$,

$$\psi(x, k) = e^{ikx} + \int_{\mathbb{R}^d} G(x - y, k)\,v(y)\,\psi(y, k)\,dy, \quad (23)$$

$$G(x, k) = e^{ikx}g(x, k),$$
$$g(x, k) = -(2\pi)^{-d}\int_{\mathbb{R}^d}\frac{e^{i\xi x}d\xi}{\xi^2 + 2k\xi}, \quad (24)$$

where $x \in \mathbb{R}^d$, $k \in \mathbb{C}^d$, $\operatorname{Im} k \ne 0$, $d \ge 3$,

One can consider (22), (23) assuming that

$$v \text{ is a sufficiently regular function on } \mathbb{R}^d \text{ with suffucient decay at infinity.} \quad (25)$$

For example, in connection with Theorem 1, we consider (22), (23) assuming that

$$v \in \mathbb{L}^\infty(B_r), \quad v \equiv 0 \text{ on } \mathbb{R} \setminus B_r. \quad (26)$$

We recall that (see [6, 8, 30, 31])

$$\left(\Delta + k^2\right)G(x, k) = \delta(x), \quad x \in \mathbb{R}^d,\ k \in \mathbb{C}^d \setminus \mathbb{R}^d, \quad (27)$$

formula (23) at fixed k is considered as an equation for

$$\psi = e^{ikx}\mu(x, k), \quad (28)$$

where μ is sought in $\mathbb{L}^\infty(\mathbb{R}^d)$; as a corollary of (23), (24), and (27), ψ satisfies (1) for $E = k^2$; h of (22) is a generalized "scattering" amplitude.

In addition, h, ψ, G in their zero energy restriction, that is for $E = 0$ were considered for the first time in [32]. The Faddeev functions h, ψ, G were, actually, rediscovered in [32].

Let

$$\Sigma_E = \left\{k \in \mathbb{C}^d : k^2 = k_1^2 + \cdots + k_d^2 = E\right\},$$
$$\Theta_E = \left\{k \in \Sigma_E, l \in \Sigma_E : \operatorname{Im} k = \operatorname{Im} l\right\}, \quad (29)$$
$$|k| = \left(|\operatorname{Re} k|^2 + |\operatorname{Im} k|^2\right)^{1/2}.$$

Let

$$v \text{ satisfy } (8), \quad \|v\|_{m,1} \le N,$$
$$\hat{v}(p) = (2\pi)^{-d} \int_{\mathbb{R}^d} e^{ipx} v(x)\, dx, \quad p \in \mathbb{R}^d, \quad (30)$$

then we have that

$$\mu(x,k) \longrightarrow 1 \quad \text{as } |k| \longrightarrow \infty \quad (31)$$

and, for any $\sigma > 1$,

$$|\mu(x,k)| + |\nabla \mu(x,k)| \le \sigma \quad \text{for } |k| \ge \lambda_1(N,m,d,r,\sigma), \quad (32)$$

where $x \in \mathbb{R}^d$, $k \in \Sigma_E$;

$$\hat{v}(p) = \lim_{\substack{(k,l) \in \Theta_E,\, k-l=p \\ |\operatorname{Im} k| = |\operatorname{Im} l| \to \infty}} h(k,l), \quad \text{for any } p \in \mathbb{R}^d, \quad (33)$$

$$|\hat{v}(p) - h(k,l)| \le \frac{c_1(m,d,r)N^2}{(E+\rho^2)^{1/2}}, \quad \text{for } (k,l) \in \Theta_E,\ p = k-l,$$
$$|\operatorname{Im} k| = |\operatorname{Im} l| = \rho, \quad E + \rho^2 \ge \lambda_2(N,m,d,r), \quad (34)$$
$$p^2 \le 4\left(E + \rho^2\right).$$

Results of the type (31), (32) go back to [32]. For more information concerning (32) see estimate (4.11) of [33]. Results of the type (33), (34) (with less precise right-hand side in (34)) go back to [6]. Estimate (34) follows, for example, from formulas (23), (22), and the estimate

$$\|\Lambda^{-s} g(k) \Lambda^{-s}\|_{\mathbb{L}^2(\mathbb{R}^d) \to \mathbb{L}^2(\mathbb{R}^d)} = O\left(|k|^{-1}\right) \quad \text{as } |k| \longrightarrow \infty,\ k \in \mathbb{C}^d \setminus \mathbb{R}^d, \quad (35)$$

for $s > 1/2$, where $g(k)$ denotes the integral operator with the Schwartz kernel $g(x-y,k)$ and Λ denotes the multiplication operator by the function $(1+|x|^2)^{1/2}$. Estimate (35) was formulated, first, in [34] for $d \ge 3$. Concerning proof of (35), see [35].

In addition, we have that

$$h_2(k,l) - h_1(k,l) = (2\pi)^{-d} \int_{\mathbb{R}^d} \psi_1(x,-l)\left(v_2(x) - v_1(x)\right)\psi_2(x,k)\, dx$$
$$\text{for } (k,l) \in \Theta_E, \quad |\operatorname{Im} k| = |\operatorname{Im} l| \ne 0, \quad \text{and } v_1, v_2 \text{ satisfying } (25), \quad (36)$$

and, under assumptions of Theorem 1,

$$|\hat{v}_1(p) - \hat{v}_2(p) - h_1(k,l) + h_2(k,l)| \le \frac{c_2(m,d,r)N\|v_1 - v_2\|_{\mathbb{L}^\infty(B_r)}}{(E+\rho^2)^{1/2}} \quad (37)$$
$$\text{for } (k,l) \in \Theta_E, \quad p = k-l, \quad |\operatorname{Im} k| = |\operatorname{Im} l| = \rho,$$
$$E + \rho^2 \ge \lambda_3(N,m,d,r), \quad p^2 \le 4\left(E+\rho^2\right),$$

where h_j, ψ_j denote h and ψ of (22) and (23) for $v = v_j$, $j = 1, 2$.

Formula (36) was given in [36]. Estimate (37) was given for example in [15].

5. Proof of Theorem 1

Let

$$\mathbb{L}_\mu^\infty\left(\mathbb{R}^d\right) = \left\{u \in \mathbb{L}^\infty\left(\mathbb{R}^d\right) : \|u\|_\mu < +\infty\right\},$$
$$\|u\|_\mu = \operatorname{ess\,sup}_{p \in \mathbb{R}^d} (1+|p|)^\mu |u(p)|, \quad \mu > 0. \quad (38)$$

Note that

$$w \in W^{m,1}\left(\mathbb{R}^d\right) \Longrightarrow \hat{w} \in \mathbb{L}_\mu^\infty\left(\mathbb{R}^d\right) \cap C\left(\mathbb{R}^d\right),$$
$$\|\hat{w}\|_\mu \le c_3(m,d)\|w\|_{m,1}, \quad \text{for } \mu = m, \quad (39)$$

where $W^{m,1}$, $\mathbb{L}_\mu^\infty$ are the spaces of (9), (38),

$$\hat{w}(p) = (2\pi)^{-d} \int_{\mathbb{R}^d} e^{ipx} w(x)\, dx, \quad p \in \mathbb{R}^d. \quad (40)$$

Using the inverse Fourier transform formula

$$w(x) = \int_{\mathbb{R}^d} e^{-ipx} \hat{w}(p)\, dp, \quad x \in \mathbb{R}^d, \quad (41)$$

we have that

$$\|v_1 - v_2\|_{\mathbb{L}^\infty(B_r)} \le \sup_{x \in \bar{B}_r} \left|\int_{\mathbb{R}^d} e^{-ipx}\left(\hat{v}_2(p) - \hat{v}_1(p)\right) dp\right| \le I_1(\kappa) + I_2(\kappa), \quad \text{for any } \kappa > 0, \quad (42)$$

where

$$I_1(\kappa) = \int_{|p| \le \kappa} |\hat{v}_2(p) - \hat{v}_1(p)|\, dp,$$
$$I_2(\kappa) = \int_{|p| \ge \kappa} |\hat{v}_2(p) - \hat{v}_1(p)|\, dp. \quad (43)$$

Using (39), we obtain that

$$|\widehat{v}_2(p) - \widehat{v}_1(p)| \le 2c_3(m,d)\,N(1+|p|)^{-m}, \quad p \in \mathbb{R}^d. \tag{44}$$

Let

$$c_4 = \int_{p\in\mathbb{R}^d,\,|p|=1} dp. \tag{45}$$

Combining (43), (44), we find that, for any $\kappa > 0$,

$$I_2(\kappa) \le 2c_3(m,d)\,Nc_4 \int_\kappa^{+\infty} \frac{dt}{t^{m-d+1}} \le \frac{2c_3(m,d)\,Nc_4}{m-d}\frac{1}{\kappa^{m-d}}. \tag{46}$$

Due to (37), we have that

$$\begin{aligned} |\widehat{v}_2(p) - \widehat{v}_1(p)| \le {} & |h_2(k,l) - h_1(k,l)| \\ & + \frac{c_2(m,d,r)\,N\|v_1 - v_2\|_{\mathbb{L}^\infty(B_r)}}{(E+\rho^2)^{1/2}}, \\ \text{for } (k,l) \in \Theta_E, \quad & p = k-l, \quad |\operatorname{Im} k| = |\operatorname{Im} l| = \rho, \\ E+\rho^2 \ge \lambda_3(N,m,d,r), \quad & p^2 \le 4(E+\rho^2). \end{aligned} \tag{47}$$

Let

$$\delta = \|S_1^+(E) - S_2^+(E)\|_{\mathbb{L}^2(\partial B_r\times\partial B_r)}. \tag{48}$$

Combining (17), (19), and (36), we get that

$$\begin{aligned} & |h_2(k,l) - h_1(k,l)| \\ & \le \delta \left\| \frac{\partial\phi_1}{\partial\nu_+} - \frac{\partial\phi_1}{\partial\nu_-} \right\|_{\mathbb{L}^2(B_r)} \left\| \frac{\partial\phi_2}{\partial\nu_+} - \frac{\partial\phi_2}{\partial\nu_-} \right\|_{\mathbb{L}^2(B_r)}, \\ & (k,l) \in \Theta_E, \quad |\operatorname{Im} k| = |\operatorname{Im} l| \neq 0, \end{aligned} \tag{49}$$

where ϕ_j, $j = 1,2$, denotes the solution of (18) with $v = v_j$, satisfying

$$\phi_j(x) = \psi_j(x,k), \quad \text{for } x \in \overline{B}_r. \tag{50}$$

Using (21), (32), and the fact that $C^1(\partial B_r) \subset \mathbb{H}^1(\partial B_r)$, we find that

$$\begin{aligned} & \left\| \frac{\partial\phi_j}{\partial\nu_+} - \frac{\partial\phi_j}{\partial\nu_-} \right\|_{\mathbb{L}^2(B_r)} \\ & \le \sigma c_5(r,d)(1+E)\ \exp(|\operatorname{Im} k|(r+1)), \\ & k \in \Sigma_E, \quad |k| \ge \lambda_1(N,m,d,r,\sigma), \quad j = 1,2. \end{aligned} \tag{51}$$

Here and bellow in this section the constant σ is the same that in (32).

Combining (49) and (51), we obtain that

$$\begin{aligned} & |h_2(k,l) - h_1(k,l)| \le c_5^2\sigma^2(1+E)^2 e^{2\rho(r+1)}\delta, \\ & \text{for } (k,l) \in \Theta_E, \quad \rho = |\operatorname{Im} k| = |\operatorname{Im} l|, \\ & E+\rho^2 \ge \lambda_1^2(N,m,d,r,\sigma). \end{aligned} \tag{52}$$

Using (47), (52), we get that

$$\begin{aligned} & |\widehat{v}_2(p) - \widehat{v}_1(p)| \le c_5^2\sigma^2(1+E)^2 e^{2\rho(r+1)}\delta \\ & \quad + \frac{c_2(m,d,r)\,N\|v_1 - v_2\|_{\mathbb{L}^\infty(B_1)}}{(E+\rho^2)^{1/2}}, \\ & p \in \mathbb{R}^d, \quad p^2 \le 4(E+\rho^2), \quad E+\rho^2 \ge \max\{\lambda_1^2, \lambda_3\}. \end{aligned} \tag{53}$$

Let

$$\varepsilon = \left(\frac{1}{2c_2(m,d,r)\,Nc_6}\right)^{1/d}, \quad c_6 = \int_{p\in\mathbb{R}^d,\,|p|\le 1} dp, \tag{54}$$

and $\lambda_4(N,m,d,r,\sigma) > 0$ be such that

$$\begin{aligned} & E+\rho^2 \ge \lambda_4(N,m,d,r,\sigma) \\ & \Longrightarrow \begin{cases} E+\rho^2 \ge \lambda_1^2(N,m,d,r,\sigma), \\ E+\rho^2 \ge \lambda_3(N,m,d,r), \\ \left(\varepsilon(E+\rho^2)^{1/2d}\right)^2 \le 4(E+\rho^2). \end{cases} \end{aligned} \tag{55}$$

Using (43), (53), we get that

$$\begin{aligned} I_1(\kappa) \le {} & c_6\kappa^d \Bigg(c_5^2\sigma^2(1+E)^2 e^{2\rho(r+1)}\delta \\ & + \frac{c_2(m,d,r)\,N\|v_1 - v_2\|_{\mathbb{L}^\infty(B_1)}}{(E+\rho^2)^{1/2}} \Bigg), \\ & \kappa > 0, \ \kappa^2 \le 4(E+\rho^2), \\ & E+\rho^2 \ge \lambda_4(N,m,d,r,\sigma). \end{aligned} \tag{56}$$

Combining (42), (46), and (56) for $\kappa = \varepsilon(E+\rho^2)^{1/2d}$ and (55), we get that

$$\begin{aligned} & \|v_1 - v_2\|_{\mathbb{L}^\infty(B_r)} \\ & \le c_7(N,m,d,r,\sigma)(1+E)^2\sqrt{E+\rho^2}\ e^{2\rho(r+1)}\delta \\ & \quad + c_8(N,m,d)\left(E+\rho^2\right)^{-(m-d)/2d} \\ & \quad + \frac{1}{2}\|v_1 - v_2\|_{\mathbb{L}^\infty(B_r)}, \\ & E+\rho^2 \ge \lambda_4(N,m,d,r,\sigma). \end{aligned} \tag{57}$$

Let $\tau' \in (0,1)$,

$$\beta = \frac{1-\tau'}{2(r+1)}, \quad \rho = \beta\ln\left(3+\delta^{-1}\right), \tag{58}$$

and $\delta_1 = \delta_1(N, m, d, \sigma, r, \tau') > 0$ be such that

$$\delta \in (0, \delta_1) \Longrightarrow \begin{cases} E + \left(\beta \ln\left(3 + \delta^{-1}\right)\right)^2 \\ \qquad \geq \lambda_4(N, m, d, r, \sigma), \\ E + \left(\beta \ln\left(3 + \delta^{-1}\right)\right)^2 \\ \qquad \leq (1 + E)\left(\beta \ln\left(3 + \delta^{-1}\right)\right)^2. \end{cases} \quad (59)$$

Then for the case when $\delta \in (0, \delta_1)$, due to (57), we have that

$$\begin{aligned} &\frac{1}{2}\|v_1 - v_2\|_{\mathbb{L}^\infty(B_r)} \\ &\quad \leq c_7(1 + E)^2\left(E + \left(\beta \ln\left(3 + \delta^{-1}\right)\right)^2\right)^{1/2}\left(3 + \delta^{-1}\right)^{2\beta(r+1)}\delta \\ &\qquad + c_8\left(E + \left(\beta \ln\left(3 + \delta^{-1}\right)\right)^2\right)^{-(m-d)/2d} \\ &\quad = c_7(1 + E)^2\left(E + \left(\beta \ln\left(3 + \delta^{-1}\right)\right)^2\right)^{1/2}(1 + 3\delta)^{1-\tau'}\delta^{\tau'} \\ &\qquad + c_8\left(E + \left(\beta \ln\left(3 + \delta^{-1}\right)\right)^2\right)^{-(m-d)/2d}. \end{aligned} \quad (60)$$

Combining (59) and (60), we obtain that for $s \in [0, s^*]$, $\tau \in (0, \tau')$ and $\delta \in (0, \delta_1)$ the following estimate holds:

$$\begin{aligned} \|v_2 - v_1\|_{L^\infty(B_r)} &\leq c_9(1 + E)^{5/2}\delta^\tau \\ &\quad + c_{10}(1 + E)^{(s-s^*)/2}\left(\ln\left(3 + \delta^{-1}\right)\right)^{-s}, \end{aligned} \quad (61)$$

where $s^* = (m - d)/d$ and $c_9, c_{10} > 0$ depend only on N, m, d, r, σ, τ', and τ.

Estimate (61) in the general case (with modified c_9 and c_{10}) follows from (61) for $\delta \leq \delta_1(N, m, d, \sigma, r, \tau')$ and the property that

$$\|v_j\|_{\mathbb{L}^\infty(B_r)} \leq c_{11}(m, d)\, N. \quad (62)$$

This completes the proof of (12).

6. Buckhgeim-Type Analogs of the Faddeev Functions

Let us identify $\mathbb{R}^2$ with $\mathbb{C}$ and use coordinates $z = x_1 + ix_2$, $\bar{z} = x_1 - ix_2$, where $(x_1, x_2) \in \mathbb{R}^2$. Following [21–23, 37], we consider the functions G_{z_0}, ψ_{z_0}, $\widetilde{\psi}_{z_0}$, δh_{z_0} going back to Buckhgeim's paper [3] and being analogs of the Faddeev functions:

$$\begin{aligned} \psi_{z_0}(z, \lambda) &= e^{\lambda(z-z_0)^2} \\ &\quad + \int_{B_r} G_{z_0}(z, \zeta, \lambda)\, v(\zeta)\, \psi_{z_0}(\zeta, \lambda)\, d\,\mathrm{Re}\,\zeta\, d\,\mathrm{Im}\,\zeta, \\ \widetilde{\psi}_{z_0}(z, \lambda) &= e^{\bar{\lambda}(\bar{z}-\bar{z}_0)^2} \\ &\quad + \int_{B_r} \overline{G_{z_0}(z, \zeta, \lambda)}\, v(\zeta)\, \widetilde{\psi}_{z_0}(\zeta, \lambda)\, d\,\mathrm{Re}\,\zeta\, d\,\mathrm{Im}\,\zeta, \end{aligned} \quad (63)$$

$$\begin{aligned} G_{z_0}(z, \zeta, \lambda) &= \frac{1}{4\pi^2}\int_{B_r} \frac{e^{-\lambda(\eta-z_0)^2+\bar{\lambda}(\bar{\eta}-\bar{z}_0)^2}\, d\,\mathrm{Re}\,\eta\, d\,\mathrm{Im}\,\eta}{(z - \eta)\left(\bar{\eta} - \bar{\zeta}\right)} \\ &\quad \times e^{\lambda(z-z_0)^2-\bar{\lambda}(\bar{\zeta}-\bar{z}_0)^2}, \\ &\quad z = x_1 + ix_2, \quad z_0 \in B_r, \quad \lambda \in \mathbb{C}, \end{aligned} \quad (64)$$

where v satisfies (13),

$$\begin{aligned} \delta h_{z_0}(\lambda) &= \int_{B_r} \widetilde{\psi}_{z_0,1}(z, -\lambda)\left(v_2(z) - v_1(z)\right) \\ &\quad \times \psi_{z_0,2}(z, \lambda)\, d\,\mathrm{Re}\,z\, d\,\mathrm{Im}\,z, \quad \lambda \in \mathbb{C}, \end{aligned} \quad (65)$$

where v_1, v_2 satisfy (13) and $\widetilde{\psi}_{z_0,1}$, $\psi_{z_0,2}$ denote $\widetilde{\psi}_{z_0}$, ψ_{z_0} of (63) for $v = v_1$ and $v = v_2$, respectively.

We recall that (see [21, 22]):

(i) the function G_{z_0} satisfies the equations

$$\begin{aligned} 4\frac{\partial^2}{\partial z \partial \bar{z}} G_{z_0}(z, \zeta, \lambda) &= \delta(z - \zeta), \\ 4\frac{\partial^2}{\partial \zeta \partial \bar{\zeta}} G_{z_0}(z, \zeta, \lambda) &= \delta(z - \zeta), \end{aligned} \quad (66)$$

where $z, z_0, \zeta \in B_r$, $\lambda \in \mathbb{C}$, and δ is the Dirac delta function;

(ii) formulas (63) at fixed z_0 and λ are considered as equations for ψ_{z_0}, $\widetilde{\psi}_{z_0}$ in $L^\infty(B_r)$;

(iii) as a corollary of (63), (64), (66), the functions ψ_{z_0}, $\widetilde{\psi}_{z_0}$ satisfy (1) in B_r for $E = 0$ and $d = 2$;

(iv) the function δh_{z_0} is similar to the right side of (36).

Let potentials $v, v_1, v_2 \in C^2(\bar{B}_r)$, and

$$\begin{aligned} &\|v\|_{C^2(\bar{B}_r)} \leq N, \qquad \|v_j\|_{C^2(\bar{B}_r)} \leq N, \quad j = 1, 2, \\ &(v_1 - v_2)|_{\partial B_r} = 0, \qquad \frac{\partial}{\partial \nu}(v_1 - v_2)\Big|_{\partial B_r} = 0, \end{aligned} \quad (67)$$

then we have that

$$\psi_{z_0}(z,\lambda) = e^{\lambda(z-z_0)^2}\mu_{z_0}(z,\lambda),$$
$$\widetilde{\psi}_{z_0}(z,\lambda) = e^{\overline{\lambda}(\overline{z}-\overline{z}_0)^2}\widetilde{\mu}_{z_0}(z,\lambda), \quad (68)$$

$$\mu_{z_0}(z,\lambda) \longrightarrow 1, \quad \widetilde{\mu}_{z_0}(z,\lambda) \longrightarrow 1 \quad \text{as } |\lambda| \longrightarrow \infty \quad (69)$$

and, for any $\sigma > 1$,

$$\left|\mu_{z_0}(z,\lambda)\right| + \left|\nabla\mu_{z_0}(z,\lambda)\right| \le \sigma, \quad (70a)$$

$$\left|\widetilde{\mu}_{z_0}(z,\lambda)\right| + \left|\nabla\widetilde{\mu}_{z_0}(z,\lambda)\right| \le \sigma, \quad (70b)$$

where $\nabla = (\partial/\partial x_1, \partial/\partial x_2)$, $z = x_1 + ix_2$, $z_0 \in B_r$, $\lambda \in \mathbb{C}$, $|\lambda| \ge \rho_1(N,r,\sigma)$;

$$v_2(z_0) - v_1(z_0) = \lim_{\lambda\to\infty}\frac{2}{\pi}|\lambda|\,\delta h_{z_0}(\lambda) \quad \text{for any } z_0 \in B_r, \quad (71)$$

$$\left|v_2(z_0) - v_1(z_0) - \frac{2}{\pi}|\lambda|\,\delta h_{z_0}(\lambda)\right| \le \frac{c_{12}(N,r)(\ln(3|\lambda|))^2}{|\lambda|^{3/4}}, \quad (72)$$
$$\text{for } z_0 \in B_r, \quad |\lambda| \ge \rho_2(N,r).$$

Formulas (68) can be considered as definitions of μ_{z_0}, $\widetilde{\mu}_{z_0}$. Formulas (69), (71) were given in [21, 22] and go back to [3]. Estimates (70a) and (70b) were proved in [33]. Estimate (72) was obtained in [21, 37].

7. Proof of Theorem 2

We suppose that $\widetilde{\psi}_{z_0,1}(\cdot,-\lambda)$, $\psi_{z_0,2}(\cdot,\lambda)$, $\delta h_{z_0}(\lambda)$ are defined as in Section 6 but with $v_j - E$ in place of v_j, $j = 1,2$. Note that functions $\widetilde{\psi}_{z_0,1}(\cdot,-\lambda)$, $\psi_{z_0,2}(\cdot,\lambda)$ satisfy (1) in B_r with $v = v_j$, $j = 1,2$, respectively. We also use the notation $N_E = N + E$. Then, using (72), we have that

$$\left|v_2(z_0) - v_1(z_0) - \frac{2}{\pi}|\lambda|\,\delta h_{z_0}(\lambda)\right| \le \frac{c_{12}(N_E,r)(\ln(3|\lambda|))^2}{|\lambda|^{3/4}}, \quad (73)$$
$$\text{for } z_0 \in B_r, \; |\lambda| \ge \rho_2(N_E,r).$$

Let

$$\delta = \left\|S_1^+(E) - S_2^+(E)\right\|_{\mathbb{L}^2(\partial B_r\times\partial B_r)}. \quad (74)$$

Combining (17), (19), and (65), we get that

$$\left|\delta h_{z_0}(\lambda)\right| \le \delta\left\|\frac{\partial\phi_1}{\partial\nu_+} - \frac{\partial\phi_1}{\partial\nu_-}\right\|_{\mathbb{L}^2(B_r)}\left\|\frac{\partial\phi_2}{\partial\nu_+} - \frac{\partial\phi_2}{\partial\nu_-}\right\|_{\mathbb{L}^2(B_r)}, \quad (75)$$
$$(k,l) \in \Theta_E, \quad |\operatorname{Im} k| = |\operatorname{Im} l| \ne 0,$$

where ϕ_j, $j = 1,2$, denotes the solution of (18) with $v = v_j$, satisfying

$$\phi_1(x) = \widetilde{\psi}_{z_0,1}(x,-\lambda),$$
$$\phi_2(x) = \psi_{z_0,2}(x,\lambda), \quad \text{for } x \in \overline{B}_r. \quad (76)$$

Using (21), (70a), and (70b) and the fact that $C^1(\partial B_r) \subset \mathbb{H}^1(\partial B_r)$, we find that

$$\left\|\frac{\partial\phi_j}{\partial\nu_+} - \frac{\partial\phi_j}{\partial\nu_-}\right\|_{\mathbb{L}^2(B_r)} \le \sigma c_{13}(r)(1+E)\exp\left(|\lambda|\left(4r^2+4r\right)\right),$$
$$\lambda \in \mathbb{C}, \quad |\lambda| \ge \rho_1(N_E,r,\sigma), \quad j = 1,2. \quad (77)$$

Here and bellow in this section the constant σ is the same that in (70a) and (70b).

Combining (75), (77), we obtain that

$$\left|\delta h_{z_0}(\lambda)\right| \le c_{14}(E,r,\sigma)\exp\left(|\lambda|\left(8r^2+8r\right)\right)\delta,$$
$$\lambda \in \mathbb{C}, \quad |\lambda| \ge \rho_1(N_E,r,\sigma). \quad (78)$$

Using (73) and (78), we get that

$$\left|v_2(z_0) - v_1(z_0)\right| \le c_{14}(E,r,\sigma)\exp\left(|\lambda|\left(8r^2+8r\right)\right)\delta + \frac{c_{12}(N_E,r)(\ln(3|\lambda|))^2}{|\lambda|^{3/4}},$$
$$z_0 \in B_r, \quad \lambda \in \mathbb{C}, \quad |\lambda| \ge \rho_3(N_E,r,\sigma) = \max\{\rho_1,\rho_2\}. \quad (79)$$

We fix some $\tau \in (0,1)$ and let

$$\beta = \frac{1-\tau}{8r^2+8r}, \quad \lambda = \beta\ln\left(3+\delta^{-1}\right), \quad (80)$$

where δ is so small that $|\lambda| \ge \rho_3(N_E,r,\sigma)$. Then due to (79), we have that

$$\begin{aligned}\|v_1 - v_2\|_{\mathbb{L}^\infty(B_r)} &\le c_{14}(E,r,\sigma)\left(3+\delta^{-1}\right)^{\beta(8r^2+8r)}\delta + c_{12}(N_E,r)\frac{\left(\ln\left(3\beta\ln\left(3+\delta^{-1}\right)\right)\right)^2}{\left(\beta\ln\left(3+\delta^{-1}\right)\right)^{3/4}} \\ &= c_{14}(E,r,\sigma)(1+3\delta)^{1-\tau}\delta^{\tau} + c_{12}(N_E,r)\beta^{-3/4}\frac{\left(\ln\left(3\beta\ln\left(3+\delta^{-1}\right)\right)\right)^2}{\left(\ln\left(3+\delta^{-1}\right)\right)^{3/4}},\end{aligned} \quad (81)$$

where τ, β, and δ are the same as in (80).

Using (81), we obtain that

$$\|v_1 - v_2\|_{\mathbb{L}^\infty(B_r)} \le c_{15}(N,E,r,\sigma)\left(\ln\left(3+\delta^{-1}\right)\right)^{-3/4} \times \left(\ln\left(3\ln\left(3+\delta^{-1}\right)\right)\right)^2, \quad (82)$$

for $\delta = \|S_1^+(E) - S_2^+(E)\|_{\mathbb{L}^2(\partial B_r \times \partial B_r)} \leq \delta_2(N_E, r, \sigma)$, where δ_2 is a sufficiently small positive constant. Estimate (82) in the general case (with modified c_{15}) follows from (82) for $\delta \leq \delta_2(N_E, r, \sigma)$ and the property that $\|v_j\|_{\mathbb{L}^\infty(B_r)} \leq N$.

This completes the proof of (14).

8. Proof of Lemma 7

In this section we assume for simplicity that $r = 1$ and therefore $\partial B_r = \mathbb{S}^{d-1}$.

We fix an orthonormal basis in $\mathbb{L}^2(\partial B_r)$:

$$\left\{f_{jp} : j \geq 0; 1 \leq p \leq p_j\right\},$$
$$f_{jp} \text{ is a spherical harmonic of degree } j, \tag{83}$$

where p_j is the dimension of the space of spherical harmonics of order j,

$$p_j = \binom{j+d-1}{d-1} - \binom{j+d-3}{d-1}, \tag{84}$$

where

$$\binom{n}{k} = \frac{n(n-1)\cdots(n-k+1)}{k!}, \quad \text{for } n \geq 0,$$
$$\binom{n}{k} = 0, \quad \text{for } n < 0. \tag{85}$$

The precise choice of f_{jp} is irrelevant for our purposes. Besides orthonormality, we only need f_{jp} to be the restriction of a homogeneous harmonic polynomial of degree j to the sphere ∂B_r and so $|x|^j f_{jp}(x \mid |x|)$ is harmonic pn $\mathbb{R}^d$. In the Sobolev spaces $\mathbb{H}^s(\partial B_r)$ the norm is defined by

$$\left\|\sum_{j,p} c_{jp} f_{jp}\right\|^2_{\mathbb{H}^s(\partial B_r)} = \sum_{j,p} (1+j)^{2s} \left|c_{jp}\right|^2. \tag{86}$$

The solution ϕ of the exterior Dirichlet problem

$$-\Delta\phi = E\phi, \quad \text{in } \mathbb{R}^d \setminus \overline{B}_r,$$
$$\lim_{|x| \to +\infty} |x|^{(d-1)/2} \left(\frac{\partial}{\partial |x|}\phi - i\sqrt{E}\phi\right) = 0, \tag{87}$$
$$\phi|_{\partial B_r} = u \in \mathbb{H}^1(\partial B_r),$$

can be expressed in the following form (see, e.g., [1, 38]):

$$\phi = \sum_{j,p} c_{jp}\phi_{jp}, \tag{88}$$

where c_{jp} are expansion coefficients of u in the basis $\{f_{jp} : j \geq 0; 1 \leq p \leq p_j\}$, and

ϕ_{jp} denotes the solution of (89) with $u = f_{jp}$,

$$\phi_{jp}(x) = h_{jp}(|x|) f_{jp}\left(\frac{x}{|x|}\right),$$
$$h_{jp}(|x|) = |x|^{-(d-2)/2} \frac{H^{(1)}_{j+((d-2)/2)}\left(\sqrt{E}|x|\right)}{H^{(1)}_{j+((d-2)/2)}\left(\sqrt{E}\right)}, \tag{89}$$

where $H^{(1)}_\mu$ is the Hankel function of the first kind. Let

$$\phi^0_{jp}(x) = |x|^{-j-d+2} f_{jp}\left(\frac{x}{|x|}\right). \tag{90}$$

Note that ϕ^0_{jp} is harmonic in $\mathbb{R}^d \setminus \{0\}$ and

$$\lim_{|x| \to +\infty} |x|^{(d-1)/2} \left(\frac{\partial}{\partial |x|}\phi^0_{jp} - i\sqrt{E}\phi^0_{jp}\right)$$
$$= 0, \quad \text{for } j + \frac{d-3}{2} > 0. \tag{91}$$

Using the Green formula and the radiation condition for ϕ_{jp}, ϕ^0_{jp}, we get that

$$\int_{\mathbb{R}^d \setminus B_r} E\phi_{jp}\phi^0_{jp} dx$$
$$= \int_{\mathbb{R}^d \setminus B_r} \left(\Delta\phi^0_{jp}\phi_{jp} - \Delta\phi_{jp}\phi^0_{jp}\right) dx$$
$$= \int_{\partial B_r} \left(\frac{\partial\phi^0_{jp}}{\partial\nu_+}\phi_{jp} - \frac{\partial\phi_{jp}}{\partial\nu_+}\phi^0_{jp}\right) dx, \quad \text{for } j + \frac{d-3}{2} > 0. \tag{92}$$

Due to (89) and (90), we have that

$$\left|\int_{\partial B_r} \frac{\partial\phi^0_{jp}}{\partial\nu_+}\phi_{jp} dx\right| = (j+d-2)\int_{\partial B_r} f^2_{jp} dx$$
$$= j + d - 2. \tag{93}$$

Using also the following property of the Hankel function of the first kind (see, e.g., [39]):

$$\left|H^{(1)}_\mu(x)\right| \text{ is a decreasing function of } x, \text{ for } x \in \mathbb{R}_+, \mu \in \mathbb{R}, \tag{94}$$

we get that

$$\begin{aligned}\left|\int_{\mathbb{R}^d\setminus B_r}\phi_{jp}\phi^0_{jp}dx\right| &= \left|\int_1^{+\infty}t^{-j-d+2}h_{jp}(t)\,t^{d-1}dt\right| \\ &= \left|\int_1^{+\infty}t^{-j-(d/2)}\frac{H^{(1)}_{j+((d-2)/2)}\left(\sqrt{E}t\right)}{H^{(1)}_{j+((d-2)/2)}\left(\sqrt{E}\right)}dt\right| \\ &\le \int_1^{+\infty}t^{-j-(d/2)}dt = \frac{1}{j+(d/2)-1} \\ &\le 2, \quad \text{for } j+\frac{d-3}{2}>0.\end{aligned} \tag{95}$$

Combining (89), (90), (92), (93), and (95), we obtain that

$$\left|\int_{\partial B_r}\frac{\partial\phi^0_{jp}}{\partial\nu_+}\phi_{jp}dx\right| = \left|\frac{h'_{jp}(r)}{h_{jp}(r)}\right| \le j+d-2+2E, \quad \text{for } j+\frac{d-3}{2}>0. \tag{96}$$

Let consider the cases when $j+((d-3)/2)\le 0$.

Case 1 ($j=0$, $d=2$). Using the property $dH^{(1)}_0(t)/dt = -H^{(1)}_1(t)$, we get that

$$\frac{h'_{jp}(r)}{h_{jp}(r)} = \sqrt{E}\frac{H^{(1)}_1\left(\sqrt{E}\right)}{H^{(1)}_0\left(\sqrt{E}\right)}. \tag{97}$$

We recall that functions $H^{(1)}_0$ and $H^{(1)}_1$ have the following asymptotic forms (see, e.g. [39]):

$$\begin{aligned} H^{(1)}_0(t) &\sim \frac{2i}{\pi}\ln\left(\frac{t}{2}\right), \quad \text{as } t\longrightarrow +0, \\ H^{(1)}_0(t) &\sim \sqrt{\frac{2}{\pi t}}e^{i(t-\pi/4)}, \quad \text{as } t\longrightarrow +\infty, \\ H^{(1)}_1(t) &\sim -\frac{i}{\pi}\left(\frac{2}{t}\right), \quad \text{as } t\longrightarrow +0, \\ H^{(1)}_1(t) &\sim \sqrt{\frac{2}{\pi t}}e^{i(t-3\pi/4)}, \quad \text{as } t\longrightarrow +\infty.\end{aligned} \tag{98}$$

Using (94) and (98), we get that for some $c>0$

$$\frac{H^{(1)}_1(t)}{H^{(1)}_0(t)} \le c\left(1+\frac{1}{t}\right). \tag{99}$$

Combining (97) and (99), we obtain that for $j=0$, $d=2$

$$\left|\frac{h'_{jp}(r)}{h_{jp}(r)}\right| \le c\left(1+\sqrt{E}\right). \tag{100}$$

Case 2 ($j=0$, $d=3$). We have that

$$H^{(1)}_{j+((d-2)/2)}(t) = \sqrt{\frac{2}{\pi t}}e^{i(t-\pi/2)}. \tag{101}$$

Using (89) and (101), we get that for $j=0$, $d=3$

$$\frac{h'_{jp}(r)}{h_{jp}(r)} = -1+i\sqrt{E}. \tag{102}$$

Combining (86)–(89), (96), (100), and (102), we get that for some constant $c'=c'(d)>0$:

$$\begin{aligned}\left\|\frac{\partial\phi}{\partial\nu_+}\bigg|_{\partial B_r}\right\|^2_{\mathbb{L}^2(\partial B_r)} &= \sum_{j,p}c^2_{jp}\left|\frac{h'_{jp}(r)}{h_{jp}(r)}\right|^2 \\ &\le c'(1+E)^2\sum_{j,p}(1+j)^2c^2_{jp}.\end{aligned} \tag{103}$$

Using (86) and (103), we obtain (21).

Acknowledgment

This work was partially supported by FCP Kadry no. 14.A18.21.0866.

References

[1] Ju. M. Berezanskiĭ, "The uniqueness theorem in the inverse problem of spectral analysis for the Schrödinger equation," *Trudy Moskovskogo Matematičeskogo Obščestva*, vol. 7, pp. 1–62, 1958 (Russian).

[2] N. V. Alekseenko, V. A. Burov, and O. D. Rumyantseva, "Solution of the three-dimensional acoustic inverse scattering problem. the modified Novikov algorithm," *Akusticheskiĭ Zhurnal*, vol. 54, no. 3, pp. 469–482, 2008 (Russian), English translation in *Acoustical Physics*, vol. 54, no. 3, pp. 407–419, 2008.

[3] A. L. Bukhgeim, "Recovering a potential from Cauchy data in the two-dimensional case," *Journal of Inverse and Ill-Posed Problems*, vol. 16, no. 1, pp. 19–33, 2008.

[4] V. A. Burov, N. V. Alekseenko, and O. D. Rumyantseva, "Multifrequency generalization of the Novikov algorithm for the two-dimensional inverse scattering problem," *Acoustical Physics*, vol. 55, no. 6, pp. 843–856, 2009.

[5] G. Eskin and J. Ralston, "Inverse scattering problem for the Schrödinger equation with magnetic potential at a fixed energy," *Communications in Mathematical Physics*, vol. 173, no. 1, pp. 199–224, 1995.

[6] R. G. Novikov and G. M. Khenkin, "The $\bar{\partial}$-equation in the multidimensional inverse scattering problem," *Uspekhi Matematicheskikh Nauk*, vol. 42, no. 3(255), pp. 93–152, 1987 (Russian), English translation in *Russian Mathematical Surveys*, vol. 42, no. 3, pp.109–180, 1987.

[7] M. I. Isaev, "Exponential instability in the inverse scattering problem on the energy interval," to appear in *Funktsional'nyi Analiz i ego Prilozheniya*, http://arxiv.org/abs/1012.5526.

[8] R. G. Novikov, "A multidimensional inverse spectral problem for the equation $-\Delta\psi+(v(x)-Eu(x))\psi=0$," *Funktsional'nyĭ

Analiz i ego Prilozheniya, vol. 22, no. 4, pp. 11–22, 1988, English translation in *Functional Analysis and Its Applications*, vol. 22, pp. 263–272, 1988.

[9] R. G. Novikov, "The inverse scattering problem at fixed energy for the three-dimensional Schrödinger equation with an exponentially decreasing potential," *Communications in Mathematical Physics*, vol. 161, no. 3, pp. 569–595, 1994.

[10] R. G. Novikov, "Approximate solution of the inverse problem of quantum scattering theory with fixed energy in dimension 2," *Trudy Matematicheskogo Instituta Imeni V. A. Steklova*, vol. 225, pp. 301–318, 1999 (Russian), English translation in *Proceedings of the Steklov Institute of Mathematic*, no. 2(225), pp. 285–302, 1999.

[11] R. G. Novikov, "The $\bar{\partial}$-approach to approximate inverse scattering at fixed energy in three dimensions," *IMRP. International Mathematics Research Papers*, no. 6, pp. 287–349, 2005.

[12] P. Stefanov, "Stability of the inverse problem in potential scattering at fixed energy," *Annales de l'Institut Fourier*, vol. 40, no. 4, pp. 867–884, 1990.

[13] A. I. Nachman, "Reconstructions from boundary measurements," *Annals of Mathematics*, vol. 128, no. 3, pp. 531–576, 1988.

[14] G. Alessandrini, "Stable determination of conductivity by boundary measurements," *Applicable Analysis*, vol. 27, no. 1–3, pp. 153–172, 1988.

[15] M. I. Isaev and R. G. Novikov, "Energy and regularity dependent stability estimates for the Gel'fand inverse problem in multidimensions," *Journal of Inverse and Ill-posed Problems*, vol. 20, no. 3, pp. 313–325, 2012.

[16] V. Isakov, "Increasing stability for the Schrödinger potential from the Dirichlet-to-Neumann map," *Discrete and Continuous Dynamical Systems. Series S*, vol. 4, no. 3, pp. 631–640, 2011.

[17] N. Mandache, "Exponential instability in an inverse problem for the Schrödinger equation," *Inverse Problems*, vol. 17, no. 5, pp. 1435–1444, 2001.

[18] R. G. Novikov, "Formulae and equations for finding scattering data from the Dirichlet-to-Neumann map with nonzero background potential," *Inverse Problems*, vol. 21, no. 1, pp. 257–270, 2005.

[19] R. G. Novikov, "An effectivization of the global reconstruction in the Gel'fand-Calderón inverse problem in three dimensions," in *Imaging Microstructures*, vol. 494 of *Contemp. Math.*, pp. 161–184, American Mathematical Society, Providence, RI, USA, 2009.

[20] R. G. Novikov, "New global stability estimates for the Gel'fand-Calderon inverse problem," *Inverse Problems*, vol. 27, no. 1, Article ID 015001, 21 pages, 2011.

[21] R. Novikov and M. Santacesaria, "A global stability estimate for the Gel'fand-Calderón inverse problem in two dimensions," *Journal of Inverse and Ill-Posed Problems*, vol. 18, no. 7, pp. 765–785, 2010.

[22] R. G. Novikov and M. Santacesaria, "Global uniqueness and reconstruction for the multi-channel Gel'fand-Calderón inverse problem in two dimensions," *Bulletin des Sciences Mathématiques*, vol. 135, no. 5, pp. 421–434, 2011.

[23] R. Novikov and M. Santacesaria, "Monochromatic reconstruction algorithms for two-dimensional multi-channel inverse problems," *International Mathematics Research Notes*, 2012.

[24] J. Sylvester and G. Uhlmann, "A global uniqueness theorem for an inverse boundary value problem," *Annals of Mathematics*, vol. 125, no. 1, pp. 153–169, 1987.

[25] P. Hähner and T. Hohage, "New stability estimates for the inverse acoustic inhomogeneous medium problem and applications," *SIAM Journal on Mathematical Analysis*, vol. 33, no. 3, pp. 670–685, 2001.

[26] M. I. Isaev and R. G. Novikov, "New global stability estimates for monochromatic inverse acoustic scattering," http://arxiv.org/abs/1210.3821.

[27] L. Beilina and M. V. Klibanov, *Approximate Global Convergence and Adaptivity for Coefficient Inverse Problems*, Springer, New York, NY, USA, 2012.

[28] M. M. Lavrent'ev, V. G. Romanov, and S. P. Shishatskii, *Ill-Posed Problems of Mathematical Physics and Analysis*, vol. 64 of *Translations of Mathematical Monographs*, American Mathematical Society, Providence, RI, USA, 1986.

[29] M. Santacesaria, "Stability estimates for an inverse problem for the Schrödinger equation at negative energy in two dimensions," *Applicable Analysis*, 2012.

[30] L. D. Faddeev, "Growing solutions of the Schrödinger equation," *Doklady Akademii Nauk SSSR*, vol. 165, no. 3, pp. 514–517, 1965 (Russian), English translation in *Soviet Physics. Doklady*, vol. 10, pp. 1033–1035, 1966.

[31] L. D. Faddeev, "The inverse problem in the quantum theory of scattering. II," *Current Problems in Mathematics*, vol. 3, pp. 93–180, 259, 1974 (Russian), English translation in *Journal of Soviet Mathematics*, vol. 5, pp. 334–396, 1976.

[32] R. Beals and R. R. Coifman, "Multidimensional inverse scatterings and nonlinear partial differential equations," in *Pseudodifferential Operators and Applications (Notre Dame, Ind., 1984)*, vol. 43 of *Proc. Sympos. Pure Math.*, pp. 45–70, American Mathematical Society, Providence, RI, USA, 1985.

[33] M. I. Isaev and R. G. Novikov, "Stability estimates for determination of potential from the impedance boundary map," *Algebra and Analysis*, 2013, http://128.84.158.119/abs/1112.3728v2.

[34] R. B. Lavine and A. I. Nachman, "On the inverse scattering transform for the n-dimensional Schrödinger operator," in *Topics in Soliton Theory and Exactly Solvable Nonlinear Equations (Oberwolfach, 1986)*, pp. 33–44, World Scientific, Singapore, 1987.

[35] R. Weder, "Generalized limiting absorption method and multidimensional inverse scattering theory," *Mathematical Methods in the Applied Sciences*, vol. 14, no. 7, pp. 509–524, 1991.

[36] R. G. Novikov, "$\bar{\partial}$-method with nonzero background potential. Application to inverse scattering for the two-dimensional acoustic equation," *Communications in Partial Differential Equations*, vol. 21, no. 3-4, pp. 597–618, 1996.

[37] M. Santacesaria, "Global stability for the multi-channel Gel'fandCalderon inverse problem in two dimensions," *Bulletin des Sciences Mathématiques*, vol. 136, no. 7, pp. 731–744, 2012.

[38] D. Colton and R. Kress, *Inverse Acoustic and Electromagnetic Scattering Theory*, vol. 93 of *Applied Mathematical Sciences*, Springer, Berlin, Germany, 2nd edition, 1998.

[39] G. N. Watson, *A Treatise on the Theory of Bessel Functions*, Cambridge University Press, Cambridge, UK, 1944.

An Order-Preserving Linear Map from Matrices to Banach $*$-Algebras and Applications

Pattrawut Chansangiam

Department of Mathematics, Faculty of Science, King Mongkut's Institute of Technology Ladkrabang, Bangkok 10520, Thailand

Correspondence should be addressed to Pattrawut Chansangiam; kcpattra@kmitl.ac.th

Academic Editor: Tepper L. Gill

We establish a theorem for which a number of absolute-value identities and inequalities in the framework of Banach $*$-algebras can be generated. To do this job, we construct an order-preserving linear map from the vector space of 2-by-2 hermitian matrices to a hermitian Banach $*$-algebras. This map can convert any suitable matrix ordering to a number of identities and inequalities in Banach $*$-algebras. Hence, we obtain a number of analogues of the well-known results in a framework of hermitian Banach $*$-algebras.

1. Introduction

There are many works involving absolute-value identities and inequalities. The famous such identities are the parallelogram law and the polarization identity. It is well known that these identities hold in the context of Euclidean spaces, normed linear spaces, and inner product spaces. On the other hand for inequalities, Bohr [1] established the classical Bohr's inequality which asserts that

$$|a+b|^2 \leqslant p|a|^2 + q|b|^2 \tag{1}$$

for complex numbers a, b and real numbers $p, q > 1$ such that $1/p + 1/q = 1$. The equality in (1) occurs if and only if $pa = qb$. Then a number of extensions and variations of absolute-value inequalities concerning Bohr's inequality were developed in various contexts by many authors. The results for complex numbers are obtained in [2–4]. The context of matrices is given in [5]. Bohr's inequality and related results were generalized to operator algebras in [6–15].

The first version of operator Bohr's inequality was established via direct computations by Hirzallah [12]. Later, Hirzallah's results were extended by the same technique in [6, 10, 14]. Zhang [15] used operator identities and inequalities for approaching operator inequalities related to (1). Recently, the idea of matrix ordering for discussing operator absolute-value inequalities appears in [8, 9, 11]. However, the results mentioned above has been proved in separate ways.

This paper consists of two main purposes. The first goal is to construct a theorem for which a number of absolute-value identities and inequalities can be generated from it. The second is to extend identities and inequalities involving absolute values to an abstract framework of Banach $*$-algebras. In Section 2, after recalling some terminology, we propose an order-preserving linear map from the set of 2-by-2 hermitian matrices to a hermitian Banach $*$-algebra. The image of this map for each 2-by-2 hermitian matrix is in the form involving absolute values. In Section 3, we apply this order preserving to elementary matrix identities and inequalities to get a number of identities and inequalities about absolute values of elements in hermitian Banach $*$-algebras.

2. An Order-Preserving Linear Map from Matrices to Banach $*$-Algebras

An element a in a Banach $*$-algebra is called *self-adjoint* if $a^* = a$. An element a which has real spectrum, that is, $\sigma(a) \subseteq \mathbb{R}$, is said to be *hermitian*. A Banach $*$-algebra is called *hermitian* if each self-adjoint element is hermitian. The class of hermitian Banach ∗-algebras includes any C^*-algebra, any group algebra of an abelian group, any group algebra of a compact group, and any measure algebra of discrete group. Throughout this paper, $\mathscr{A}$ denotes a hermitian Banach ∗-algebra.

Every hermitian Banach ∗-algebra is equipped with a natural order structure as follows. Given self-adjoint elements $a, b \in \mathscr{A}$, the relation $a \leqslant b$ means that $b-a$ is self-adjoint and $\sigma(b-a) \subseteq [0,\infty)$. Then the relation "$\leqslant$" forms a partial order on the real vector space of self-adjoint elements in $\mathscr{A}$. The set of $a \in \mathscr{A}$ such that $a \geqslant 0$ forms a positive cone in $\mathscr{A}$ (see [16, Lemma 41.4]), that is, if $a, b \in \mathscr{A}$ are such that $a, b \geqslant 0$, then $\alpha a + \beta b \geqslant 0$ for any $\alpha, \beta \in [0,\infty)$.

The Shirali-Ford Theorem [17, Theorem 1] assures that $a^*a \geqslant 0$ for any $a \in \mathscr{A}$. Then the *absolute value* of $a \in \mathscr{A}$ is defined to be $(a^*a)^{1/2}$. By the spectral mapping theorem, $\sigma(|a|) \subseteq [0,\infty)$ and hence $|a| \geqslant 0$ for every $a \in \mathscr{A}$. Note that $|a| = 0$ if and only if $a = 0$ by [18, Lemma 3].

Denote by $H_2(\mathbb{C})$ the hermitian Banach ∗-subalgebra of 2-by-2 hermitian matrices of the hermitian Banach ∗-algebra $M_2(\mathbb{C})$ of 2-by-2 complex matrices. The natural order structure on $H_2(\mathbb{C})$ is called the Löwner partial order: the relation $A \leqslant B$ in $H_2(\mathbb{C})$ means that $B - A$ is a positive semidefinite matrix, that is, a hermitian matrix with nonnegative eigenvalues.

Theorem 1. *Let $a, b \in \mathscr{A}$. The map $\Psi : M_2(\mathbb{C}) \to \mathscr{A}$ given by*

$$\begin{pmatrix} \alpha & \beta \\ \gamma & \mu \end{pmatrix} \longmapsto \alpha|a|^2 + \beta a^*b + \gamma b^*a + \mu|b|^2, \qquad (2)$$

for $\alpha, \beta, \gamma, \mu \in \mathbb{C}$, is $\mathbb{C}$-linear. The map $\Phi : H_2(\mathbb{C}) \to \mathscr{A}$ given by

$$\begin{pmatrix} \alpha & \beta \\ \overline{\beta} & \gamma \end{pmatrix} \longmapsto \alpha|a|^2 + \beta a^*b + \overline{\beta} b^*a + \gamma|b|^2, \qquad (3)$$

for $\alpha, \gamma \in \mathbb{R}$ and $\beta \in \mathbb{C}$, is $\mathbb{R}$-linear, positive, and order-preserving. The positivity of Φ means $A \geqslant 0$ in $H_2(\mathbb{C})$ implies $\Phi(A) \geqslant 0$ in $\mathscr{A}$. The order preserving of Φ means $A \geqslant B$ in $H_2(\mathbb{C})$ implies $\Phi(A) \geqslant \Phi(B)$ in $\mathscr{A}$.

Proof. The $\mathbb{C}$-linearity of Ψ and the $\mathbb{R}$-linearity of Φ are clear. For the positivity of Φ, consider $\alpha, \gamma \geqslant 0$ and $\beta \in \mathbb{C}$ such that $\alpha\gamma \geqslant |\beta|^2$. If $\beta = 0$, we are done. If $\beta \neq 0$, then $\alpha > 0$ and $\gamma > 0$. Set $\lambda = \alpha\gamma - |\beta|^2$. Since $\alpha = (\lambda + |\beta|^2)/\gamma$, it follows that

$$\begin{aligned} &\alpha|a|^2 + \beta a^*b + \overline{\beta} b^*a + \gamma|b|^2 \\ &= \frac{\lambda + |\beta|^2}{\gamma}|a|^2 + \beta a^*b + \overline{\beta} b^*a + \gamma|b|^2 \\ &= \frac{\lambda}{\gamma}|a|^2 + \left|\frac{\beta}{\sqrt{\gamma}}a + \sqrt{\gamma}b\right|^2 \\ &\geqslant 0. \end{aligned} \qquad (4)$$

So, Φ is positive. Now, if $A \leqslant B$ in $H_2(\mathbb{C})$, it follows from the linearity and positivity of Φ that $B - A \geqslant 0$ which implies $\Phi(B) - \Phi(A) = \Phi(B - A) \geqslant \Phi(0) = 0$, that is, $\Phi(A) \leqslant \Phi(B)$. □

The most useful form of Φ in this paper is

$$\Phi\begin{pmatrix} |\alpha|^2 & \overline{\alpha}\beta \\ \alpha\overline{\beta} & |\beta|^2 \end{pmatrix} = |\alpha a + \beta b|^2. \qquad (5)$$

The next theorem is very useful to get a necessity and sufficiency condition for the equality case in the later discussions.

Theorem 2. *Let a, b be nonzero elements in $\mathscr{A}$. For each $\alpha, \gamma \geqslant 0$ and $\beta \in \mathbb{C}$ such that $\alpha\gamma \geqslant |\beta|^2$, the equation*

$$\alpha|a|^2 + \beta a^*b + \overline{\beta} b^*a + \gamma|b|^2 = 0 \qquad (6)$$

holds if and only if either $\alpha = \beta = \gamma = 0$ or $\alpha\gamma = |\beta|^2 \neq 0$ and $\alpha a + \beta b = 0$. In particular, the restriction of Φ in the previous theorem to the 2-by-2 positive semidefinite matrices satisfies

$$\begin{aligned} \Phi^{-1}(\{0\}) = &\left\{\begin{pmatrix} \alpha & \beta \\ \overline{\beta} & \dfrac{|\beta|^2}{\alpha} \end{pmatrix} : \alpha a + \beta b = 0, \alpha > 0, \ \beta \in \mathbb{C}\right\} \\ &\cup \left\{\begin{pmatrix} 0 & 0 \\ 0 & 0 \end{pmatrix}\right\}. \end{aligned} \qquad (7)$$

Proof. If either $\alpha = \beta = \gamma = 0$ or $\alpha\gamma = \beta^2$ and $\alpha a + \beta b = 0$ holds, then (6) holds. Suppose now that (6) holds for $\alpha, \gamma \geqslant 0$ and $\beta \in \mathbb{C}$ such that $\alpha\gamma \geqslant |\beta|^2$. If $\alpha = 0$, then $\beta = 0$ and $0 = \gamma|b|^2$, that is, $\gamma = 0$. Consider the case $\alpha > 0$. If $\gamma = 0$, we get $\beta = 0$ and then $a = 0$, a contradiction. Now for $\gamma > 0$, let $\lambda = \alpha\gamma - |\beta|^2$. We have $\lambda \geqslant 0$ and $\gamma = (\lambda + |\beta|^2)/\alpha$. Hence

$$\begin{aligned} 0 &= \alpha|a|^2 + \beta a^*b + \overline{\beta} b^*a + \gamma|b|^2 \\ &= \alpha|a|^2 + \beta a^*b + \overline{\beta} b^*a + \frac{\lambda + |\beta|^2}{\alpha}|b|^2 \\ &= \left|\sqrt{\alpha}a + \frac{\beta}{\sqrt{\alpha}}b\right|^2 + \frac{\lambda}{\alpha}|b|^2. \end{aligned} \qquad (8)$$

This forces $|\sqrt{\alpha}a + (\beta/\sqrt{\alpha})b| = (\lambda/\alpha)|b| = 0$ and hence $\alpha a + \beta b = 0$ and $\alpha\gamma = |\beta|^2$. □

3. Applications

The $\mathbb{R}$-linearity and order preserving of the map Φ and the $\mathbb{C}$-linearity of Ψ are used in order to obtain a number of identities and inequalities from suitable matrix identities and inequalities. We give some applications of Theorem 1 as follows:

(1) parallelogram law and its generalizations,

(2) polarization identity and its generalizations,

(3) Bohr's inequality and its reverse,

(4) generalizations of Bohr's inequality,

(5) related absolute-value identities and inequalities.

Let $\mathscr{A}$ be a hermitian Banach $*$-algebra.

Corollary 3. *Let $a, b \in \mathscr{A}$ and $n \in \mathbb{N}$. If ω is an nth-root of unity (i.e., $\omega \in \mathbb{C} \setminus \{1\}$ and $\omega^n = 1$), then*

$$|a|^2 + |b|^2 = \frac{1}{n}\sum_{k=1}^{n} \left|a + \omega^k b\right|^2. \tag{9}$$

In particular, the usual parallelogram law holds:

$$2|a|^2 + 2|b|^2 = |a - b|^2 + |a + b|^2. \tag{10}$$

Proof. Since ω is an nth-root of unity, $\omega^{n-1} + \omega^{n-2} + \cdots + 1 = 0$ and $|\omega| = 1$. Then a computation shows that

$$\begin{aligned} &\begin{pmatrix} 1 & \omega \\ \overline{\omega} & |\omega|^2 \end{pmatrix} + \begin{pmatrix} 1 & \omega^2 \\ \overline{\omega}^2 & |\omega|^4 \end{pmatrix} + \cdots + \begin{pmatrix} 1 & \omega^n \\ \overline{\omega}^n & |\omega|^{2n} \end{pmatrix} \\ &= \begin{pmatrix} n & \omega + \omega^2 + \cdots + \omega^n \\ \overline{\omega} + \overline{\omega}^2 + \cdots + \overline{\omega}^n & |\omega|^2 + |\omega|^4 + \cdots + |\omega|^{2n} \end{pmatrix} \\ &= \begin{pmatrix} n & 0 \\ 0 & n \end{pmatrix}. \end{aligned} \tag{11}$$

By linearity of Ψ or Φ, we get (9). To get a usual parallelogram law, take $n = 2$ and $\omega = -1$. □

Corollary 4. *Let $a, b \in \mathscr{A}$. Then for any $t \in \mathbb{R} \setminus \{0\}$,*

$$|a + b|^2 + \frac{1}{t}|ta - b|^2 = (1 + t)\,|a|^2 + \left(1 + \frac{1}{t}\right)|b|^2. \tag{12}$$

In particular, for $p, q \in \mathbb{R}$ such that $p, q \neq 1$, $pq > 0$ and $1/p + 1/q = 1$,

$$|a + b|^2 + \left|\sqrt{\frac{p}{q}}a - \sqrt{\frac{q}{p}}b\right|^2 = p|a|^2 + q|b|^2. \tag{13}$$

Proof. Equation (12) is done by applying Ψ or Φ to a matrix identity

$$\begin{pmatrix} 1 & 1 \\ 1 & 1 \end{pmatrix} + \frac{1}{t}\begin{pmatrix} t^2 & -t \\ -t & 1 \end{pmatrix} = \begin{pmatrix} 1 + t & 0 \\ 0 & 1 + \frac{1}{t} \end{pmatrix}. \tag{14}$$

To get (13), choose $t = p/q > 0$ which implies $1 + t = p$ and $1 + 1/t = q$ from the condition $1/p + 1/q = 1$. □

The identities (12) and (13) become the parallelogram law when $t = 1$ and $p = q = 2$, respectively. The identities (12) and (13) for Hilbert space operators are provided in [11, Theorem 4.1] and [15, Theorem 2], respectively. The identity (13) can be stated equivalently that for any $0 \leqslant \alpha \leqslant 1$,

$$|\alpha a + (1 - \alpha)\,b|^2 + \alpha\,(1 - \alpha)\,|a - b|^2 = \alpha|a|^2 + (1 - \alpha)\,|b|^2. \tag{15}$$

Corollary 5. *Let $a, b \in \mathscr{A}$ and $n \in \mathbb{N}$. If ω is an nth -root of unity, then*

$$b^* a = \frac{1}{n}\sum_{k=1}^{n} \omega^k \left|a + \omega^k b\right|^2. \tag{16}$$

In particular, the usual polarization identity holds when we denote $\langle a, b\rangle = b^* a$ (e.g., in $\mathbb{C}^n$):

$$\langle a, b\rangle = \frac{1}{4}\sum_{k=1}^{4} i^k \left|a + i^k b\right|^2. \tag{17}$$

Proof. Since ω is an nth-root of unity, $\omega^{n-1} + \omega^{n-2} + \cdots + 1 = 0$ and $|\omega| = 1$. Note that ω^2 is also an nth-root of unity and hence $1 + \omega^2 + \omega^4 + \cdots + \omega^{2n-2} = 0$. Then a computation shows that

$$\begin{aligned} &\omega\begin{pmatrix} 1 & \omega \\ \overline{\omega} & |\omega|^2 \end{pmatrix} + \omega^2\begin{pmatrix} 1 & \omega^2 \\ \overline{\omega}^2 & |\omega|^4 \end{pmatrix} + \cdots + \omega^n\begin{pmatrix} 1 & \omega^n \\ \overline{\omega}^n & |\omega|^{2n} \end{pmatrix} \\ &= \begin{pmatrix} \omega + \omega^2 + \cdots + \omega^n & \omega^2 + \omega^4 + \cdots + \omega^{2n} \\ \omega\overline{\omega} + \omega^2\overline{\omega}^2 + \cdots + \omega^n\overline{\omega}^n & \omega|\omega|^2 + \omega^2|\omega|^4 + \cdots + \omega^n|\omega|^{2n} \end{pmatrix} \\ &= \begin{pmatrix} \omega\left(1 + \omega + \cdots + \omega^{n-1}\right) & \omega^2\left(1 + \omega^2 + \omega^4 + \cdots + \omega^{2n-2}\right) \\ |\omega|^2 + |\omega|^4 + \cdots + |\omega|^{2n} & \omega\left(1 + \omega + \cdots + \omega^{n-1}\right) \end{pmatrix} \\ &= \begin{pmatrix} 0 & 0 \\ n & 0 \end{pmatrix}. \end{aligned} \tag{18}$$

By $\mathbb{C}$-linearity of Ψ, we get (16). To get a usual polarization identity, take $n = 4$ and $\omega = i$. □

Next, consider Bohr's inequality and its extensions. The operator version of Bohr's inequality is first proved in [12, Corallary 1]. The following result gives a generalization of Bohr's inequality and also includes its reverse.

Corollary 6. *Let $a, b \in \mathscr{A}$ and $p, q \in \mathbb{R}$ such that $1/p + 1/q = 1$.*

(i) *Bohr's inequality: if $p, q > 1$, then*

$$|a + b|^2 \leqslant p|a|^2 + q|b|^2 \tag{19}$$

with equality if and only if $pa = qb$ (i.e., $b = (p - 1)a$ or $a = (q - 1)b$).

(ii) *Reverse Bohr's inequality: if $p, q < 1$, then*

$$|a + b|^2 \geqslant p|a|^2 + q|b|^2 \tag{20}$$

with equality if and only if $pa = qb$.

Proof. Let $p, q > 1$ be such that $1/p + 1/q = 1$. Then $(p - 1)(q - 1) = 1$ which implies

$$\begin{pmatrix} 1 & 1 \\ 1 & 1 \end{pmatrix} \leqslant \begin{pmatrix} p & 0 \\ 0 & q \end{pmatrix}. \tag{21}$$

Since Φ is order preserving, we obtain the Bohr's inequality (19). By Theorem 2, $(p - 1)|a|^2 - a^* b - b^* a + (q - 1)|b|^2 = 0$ if and only if $(p - 1)a - b = 0$, that is, $b = (p - 1)a$ (note that $p, q \neq 1$). Part (ii) is similarly proven. □

Corollary 7. *Let $a, b \in \mathcal{A}$ and p, q real numbers such that $1/p + 1/q = 1$.*

(i) *If $p < 1$, then*

$$|a+b|^2 + |(p-1)a-b|^2 \geqslant p|a|^2 + q|b|^2, \quad |a+b|^2 + |a-(q-1)b|^2 \geqslant p|a|^2 + q|b|^2, \tag{22}$$

with equality if and only if $b = (p-1)a$.

(ii) *If $1 < p \leqslant 2$, then*

$$|a+b|^2 + |(p-1)a-b|^2 \leqslant p|a|^2 + q|b|^2, \tag{23}$$

$$|a+b|^2 + |a-(q-1)b|^2 \geqslant p|a|^2 + q|b|^2, \tag{24}$$

with equality if and only if $p = q = 2$ or $b = (p-1)a$.

(iii) *If $p > 2$, then*

$$|a+b|^2 + |(p-1)a-b|^2 \geqslant p|a|^2 + q|b|^2, \quad |a+b|^2 + |a-(q-1)b|^2 \leqslant p|a|^2 + q|b|^2, \tag{25}$$

with equality if and only if $b = (p-1)a$.

Proof. Assume $1 < p \leqslant 2$. The inequality (23) is obtained by applying the order-preserving Φ to a matrix ordering

$$\begin{pmatrix} 1 & 1 \\ 1 & 1 \end{pmatrix} + \begin{pmatrix} (p-1)^2 & 1-p \\ 1-p & 1 \end{pmatrix} = \begin{pmatrix} p^2-2p+2 & 2-p \\ 2-p & 2 \end{pmatrix} \leqslant \begin{pmatrix} p & 0 \\ 0 & q \end{pmatrix}. \tag{26}$$

By Theorem 2, $(p^2 - 2p + 2 - p)|a|^2 + (2-p)a^*b + (2-p)b^*a + (2-q)|b|^2 = 0$ holds if and only if either $p^2 - 3p - 2 = p - 2 = q - 2 = 0$ or $(p-1)(p-2)a + (2-p)b = 0$ and $(p^2-3p-2)(2-q) = (2-p)^2$. The first case is $p = q = 2$. The latter case is $b = (p-1)a$ since $(p^2-3p-2)(2-q) = (2-p)^2$ always holds from the fact that $p + q = pq$.

Now, a matrix ordering

$$\begin{pmatrix} 1 & 1 \\ 1 & 1 \end{pmatrix} + \begin{pmatrix} 1 & 1-q \\ 1-q & (q-1)^2 \end{pmatrix} = \begin{pmatrix} 2 & 2-q \\ 2-q & q^2-2q+2 \end{pmatrix} \geqslant \begin{pmatrix} p & 0 \\ 0 & q \end{pmatrix} \tag{27}$$

yields (24) via the map Φ. The proofs of others results are similar to that one. □

The analogue results for the case of operators on a Hilbert space are obtained in [10] (cf. Theorem 2, Theorem 1 and Corollary 1 in [10], resp.). The next result generalizes [11, Theorem 3.2].

Corollary 8. *Let $a, b \in \mathcal{A}$,*

(a) *if $0 < t \leqslant 1$, then $|a+b|^2 + |ta-b|^2 \leqslant (1+t)|a|^2 + (1+1/t)|b|^2$ with equality if and only if $t = 1$ or $b = ta$,*

(b) *if $t < 0$ or $t \geqslant 1$, then $|a+b|^2 + |ta-b|^2 \geqslant (1+t)|a|^2 + (1+1/t)|b|^2$ with equality if and only if $t = 1$ or $b = ta$.*

Proof. Assume $0 < t \leqslant 1$. The order-preserving Φ brings a matrix inequality

$$\begin{pmatrix} 1 & 1 \\ 1 & 1 \end{pmatrix} + \begin{pmatrix} t^2 & -t \\ -t & 1 \end{pmatrix} \leqslant \begin{pmatrix} 1+t & 0 \\ 0 & 1+\dfrac{1}{t} \end{pmatrix} \tag{28}$$

to the desired inequality in $\mathcal{A}$. This inequality becomes an equality if and only if

$$(t-1)\left(t|a|^2 - a^*b - b^*a + \frac{1}{t}|b|^2\right) = 0 \tag{29}$$

which is equivalent to the condition $t = 1$ or $ta - b = 0$ by Theorem 2. The proof of (b) is similar to that one. □

In 2009, generalized Bohr's inequality and its reverse for operators acting on a Hilbert space are done in [9, Theorem 9]. The following results give analogues results in the framework of Banach $*$-algebras.

Corollary 9. *Let $a, b \in \mathcal{A}$ and $s, t \in \mathbb{C}$, $p, q \in \mathbb{R} \setminus \{0\}$ such that $|s|^2/p + |t|^2/q \leqslant 1$.*

(a) *Generalized Bohr's inequality: if $|s|^2 \leqslant p$ and $|t|^2 \leqslant q$, then*

$$|sa + tb|^2 \leqslant p|a|^2 + q|b|^2. \tag{30}$$

(b) *Generalized reverse Bohr's inequality: if $|s|^2 \geqslant p$ and $|t|^2 \geqslant q$, then*

$$|sa + tb|^2 \geqslant p|a|^2 + q|b|^2. \tag{31}$$

In both cases, equality holds if and only if one of the following occurs:

(i) *$a = b = 0$,*

(ii) *$a = 0, b \neq 0$, and $q = |t|^2$,*

(iii) *$a \neq 0, b = 0$, and $p = |s|^2$,*

(iv) *$a \neq 0, b \neq 0, (p - |s|^2)a = \bar{s}tb$, and $|st| = \sqrt{(p-|s|^2)(q-|t|^2)}$ (equivalently, $a \neq 0$, $b \neq 0$, $s\bar{t}a = (q-|t|^2)b$ and $|st| = \sqrt{(p-|s|^2)(q-|t|^2)}$).*

Proof. The proofs of (a) and (b) are similar. For (a) assume $|s|^2 \leqslant p$ and $|t|^2 \leqslant q$. The conditions $|s|^2 - p \leqslant 0, |t|^2 - q \leqslant 0$ and $(p-|s|^2)(q-|t|^2) = |st|^2$ imply

$$\begin{pmatrix} |s|^2 & \bar{s}t \\ s\bar{t} & |t|^2 \end{pmatrix} \leqslant \begin{pmatrix} p & 0 \\ 0 & q \end{pmatrix}. \tag{32}$$

So, by passing through the order-preserving Φ, we obtain (30). When $a \neq 0$ and $b \neq 0$, the equality holds by Theorem 2 if and only if either $p - |s|^2 = \bar{s}t = q - |t|^2 = 0$ or $(p-|s|^2)a = \bar{s}tb$ and $(p-|s|^2)(q-|t|^2) = |st|^2$. The first case is impossible. The latter case is equivalent to $s\bar{t}a = (q-|t|^2)b$ and $|st| = \sqrt{(p-s|^2)(q-t|^2)})$. For $a = 0$ and $b \neq 0$, it forces $q = |t|^2$. For $a \neq 0$ and $b = 0$, it forces $p = |s|^2$. □

Corollary 10. *For any $a, b \in \mathscr{A}$, we have*

$$\pm(a^*b + b^*a) \leqslant \frac{1}{2}|a \pm b|^2 \leqslant |a|^2 + |b|^2 \qquad (33)$$

and each equality holds if and only if $a = \pm b$.

Proof. The inequality (33) follows via applying Φ to a matrix inequality

$$\begin{pmatrix} 0 & \pm 1 \\ \pm 1 & 0 \end{pmatrix} \leqslant \frac{1}{2}\begin{pmatrix} 1 & \pm 1 \\ \pm 1 & 1 \end{pmatrix} \leqslant \begin{pmatrix} 1 & 0 \\ 0 & 1 \end{pmatrix}. \qquad (34)$$

Each equality holds if and only if $|a|^2 \mp a^*b \mp b^*a + |b|^2 = 0$. By Theorem 2, it holds if and only if $a = \pm b$. □

In particular, for any $a, b \in \mathscr{A}$ and $t \in \mathbb{C} \setminus \{0\}$

$$\pm(a^*b + b^*a) \leqslant \frac{1}{2}\left|ta \pm \frac{1}{t}b\right|^2 \leqslant |t|^2|a|^2 + \frac{1}{|t|^2}|b|^2, \qquad (35)$$

with each equality holds if and only if $b = \pm t^2 a$. The inequalities of this form for operators appear in [15, Theorem 3].

We finally comment that the order-preserving Φ can generate many absolute-value identities and inequalities from any 2-by-2 matrix inequalities.

References

[1] H. Bohr, "Zur Theorie der Fastperiodischen Funktionen I," *Acta Math*, vol. 45, pp. 29–127, 1924.

[2] D. Mitrinovic, *Analytic Inequalities*, Springer-Verlag, New York, NY, USA, 1970.

[3] D. Mitrinovic, J. Pečarić, and A. Fink, *Classical and New Inequalities in Analysis*, Kluwer Academic, Dordrecht, The Netherlands, 1993.

[4] P. Vasić and J. Kečkić, "Some inequalities for complex numbers," *Mathematica Balkanica*, vol. 1, pp. 282–286, 1971.

[5] H. Bergström, "A triangle-inequality for matrices," in *Proceedings of the Den 11-te Skandinaviske Matematikerkongress, 1949*, pp. 264–267, Trondheim, Oslo, 1952.

[6] S. Abramovich, J. Barić, and J. Pečarić, "A new proof of an inequality of Bohr for Hilbert space operators," *Linear Algebra and Its Applications*, vol. 430, no. 4, pp. 1432–1435, 2009.

[7] S. Abramovich, J. Barić, and J. Pečarić, "Superquadracity, Bohr's inequality and deviation from a mean value," *Australian Journal of Mathematical Analysis and Applications*, vol. 7, no. 1, pp. 1–9, 2010.

[8] P. Chansangiam, "Bohr inequalities in C^*-algebras," *ScienceAsia*, vol. 36, no. 4, pp. 326–332, 2010.

[9] P. Chansangiam, P. Hemchote, and P. Pantaragphong, "Generalizations of Bohr inequality for Hilbert space operators," *Journal of Mathematical Analysis and Applications*, vol. 356, no. 2, pp. 525–536, 2009.

[10] W. S. Cheung and J. Pečarić, "Bohr's inequalities for Hilbert space operators," *Journal of Mathematical Analysis and Applications*, vol. 323, no. 1, pp. 403–412, 2006.

[11] M. Fujii and H. Zuo, "Matrix order in Bohr inequality for operators," *Banach Journal of Mathematical Analysis*, vol. 4, no. 1, pp. 21–27, 2010.

[12] O. Hirzallah, "Non-commutative operator Bohr inequality," *Journal of Mathematical Analysis and Applications*, vol. 282, no. 2, pp. 578–583, 2003.

[13] M. Moslehian, J. Pečarić, and I. Perić, "An operator extension of Bohr's inequality," *Bulletin of the Iranian Mathematical Society*, vol. 35, pp. 77–84, 2009.

[14] M. Moslehian and R. Rajic, "Generalizations of Bohr's inequality in Hilbert C^*-modules," *Linear Multilinear Algebra*, vol. 58, pp. 323–331, 2010.

[15] F. Zhang, "On the Bohr inequality of operators," *Journal of Mathematical Analysis and Applications*, vol. 333, no. 2, pp. 1264–1271, 2007.

[16] F. Bonsall and J. Duncan, *Complete Normed Algebras*, Springer-Verlag, New York, NY, USA, 1973.

[17] S. Shirali and J. Ford, "Symmetry in complex involutory Banach algebras II," *Duke Mathematical Journal*, vol. 37, no. 2, pp. 275–280, 1970.

[18] B. Sims, "A characterization of a Banach-star-algebras with numerical range," *Bulletin of the Australian Mathematical Society*, vol. 4, no. 2, pp. 193–200, 1971.

Permissions

The contributors of this book come from diverse backgrounds, making this book a truly international effort. This book will bring forth new frontiers with its revolutionizing research information and detailed analysis of the nascent developments around the world.

We would like to thank all the contributing authors for lending their expertise to make the book truly unique. They have played a crucial role in the development of this book. Without their invaluable contributions this book wouldn't have been possible. They have made vital efforts to compile up to date information on the varied aspects of this subject to make this book a valuable addition to the collection of many professionals and students.

This book was conceptualized with the vision of imparting up-to-date information and advanced data in this field. To ensure the same, a matchless editorial board was set up. Every individual on the board went through rigorous rounds of assessment to prove their worth. After which they invested a large part of their time researching and compiling the most relevant data for our readers. Conferences and sessions were held from time to time between the editorial board and the contributing authors to present the data in the most comprehensible form. The editorial team has worked tirelessly to provide valuable and valid information to help people across the globe.

Every chapter published in this book has been scrutinized by our experts. Their significance has been extensively debated. The topics covered herein carry significant findings which will fuel the growth of the discipline. They may even be implemented as practical applications or may be referred to as a beginning point for another development. Chapters in this book were first published by Hindawi Publishing Corporation; hereby published with permission under the Creative Commons Attribution License or equivalent.

The editorial board has been involved in producing this book since its inception. They have spent rigorous hours researching and exploring the diverse topics which have resulted in the successful publishing of this book. They have passed on their knowledge of decades through this book. To expedite this challenging task, the publisher supported the team at every step. A small team of assistant editors was also appointed to further simplify the editing procedure and attain best results for the readers.

Our editorial team has been hand-picked from every corner of the world. Their multi-ethnicity adds dynamic inputs to the discussions which result in innovative outcomes. These outcomes are then further discussed with the researchers and contributors who give their valuable feedback and opinion regarding the same. The feedback is then collaborated with the researches and they are edited in a comprehensive manner to aid the understanding of the subject.

Apart from the editorial board, the designing team has also invested a significant amount of their time in understanding the subject and creating the most relevant covers. They scrutinized every image to scout for the most suitable representation of the subject and create an appropriate cover for the book.

The publishing team has been involved in this book since its early stages. They were actively engaged in every process, be it collecting the data, connecting with the contributors or procuring relevant information. The team has been an ardent support to the editorial, designing and production team. Their endless efforts to recruit the best for this project, has resulted in the accomplishment of this book. They are a veteran in the field of academics and their pool of knowledge is as vast as their experience in printing. Their expertise and guidance has proved useful at every step. Their uncompromising quality standards have made this book an exceptional effort. Their encouragement from time to time has been an inspiration for everyone.

The publisher and the editorial board hope that this book will prove to be a valuable piece of knowledge for researchers, students, practitioners and scholars across the globe.

List of Contributors

Sunil Kumar Singh
Department of Mathematics, Rajiv Gandhi University, Doimukh, Arunachal Pradesh 791112, India

Arsen Pskhu
Scientific Research Institute of Applied Mathematics and Automation, Kabardino-Balkar Scientific Centre, Russian Academy of Sciences, 89A Shortanov Street, Nalchik 360000, Russia

Zehra Pinar and Turgut Özis
Department of Mathematics, Science Faculty, Ege University, Bornova, 35100 ˙Izmir, Turkey

George Voutsadakis
School of Mathematics and Computer Science, Lake Superior State University, Sault Sainte Marie, MI 49783, USA

Morten Nielsen
Department of Mathematical Sciences, Aalborg University, Fredrik Bajers Vej 7G, 9220 Aalborg East, Denmark

Marwan Alquran, Mahmoud Mohammad and Ahmad Ababneh
Department of Mathematics and Statistics, Jordan University of Science and Technology, Irbid 22110, Jordan

Xiao-Bao Shu
Department of Mathematics, Hunan University, Changsha 410082, China

Fei Xu
Department of Mathematics, Wilfrid Laurier University, Waterloo, ON, Canada N2L 3C5

Hatem Mejjaoli
Department of Mathematics, College of Sciences, Taibah University, P.O. Box 30002, Al Madinah AL Munawara, Saudi Arabia

Ahmedou Ould Ahmed Salem
Department of Mathematics, College of Sciences, King Khalid University, Mohayil, Saudi Arabia

José Antonio Belinchón
Departamento Física Atómica, Molecular y Nuclear, Universidad Complutense de Madrid, 28040 Madrid, Spain

Michael S. Milgram
Consulting Physicist, Geometrics Unlimited, Ltd., P.O. Box 1484, Deep River, ON, Canada K0J 1P0

Yusuf Pandir and Halime Ulusoy
Department of Mathematics, Faculty of Science, Bozok University, 66100 Yozgat, Turkey

Suhel Ahmad Khan
Department of Mathematics, BITS-Pilani, Dubai Campus, P.O. Box 345055, Dubai, UAE

Naeem Ahmad
Department of Mathematics, Al-Jouf University, P.O. Box 2014, Skaka, Saudi Arabia

Peixin Zhao
Department of Mathematics, Hechi University, Yizhou, Guangxi 546300, China

Xiao-Guang Lv and Ting-Zhu Huang
School of Mathematical Sciences, University of Electronic Science and Technology of China, Chengdu, Sichuan 611731, China

Devendra Kumar
Department of Mathematics, Research and Post Graduate Studies, M. M. H. College, Model Town, Ghaziabad, Uttar Pradesh 20/001, India

Bin-Chao Deng and Tong Chen
School of Management, Tianjin University, Tianjin 300072, China

Anna Pascoletti and Fabio Zanolin
Dipartimento di Matematica e Informatica, Universit`a di Udine, Via delle Scienze 206, 33100 Udine, Italy

Le Jiang
School of Mathematical Sciences, University of Electronic Science and Technology of China, Chengdu, Sichuan 610054, China
School of Science, Huaihai Institute of Technology, Lianyungang, Jiangsu 222005, China

Jin Huang and Jun Liu
School of Mathematical Sciences, University of Electronic Science and Technology of China, Chengdu, Sichuan 610054, China

Xiao-Guang Lv
School of Science, Huaihai Institute of Technology, Lianyungang, Jiangsu 222005, China

Szymon Plewik and Marta Walczyńska
Institute of Mathematics, University of Silesia, ul. Bankowa 14, 40-007 Katowice, Poland

Satish Shukla
Department of Applied Mathematics, Shri Vaishnav Institute of Technology and Science, Gram Baroli, Sanwer Road, Indore 453331, India

Stojan Radenovic and Slaviša Pantelic
Faculty of Mechanical Engineering, University of Belgrade, Kraljice Marije 16, 11120 Beograd, Serbia

Boubakari Ibrahimou
Department of Public Health, Western Kentucky University, 1906 College Heights Boulevard, Bowling Green, KY 42104,USA

Asao Arai
Department of Mathematics, Hokkaido University, Sapporo 060-0810, Japan

Ugur Ulusu and Fatih Nuray
Department of Mathematics, Faculty of Science and Literature, Afyon Kocatepe University, 03200 Afyonkarahisar, Turkey

Kun Wang and Zhilin Yang
Department of Mathematics, Qingdao Technological University, Fushun Road No. 11, Quingdao, Shandong 266033, China

M. I. Isaev
Centre de Mathématiques Appliquées, Ecole Polytechnique, 91128 Palaiseau, France

Pattrawut Chansangiam
Department of Mathematics, Faculty of Science, King Mongkut's Institute of Technology Ladkrabang, Bangkok 10520, Thailand